MUTAGENESIS

Benchmark Papers in Genetics

Series Editor: David L. Jameson
University of Houston

PUBLISHED VOLUMES AND VOLUMES IN PREPARATION

GENETICS AND SOCIAL STRUCTURE / *Paul A. Ballonoff*
GENES AND PROTEINS / *Robert P. Wagner*
DEMOGRAPHIC GENETICS / *Kenneth M. Weiss and Paul A. Ballonoff*
MUTAGENESIS / *John W. Drake and Robert E. Koch*
EUGENICS: Then and Now / *Carl Jay Bajema*
POPULATION GENETICS / *James F. Crow and Carter Denniston*
MEDICAL GENETICS / *William Jack Schull*
ECOLOGICAL GENETICS / *W. W. Anderson*
QUANTITATIVE GENETICS / *R. E. Comstock*
GENETIC RECOMBINATION / *Rollin Hotchkiss*
REGULATION GENETICS / *Werner K. Maas*
PLANT BREEDING / *D. F. Matzinger*
DEVELOPMENTAL GENETICS / *Antonie W. Blackler and Richard Hallberg*
CYTOGENETICS / *Ronald L. Phillips and Charles H. Burnham*

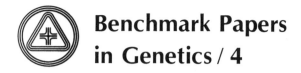

**Benchmark Papers
in Genetics / 4**

A BENCHMARK® Books Series

MUTAGENESIS

Edited by

**John W. Drake
and Robert E. Koch**
University of Illinois at Urbana-Champaign

**Dowden, Hutchinson
& Ross, Inc.**
STROUDSBURG, PENNSYLVANIA

Distributed by
HALSTED
PRESS

A Division of
John Wiley & Sons, Inc.

Exclusive Distributor: **Halsted Press**
A Division of John Wiley & Sons, Inc.

ACKNOWLEDGMENTS AND PERMISSIONS

ACKNOWLEDGMENTS

AMERICAN ASSOCIATION FOR THE ADVANCEMENT OF SCIENCE—*Science*
Artificial Transmutation of the Gene
Radiation Dose Rate and Mutation Frequency

GENETIC SOCIETY OF AMERICA—*Genetics*
Cryptic Mutants of Bacteriophage T4
The Effect of X-Rays upon Mutation of the Gene *A* in Maize
Ethyl Methanesulfonate-Induced Reversion of Bacteriophage T4rII Mutants

NATIONAL ACADEMY OF SCIENCES—*Proceedings of the National Academy of Sciences*
The Chemical and Mutagenic Specificity of Hydroxylamine
The Difference Between Spontaneous and Base-Analogue Induced Mutations of Phage T4
Induction Kinetics and Genetic Analysis of X-Ray-Induced Mutations in the *ad-3* Region of *Neurospora crassa*
The Influence of Neighboring Base Pairs upon Base-Pair Substitution Mutation Rates
On the Topography of the Genetic Fine Structure
Reversal of Mutant Phenotypes by 5-Fluorouracil: An Approach to Nucleotide Sequences in Messenger-RNA
The Structure of the DNA–Acridine Complex
The Unusual Mutagenic Specificity of an *E. coli* Mutator Gene

PERMISSIONS

The following papers have been reprinted with the permission of the authors and copyright holders.

ACADEMIC PRESS, INC.
Biochemical and Biophysical Research Communications
The Base Specificity of Mutation Induced by Nitrous Acid in Phage T2
Mutagenic DNA Polymerase
Virology
Mutagenesis in Phages φX174 and T4 and Properties of the Genetic Material

ACADEMIC PRESS INC. (LONDON) LTD.—*Journal of Molecular Biology*
The Efficiency of Induction of Mutations by Hydroxylamine
The Theory of Mutagenesis

AMERICAN ASSOCIATION FOR THE ADVANCEMENT OF SCIENCE—*Science*
Mutagenic Effects of Hydroxylamine *in vivo*

Acknowledgments and Permissions

THE AMERICAN SOCIETY OF BIOLOGICAL CHEMISTS, INC.—*The Journal of Biological Chemistry*
Studies on the Biochemical Basis of Spontaneous Mutation: I. A Comparison of the Deoxyribonucleic Acid Polymerases of Mutator, Antimutator, and Wild Type Strains of Bacteriophage T4

ASP BIOLOGICAL AND MEDICAL PRESS B.V.—*Mutation Research*
The Mutability Toward Ultraviolet Light of Recombination-Deficient Strains of *Escherichia coli*

THE BIOLOGICAL LABORATORY, LONG ISLAND BIOLOGICAL ASSOCIATION—*Cold Spring Harbor Symposia on Quantitative Biology*
The Structure of DNA

COLD SPRING HARBOR LABORATORY—*Cold Spring Harbor Symposia on Quantitative Biology*
Antimutagenic DNA Polymerases of Bacteriophage T4
Frameshift Mutagenesis in *Salmonella*
Frameshift Mutations and the Genetic Code

MACMILLAN JOURNALS LTD.—*Nature*
Comparative Rates of Spontaneous Mutation
General Nature of the Genetic Code for Proteins
Genetical Implications of the Structure of Deoxyribonucleic Acid
Genetic Code: The "Nonsense" Triplets for Chain Termination and Their Suppression
Molecular Basis of a Mutational Hot Spot in the Lysozyme Gene of Bacteriophage T4
Possible Relevance of O-6 Alkylation of Deoxyguanosine to the Mutagenicity and Carcinogenicity of Nitrosamines and Nitrosamides
Production of Mutants of Tobacco Mosaic Virus by Chemical Alteration of its Ribonucleic Acid *in vitro*
Uniformity of Radiation-Induced Mutation Rates Among Different Species

ROYAL SOCIETY OF EDINBURGH—*Proceedings of the Royal Society of Edinburgh*
The Induction by Mustard Gas of Chromosomal Instabilities in *Drosophila melanogaster*
The Production of Mutations by Chemical Substances

ROYAL SOCIETY OF LONDON—*Proceedings of the Royal Society (London)*
The Influence of Radiomimetic Substances on Deoxyribonucleic Acid Synthesis and Function Studied in *Escherichia coli*/Phage Systems: III. Mutation of T2 Bacteriophage as a Consequence of Alkylation *in vitro*: The Uniqueness of Ethylation

THE SCIENCE COUNCIL OF JAPAN—*Proceedings of the XII International Congress on Genetics*
Mutagenicity Versus Radiosensitivity in *Escherichia coli*
The Role of DNA Repair and Recombination in Mutagenesis

SERIES EDITOR'S PREFACE

The study of any discipline assumes mastery of the literature of the subject. In many branches of science, even one as new as genetics, the expansion of knowledge has been so rapid that there is little hope of learning of the development of all phases of the subject. The student has difficulty mastering the textbook, the young scholar must tend to the literature near his own research, the young instructor barely finds time to expand his horizons to meet his class preparation requirements, the monographer copes with a wider literature but usually from a specialized viewpoint, and the textbook author is forced to cover much the same materials as previous and competing texts to respond to the user's needs and abilities.

Few publishers have the dedication to scholarship to serve primarily the limited market of advanced studies. The opportunity to assist professionals at all stages of their careers has been recognized by the publishers of the Benchmark series and by a distinguished group of Benchmark Volume Editors knowledgeable in specific aspects of the literature of genetics. Some have contributed heavily to the development of that literature, some have studied with the early scholars, and some have developed and are in the process of developing entirely new fields of genetic knowledge. In many cases, the judgments of the editors become a historical document that records their opinion of the important steps in the development of the subject. These editors have selected papers and portions of papers that demonstrate both the development of knowledge and the atmosphere in which that knowledge was developed. There is no substitute for reading great papers. Here you can learn how questions are asked, how they are approached, and how difficult and essential it is to obtain definitive answers and clear writing. My own pleasure in working with this distinguished panel is exceeded only by the considerable pleasure of reading their remarks and their selections. Their dedication and wisdom are impressive.

The source of genetic variation lies in the errors of replication. Thus studies, such as *Mutagenesis*, which analyze the nature of the error process lie at the very heart of genetics. Studies of spontaneous mutation, induced mutations, and resistance to mutation provide insight to the underlying genetic mechanisms while they develop a basis for the control of environmental mutagens. This collection of benchmark papers is a link between active and expanding basic studies and the necessary vital application of those studies. With the publication of this volume, Drake and Koch have provided a service to a broad range of scholars and practitioners.

DAVID L. JAMESON

CONTENTS

ix

Contents

PART III: FRAMESHIFT MUTAGENESIS

PART IV: MISREPAIR MUTAGENESIS

Contents

CONTENTS BY AUTHOR

MUTAGENESIS

INTRODUCTION

Infidelity has always fascinated geneticists. The genetically most important infidelities occur when genes mutate, and many early-twentieth-century geneticists hoped to unravel the nature of the gene by studying its mutants and their origins. They were frustrated for over two decades, however, by an inability to induce mutations by artificial means. Evolutionary geneticists, both the neo-Darwinians and more recently the non-Darwinians, continued to emphasize the overwhelming importance of mutation as the process that generates raw genetic variation, which in turn is manipulated by evolutionary forces (whether these be selective or random in nature). Within the past two decades, molecular geneticists have devoted considerable enthusiasm to working out specific mechanisms of mutation, and, in doing so, have also continued to do what numerous geneticists have always done, to use the products of mutagenesis to attack exciting contemporary problems.

Although Charles Darwin directed relatively few of his efforts to deducing the origins of organismal variation, he placed this variation unequivocally at the center of his theories of evolution. The first worker to study mutation comprehensively (and also, in 1901, to coin the word itself) was Hugo de Vries. Curiously, the highly variable organism he so intensively studied, *Oenothera* (the evening primrose), "mutates" by a very atypical mechanism. Most *Oenothera* populations maintain high frequencies of balanced translocations, sometimes involving the entire chromosome complement. The resulting anomalies of meiosis, which include frequent chromosome breakage and anomalous rejoining events, generate large numbers of somatic and germ line chromosomal mutations, but, as far as is known, do not produce unusually high frequencies of point mutations.

1

The Morgan school of *Drosophila* geneticists rose to preeminence in the first three decades of this century through their demonstration of the critical role that mutant genes necessarily play in genetic analysis. Except in the still relatively rare circumstances, where specific protein or RNA species can be isolated and characterized, the very existence of a gene can be demonstrated only by recovering one of its mutant alleles. Nearly all experimental studies of genetic systems begin with the isolation of mutants, and the practicing student of mutagenesis is continually beseeched to suggest methods for inducing and selecting desirable types of mutants. The art of artificial mutagenesis was initiated by H. J. Muller, who, in 1927, demonstrated the induction of sex-linked recessive lethal mutations in *Drosophila melanogaster* by X-irradiation. A large literature describing X-ray-induced mutagenesis in diverse organisms accumulated rather rapidly over the next dozen years, but numerous attempts to induce mutations by other means always produced results that were at best highly equivocal.

This situation was dramatically reversed by Charlotte Auerbach's discovery, under very curious circumstances, of the mutagenicity of the infamous weapon, mustard gas. This compound, and numerous related alkylating agents, were soon studied as intensively as X-rays. Only a decade later, however, yet another major breakthrough occurred, the promulgation of the Watson–Crick structure of DNA. A considerable number of very simple, and simply acting, chemical mutagens were soon discovered, and the conceptual bases of point mutagenesis were rapidly formulated by Ernst Freese, Sydney Brenner, and Francis Crick.

The elucidation of basic mutagenic mechanisms, with which this collection of papers is mainly concerned, has been a source of great excitement and pleasure to many workers. Serendipity has played its usual crucial role, with the rewards being frequently seized by those most sensitive to the potential significance of anomalous experimental results. While advances in laboratory technology have frequently proved indispensable, imagination has played an even more important role in solving mutagenic puzzles, mainly because of the intrinsic rarity of mutational events. The typical mutational event occurs sufficiently infrequently to render its straightforward biochemical analysis virtually impossible. Only the products of mutagenesis, the mutant clones themselves, can be sufficiently magnified by growth to become susceptible to biochemical analysis. As a result, deduction based jointly upon integrated genetic and biochemical results has usually been required to understand mutagenic mechanisms. This necessarily combined approach, plus the fact that it is frequently far less expensive to study mutagenesis than other genetic phenomena (at least in microbes), also helps to make the study of mutagenesis such great fun.

Recent advances in our understanding of mutation rates in numerous organisms, and particularly in man, have added a new impetus to studies in mutagenesis. It now seems very likely that the human mutation rate is very large indeed, so large that most individuals probably harbor not only numerous (deleterious) mutant alleles inherited from relatively recent ancestors, but also an average of about one newly arisen deleterious mutation. The resulting "mutational load" (a phrase originated in 1950 by H. J. Muller) probably represents a serious and continuing challenge to human genetic health. As a result, a whole new branch of medicine—medical genetics and its allied art, genetic counseling—has come into existence, and the number of even relatively simple clinical syndromes now exceeds 2,000. A concomitant concern is now developing about "environmental mutagenesis," the presence in the environment of mutagens, many of which (chemicals for the most part) were never before present. The control of environmental mutagenesis now engages the attention of numerous scientists, including geneticists, toxicologists, pharmacologists, and environmentalists generally.

In addition to making the solution of mutagenic mechanisms a tricky process, the rarity of mutational events has also determined to a large extent the choice of experimental organisms. It is much easier to screen 1 million or 1 billion microbes for mutant individuals (particularly when one uses sophisticated selective schemes) than it is to screen an equal number of fruit flies; and it is vastly easier to screen microbes than to screen an equal number of mice or men. Microbes have therefore long constituted the organisms of choice in this area of research. The continuing necessity to understand mutagenic mechanisms in higher eucaryotes has forced us to study these experimentally awkward organisms, but our understanding of basic molecular mechanisms in fruit flies, mice, and men remains very thin indeed. This collection of papers is therefore heavily weighted in favor of very small organisms, among which bacteriophage T4 tends to predominate.

Our choice of papers may reflect in part our personal research preoccupations with T4, but it is nevertheless unequivocally clear that most of what we know, or claim to know, about mutagenic mechanisms at the molecular level was first worked out in laboratories employing the talents of this very serviceable organism. In considering our choice of papers, however, we found it desirable to include studies employing many of the other organisms favored by students of mutation, including maize, mice, *Drosophila*, *Neurospora*, yeasts, *Salmonella typhimurium* and *Escherichia coli*, and even other viruses. Indeed, a few of these studies have abandoned living organisms altogether and employ only DNA molecules.

Several topics have been deliberately slighted in this collection, being of only peripheral relevance to our central topic of mutagenic mechanisms. These include certain types of evolutionary studies, numerous studies of mutant gene products (except when these have contributed directly to an understanding of mutation itself), most aspects of genetic suppression, and applied mutagenesis, such as the technology of producing mutant variants of economically important organisms or the whole arena of environmental mutagenesis. A number of works have appeared, however, that describe in far greater breadth than is possible in this collection both mutagenic mechanisms and diverse procedures for detecting mutations. The more recent of these are Auerbach and Kilbey (1971), Drake (1969, 1970), and Hollaender (1971, 1973).

The cross section of mutagenesis research represented in this collection has as its most recent entry a paper from the year 1973 (and only four more from the rest of the 1970s). It is interesting to speculate about the papers that might have been included had we dallied yet another decade. It seems likely that ongoing studies of the enzymatic details of DNA replication and repair will reveal important aspects of the mechanisms by which fidelity is optimized during DNA synthesis, and why certain types of postreplication repair systems are so very error prone. A definitive resolution will no doubt be achieved of the current controversy (not directly represented by any of the papers in this collection) concerning whether variants selected from populations of cultured mammalian cells represent true mutations or epigenetic changes akin to differentiation, but not necessarily resulting from altered DNA base pair sequences. It may also turn out that completely novel mechanisms of mutation will be discovered in eucaryotes. Finally, and of particular importance to those who see the understanding of man himself as the ultimate goal of all biological research, the possibility continues to exist that ways will be discovered to control human mutation rates. It is extremely difficult at the present time to imagine how this might be accomplished.

REFERENCES

Auerbach, C., and B. J. Kilbey. 1971. Mutation in eucaryotes, *Ann. Rev. Genetics*, *5:*163.

Drake, J. W. 1969. Mutagenic mechanisms. *Ann Rev. Genetics, 3:* 247.

———. 1970. *The Molecular Basis of Mutation.* Holden-Day, San Francisco.

Hollaender, A. (ed.). *Chemical Mutagens. Principles and Methods for Their Detection,* Vols. 1 and 2 (1971) and 3 (1973). Plenum Press, New York.

Part I

IONIZING RADIATION MUTAGENESIS

Editors' Comments
on Papers 1 Through 5

IONIZING RADIATION MUTAGENESIS

Throughout most of the first half of this century, many geneticists were deeply involved in resolving the nature of the gene. During the first four decades, for most practical purposes the gene was simply a mysterious point entity, localizable by mapping procedures to a particular site on a particular chromosome, but not further subdivisible. Both theoretical considerations and the existence of multiple phenotypically distinguishable alleles of single genes, however, argued strongly that the gene possessed a complex internal structure. Many workers, H. J. Muller in particular, believed that the gene's internal structure might be deduced if methods could be devised to mutate genes at will, rather than relying on the relatively rare occurrence of spontaneous mutations. Muller's contribution was twofold. The first, described as a rather brief announcement in Paper 1, demonstrated that the highly energetic and localized releases of energy resulting from X-irradiation produced unequivocal mutagenic responses in the then favorite organism for the study of chromosome mechanics, *Drosophila melanogaster*. Second, to achieve this end, he devised an ingenious method for simultaneously scoring mutations at a large number of gene loci, a number now known to be on the order of 1,000.

Since Muller's method is not adequately described in Paper 1, but remains in widespread use with only relatively minor modifications (as, for instance, in Paper 7), we shall describe its essentials here. The method measures the frequency of sex-linked (X chromosome) recessive lethal mutations, which are not readily detected by conventional screening techniques. It employs a special mutant X chromosome, called *ClB*, which contains three important components: (1) an inversion (*C*), which effectively abolishes recombination between a *ClB* and a normal chromosome, (2) a recessive lethal mutation (*l*), which kills all hemizygous males and homozygous females early in development, and (3) a dominant eye-morphology mutation *Bar (B)*, which enables heterozygous X^{ClB}/X females to be recognized visually. To detect mutations, normal X/Y males are irradiated; we shall designate the irradiated X chromosome as X^*. These treated males are then crossed with X^{ClB}/X females, and four progeny classes result: X^{ClB}/X^* (*Bar* females carrying the irradiated X chromosome), X^{ClB}/Y (dead males), X/X^* (normal females), and X/Y (normal males). The *Bar* females and the surviving males are then mated with each other, with each mating pair placed in a separate vial (of which there will be hundreds or thousands). The four resulting progeny classes of the next generation will therefore be X^{ClB}/X, X^{ClB}/Y (dies), X^*/X, and X^*/Y. However, if X^* harbors a newly induced recessive lethal mutation, the X^*/Y males also die. Each vial totally lacking male progeny therefore signals an induced mutation, which may, if desired, be recovered from among the normal females in the same vial. The spontaneous mutation rate in this system is typically about 0.2 percent per sexual generation, but this frequency can readily be increased by as much as two orders of magnitude by X-irradiation.

A number of lines of evidence, obtained mainly in the 1930s and 1940s, suggested that X-irradiation might not, after all, be such a useful probe of the gene's interior as had been anticipated, since it might induce mainly deficiency or deletion mutations that remove a gene altogether, rather than simply modifying it. Very often, for instance, induced mutants of a particular gene seemed to simultaneously affect other nearby loci, in contrast to other alleles of that gene which had arisen spontaneously. In Paper 2, Stadler and Roman described an extreme case of this type and demonstrated that virtually all X-ray-induced mutations in maize are deficiencies. It also seems likely that a majority of X-ray-induced mutations in *Drosophila* are deficiencies (Muller, 1954), but it is now known that X-irradiation is quite capable of inducing point mutations as well, at least in *Neurospora* (see Paper 4) and in phage T4 (M. Conkling and J. W. Drake, unpublished results). (The term *point mu-*

tation as used above refers specifically to a lesion that alters only one or at most a few base pairs. It should not be confused with the same term as it was used in Paper 2 at a time when the structure of the gene was largely unknown. The earlier usage did not distinguish between what we now call a point mutation and a deletion or deficiency that affected or fell within only a single gene.)

With the advent of atomic weaponry and the associated radioactive fallout, as well as with the indiscriminate use of medical X-rays, radiation geneticists began to become acutely aware of the potential hazards to man that might accrue from artificial sources of ionizing radiations. One result was the initiation of an extensive program to study the genetic effects of X-irradiation in mice; in this instance, mutations were scored at specific loci rather than along an entire chromosome. Paper 3 describes one such study from the Russell group at Oak Ridge. This report was notable in that it demonstrated a very considerable effect of dose *rate* on the mutation rate per incident rad. This observation was one of the first to hint at the existence of a phenomenon that is now known to be crucial to many aspects of induced mutation, DNA *repair*. It is now clear that repair can either reduce *or* enhance mutagenic efficiencies, depending upon the mutagen, the type of repair system involved, and the conditions of treatment (see, for instance, Papers 27 through 29).

The development of fine-scale genetics (see Paper 30), through which genetic lesions can be localized within genes, allowed the first truly detailed analysis of the molecular dimensions of X-ray-induced mutations to be performed. Although X-ray-induced mutagenesis in some organisms, including both *Drosophila* and phage T4, is essentially directly proportional to dose, the kinetics of ionizing radiation mutagenesis are sometimes clearly complex in other organisms. Complex kinetics were first comprehensively analyzed by the de Serres group at Oak Ridge using *Neurospora crassa*, as described by Webber and de Serres in Paper 4. This study, which employed high dose rates, revealed that the induction of point mutations is directly proportional to the X-ray dose, whereas the induction of deficiencies followed dose-squared or two-hit kinetics. Later, however, this same group also showed that, when the dose rate was reduced 100-fold, deletions were induced at a substantially lower rate per unit of radiation; their rate of appearance became much more single hit in character (de Serres et al., 1967). Simple results like these, plus continuing advances in our understanding of the diverse roles of repair in mutagenesis, now provide a reasonable conceptual basis for interpreting the more complex types of does dependencies with which mutations sometimes arise in response to a multitude of chemical mutagens.

Growing concern about environmental mutagenesis, particularly in relation to the proliferation of commercial nuclear power plants, prompted the U. S. Atomic Energy Commission to organize a 1973 workshop to consider, among other things, how to extrapolate induced mutation rates from experimental animals to man. A literature search stimulated by the workshop debates produced the remarkable discovery, described by Abrahamson et al. in Paper 5: specific locus (e.g., per gene) mutation rates per rad of ionizing radiation are directly proportional to the total amount of DNA per haploid genome complement. Thus a 1,000-fold increase in the amount of DNA per cell is associated with a 1,000-fold increase in the induced mutation rate per gene (more specifically, per cistron or complementation unit) per rad. This relationship is of dual significance. First, it provides a useful empirical rationale for extrapolating mutation rates from organisms such as *Neurospora, Drosophila,* and the mouse to man. Second, it offers a novel type of support for the currently popular concept that the number of genes does not increase very greatly as one progresses from the lowest eucaryotes (yeasts) to the highest (such as man and some plants), whereas the size of the average gene may increase very greatly. Gene size would increase, not from an increase in the size of an average protein, which remains relatively constant in all organisms, but from an increase in the number of regulatory DNA base sequences associated with each protein-encoding sequence.

In further support of this notion, M. Conkling and J. W. Drake have shown (unpublished results) that the mutation rate per locus per rad of ionizing radiation is essentially constant from phage T4 through *E. coli* (the *E. coli* point in Paper 5 is mildly in error) through the yeasts. It is also possible that deletion induction may make a large contribution to the points higher up on the curve shown in Figure 1 of Paper 5, that deletions may originate from radiation lesions lying well outside of the target gene itself, and that the slope of the figure may therefore be less steep than shown. If true, this could imply that higher organisms may have more as well as larger genes compared to more simple organisms. The generality of this observation is indicated by another literature search, which revealed that a chemical mutagen, ethyl methanesulfonate (see Papers 19, 20, and 21), produces a similar relationship (J. A. Heddle, personal communication). Curiously, however, spontaneous mutation rates do *not* appear to follow this relationship.

REFERENCES

de Serres, F. J., H. V. Malling, and B. B. Webber. 1967. Dose-rate effects on inactivation and mutation induction in *Neurospora crassa. Brookhaven Symp. Biol., 20:* 56.

Muller, H. J. 1954. The manner of production of mutations by radiation. *Radiation Biol., 1:* 475 (see especially pp. 496–507).

Reprinted from *Science*, **66**(1699), 84–87 (1927)

ARTIFICIAL TRANSMUTATION OF THE GENE

H. J. Muller

University of Texas

MOST modern geneticists will agree that gene mutations form the chief basis of organic evolution, and therefore of most of the complexities of living things. Unfortunately for the geneticists, however, the study of these mutations, and, through them, of the genes themselves, has heretofore been very seriously hampered by the extreme infrequency of their occurrence under ordinary conditions, and by the general unsuccessfulness of attempts to modify decidedly, and in a sure and detectable way, this sluggish "natural" mutation rate. Modification of the innate nature of organisms, for more directly utilitarian purposes, has of course been subject to these same restrictions, and the practical breeder has hence been compelled to remain content with the mere making of recombinations of the material already at hand, providentially supplemented, on rare and isolated occasions, by an unexpected mutational windfall. To these circumstances are due the wide-spread desire on the part of biologists to gain some measure of control over the hereditary changes within the genes.

It has been repeatedly reported that germinal changes, presumably mutational, could be induced by X or radium rays, but, as in the case of the similar published claims involving other agents (alcohol, lead, antibodies, etc.), the work has been done in such a way that the meaning of the data, as analyzed from a modern genetic standpoint, has been highly disputatious at best; moreover, what were apparently the clearest cases have given negative or contrary results on repetition. Nevertheless, on theoretical grounds, it has appeared to the present writer that radiations of short wave length should be especially promising for the production of mutational changes, and for this and other reasons a series of experiments concerned with this problem has been undertaken during the past year on the fruit fly, *Drosophila melanogaster*, in an attempt to provide critical data. The well-known favorableness of this species for genetic study, and the special methods evolved during the writer's eight years' intensive work on its mutation rate (including the work on temperature, to be referred to later), have finally made possible the finding of some decisive effects, consequent upon the application of X-rays. The effects here referred to are truly mutational, and not to be confused with the well-known effects of X-rays upon the distribution of the chromatin, expressed by non-disjunction, non-inherited crossover modifications, etc. In the present condensed digest of the work, only the broad facts and conclusions therefrom, and some of the problems raised, can be presented, without any details of the genetic methods employed, or of the individual results obtained.

It has been found quite conclusively that treatment of the sperm with relatively heavy doses of X-rays induces the occurrence of true "gene mutations" in a high proportion of the treated germ cells. Several hundred mutants have been obtained in this way in a short time and considerably more than a hundred of the mutant genes have been followed through three, four or more generations. They are (nearly all of them, at any rate) stable in their inheritance, and most of them behave in the manner typical of the Mendelian chromosomal mutant genes found in organisms generally. The nature of the crosses was such as to be much more favorable for the detection of mutations in the X-chromosomes than in the other chromosomes, so that most of the mutant genes dealt with were sex-linked; there was, however, ample proof that mutations were occurring similarly throughout the chromatin. When the heaviest treatment was given to the sperm, about a seventh of the offspring that hatched from them and bred contained individually detectable mutations in their treated X-chromosome. Since the X forms about one fourth of the haploid chromatin, then, if we assume an equal rate of mutation in all the chromosomes (per unit of their length), it follows that almost "every other one" of the sperm cells capable of producing a fertile adult contained an "individually detectable" mutation in some chromosome or other. Thousands of untreated parent flies were bred as controls in the same way as the treated ones. Comparison of the mutation rates under the two sets of conditions showed that the heavy treatment had caused a rise of about fifteen thousand per cent. in the mutation rate over that in the untreated germ cells.

Regarding the types of mutations produced, it was found that, as was to have been expected both on theoretical grounds and on the basis of the previous mutation studies of Altenburg and the writer, the lethals (recessive for the lethal effect, though some were dominant for visible effects) greatly outnumbered the non-lethals producing a visible morphological abnormality. There were some "semi-lethals" also (defining these as mutants having a viability ordinarily between about 0.5 per cent. and 10 per cent. of the normal), but, fortunately for the use of lethals as an index of mutation rate, these were not nearly so numerous as the lethals. The elusive class of "invisible" mutations that caused an even lesser reduction of viability, not readily confusable with lethals, appeared larger than that of the semi-lethals, but they were not subjected to study. In addition, it was also possible to obtain evidence in these experiments for the first time, of the occurrence of dominant lethal genetic changes, both in the X and in the other chromosomes. Since the zygotes receiving these never developed to maturity, such lethals could not be detected individually, but their number was so great that through egg counts and effects on the sex ratio evidence could be obtained of them *en masse*. It was found that their numbers are of the same order of magnitude as those of the recessive lethals. The "partial sterility" of treated males is, to an appreciable extent at least, caused by these dominant lethals. Another abundant class of mutations not previously recognized was found to be those which, when heterozygous, cause sterility but produce no detectable change in appearance; these too occur in numbers rather similar to those of the recessive lethals, and they may hereafter afford one of the readiest indices of the general mutation rate, when this is high. The sterility thus caused, occurring as it does in the offspring of the treated individuals, is of course a separate phenomenon from the "partial sterility" of the treated individuals themselves, caused by the dominant lethals.

In the statement that the proportion of "individually detectable mutations" was about one seventh for the X, and therefore nearly one half for all the chromatin, only the recessive lethals and semi-lethals and the "visible" mutants were referred to. If the dominant lethals, the dominant and recessive sterility genes and the "invisible" genes that merely reduce (or otherwise affect) viability or fertility had been taken into account, the percentage of mutants given would

have been far higher, and it is accordingly evident that in reality the great majority of the treated sperm cells contained mutations of some kind or other. It appears that the rate of gene mutation after X-ray treatment is high enough, in proportion to the total number of genes, so that it will be practicable to study it even in the case of individual loci, in an attack on problems of allelomorphism, etc.

Returning to a consideration of the induced mutations that produced visible effects, it is to be noted that the conditions of the present experiment allowed the detection of many which approached or overlapped the normal type to such an extent that ordinarily they would have escaped observation, and definite evidence was thus obtained of the relatively high frequency of such changes here, as compared with the more conspicuous ones. The belief has several times been expressed in the *Drosophila* literature that this holds true in the case of "natural" mutations in this organism, but it has been founded only on "general impressions"; Baur, however, has demonstrated the truth of it in *Antirrhinum*. On the whole, the visible mutations caused by raying were found to be similar, in their general characteristics, to those previously detected in non-rayed material in the extensive observations on visible mutations in *Drosophila* carried out by Bridges and others. A considerable proportion of the induced visible mutations were, it is true, in loci in which mutation apparently had never been observed before, and some of these involved morphological effects of a sort not exactly like any seen previously (*e.g.*, "splotched wing," "sex-combless," etc.), but, on the other hand, there were also numerous repetitions of mutations previously known. In fact, the majority of the well-known mutations in the X-chromosome of *Drosophila melanogaster*, such as "white eye," "miniature wing," "forked bristles," etc., were reobtained, some of them several times. Among the visible mutations found, the great majority were recessive, yet there was a considerable "sprinkling" of dominants, just as in other work. All in all, then, there can be no doubt that many, at least, of the changes produced by X-rays are of just the same kind as the "gene mutations" which are obtained, with so much greater rarity, without such treatment, and which we believe furnish the building blocks of evolution.

In addition to the gene mutations, it was found that there is also caused by X-ray treatment a high proportion of rearrangements in the linear order of the genes. This was evidenced in general by the frequent inherited disturbances in crossover frequency (at least 3 per cent. were detected in the X-chromosome alone, many accompanied but some unaccompanied by lethal effects), and evidenced specifically by various cases that were proved in other ways to involve inversions, "deficiencies," fragmentations, translocations, etc., of portions of a chromosome. These cases are making possible attacks on a number of genetic problems otherwise difficult of approach.

The transmuting action of X-rays on the genes is not confined to the sperm cells, for treatment of the unfertilized females causes mutations about as readily as treatment of the males. The effect is produced both on oöcytes and early oögonia. It should be noted especially that, as in mammals, X-rays (in the doses used) cause a period of extreme infertility, which commences soon after treatment and later is partially recovered from. It can be stated positively that the return of fertility does not mean that the new crop of eggs is unaffected, for these, like those mature eggs that managed to survive, were found in the present experiments to contain a high proportion of mutant genes (chiefly lethals, as usual). The practice, common in current X-ray therapy, of giving treatments that do not certainly result in permanent sterilization, has been defended chiefly on the ground of a purely theoretical conception that eggs produced after the return of fertility must necessarily represent "uninjured" tissue. As this presumption is hereby demonstrated to be faulty it would seem incumbent for medical practice to be modified accordingly, at least until genetically sound experimentation upon mammals can be shown to yield results of a decisively negative character. Such work upon mammals would involve a highly elaborate undertaking, as compared with the above experiments on flies.

From the standpoint of biological theory, the chief interest of the present experiments lies in their bearing on the problems of the composition and behavior of chromosomes and genes. Through special genetic methods it has been possible to obtain some information concerning the manner of distribution of the transmuted genes amongst the cells of the first and later zygote generations following treatment. It is found that the mutation does not usually involve a permanent alteration of all of the gene substance present at a given chromosome locus at the time of treatment, but either affects in this way only a portion of that substance, or else occurs subsequently, as an after-effect, in only one of two or more descendant genes derived from the treated gene. An extensive series of experiments, now in project, will be necessary for deciding conclusively between these two possibilities, but such evidence as is already at hand speaks rather in favor of the former. This would imply a somewhat compound structure for the gene (or chromosome as a whole) in the sperm cell. On the other hand, the mutated tissue is distributed in a manner that seems inconsistent with a general applicability of the theory of "gene elements" first sug-

gested by Anderson in connection with variegated pericarp in maize, then taken up by Eyster, and recently reenforced by Demerec in *Drosophila virilis*.

A precociously doubled (or further multiplied) condition of the chromosomes (in "preparation" for later mitoses) is all that is necessary to account for the above-mentioned *fractional effect* of X-rays on a given locus; but the theory of a divided condition of each gene, into a number of (originally identical) "elements" that can become separated somewhat indeterminately at mitosis, would lead to expectations different from the results that have been obtained in the present work. It should, on that theory, often have been found here, as in the variegated corn and the eversporting races of *D. virilis*, that mutated tissue gives rise to normal by frequent "reverse mutation"; moreover, treated tissues not at first showing a mutation might frequently give rise to one, through a "sorting out" of diverse elements, several generations after treatment. Neither of these effects was found. As has been mentioned, the mutants were found to be stable through several generations, in the great majority of cases at least. Hundreds of non-mutated descendants of treated germ cells, also, were carried through several generations, without evidence appearing of the production of mutations in generations subsequent to the first. Larger numbers will be desirable here, however, and further experiments of a different type have also been planned in the attack on this problem of gene structure, which probably can be answered definitely.

Certain of the above points which have already been determined, especially that of the fractional effect of X-rays, taken in conjunction with that of the production of dominant lethals, seem to give a clue to the especially destructive action of X-rays on tissues in which, as in cancer, embryonic and epidermal tissues, the cells undergo repeated divisions (though the operation of additional factors, *e.g.*, abnormal mitoses, tending towards the same result, is not thereby precluded); moreover, the converse effect of X-rays, in occasionally producing cancer, may also be associated with their action in producing mutations. It would be premature, however, at this time to consider in detail the various X-ray effects previously considered as "physiological," which may now receive a possible interpretation in terms of the gene-transmuting property of X-rays; we may more appropriately confine ourselves here to matters which can more strictly be demonstrated to be genetic.

Further facts concerning the nature of the gene may emerge from a study of the comparative effects of varied dosages of X-rays, and of X-rays administered at different points in the life cycle and under varied conditions. In the experiments herein reported, several different dosages were made use of, and while the figures are not yet quite conclusive they make it probable that, within the limits used, the number of recessive lethals does not vary directly with the X-ray energy absorbed, but more nearly with the square root of the latter. Should this lack of exact proportionality be confirmed, then, as Dr. Irving Langmuir has pointed out to me, we should have to conclude that these mutations are not caused directly by single quanta of X-ray energy that happen to be absorbed at some critical spot. If the transmuting effect were thus relatively indirect there would be a greater likelihood of its being influenceable by other physico-chemical agencies as well, but our problems would tend to become more complicated. There is, however, some danger in using the total of lethal mutations produced by X-rays as an index of gene mutations occurring in single loci, for some lethals, involving changes in crossover frequency, are probably associated with rearrangements of chromosome regions, and such changes would be much less likely than "point mutations" to depend on single quanta. A re-examination of the effect of different dosages must therefore be carried out, in which the different types of mutations are clearly distinguished from one another. When this question is settled, for a wide range of dosages and developmental stages, we shall also be in a position to decide whether or not the minute amounts of gamma radiation present in nature cause the ordinary mutations which occur in wild and in cultivated organisms in the absence of artificially administered X-ray treatment.

As a beginning in the study of the effect of varying other conditions, upon the frequency of the mutations produced by X-rays, a comparison has been made between the mutation frequencies following the raying of sperm in the male and in the female receptacles, and from germ cells that were in different portions of the male genital system at the time of raying. No decisive differences have been observed. It is found, in addition, that aging the sperm after treatment, before fertilization, causes no noticeable alteration in the frequency of detectable mutations. Therefore the death rate of the mutant sperm is no higher than that of the unaffected ones; moreover, the mutations can not be regarded as secondary effects of any semi-lethal physiological changes which might be supposed to have occurred more intensely in some ("more highly susceptible") spermatozoa than in others.

Despite the "negative results" just mentioned, however, it is already certain that differences in X-ray influences, by themselves, are not sufficient to account for all variations in mutation frequency, for the present X-ray work comes on the heels of the determination of mutation rate being dependent upon tempera-

ture (work as yet unpublished). This relation had first been made probable by work of Altenburg and the writer in 1918, but was not finally established until the completion of some experiments in 1926. These gave the first definite evidence that gene mutation may be to any extent controllable, but the magnitude of the heat effect, being similar to that found for chemical reactions in general, is too small, in connection with the almost imperceptible "natural" mutation rate, for it, by itself, to provide a powerful tool in the mutation study. The result, however, is enough to indicate that various factors besides X-rays probably do affect the composition of the gene, and that the measurement of their effects, at least when in combination with X-rays, will be practicable. Thus we may hope that problems of the composition and behavior of the gene can shortly be approached from various new angles, and new handles found for their investigation, so that it will be legitimate to speak of the subject of "gene physiology," at least, if not of gene physics and chemistry.

In conclusion, the attention of those working along classical genetic lines may be drawn to the opportunity, afforded them by the use of X-rays, of creating in their chosen organisms a series of artificial races for use in the study of genetic and "phaenogenetic" phenomena. If, as seems likely on general considerations, the effect is common to most organisms, it should be possible to produce, "to order," enough mutations to furnish respectable genetic maps, in their selected species, and, by the use of the mapped genes, to analyze the aberrant chromosome phenomena simultaneously obtained. Similarly, for the practical breeder, it is hoped that the method will ultimately prove useful. The time is not ripe to discuss here such possibilities with reference to the human species.

The writer takes pleasure in acknowledging his sincere appreciation of the cooperation of Dr. Dalton Richardson, Roentgenologist, of Austin, Texas, in the work of administering the X-ray treatments.

13

2

Reprinted from *Genetics*, *33*, 273–303 (May 1948)

THE EFFECT OF X-RAYS UPON MUTATION OF THE GENE *A* IN MAIZE*

L. J. STADLER AND HERSCHEL ROMAN

University of Missouri and U. S. Department of Agriculture, Columbia, Missouri,[1]
and University of Washington, Seattle, Washington

Received February 18, 1948

THE experimental evidence indicating that gene mutation is induced by X-rays comes almost wholly from experiments on the general mutation rate; that is, from experiments in which the total frequency of detectable mutations at all loci is compared in treated and untreated cultures. "Gene mutations" as experimentally identified in such studies, include all variations inherited as if due to a change in a single gene. If it may be assumed that such variations can arise only from the transformation of a gene to an alternative form, these experiments prove the effect of the treatment upon some process of genic evolution.

It has long been evident that this assumption is not valid. Various extra-genic alterations may produce effects which are distinguishable only with great difficulty, if at all, from the effects expected from qualitative change in the hereditary unit. These include alterations involving loss, reduplication, and rearrangement of unaltered genes. To avoid ambiguity, the non-committal term "point mutation" will therefore be used to designate apparent gene mutations as experimentally identified, and the term "gene mutation" will be used only to designate the hypothetical conversion of a gene to an allelic form.

The inference that gene mutation is induced by X-rays is founded upon the assumption that some among the induced point mutations are in fact due to intragenic change, although others may be shown to be due to extra-genic alterations simulating gene mutation. But there are no general criteria by which the observed mutations occurring at miscellaneous loci may be individually identified as intra-genic or extra-genic. Consequently it is not possible to determine the effect of X-ray treatment upon gene mutation by experimental modification of the general mutation rate.

The alternative is to study the mutational behavior of selected genes especially suited to the purpose, both in spontaneous mutation and in mutation induced by various treatments, in the hope of developing for these genes a sounder basis for interpretation. For this purpose the loci *A* and *R* in maize provide favorable material, for reasons which have been stated in an earlier paper (STADLER 1941).

* The cost of the accompanying illustrations is paid by the GALTON AND MENDEL MEMORIAL FUND.

[1] Cooperative investigations of the Division of Cereal Crops and Diseases, Bureau of Plant Industry, U. S. DEPARTMENT OF AGRICULTURE, and DEPARTMENT OF FIELD CROPS, UNIVERSITY OF MISSOURI. Missouri Agricultural Experiment Station Journal Series No. 1083.

The present study is concerned with the nature of X-ray-induced mutations at the A locus. A parallel study of ultraviolet-induced mutations at the A locus will be published shortly. These studies were made several years ago and the summarized results have been published in abstract form (STADLER and ROMAN 1943), but various delays incidental to the war have prevented the completion and fuller publication of the work. Meanwhile, certain other studies concerned with the general problem have been reported (STADLER and FOGEL 1943, 1945; STADLER 1944, 1946, FOGEL 1946a, b; LAUGHNAN 1946).

EXPERIMENTAL MATERIALS AND METHODS

The gene A, in the presence of appropriate complementary genes, affects the pigmentation of various tissues of the maize plant. In the presence of A_2, B, Pl, C, R, Pr, and P, the phenotypic effect of A may be represented as follows:

GENOTYPE	ALEURONE COLOR	PLANT COLOR	PERICARP COLOR
$A A, A a$	Purple	Purple	Red
$a a$	Colorless	Brown	Brown

The purple color of the aleurone and plant tissues is due to an anthocyanin pigment, while the red color of the pericarp is due to an unknown pigment, not an anthocyanin. Various additional alleles of A have been reported, including some with effects apparently intermediate between those of A and a (EMERSON and ANDERSON 1932; RHOADES 1941).

The populations from which the mutants here described were selected were F_1 progenies produced by the cross $aa \times AA$, the pollen of the male parent being X-rayed just before pollination. Plants showing loss of the A effect, whether by mutation, deficiency, or other types of genetic alteration, will be referred to as "A losses." These are readily identified in F_1, A loss in the endosperm resulting in a seed with colorless aleurone, A loss in the embryo in a seedling devoid of anthocyanin. Endosperm and embryo effects are produced independently in maize pollen irradiated at maturity, since the second microspore division occurs several days before the pollen is mature. Spontaneous mutations of A to a, occurring previous to gene reduplication for this division, would result in A loss in both endosperm and embryo. As determined from the frequency of mutation affecting endosperm and embryo together, the frequency of spontaneous mutation of A is low (STADLER 1941 and unpubl.), and no spontaneous mutations would be expected in a population equal to the total number of plants examined in this study.

With X-ray treatment of mature pollen, loss of the A effect is not uncommon, its frequency varying with the X-ray dose. A losses in the endosperm cannot be tested to discriminate between mutation and deficiency. The anthocyanin-free seedlings of the F_1, which include all A losses in the embryo, may include also maternal haploids or diploids resulting from development of the embryo without fertilization. The plants may be grown for further study, to distinguish these classes, and such alterations as are not lethal to the gametophyte may be carried on to subsequent generations for more critical analysis.

Haploids which survive the seedling stage are readily recognizable by their distinctive growth habit, reduced cell-size, and almost complete pollen abor-

15

tion, as well as by their wholly maternal phenotype. Maternal diploids would be recognizable only by their maternal phenotype. They would presumably be normal in development, and would be distinguishable from self-contaminations only by their failure to show the maternal phenotype in the endosperm of the seed from which the affected plant was grown. In cultures so marked as to permit identification of both maternal haploids and maternal diploids, more than 30 maternal haploids, but no maternal diploids, were found. The frequency of maternal haploids varies rather widely in different stocks used as female parent, and is not materially affected by X-ray treatment of the pollen parent, except for the slight increase in relative frequency expected from the elimination of a portion of the diploid progeny through lethal effects of the treatment.

The great majority of the *A* losses found were distinctly defective plants among which those which survived to flowering had at least 50 percent of the pollen aborted. Among the remainder, with approximately normal plant development, similar segregation of aborted pollen is found in almost all cases. In a few of these plants the segregating defective pollen was not aborted, but was distinctly subnormal in size and in starch content at maturity. Many of these plants show segregation for two or more factors producing defective pollen, and in some of these there is segregation for both aborted and subnormal pollen. Various grades of defective pollen development are recognizable in iodine-stained pollen specimens, and specific grades are associated with specific deficiencies. These range form aborted pollen containing little or no starch, characteristic of most known deficiencies, to types only slightly subnormal in size and starch content. In a few cases the pollen is wholly normal in appearance but fails to accomplish fertilization in competition with normal pollen. McClintock (1944) has described minute terminal deficiencies in chromosome 9 with pollen visibly normal and capable of normal functioning in competition with non-deficient pollen. The occurence of segregating defective pollen serves as a rather sensitive detector of segmental deficiency, but deficiencies without detectable effect upon the pollen may occur.

The purpose of the experiment was to select from among the *A* losses induced by X-ray treatment those least subject to the suspicion of deficiency, in order to determine whether these were in fact experimentally indistinguishable from gene mutation of *A*. If *A* losses free from pollen effect occur, a certain proportion of these may be missed in the F_1 plants because of the coincidental occurrence of alterations at other loci resulting in defective pollen segregation. The proportion of such cases would increase with increasing dosage, and would be roughly equal to the proportion of pollen-segregating plants in the entire progeny. For this reason relatively low X-ray doses (usually 400 r) were used at first, but when it became evident that a very large sample of *A* losses would be needed, doses of 1000 r were used. Among a total of 126 *A* losses found in these progenies, 76 failed to survive to flowering or were so defective as to give no pollen specimen, 49 showed segregation for aborted pollen, or for aborted and subnormal pollen, and one showed segregation for subnormal pollen only. None was a normal plant free from defective pollen segregation. Since the pro-

16

portion of A losses free from gametophytic effect was obviously extremely low, various progenies produced for other purposes were used as a supplementary source of A losses. These were mostly from doses of 1000 r, though some with higher doses were included. They included progenies not so marked as to permit regular identification of haploids, and in some cases the genotype was such as to give anthocyanin-free seedlings by R loss as well as A loss. Among the anthocyanin-free plants which did not survive the seedling stage the R losses and in some cases the haploids could not be distinguished from the A losses. The inclusion of this supplementary material made it impossible to determine accurately the total number of A losses observed, but it did result in the finding of two A losses with no visible effect upon the pollen. A total of 543 anthocyanin-free seedlings was observed in the course of the experiment. On the basis of the relative frequency of haploids and R losses in cultures properly marked for their identification, it is estimated that about 415 of these plants were A losses.

THE MUTANTS

The genetic alterations responsible for the two A loss plants with normal pollen were designated respectively a-$X1$ and a-$X2$, and were extracted for further study. In addition there was included for comparison the genetic alteration extracted from one of the A losses with segregating subnormal pollen. This was designated a-$X3$.

Alteration a-$X1$ occurred in a plant of normal vigor in an F_1 progeny of the cross $a\ A_2\ R^g\ C\ b\ pl$-$y\ pr\ wx\ p \times A\ A_2\ r^{ch}\ C\ b\ pl$-$y\ Pr\ Wx\ p$, in which the pollen was irradiated (dose 1000 r) just before pollination. The genotype was such as to give green plants by either A loss or r^{ch} loss. The ear, pollinated by a^p $Dt\ Dt$, yielded a full seed set, showing that the alteration was haplo-viable and transmissible through female germ cells. All of the seeds were pale, showing that the case was due to A loss rather than r^{ch} loss. Somewhat less than half of these seeds showed one or more dots, and the remainder were dotless. The plants grown from these seeds were crossed on and by an a-tester ($a\ A_2\ C\ R$). In crosses on the a tester stock all of the plants from dotted seeds gave approximately equal transmission of pale and colorless, while all but one of the plants from dotless seeds showed lowered transmission of the colorless seed type. Subsequent tests confirmed this indication, showing that the a "allele" derived from the treated parent was distinguished from the standard a allele by failure to respond to Dt (in any dosage) and by consistently lowered transmission through male germ cells.

Alteration a-$X2$ was found in F_1 of the cross $a\ A_2\ C\ R^g\ b\ pl\ p \times A\ A_2\ R^r\ C$ $b\ Pl\ p$, with treatment of the parental pollen at a dose of about 900 r just before pollination. The treatment was applied in the field with a mobile X-ray unit not calibrated in r units. The dose is estimated from the relative frequency of genetic effects commonly produced by this treatment, in comparison to the frequency of similar effects produced by treatments at measured doses. The plant was normal in vigor and produced two ears, both with approximately a full set of seed. Pollinations similar to those described for a-$X1$ gave similar results, except for a greater reduction in transmission through male germ cells

17

and an indication in some ears of possible reduced transmission through female germ cells.

Alteration a-$X3$ occurred in the F_1 of the cross a A_2 C-Wx r^r j $lg \times A$ A_2 c-wx R^g pr su y, in which both the parental ear and the parental tassel were X-rayed just before pollination. This treatment also was applied in the field. The dose applied to the tassel was approximately 1200 r. The original F_1 plant was normal in development, but about 50 percent of its pollen was clearly subnormal in size and starch content. A similar subnormal pollen type, associated with an R deficiency (DfX-1) has been illustrated and described in some detail in an earlier paper (STADLER 1933). The ear of this plant did not produce a full seed set, but in parts of the ear there seemed to be definitely more than 50 percent seed set. Cultures grown from the seed in these sections yielded a few plants with pollen segregation similar to that observed in the F_1 plant, and these were crossed with various testers to provide material for further study. Pollinations similar to those described for a-$X1$ showed that the subnormal pollen segregation is regularly associated with a-$X3$, and that there is no transmission of the alteration through male germ cells. The subnormal pollen is slightly less defective than in Df X-1, but is always clearly detectable. Transmission through female germ cells also is distinctly reduced. Its frequency is readily shown in pollinations of the type A/a-$X3 \times a^p$, in which the seeds heterozygous for a-$X3$ are pale, while their normal sibs are fully colored. The pale seeds in this pollination are usually less than one fourth as numerous as the colorless seeds, and in general are slightly smaller. The difference in seed size is slight, and would not be detectable in the shelled seed, because of difference in seed size due to position of the seed on the ear. The reality of the difference may be demonstrated objectively in ears of the type A/a-$X3 \times A$, in which the two classes are indistinguishable in color-phenotype. Small seeds, carefully selected by comparison of seed size in neighboring seeds, give with few exceptions plants showing subnormal pollen segregation. The plants grown from the small seeds are not visibly inferior to their sibs, either in seedling growth or in development at later stages.

All three alterations occurred in cultures lacking the complementary factors, B, Pl, and P. In subsequent extractions it was found, for all three alterations, that plants of genotype a/a-X B Pl are brown, like a/a B Pl, and plants of genotype a/a-X P have brown pericarp like a/a P. Several objective tests were made in an attempt to distinguish a/a-$X3$ from a/a sibs on the basis of plant color phenotype at various stages of development, but these attempts were wholly unsuccessful.

GAMETOPHYTE VIABILITY

In crosses of the type A/a-X $\times a^p$, the production of pale seeds demonstrably a-X/a^p in genotype proves that the a-X alteration is viable in the haploid female gametophyte, for these seeds can be produced only by the normal functioning of the a-X gametophyte. Similarly, the production of a^p/a-X seeds in crosses of the type $a^p \times a/a$-X proves that the alteration is viable in the male gametophyte. On the basis of these tests it is clear that a-$X1$ and a-$X2$ are viable in both male and female gametophytes, and that a-$X3$ is viable at least in female gametophytes. The failure of male transmission of a-$X3$ does not necessarily mean that pollen bearing this alteration is inviable, since it could result from the slower pollen tube growth of this type in competition with pollen bearing the standard allele a, as well as from failure of the pollen to germinate.

The pollinations made to determine haplo-viability show also the relative frequency of fertilization accomplished by a-X gametophytes versus gametophytes bearing a standard A or a^p allele. Differences in female transmission between two contrasted types presumably result only from differences in the proportion effectively fertilized or differences in survival of the heterozygotes produced. Differences in male transmission may result from these causes also, but in addition the factor of competition in pollen-tube growth may cause large inequalities in the number of ovules which may be fertilized by the two types of sperms. Trials of male transmission frequency are complicated by the fact that there are many unidentified pollen tube growth factors in maize, so that 1:1 transmission through male germ cells is not necessarily to be expected in the case of two alleles which are themselves without effect upon gametophyte functioning. A significant deviation from equality may be the result of other factors linked to those under study. For more critical evidence of the effect, it is desirable to make comparisons of male transmission between sibs secured from a cross of the type $A \times a$-X/a, the transmission tests being crosses of a number of the progeny plants, including A/a-X and A/a individuals, upon an a^{dl}-$tester$ (a^{dl} A_2 C R Dt Dt). The symbol a^{dl} represents an allele similar to a except for failure to respond to Dt. Colorless seeds in the crosses by A/a are dotted; those in crosses by A/a-X are dotless. The relative frequency of colored and colorless seeds in the crosses by A/a plants establishes a norm of transmission of the A chromosome for comparison with that shown in crosses by the A/a-X plants.

Trials of this type were made in various extracted stocks of both a-$X1$ and a-$X2$, and trials of female transmission in several extracted stocks of each of the three alterations. Alteration a-$X1$ regularly shows reduced male transmission, but the extent of reduction varies in different cultures. For example, in tests of a-$X1/a^p$ plants, in comparison with a/a^p sibs, seven a-$X1/a^p$ plants of one culture yielded a total of 803 colorless and 1390 pale seeds (58 percent of normal transmission), while in a closely related culture tested at the same time five plants tested yielded 317 colorless and 742 pale (43 percent of normal transmission). The individual plants did not differ significantly from the averages stated, and the a/a^p sibs in both cultures gave normal transmission. Occasional plants heterozygous for a-$X1$ have been found to give less than 25 percent normal transmission, but ordinarily male transmission of this alteration is not far from 50 percent of normal. Female transmission is usually equal to normal, with occasional exceptions showing slightly lowered transmission.

With a-$X2$, male transmission is commonly about 25 percent of normal, and female transmission about 65 percent of normal, again with occasional rather wide variations. One culture, in which seven plants of A/a-$X2$ and six plants of A/a were all tested for both male and female transmission, gave a total fre quency of 1334 A and 398 a-$X2$ in male transmission (30 percent normal) and

931 *A* and 661 *a-X2* in female transmission (71 percent normal). Differences between individual plants were slight and transmission was normal in the *A/a* sibs. In other cultures male transmission has frequently been as low as 15 percent and occasionally as low as 5 percent. For example, in one culture in which three *a^p/a-X2* plants were tested rather extensively for male transmission, the total frequency was 1588 *a^p* and 72 *a-X2* (4.5 percent normal).

With *a-X3*, male transmission has never been found, and female transmission is always low, ordinarily about 15–35 percent. Individual ears with no transmission of the alteration are not rare, but usually at least a few *a-X3* seeds are produced, and occasionally female transmission approaches 50 percent.

VIABILITY OF HOMOZYGOTES AND COMPOUNDS

When *A/a-X1* or *a^p/a-X1* plants are selfed or intercrossed, the homozygote *a-X1/a-X1* should be recognizable by absence of anthocyanin in the aleurone and plant tissues, and should occur with an average frequency of about 17 percent (assuming normal female transmission and 50 percent male transmission of *a-X1*). Selfed or intercrossed ears of these genotypes yield no colorless seeds at all. The absence of colorless seeds cannot be accounted for by the assumption that the phenotype of *A-X1/a-X1* is other than as expected, for all of the seeds from *A/a-X1* selfs are full-colored, and all of the seeds from *a^p/a-X1* selfs are pale. The plan s grown from these seeds show the corresponding plant color phenotype, and at least half of these plants are found to be heterozygous for *a-X1*. In many of these pollinations, the same pollen sample used in selfing was used also in outcrosses on an *a* tester, to establish the level of male transmission of *a-X1*, and sibs of the plants selfed were pollinated by an *a* tester to establish female transmission of *a-X1* in the cultures used. Male and female transmission tests of *a-X1* in the outcrosses gave results in agreement with the tests described in the preceding section. Assuming random fertilisation, the absense of colorless seeds thus shows that the homozygote is inviable. There was no class of aborted or defective seeds, or of empty pericarps, which could be considered to represent the ovules in which the combination occurred.

With *a-X2* the proportion of homozygotes expected in selfs and intercrosses is lower, due to the lower male and female transmission of this alteration. With female transmission at 65 percent and male transmission at 25 percent of normal, the expected average frequency of homozygotes would be about 8 percent. A large number of selfs and intercrosses of *a-X2* heterozygotes, including several with simultaneous tests of male and female transmission, failed to yield a single colorless seed, and it is clear that in this case also the absence of colorless seeds is due to the inviability of the homozygote.

The viability of *a-X3* homozygotes cannot be tested in this manner, because of the lack of male transmission.

The viability of the compound *a-X1/a-X2* is readily tested by crosses of the type *a^p/a-X2 × a^p/a-X1*. The expected frequency of colorless seeds representing the compound is about 13 percent, or for the reciprocal cross about 10 percent. Numerous crosses of these types were made, and all of the seeds produced were pale, except for one colorless seed. The plant grown from this seed proved

20

to be heterozygous for a dottable a. The seed thus does not represent the com pound, and presumable was the result of pollen contamination.

Viability of the compounds a-$X1/a$-$X3$ and a-$X2/a$-$X3$ may be tested similarly, but because of the low and variable female transmission of a-$X3$ the tests must be made on a fairly large number of ears of the a-$X3$ heterozygote. Crosses of the type a^p/a-$X3 \times a^p/a$-$X1$ were made on 11 ears, which produced a total of 985 seeds, all pale. With average transmission of the a-X alterations concerned, and with normal viability of the compound, the expected frequency of colorless seeds would be 55 (5.6 percent). Similarly, 13 ears of the cross a^p/a-$X3 \times a^p/a$-$X2$ yielded a total of 1130 seeds, all of which were pale except two colorless seeds which proved to be due to pollen contamination. With average transmission rates the expected frequency of colorless seeds would be 37 (3.3 percent). These crosses indicate beyond reasonable doubt that the compounds in all combinations, like the homozygotes, are zygotically lethal.[1]

EFFECT ON CROSSING OVER

Short intercalary deficiencies in maize sometimes reduce crossing over much more than would be expected from the length of the deficient segment. For example (STADLER 1935) *Df 5-1*, a deficiency of a short intercalary segment of the longer arm of chromosome 5, includes the locus V_3, and not the neighboring loci *Bt*, to the left, and *Bv*, to the right. Its genetic length is therefore less than 3.8 crossover units, the map distance from *Bt* to *Bv*, The deficiency is haplo-viable in the female gemetophyte, and in heterozygous plants it is cytologically detectable at pachytene by the occurrence of a short "buckle" in chromosome 5, usually located on the longer arm near the centromere. In plants heterozygous for *Df 5-1*, crossing over in the region between *Bm* and *Bv* (the region including the deficiency) is almost completely inhibited, and crossing over in the adjacent region between *Bv* and *Pr* is greatly reduced. The total reduction in crossing over is considerably greater than the map length of the deficiency.

Presumably, the reason for the reduction in crossing over outside of the deficient region is the strong tendency toward non-homologous pairing in the chromosomes of maize. McCLINTOCK (1933) has shown, in plants heterozygous for various chromosomal derangements, the frequent occurrence of close pachytene pairing of non-homologous regions. In plants heterozygous for *Df 5-1*, the buckle observed in the chromosome 5 pair at pachytene is formed by a segment of the non-deficient homolog equal in length to the segment missing from the deficient homolog. If pairing were always between homologous regions, the position of the buckle would be constant and would mark the location of the deficiency. In fact, however (STADLER 1935), the position of the buckle varies widely, and in extreme instances it may occur on the distal half of the longer arm, or in the proximal region of the shorter arm. When the position of the buckle is not identical with the locus of the deficiency, the chromosome re-

[1] Note added in proof: RHOADES (Maize News Letter, Mar. 1948), using a-$X2$ in combination with a gametophyte factor ga distal to et, finds no production of homozygous a-$X2$ seeds on selfed ears of the heterozygote ga/a-$X2$. Here the ratio expected is increased to about 30 percent, and the absence of visible sterility on the ear suggests that selective fertilization may be involved.

gions paired between these two points must be non-homologous. In cases in which the buckle is found on the shorter arm, the non-homology of the paired region is cytologically evident, for the centromeres of the two chromosomes are separated by a distance approximately equal to the length of the buckled segment.

If non-homologous pairing is a major cause of the reduction in crossing over, it may be expected that pronounced reduction in crossing over may usually be found with short intercalary deficiencies in maize. The cytological observations described above are typical of the observed behavior of short intercalary deficiencies. Deficiencies too short to cause the formation of a visible buckle, though they would not permit its cytological demonstration, could result in non-homologous pairing which might extend over a considerable length of chromosome and thus might lead to distinct reduction in crossing over.

The effect of the *a-X* alterations upon crossing over can be determined only with rather distant or otherwise unfavorable marker genes. The only known marker distal to *A* is *et*, an X-ray-induced mutant located about 12 units from *A* (STADLER 1940). The phenotypic effect of *et* is shown by both endosperm and seedling, the endosperm surface being characteristically scarred (etched), and the seedling being virescent. The endosperm effect is usually distinct enough for positive separation, though it may sometimes be confused with scarred endosperm resulting from the action of other genes or occurring sporadically with no simple genetic basis. In some matings the endosperm effect is very extreme, and many of the etched seeds are small and inviable. The virescent character of the seedling is always clear and readily separable, except at unusually high temperatures. Cultures which give a clear etched endosperm separation always give virescent seedlings from all etched seeds, and in ears with endosperm separation doubtful, the seedling separation is useful as a check. The alteration is readily transmitted through both male and female germ cells, though sometimes in less than normal ratio.

The nearest useful locus proximal to *A* is *lg₂*, about 34 crossover units distant. The gene *na* is probably a few units closer to *A* but is not well suited to seedling detection. The effect of *lg₂* is clear in the seedling stage, permitting positive classification in the rather large numbers needed for the comparison.

The effect of the *a-X* alterations upon crossing over in the regions *Et-A* and *A-Lg₂* was determined from sib plants produced in crosses of the type *Et a-X Lg₂/Et a Lg₂×et A lg₂*. This cross yields plants of two types, *Et a-X Lg₂/et A lg₂* and *Et a Lg₂/et A lg₂*. Crossover frequency was determined in plants of both types, by pollinating them by *et a lg₂*. The results are shown in table 1. The relationship of the plants tested is indicated by the culture numbers. For example, the effect of *a-X1* was tested in two cultures, 49:763 (including two *a-X1/A* plants and three *a/A* plants tested) and 51:14 (including three *a-X1/A* and three *a/A* plants tested). The cultures 49:763 and 51:14 were progenies of two different crosses of the type *Et a-X Lg₂/Et a lg₂×et A Lg₂*.

With *a-X1* there is no consistent indication of crossover modification, although the average frequency of crossing over in the *Et-A* interval is slightly lower than in the normal sibs tested.

With *a-X2*, there is a pronounced and consistent reduction in crossover fre-

22

Table 1

Effect of a-X1, a-X2, and a-X3 on crossing over. Crossover frequency in female gametes in backcrosses.

F₁ GENOTYPE	NON-CROSSOVERS		CROSSOVERS						CROSSOVER %	
			REG. 1		REG. 2		REG. 1, 2		REG. 1	REG. 2
a-X1										
et A lg/Et a-X1 Lg										
49:763-3	44	57	7	8	22	26	2	0	10.2	30.1
49:763-12	106	85	11	16	36	52	4	2	10.6	30.1
51:14-6	98	110	10	14	55	48	6	5	10.1	33.0
51:14-13	88	70	19	20	41	47	2	2	14.9	31.8
51:14-19	69	67	13	13	21	14	2	9	17.8	22.1
Total	405	389	60	71	175	187	16	18	12.5	30.0
et A lg/Et a Lg										
49:763-4	80	81	25	24	26	45	2	1	18.3	26.1
49:763-9	81	82	19	16	36	39	3	2	14.4	28.8
49:763-10	42	61	13	10	28	26	2	6	16.5	33.0
51:14-5	72	70	17	16	35	29	0	4	15.2	28.0
51:14-9	93	85	16	16	36	58	4	8	13.9	33.5
51:14-20	82	62	6	18	28	31	4	3	13.3	28.2
Total	450	441	96	100	189	228	15	24	15.2	29.6
a-X2										
et A lg/Et a-X2 Lg										
49:774-7	86	62	0	5	26	28	0	3	3.8	27.1
49:774-8	43	60	2	7	12	9	0	0	6.8	15.8
51:15-2	63	110	2	9	12	25	1	1	5.8	17.5
51:15-4	21	19	0	3	3	6	0	0	5.8	17.3
51:15-11	62	82	1	4	13	35	1	1	3.5	25.1
51:15-13	48	64	1	8	16	9	1	3	8.7	19.3
51:15-14	57	90	2	1	14	25	0	0	1.6	20.6
51:15-17	68	97	4	9	13	34	0	1	6.2	21.2
51:15-18	45	92	2	5	10	18	1	0	4.6	16.8
Total	493	676	14	51	119	189	4	9	5.0	20.6
et A lg/Et a Lg										
49:774-9	16	54	3	10	15	31	1	0	10.8	36.2
49:774-11	28	67	7	11	13	28	1	2	13.4	28.0
49:774-14	40	42	7	15	19	32	1	4	16.9	35.0
51:15-3	55	100	7	19	35	52	2	1	10.7	33.2
51:15-15	60	113	16	12	31	77	2	6	11.4	36.6
51:15-19	55	120	10	19	25	65	1	2	10.8	31.3
Total	254	496	50	86	138	285	8	15	11.9	33.5
a-X3										
et A lg/Et a-X3 Lg										
49:751.1-1	93	7	0	2	37	4	0	0	1.4	28.7
49:751.1-3	122	25	0	1	45	12	0	0	0.5	27.8
Total	215	32	0	3	82	16	0	0	0.9	28.2
et A lg/Et a Lg										
49:751.2-1	110	109	20	17	52	53	2	0	10.7	29.5
49:751.2-3	19	101	9	6	44	20	0	14	13.6	36.6
49:751.2-7	41	51	6	8	30	32	1	4	11.0	38.7
49:751.2-9	85	77	17	13	27	42	7	2	14.4	28.9
Total	255	338	52	44	153	147	10	20	12.4	32.4

quency, both in the *Et-A* interval, to the left of the alteration, and in the *A-Lg₂* interval, to the right. There is considerable variation in the extent of reduction of crossing over in different plants, but in each of the nine *a-X2/A* plants tested, for both of the segments covered, the crossover frequency is lower than in any of the six *a/A* plants tested.

With *a-X3*, only two plants heterozygous for the alteration are available in the test for crossover effects in the entire *Et-Lg₂* region, since in the tests made in the later series it was necessary to use cultures not marked at the *Lg₂* locus.

TABLE 2

Effect of a-X3 on crossing over (Et-A segment). Crossover frequency in female gametes in backcrosses.

F₁ GENOTYPE	NON-CROSSOVERS		CROSSOVERS		CROSSOVER PERCENT
	et A	*Et a(−X3)*	*et a(−X3)*	*et A*	
et A/Et a-X3					
49:751.1–1	130	11	0	2	1.4
49:751.1–3	167	37	0	1	0.5
51:11.1–1	30	24	0	3	5.3
51:11.1–10	91	64	1	6	4.3
51:11.2–8	64	63	0	10	7.3
51:12.1–12	68	19	0	4	4.4
51:12.1–14	71	10	0	8	9.0
Total	621	228	1	34	4.0
et A/Et a					
49:751.2–1	162	162	22	17	10.8
49:751.2–3	63	121	9	20	13.6
49:751.2–7	71	83	7	12	11.0
49:751.2–9	112	119	24	15	14.4
51:11.1–11	95	127	14	8	9.0
51:11.2–6	71	141	9	23	13.1
51:11.2–22	120	133	12	18	10.6
51:12.1–11	66	136	23	17	16.5
51:12.1–13	86	176	10	16	9.0
Total	846	1,198	130	146	11.9

The data on crossing over in the *Et-A* region in these tests will be given presently. The data for the two plants marked at all three loci show pronounced reduction of crossing over in the *Et-A* region, but only slight and insignificant reduction in the *A-Lg₂* region.

Additional data for *a-X3*, on crossing over in the *Et-A* region, are given in table 2. The endosperm classification was checked by seedling classification in all cases, and seeds which failed to germinate are omitted. The data are therefore comparable with those for the *Et-A* region in table 1. Data on the *a-X3* effect from table 1 are included in table 2 for comparison.

The results show a pronounced reduction in crossing over in plants heterozygous for a-X3, similar to that observed with a-X2, and like the latter vary-ng rather widely in different F_1 plants tested.

It should be noted that there are large inequalities in the frequency of cor-responding classes in these backcross progenies. This is evident not only in the progenies of A/a X plants, in which inequality is expected from the lowered female transmission of the various a-X alterations, but also in the progenies of A/a plants, in which this factor would not enter. The marker gene et shows distinctly reduced viability in the families used in testing the crossover effect of a-X2 and a-X3, in which the progenies of A/a plants used as controls show Et:et ratios of 1.00:0.51 and 1.00:0.73 respectively. The reduced frequency of

TABLE 3

Effect of a-X1, a-X2, and a-X3 on crossing over (ET-A segment). Data from totalled seed counts. (Crossover frequency in female gametes in backcrosses.)

F₁ GENOTYPE	NUMBER OF CUL-TURES	NON-CROSSOVERS		CROSSOVERS		CROSS-OVER percent	Et:et RATIO 1.00:	A:a(-X) RATIO 1.00:
		et A	Et a(-X)	et a(-X)	Et A			
For a-X1								
et A/Et a-X1	5	598	597	78	90	12.3	0.98	0.98
et A/Et a	6	676	683	114	132	15.3	0.97	0.99
For a-X2								
et A/Et a-X2	9	737	892	27	61	5.1	0.80	1.15
et A/Et a	6	484	798	61	108	11.7	0.60	1.45
For a-X3								
et A/Et a-X3	7	917	261	8	37	3.7	0.32	0.28
et A/Et a	9	1,036	1,261	166	153	12.2	0.85	0.83

et is due in part to lowered female transmission and in part to lowered germi-nation of et seeds in some of the cultures. Both tendencies vary widely in dif-ferent families; note for example that the Et:et ratio in progenies of A/a plants used as controls for a-X1 effects is 1.00:0.95.

Since Et and A are both identified by seed as well as seedling effects, in-equalities due to differences in germination may be excluded from data on the Et-A region by using the seed counts rather than the seedling counts. All of the data relating to Et-A crossing over in tables 1 and 2 were recalculated in terms of the total population of seeds produced. Since the seedling results cor-rected certain errors of seed classification for et, the corrected figures were used, and in the classes affected the same proportionate correction was applied to the seeds which failed to germinate. This recalculation made slight changes in the crossover frequencies observed in individual cultures, but made no ma-terial change in the indications of effects on crossing over. In the case of both a-X2 and a-X3, in the data from total seed counts, crossover frequency was lower in every A/a-X progeny than in any of the control A/a progenies. The effect upon crossover frequency in the totalled data for each group is shown in table 3.

It is evident that marked inequalities of the *Et* and *et* classes remain in the families used for the *a-X2* and *a-X3* comparisons, with a reduction of female transmission of *et* to about 60 percent and 85 percent respectively in the cultures from the *et A/Et a* control plants. In the control cultures this does not distort the crossover frequency, since it affects the crossover and non-crossover classes equally. But in the cultures segregating also for an *a-X* allele of re duced viability, the effect upon these two classes is necessarily unequal. Is this factor, rather than an actual decrease in crossing over, the cause of the reduced frequency of crossover gametes observed in the *A/a-X* plants?

Assume, for example, a crossover frequency of 12 percent in the *et A/Et a-X2* plants, as in their control sibs. In an ear with a potential yield of 200 seeds, we expect 88 gametes of each of the non-crossover types and 12 of each of the crossover types. Assuming 60 percent survival for *et*, the frequency of the *et A* class is reduced from 88 to 53. The frequency of the *Et a-X2* class will be similarly reduced by lower female transmission of *a-X2*, but the extent of this reduction varies in different cultures. We may estimate the reduction in these plants directly by comparing the *a-X* frequency observed with that expected if *a-X* had no effect on viability. With 12 percent crossing over and with full survival of *a-X2*, these cultures would be expected to have a ratio for *A:a-X* of 1.00:1.47, the reduced frequency of *A* being due to the disproportionate elimination of *A* individuals by the action of *et*. Actually the ears had a ratio for *A:a* of 1.00:1.15, indicating survival of *a-X2* at about 78 percent of normal. The expected frequencies of the four classes (with no interaction of viability effects) would thus be

	et A	*Et a-X*	*et a-X*	*Et A*	TOTAL	CROSSOVERS	% OF CROSSOVERS
Gamete ratio	88	88	12	12	200	24	12.0
Survival rate	.60	.78	.47 (.60×.78)	1.00			
Expected frequency	53	69	6	12	140	18	12.9

The effect is thus to increase slightly the apparent frequency of crossing over. With lower survival of gametes, such as might occur with *a-X3*, the error would be increased materially, as shown below:

	et A	*Et a-X*	*et a-X*	*Et A*	TOTAL	CROSSOVERS	% OF CROSSOVERS
Gametes	88	88	12	12	200	24	12.0
Survival rate	.60	.25	.15	1.00			
Expected	53	22	2	12	89	14	15.7

It is clear that this error in estimating crossover frequency, when the viability factors are in repulsion-phase, must always be in the plus direc ion. The apparent increase in crossover frequency is due to an unbalanced reduction in survival of the various classes of gametes, and while both non-crossover classes are reduced in survival, only one of the two crossover classes is so affected.

The illustrations given assume no interaction of the viability factors. For example, in the first illustration, it is assumed that of the 12 gametes of geno-

26

type *et a-X2* expected, 40 percent would be eliminated by the action of *et* and of the remainder, 22 percent would be eliminated by the action of *A-X2*, It would not be surprising if the survival of *et a-X2* gametes were less than the expected 47 percent, since the effect of both viability factors in the same individual might be greater than would be anticipated from their effects in different individuals. But even if the cumulative effect were much increased, it could not result in any large decrease in apparent crossovers compared with the control, for it could only affect individuals in the single class which provides only a minority of the observed crossovers. At its extreme, in the example given, with complete elimination of this class, it would give an apparent reduction of crossing over from 12 percent to 8.6 percent. In the second illustration, with lower survival, the maximum decrease possible would not be sufficient to balance the spurious increase in observed crossovers, and the results with complete elimination of the *et a-X* class would still indicate an increase in crossovers over the control. Obviously, the reduction of crossovers to less than half the control frequency, as found with both *a-X2* and *a-X3*, could not be due to these sources of error.

The experimental results (table 3) show further that the *et a-X* class was not eliminated; it provides 27 of the 88 crossovers observed with *a-X2* and 8 of the 45 with *a-X3*. Is this inequality of the crossover classes, and the observed inequality of the non-crossover classes, reasonably accounted for on the basis of viability differences involved?

Consider first the experimental data for *a-X2* (table 3). The survival ratio for *et* in the families used is 0.60, as indicated by the control cultures. With 60 percent survival of *et* in the *et A/Et a-X2* cultures, and with full survival of *a-X2*, these ears should have produced 1228 *Et a-X2* seeds (737/0.60) and 37 *et a-X2* seeds (61×0.60). Actually they produced 892 *Et a-X2* and 27 *et a-X2* seeds, indicating survival of about 73 percent in both classes. In other words, on the assumption of 60 percent survival for *et*, 73 percent survival for *a-X2*, and no interaction of viability effects, a gametic population of 2456 non-crossovers and 122 crossovers would have given the results observed. This represents an actual crossover percentage of 4.7, though among the surviving individuals crossovers would constitute 5.1 percent.

A similar calculation for the *et A/Et a-X3* cultures is shown below:

	et A	*Et a-X3*	*et a-X3*	*Et A*
Observed (table 3)	917	261	8	37
Expected with full survival of *a-X3*	917	1079	31	37
Indicated survival of *a-X3*		24%	26%	
Indicated gametic ratio	1079	1079	37	37

Here also the inequality of the crossover classes is well accounted for by the relative viability of the genotypes concerned, and the crossover frequency indicated, 3.3 percent, is a little lower than that shown by the percentage of crossovers among survivors.

Thus, in spite of the rather unfavorable marker genes necessarily used, the data show pronounced effects of *a-X2* and *a-X3* on the frequency of crossing over. The effect is unexpectedly large, *a-X2* heterozygotes showing a reduction of about 20 crossover units in the *Et-Lg₂* segment. In *a-X3* heterozygotes, which were adequately tested only in the *Et-A* segment, the reduction in frequency of crossing over in this region was even greater than that shown by *a-X2*.

CYTOLOGICAL OBSERVATIONS

The pronounced effect of *a-X2* and *a-X3* upon crossing over suggests an effect of these alterations on chromosome pairing in the region of the *A* locus. The additional effect of *a-X3* upon pollen development, and its sharply reduced female transmission, suggests further deficiency at this locus. Cytological examinations were made therefore of plants heterozygous for *a-X3*, in the pachytene stage of the microsporocyte.

The *A* locus is in the terminal one-fifth of the long arm of chromosome 3 (MCCLINTOCK 1931). This region was carefully examined in two types of material. In the first of these, the normal chromosome 3 carried a distinctive knob, near the midpoint of the long arm, which facilitated identification of the chromosome. Repeated observations failed to show any consistent abnormality in the region concerned. However, the material was not well suited for the detection of small alterations since the knob itself, in heterozygous condition, was responsible for some looseness and the frequent occurrence of buckles, usually in the vicinity of the knob but sometimes also in adjacent regions. More critical material was obtained from plants heterozygous for *a-X3* but lacking the knob. Several pachytene configurations were examined in which chromosome 3 was positively identified. In these there was no suggestion of a deficiency buckle or an inversion loop in the region of the *A* locus; the strands were as closely paired in this region as in other regions of the chromosomes.

The examinations made show that no reciprocal translocation is present, and no deficiency, insertion, or inversion long enough to interfere with visually normal pairing of the affected chromosome with its normal homolog. However, they do not exclude the possibility of a minute alteration that might be detectable only if located in a region most favorable for cytological study. All three *a-X* alerations, if due to deficiency, must be intercalary since the gene *Et*, located distal to *A* and present in the stocks originally treated, was not lost in any of the alterations. It is doubtful that losses as small as the minute terminal deficiences of chromosome 9, recently described by MCCLINTOCK (1944), would be cytologically detectable if they occupied an intercalary position.

EVIDENCE FROM THE USE OF AN UNSTABLE DUPLICATION

The unstable duplication used in the experiments now to be described was secured indirectly from a chromosomal aberration found in a progeny from ultraviolet-treated pollen. A stock carrying *Aᵇ* was irradiated and crossed on an *a* tester stock, in order to secure *A* losses from ultraviolet treatment for study in comparison with those from X-ray treatment. One of the plants of this progeny was a sectorial chimera, in which most of the plant showed the *a* phenotype but a sector of considerable size was of *Aᵇ* phenotype. Both sectors extended into the tassel, and pollinations made from *Aᵇ* anthers upon *a* tester ears showed some transmission of A^b (39 colored seeds+7 colored-colorless mosaics+260 colorless).

The colored seeds yielded colored plants, typical of A^b. A few of these were found to show small sectors of a tissue. The colored-colorless mosaic seeds regularly gave plants variegated for A^b and a tissue, suggestive of the ring-chromosome variegation types described by McCLINTOCK (1938). A typical plant of this sort is shown in fig. 1, A. Cytological examination of these plants showed at pachytene a short supernumerary fragment in addition to the normal complement of ten pairs of chromosomes. This is designated Dp $3a$, and will be referred to in this paper simply as Dp. The fragment was never seen as an open ring though it may have been a collapsed ring. Cytological study of the aberration was undertaken by DR. HELEN V. CROUSE, and will be reported separately.

This unstable duplication covering the A locus provides an opportunity for further study of the a-X alterations. With the addition of this duplication, the homozygotes and compounds of the a-X alterations may be shielded from the zygotic lethal effect, and, if these genotypes are viable in any tissue, the subsequent loss or diminution of the duplication fragment might make it possible to observe their effects.

Crosses were made with all three a-X alterations, according to the scheme illustrated below in the case of a-$X1$:

(1) a-$X1/a^p \times a$ a, Dp.

Most of the seeds produced by this pollination are pale or colorless. There are in addition a considerable number of seeds showing full color due to A^b, but all of these are endosperm mosaics. In about half of these mosaics the sector lacking full color is pale; in the remainder it is colorless. The latter class represents the genotype a-$X1/a$, Dp. The plants grown from these seeds show a sectorial phenotype similar to a a, Dp (fig. 1, A). A seed showing a typical endosperm mosaic of this class is shown in fig. 3, A.

(2) a-$X1/a^p \times a$-$X1/a$, Dp.

This cross similarly yields pale and colorless seeds together with colored seeds showing endosperm mosiacs. Among the latter a new type appears, in which the region lacking color, instead of being healthy endosperm tissue distinguished only by pigmentation of the aleurone layer, is shriveled, degenerate tissue. An example is shown in fig. 3, B.

Seeds of this type are a-$X1/a$-$X1$, Dp. Plants grown from these seeds are variegated, but are of a type very different from their a $X1/a$ (or a^p), Dp sibs which resemble the plant pictured in fig. 1, A. A typical plant of a-$X1/a$-$X1$, Dp is pictured in fig. 1, B. The sectors are all relatively small, and in general they are marked by absence of chlorophyll rather than absence of anthocyanin. The plants are distinctly defective in development, but they usually survive to maturity, and produce both tassels and ears. Flowering is delayed, and seed yield is small.

(3) a-$X1/a^p \times a$-$X1/a$-$X1$, Dp.

Crosses of this type are used to produce the *a-X/a-X, Dp* types in quantity, since all of the seeds produced are colored seed mosaics. The mosaics not showing degenerate sectors are mosaic for pale, not colorless.

Compounds are produced by appropriate substitution of the various *a-X* alterations in crosses (2) and (3). The chief difficulty in extracting the types desired comes from the low transmission of *a-X3*, and to a lesser extent of *a-X2*, in the heterozygotes lacking the duplication. Both alterations are readily transmitted through pollen from plants carrying the duplication. Successful extraction of *a-X2* or *a-X3* in cross (2) thus makes the production of their compounds with *a-X1* a simple matter, but the extraction of the homozygotes requires further use in cross (3) of the non-duplication stocks of low transmission. It was ultimately possible to extract all three homozygotes and all three compounds. The types *a-X3/a-X3, Dp* and *a-X2/a-X3, Dp* were first secured in crosses using *Dp* stocks as both male and female parents, a procedure which is undesirable for some purposes because of the possible inclusion in the progeny of individuals inheriting a duplication from both parents. The two types were later obtained also from crosses analogous to those of the scheme outlined above.

Dp-bearing plants homozygous for *a-X1*, and those representing the *a-X1/a-X2* and *a-X1/a-X3* compounds, were produced by crossing the appropriate types on *a-X1/a^p* (cross-type (3) above). Thus in each of the three crosses, the *a-X/a-X* types desired were produced in a progeny which included sib plants of *a-X/a^p* for comparison. In each case, the *a-X/a-X* plants were distinctly smaller and in general later than their *a-X/a^p* sibs, and were characterized invariably by the occurrence of the characteristic chlorophyll-lacking sectors described above and pictured in fig. 1, B. The variegation type observed showed no consistent differences in the three *a-X/a-X* genotypes included.

These sectors, marked by absence of chlorophyll, in general gave no indication of loss of *A*, for the anthocyanin pigmentation within the sector was in almost all cases unaffected. Examination for anthocyanin effects cannot be made effectively in leaf blade tissue, since little anthocyanin is produced in this tissue. Many of the sectors identified in the blade extend into the leaf sheath, in which, with the proper complementary factors, anthocyanin pigmentation is strong. For pigmentation of the sheaths above the base, the complementary factor *B* is essential, and unless the complementary factor *Pl* is also present, pigmentation is absent in the portion of the sheath shielded from light by overlying tissue. Critical examination for anthocyanin effects in the sectors can therefore be made only in *B Pl* plants, and in some of the sectors of *B pl* plants. Most of the sectorial plants carried *B* and a substantial proportion had *Pl* also.

Among these, close examination was made for sectors lacking anthocyanin, with or without associated lack of chlorophyll. In plants heavily pigmented with anthocyanin in the sheath, small sectors lacking chlorophyll cannot always be detected in sheath tissue. The broader sectors found in the leaf blade, clearly marked by absence of chlorophyll, may usually be traced back

A

31

B

FIGURE 1. Characteristic variegated plant types. A. *a/a, Dp*. Note broad to narrow sectors lacking anthocyanin. There are no sectors lacking chlorophyll. B. *a-X1/a-X1, Dp*. Note narrowed sectors lacking chlorophyll. Very few of these sectors lack anthocyanin.

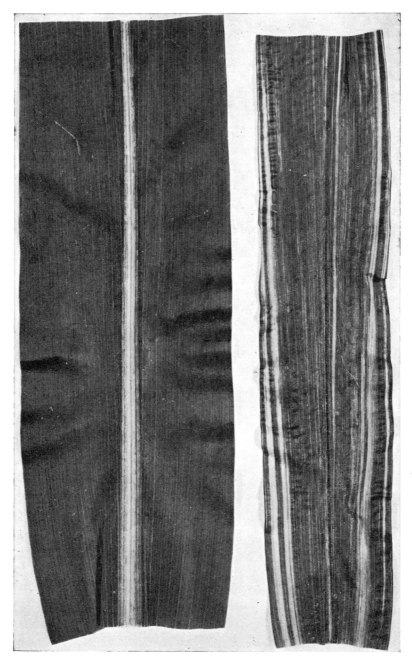

A   B

FIGURE 2.—Leaves of plants shown in Fig. 1. A. *a/a, Dp.* No chlorophyll sectors.
B. *a-X1/a-X1, Dp.* Numerous chlorophyll sectors.

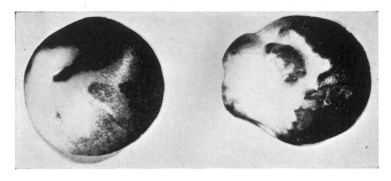

A   B

FIGURE 3. Characteristic endosperm mosaics. A. *a/a, Dp.* The mosaic sector is characterized only by the absence of anthocyanin pigmentation. B. *a-X1/a-X1, Dp.* The mosaic sector is characterized by degenerate endosperm tissue.

to the sheath, where even in the presence of strong anthocyanin the sector is recognizable by its lighter color resulting from the absence of chlorophyll. Such sectors are reasonably numerous, a single leaf sometimes providing a dozen or more. The counts were always made from only one leaf of any plant, to avoid including sectors of common origin in different counts.

It was found that in a large majority of the sectors lacking chlorophyll, anthocyanin pigmentation was present and apparently wholly normal, in comparison with adjoining tissue outside the sector. Much less frequently

A   B

FIGURE 4. Sib plants in F₁ of *a-X3/aᵖ⨯a-X1/a-X3, Dp.* A. *a-X1/a-X3,*
Dp. B. *a-X3/a-X3, Dp.*

A   B

FIGURE 5. Leaf blades of plants shown in Fig. 4. A. *a-X1/a-X3, Dp*. B. *a-X3/a-X3, Dp*.

sectors lacking anthocyanin as well as chlorophyll were found. Usually such sectors were adjoined on one or on both sides by sectors lacking chlorophyll only, as if the sector characterized by loss of both anthocyanin and chlorophyll occurred within a larger sector initiated by loss of chlorophyll only. There were, however, other cases in which sectors showing loss of both anthocyanin and chlorophyll occurred without adjoining sectors showing loss of chlorophyll only.

In rare instances sectors lacking anthocyanin and not lacking chlorophyll occurred. The clearest case of this kind, observed in an *a-X1/a-X3, Dp* plant, was a sector about four mm in width, devoid of anthocyanin and clearly normal in chlorophyll. It was adjoined on one side by normal tissue with deep anthocyanin pigmentation, with a sharp line of demarcation between the pigmented and non-pigmented tissues. On the other side it was adjoined by a sector about six mm wide, in which chlorophyll was lacking. Within the latter sector there were several streaks of anthocyanin-bearing tissue, but otherwise anthocyanin was lacking in this sector also. Among sectors lacking anthocyanin, both in cases with and without chlorophyll, there is sometimes clear brown pigmentation similar to that found in a *a B Pl* plants.

These observations suggest that the sectors observed in the *a-X/a-X* plants result not from the loss of the duplication as a whole, but from losses of a portion only. If the *a-X* alterations represent deficiencies which are effectively cell-lethal, this is the result to be expected. Losses of the duplication as a whole, occurring in the course of development, would only eliminate the cell progeny of occasional cells of growing meristems. This might result in the reduced growth and distortion observed in the *a-X/a-X* plants as compared to their *a-X/a^p* sibs, but could not produce sectors of phenotypically distinguishable tissue. But if the duplication is capable of partial loss in mitosis, such as the diminution in ring chromosomes demonstrated by McClintock (1938), and if the cell-lethal effect of the *a-X* alterations is due to a locus different from the *A* locus or the chlorophyll-essential locus (here designated *W*), then partial losses could occur which would given viable sectors lacking chlorophyll or anthocyanin or both. The relative frequency of such sectors would depend upon the sequence of the genes concerned, relative to the centromere of the duplication fragment. The relatively low frequency of sectors lacking both chlorophyll and anthocyanin, and the fact that there are usually found adjoining sectors lacking chlorophyll only, suggests that a factor essential for cell-viability may be so located that losses of a segment including both *A* and *W* would ordinarily include the cell-viability factor, but a loss involving one of the two loci might be followed by a loss involving the other, without including the viability factor.

It should be possible, by detailed study of the cytological behavior of the fragment and of the relations of the various genetic effects in the sectors, to locate these factors more definitely, and to distinguish between the various viability effects involved. For our present purpose, the essential point is the demonstration that all three of the *a-X* alterations are characterized by the absence of distinct and separable genetic effects. This is evident from the varying types of sectors observed.

The phenotypic similarity of *a-X1/a-X1, Dp* plants to the compounds *a-X1/a-X2, Dp* and *a-X1/a-X3, Dp* is best tested in the progeny of crosses of the type *a-X1/a^p × a-X1/a-X2* (or *a-X3*), *Dp*. Here the seeds with degenerate sectors include two types, *a-X1/a-X1, Dp* and *a-X1/a-X2* (or *a-X3*), *Dp*. There was little basis in the appearance of the plants for any attempted classification of the two genotypes. In the case of *a-X1/a-X1, Dp* versus *a-X1/a-X3, Dp*, attempted classifications may be checked at flowering by pollen examination, for the plants of the latter class show about 40 percent subnormal pollen due to the effect of *a-X3* in the absence of *Dp*. These trials showed no relation of the attempted classifications to the presence or absence of *a-X3*. The three

Dp-bearing a-X/a-X genotypes including a-$X1$ thus appear to be phenotypically indistinguishable.

Of the three remaining Dp-bearing genotypes, a-$X2$/a-$X2$, a-$X2$/a-$X3$, and a-$X3$/a-$X3$, the first two also are similar to the a-$X1$ homozygote and compounds. These types, obtained in various progenies, were always recognizable as defective plants with chlorophyll-lacking sectors, similar to those previously described. The third, a-$X3$/a-$X3$, Dp, is distinctly different. This may be illustrated in the progeny of the cross a-$X3$/$a^p \times a$-$X1$ (or a-$X2$)/a-$X3$, Dp. Because of low transmission of a-$X3$ by the female parent, this cross yields only a few seeds of the desired types per ear, but by pollinating several ears of a-$X3$/a^p by pollen from a single plant, small but adequate populations were secured. The results are as follows:

The seeds produced are mostly colored-pale mosaics. The remainder include some with clear degenerate sectors and some with no clear sectors, but usually with a slight pitting of the surface. In some cases no pitting is detectable, and the seed appears fully colored.

Plants grown from the seeds with degenerate sectors are typical a-X/a-X sectorial plants of the type previously described, and are indistinguishable from those previously found with a-$X1$ or a-$X2$. These plants show the segregation of subnormal pollen associated with a-$X3$, but in proportions below 50 percent, as in the a-$X3$ heterozygotes previously reported. They are thus identified as the a-$X1$ (or a-$X2$)/a-$X3$, Dp class expected.

Plants grown from the pitted seeds or from the wholly colored seeds are of a new type, regularly defective in comparison with their a-X/a^p, Dp sibs and usually more defective than their a-$X1$ (or a-$X2$)/$a X3$, Dp sibs. These plants usually show no sectors except for numerous minute streaks of chlorophyll-lacking tissue. Occasionally a small sector similar to those found in the other a-X/a-X types occurs, but ordinarily the plant is distinguishable only by its fine-streaked leaf blades and its generally defective development. A considerable proportion of these plants fail to survive to the flowering stage, but the majority survive to maturity and usually produce a few seeds. These plants show subnormal pollen in proportions approximating 80–85 percent, or subnormal and defective pollen approximating this total. They are thus identified as the a-$X3$/a-$X3$, Dp class expected. The contrasted types of a-X/a-X, Dp plants are illustrated in figs. 4 and 5.

Thus a-$X3$ must differ from a-$X1$ and a-$X2$ in some viability factor or factors concerned in the development of the sectors produced as a result of loss or diminution of the duplication fragment. If it be assumed that a-$X3$ involves deficiency of a viability factor or factors not involved in the other a-X alterations, it does not necessarily follow that sector survival would be appreciably affected, for the evidence indicates that the sectors observed with a-$X1$ and a-$X2$ result only from partial loss of the fragment. Such partial losses would not necessarily "uncover" the viability deficiency peculiar to a-$X3$. The observed types however indicate that this deficiency is so located that the more frequent types of partial loss which are capable of producing sectors in the usual a-X/a-X combinations do "uncover" it.

37

The interpretation applied may best be illustrated by considering the various classes of progeny in a cross of the type $a\text{-}X3/a^p \times a\text{-}X1/a\text{-}X3, Dp$. The progeny includes four genotypes, which we may designate for convenience as classes I, II, III and IV as follows:

GENOTYPE

I	$a\text{-}X1/a^p, Dp$
II	$a\text{-}X3/a^p, Dp$
III	$a\text{-}X1/a\text{-}X3, Dp$
IV	$a\text{-}X3/a\text{-}X3, Dp$

All classes of the progeny include the same Dp fragment (or, assuming that the fragment is constantly changing, equivalent samples from the mixture of Dp types present in the pollen grains of the plant used as male parent). The fragments present in the progeny plants therefore must undergo elimination and partial loss of the same kinds in plants of the four classes. The total effects of such losses are best indicated by classes I and II, for these carry a chromosome 3 free from viability defects (the a^p-bearing chromosome). Any loss of the duplication, in whole or in part, therefore will be without effect upon subsequent development of the affected tissue. In these classes, visible sectors result only from losses which include the A locus.

The frequent occurrence of large sectors lacking A^b in classes I and II shows that the Dp-fragment is undergoing frequent loss in development, including losses at early stages of development. These losses may be entire or partial, so long as they include the A locus. Presumably they include chiefly cases of loss of the entire fragment.

The nature of the sectors observed in class III plants shows that few, if any, such losses, when they occur in plants of this genotype, result in the production of visible sectors. Sectors lacking A^b are extremely rare. Since there is no reason to assume that these losses do not occur, we must postulate a genotype for the $a\text{-}X$ alterations involved which prevents the development of these sectors. A factor (or factors) essential for viability, present in the normal chromosome 3 in the region covered by $Dp\ 3a$, must therefore be lacking in both $a\text{-}X1$ and $a\text{-}X3$. Since $a\text{-}X2$ compounds show the same behavior, it must be lacking also in $a\text{-}X2$. The location of this factor is such as to make possible its preservation when W and A are eliminated in successive losses, though it is commonly lost when W and A are lost simultaneously. The sectors involving loss of W and not of A, which make up the great majority of the sectors observed in plants of class III, result from a type of loss not detectable in classes I and II, since these classes are not genetically marked at the W locus.

An alternative considered in the interpretation of the variegation observed in the plants of class III was the possibility that the sectors observed are in fact deficient for A^b, in spite of the fact that they show anthocyanin. It might be supposed that the boundary of a sector deficient for A would not be sharply marked by absence of anthocyanin, since cells at the margin of the sector might include the pigment by reason of diffusion of anthocyanin or precursor substances from the adjoining non-deficient tissue. If the sectors were narrow enough this might result in apparently uniform anthocyanin pigmentation throughout the sector. On this basis the sectors found in class III plants could be considered the result of the same losses as those found in classes I and II, and the loss responsible in both cases could be loss of the entire fragment. The reduction in size of sector might be considered to be the result of reduced growth in the a-X/a-X tissue, and the difference in phenotype the result of an effect upon chlorophyll development involved in the a-X alterations. It would still be necessary to postulate effects of the a-X alterations upon anthocyanin, chlorophyll, and growth, but no separation of these effects would be demonstrated by the variegation patterns observed.

This possibility is not supported by histological examination of the tissues within and adjoining the sectors. It requires the assumption that a very broad margin must be subject to pigmentation by diffusion, for some of the chlorophyll deficient sectors are eight to ten mm wide, and are uniformly colored by anthocyanin throughout. It is contradicted by the sharpness of the margins of the exceptional types of sector in which anthocyanin, or both anthocyanin and chlorophyll, are deficient, and especially by the fact that this margin is sharp on a side adjoining normal tissue as well as on a side adjoining a chlorophyll-deficient sector. We are forced to conclude that the sectors involving loss of chlorophyll alone represent a second type of loss, distinct from that involving loss of anthocyanin in classes I and II, and occurring frequently enough to produce a large number of sectors on each leaf.

The disappearance, or extreme reduction in size, of these sectors in plants of class IV, similarly shows an additional viability effect in plants of genotype a-$X3$/a-$X3$. Alterations in the Dp fragment of the "second type" mentioned in the preceding paragraph, when they occur in plants of a-$X3$/a-$X3$, Dp genotype, fail to result in sectors comparable to those found in other a-X/a-X, Dp plants. The chlorophyll-essential factor W must be deficient in a-$X3$ as well as in a-$X1$ and a-$X2$; otherwise the a-$X3$ compounds a-$X1$/a-$X3$ and a-$X2$/a-$X3$ could not show the chlorophyll-deficient sectors. The absence of these sectors in the a-$X3$/a-$X3$, Dp plants must therefore be due to their failure to develop, through the action of some viability-affecting factor, or factors, distinguishing a-$X3$ from a-$X1$ and a-$X2$. The factor must be so located that its normal homolog tends to be lost from the duplication when W is lost.

The evidence secured through the use of the unstable duplication thus shows that the a-X alterations studied involve the loss of several genetically separable effects observable in somatic development, aside from any additional factors which may be involved in their various effects upon gametophytic survival and pollen development. All three of the alterations involve the loss of effects essential for anthocyanin production, chlorophyll production, and somatic viability. The results of these losses may be suppressed in different combinations by the different derivatives of the Dp fragment which are produced in the course of development of a Dp-bearing plant. All three effects are suppressed by the "intact" Dp fragment, which is maintained in a-X/a-X, Dp stocks and thus shielded against losses which would make it useless for covering the *a-X* alterations. The viability and anthocyanin effects are suppressed without suppression of the chlorophyll effect by one type of variant; the viability and chlorophyll effects are suppressed without suppression of the anthocyanin effect by another; and the viability effect is suppressed without suppression of the anthocyanin or chlorophyll effect by a third. Each of these effects may thus be separated from the other two. In addition, *a-X3*, by similar evidence, is shown to involve loss of an additional factor or factors essential for somatic development. This demonstration of the loss of several genetically separable factors in each of the induced alterations constitutes genetic proof of deficiency.

DISCUSSION

This study of the losses of *A* action induced by X-ray treatment gives no evidence of an effect of the treatment upon gene mutation. It shows that the segmental deficiencies induced in maize by X-ray treatment range to lengths so short as to be distinguishable from gene mutations only in exceptionally favorable material. Among a large sample of "*A* losses" the two selected for detailed study as most nearly approximating the expected effect of gene mutation proved to be deficiencies, distinguished only by the relatively minute length of the segment involved.

The background for the interpretation of these results may be briefly outlined as follows: X-ray treatment, as shown by earlier experiments concerned with the general mutation rate, results in a substantial increase in the frequency of point mutations. These are identified, in experiments with maize, as new Mendelizing characters observed in the F_2 of crosses in which one or both of the parents were X-rayed. The X-ray-induced point mutations found in maize prove to be wholly recessive in all cases. They include cases in which the mutant segregates in the expected 3:1 ratio, and also cases in which the ratio is reduced to varying degrees, usually by reduced transmission through male germ cells. The same X-ray treatments produce in large numbers chromosomal derangements of various sorts. There is no appreciable tendency of the induced mutations to occur at points of chromosome interchange, a feature in contrast with the results of comparable experiments with Drosophila. The primary question is whether the point mutations observed represent in the main the result of gene mutations or the result of extra-genic alterations. If the results indicate the latter to be true, a second question arises: Do the X-ray-induced mutations include any cases in which gene mutation must be assumed to have occurred?

The presumption that the point mutations in general represent gene mutations is based upon the assumption that deficiencies would ordinarily be lethal to the gametophyte, or at any rate would be so weakened in haplo phase as to fail in transmission through pollen. If this were so their recessive effects could not appear in F_2. This assumption becomes more and more questionable as cases are found of demonstrable deficiencies transmitted through the gametophyte generation. Deficiencies now known range in gametophytic effect from total abortion of both male and female gametophyte through successive levels of development ranging to apparently normal development and normal functioning. Moreover the cases demonstrated as deficiencies range to the limit of cytological or genetic identification. The assumption that deficiencies do not occur below these limits of detectability is wholly unwarranted.

In selecting individual cases for detailed study as possible gene-mutations, it was assumed that the action of A is not essential for normal gametophyte development, and therefore that any deleterious effects of an observed A loss upon gametophyte development indicate that other loci are also involved. This assumption was based upon the normal gametophyte development found in all known A alleles, including several spontaneous mutations observed from the first generation following their occurrence. The assumption is confirmed by the results of this study, showing two deficiencies of A which permit normal development of both the male and the female gametophyte.

Among the population of A losses observed in F_1, then, we may expect that the major deficiencies including the A locus will have visible effects upon gametophyte development, though there may also be minor deficiencies with no detectable gametophytic effect. The A losses due to alterations with visible gametophytic effects will include most of the deficiencies, and may include losses of A effect due to other extra-genic alterations, but will not include any of the A losses due wholly to gene mutation. The A losses without visible gametophytic effects will include all of the gene mutations (except such as might occur from a single alteration involving both gene mutation and chromosomal derangement), and may include also some of the deficiencies and other extra-genic alterations.

Unfortunately, it is not possible in general to determine from inspection of the mutant individuals observed in F_1 whether or not the alteration concerned has gametophytic effects. The A loss observed is induced by X-ray treatment, and the coincidental occurrence of an independent chromosomal derangement induced by the same treatment may produce in the same F_1 plant a gametophytic effect not due to the alteration involving A. It is technically impossible to extract the A alteration in each case for individual study, not only because of the labor involved, but because in many cases the single plant concerned is sterile. The possibility remains therefore that among the population of A losses observed there were instances of gene mutation of A, which by reason of coincident but independent chromosomal alterations were accompanied by gametophytic or developmental defects.

The nature of the deficiencies identified in a-$X1$, a-$X2$, and a-$X3$ is of interest in relation to the question of distinguishing between deficiencies and gene mutations among the point mutations at miscellaneous loci, which are induced in maize by X-ray treatment. It is clear that deficiencies including three or more gene loci may be too minute for cytological detection. In spite of the extremely sensitive reaction of crossover frequency to minute deficiency in maize, the results with a-$X1$ show that such deficiencies may occur without pronounced effect upon crossover frequency. Genetic detection of the deficiencies in this case was facilitated by the fact that loci affecting chlorophyll development and somatic viability occur very close to the A locus, but there is of course no reason to assume that such convenient linkages would occur with another gene similarly studied. There may be many genes with phenotypic effects, which are so located that deficiencies of comparable length could remove them without removing other detectable genes. Such deficiencies would be indistinguishable from gene mutations.

The somatic viability locus deficient in all three *a-X* alterations studied is apparently close enough to *A* to ensure that point mutations of *A* would be extremely infrequent in experiments of the usual type on X-ray-induced mutation of miscellaneous genes. So far as we know, no X-ray-induced mutations of *A* have previously been found. None of the three *a-X* alterations here described would appear in F₂ in such an experiment, since all are zygotically lethal, due to the associated viability factor. Although the loci yielding mutations under X-ray treatment appear to be a random sample, there are of course many known genes for which X-ray induced mutants have not been found. It is not improbable that with significant comparisons different loci would show large differences in susceptibility to X-ray-induced mutation, due to this cause.

If no viability factor were included in the *a-X* deficiency-segments, the alterations would segregate in F₂ as mutant alleles of *A* with effects upon both anthocyanin development and chlorophyll development. Such alterations with diverse effects are commonly regarded as gene mutations, particularly if they occur at loci not already identified by a consistent group of alleles. The X-ray induced mutant *et*, described in connection with the crossover experiments in this paper, is perhaps an analogous case. It affects two characters not obviously related, scarring of the endosperm surface and chlorophyll development in the seedling stage. The homozygote is fully viable and fertile, pollen development is normal, and the mutant is transmitted through both male and female gametophytes.

The three deficiencies designated *a-X1*, *a-X2*, and *a-X3* apparently represent segmental losses of different extent in each case. *Df a-X3* must be assumed to include some locus or loci not involved in the others, to account for its visible effect on pollen development and its much more extreme reduction in gametophyte survival. The additional somatic viability factor or factors involved in *a-X3*, as indicated by the studies with the unstable duplication, may or may not represent the same loci. Similarly *a-X2* must differ from *a-X1*, to account for its much more pronounced effect upon crossing over, as well as its consistently lower gametophytic transmission.

Since these three deficiencies represent breaks at different points in the normal gene sequence, it is reasonable to infer that a larger sample would include deficiencies of smaller extent. A deficiency including the *A* locus and not including the viability and chlorophyll-essential loci here involved would be viable in homozygotes and would be identical in phenotypic effect with the standard recessive allele *a*.

SUMMARY

1. X-ray-induced losses of the phenotypic effect of *A* (here designated "A losses") were identified in F₁ populations produced by the use of X-rayed

pollen. Among about 415 *A* losses observed, only two were normal plants free from segregating pollen defects.

2. The induced germinal alterations responsible were extracted for study in these two cases (*a-X1* and *a-X2*). The alteration involved in an *A* loss showing segregation of subnormal but not aborted pollen was also extracted for comparison (*a-X3*).

3. All three alterations are haplo-viable but with reduced transmission through the gametophyte generation. *a-X1* is usually transmitted with normal frequency through female gametophytes, and with about 50 percent of normal frequency through male gametophytes. *a-X2* usually shows about 65 percent of normal female transmission and about 25 percent of normal male transmission. *a-X3* usually gives about 20 percent of normal female transmission, and is not at all transmitted through pollen.

4. Homozygotes of *a-X1* and *a-X2*, and the compounds of *a-X1*, *a-X2*, and *a-X3* in all three possible combinations, are zygotically lethal.[1]

5. Plants heterozygous for *a-X2* and *a-X3* show greatly reduced crossing over in the adjacent region of the chromosome, while plants heterozygous for *a-X1* show approximately normal crossing over in the same region.

6. Cytological examination at pachytene in plants heterozygous for *a-X2* and *a-X3* shows no visible deficiency, intercalation, or inversion, and chromosome pairing in the region involved is not visibly abnormal.

7. By the use of an unstable duplication, *Dp 3a*, covering the region concerned, the various *a-X/a-X* homozygotes and compounds were produced, and the sectors resulting from elimination or partial loss of the *Dp* fragment were studied for evidence concerning the factors involved in the various *a-X* alterations. This study showed that all three alterations are deficient for factors affecting anthocyanin production, chlorophyll production, and somatic viability, and that *a-X3* is deficient for an additional factor or factors affecting somatic viability.

LITERATURE CITED

EMERSON, R. A., and E. G. ANDERSON, 1932 The *A* series of allelomorphs in relation to pigmentation in maize. Genetics **17**: 503–509.

FOGEL, SEYMOUR, 1946a Gene action and histological specificity of pigmentation patterns of certain *R* alleles. Genetics **31**: 215–216.

    1946b Gene action and the course of anthocyanin synthesis in certain *R* alleles. Genetics **31**: 215–216.

LAUGHNAN, JOHN R., 1946 Chemical studies concerned with the action of the gene *A₁* in maize. Genetics **31**: 222.

McCLINTOCK, BARBARA, 1931 Cytological observations of deficiencies involving known genes, translocations and an inversion in *Zea mays*. Missouri Agric. Exp. Sta. Res. Bul. 163, pp. 1–30.

    1933 The association of non-homologous parts of chromosomes in the midprophase of meiosis in *Zea mays*. Z. Zellf. mik. Anat. **19**: 191–237.

    1938 The production of homozygous deficient tissues with mutant characteristics by means of the aberrant mitotic behavior of ring-shaped chromosomes. Genetics **23**: 315–376.

    1944 The relation of homozygous deficiencies to mutations and allelic series in maize. Genetics **29**: 478–502.

[1] See footnote, p. 280.

RHOADES, M. M., 1941 The genetic control of mutability in maize. Cold Spring Harbor Symp. Quant. Biol. **9**: 138–144.

STADLER, L. J., 1933 On the genetic nature of induced mutations in plants. II. A haplo-viable deficiency in maize. Missouri Agric. Exp. Sta. Res. Bul. 204. pp. 1–29.

1935 Genetic behavior of a haplo-viable internal deficiency in maize. (Abst.) Am. Nat. **69**: 56–57.

1940 Etched endosperm-virescent seedling. Maize Genetics Cooperation News Letter **1940**: 26.

1941 The comparison of ultraviolet and X-ray effects on mutation. Cold Spring Harbor Symp. Quant. Biol. **9**: 168–177.

1942 Some observations on gene variability and spontanous mutation. The Spragg Memorial Lectures on Plant Breeding (third series). Mich. State Coll., East Lansing.

1944 The effect of X-rays upon dominant mutation in maize. Proc. Nat. Acad. Sci. **30**: 123–128.

1946. Spontaneous mutation at the *R* locus in maize. I. The aleurone-color and plant-color effects. Genetics **31**: 377–394.

STADLER, L. J., and SEYMOUR FOGEL, 1943 Gene variability in maize. I. Some alleles of *R* (*R*r series). Genetics **28**: 90–91.

1945 Gene variability in maize. II. The action of certain *R* alleles. Genetics **30**: 23–24.

STADLER, L. J., and HERSCHEL ROMAN, 1943 The genetic nature of X-ray and ultraviolet induced mutations affecting the gene *A* in maize. Genetics **28**: 91.

44

3

Reprinted from *Science,* **128**(3338), 1546–1550 (1958)

Radiation Dose Rate and Mutation Frequency

The frequency of radiation-induced mutations is not, as the classical view holds, independent of dose rate.

W. L. Russell, Liane Brauch Russell, Elizabeth M. Kelly

It is usually considered to be one of the basic tenets of radiation genetics that variation in radiation intensity—that is, dose rate—does not affect mutation rate. However, the experimental results upon which this conclusion is based were obtained only from certain cell stages, particularly *Drosophila* spermatozoa. The bulk of the radiation dose causing genetic hazards in man will be accumulated not in spermatozoa but in spermatogonia and oocytes. It was therefore of both practical and fundamental importance to question whether mutation rates observed following irradiation of these cell stages would also prove to be independent of radiation intensity.

Two major considerations that prompted such a question, in the face of the general acceptance of the absence of a radiation intensity effect on induced mutation rate, may be outlined. First, there has been increasing evidence that induction of mutation may not be as direct an action as had often been supposed, and that the mutation process in the gene may not be entirely independent of variation in its cellular environment. Consequently,

The authors are on the staff of the Biology Division of Oak Ridge National Laboratory, Oak Ridge, Tenn.

there was room for speculation that even though the mutation process in spermatozoa is apparently independent of dose rate, it might not be so in metabolically active cells like spermatogonia. Second, it was reasoned that even if the actual mutation process in spermatogonia should prove to be, as in spermatozoa, independent of radiation intensity, nevertheless the mutation rate, as measured by mutations transmitted to the offspring, might still be dependent on dose rate, because of cell selection due to killing or other interference with the dynamics of the cycle of the seminiferous epithelium (*1, 2*).

With these two considerations in mind, experiments to determine mutation rates induced by chronic gamma irradiation in spermatogonia in mice were started. The first data from these experiments, and a comparison of them with mutation rates obtained earlier with acute x-irradiation, were presented at the April 1958 annual meeting of the National Academy of Sciences (*1*). They had been submitted earlier for a publication still in press (*2*), and they have also been discussed briefly elsewhere (*3*). The results showed a much lower mutation rate from chronic gamma than from acute x-irradiation. It

was pointed out that, without further analysis, it could not be definitely decided whether the difference was attributable to intensity or to quality of radiation (although the latter seemed unlikely in view of the magnitude of the effect), and whether it was the mutation process itself that was involved or some secondary process, such as cell selection.

Since the time of the early reports, the data have been approximately doubled. Also, a number of new experiments, undertaken specifically for the purpose of analyzing the observed effect, have already thrown additional light on the problem. Because of the wide interest in this field, the present interim report has been prepared, bringing tabulation of the spermatogonia results up to date and presenting preliminary results from the new experiments.

Chronic Gamma Irradiation of Spermatogonia

Young mature male mice were exposed, in polystyrene cages of 3.0 to 3.5-millimeter wall thickness (more than adequate for secondary electron equilibrium), to a 5-curie Cs^{137} source. Dose rate was regulated by distance. Exposure was continuous (except for occasional interruptions of a few minutes) until the total dose had been accumulated (*4*). The males were mated to test females (see below) immediately following removal from the radiation field. However, only mutations induced in spermatogonia are considered in this section of this article. Unirradiated males were tested simultaneously with the irradiated.

Mutation rates were determined by the specific locus method. Irradiated and control males are mated to females homozygous for seven autosomal recessive visibles. The offspring are then examined for mutations at the seven loci. Details of the experimental procedure have been described earlier (*5*).

The results from the chronic gamma irradiation experiments are given in

Table 1. The mutations listed in the table have not yet all been tested for allelism. However, classification by phenotype has proved remarkably reliable in our experience with well over 100 tested mutants at these loci, so there is little likelihood of error.

For comparison with the chronic gamma irradiation data listed in Table 1, a summary is presented in Table 2 of the results of three of our acute x-ray experiments (2). The radiation intensity in these experiments was approximately 80 to 90 roentgens per minute.

Results from the chronic gamma and acute x-ray experiments are compared in Fig. 1. All the points for the chronic gamma-ray mutation rate curve are considerably below the acute x-ray curve. However, a comparison of the two sets of results over the whole range of doses cannot be reduced to a simple statistical test of significance because the mutation rate curve following acute x-irradiation shows a clear departure from linearity, already discussed elsewhere (2, 3, 6), while the present mutation rate data from chronic gamma irradiation show no evidence of a similar departure. Three statistical tests have been made (7) which attempt to avoid this difficulty in different ways.

In view of the possible special reasons for the departure of the acute x-ray curve from linearity (the drop in the mutation rate at the 1000-roentgen dose being attributed to cell selection), one test of the significance of the difference between chronic gamma and acute x-radiation induced mutation rates was made with the 1000-roentgen x-ray point excluded. The two sets of data, with a combined control point, were fitted simultaneously to two straight lines by the method of least squares, with weights based on the Poisson assumption. The ratio of the slopes is 4.1 (95-percent confidence interval 2.36, 12.5), and the slopes differ significantly ($P < 1 \times 10^{-9}$). A similar test, but one that excludes both the acute x-ray 1000-roentgen point and the chronic gamma-ray 861-roentgen point, also yields a significant difference ($P < 1 \times 10^{-7}$). The third statistical test was made between just two points. In view of the lack of data at closely comparable doses in the lower part of the dose range, and because of the presumed complexity at the 1000-roentgen x-ray point, the two points that seemed to offer the most meaningful single comparison were the 600-roentgen point for acute x-rays and the 516-roentgen point for chronic gamma

Table 1. Mutations at specific loci induced in spermatogonia of mice by chronic gamma irradiation.

Dose (r)	Intensity (r/wk)	Offspring (No.)	Mutations at 7 loci (No.)	Mean No. of mutations per locus, per gamete ($\times 10^6$)
0		105,403	8	1.08
86	10	48,500	6	1.77
516	90	20,752	4	2.75
861	90	20,993	9	6.12

Table 2. Mutations at specific loci induced in spermatogonia of mice by acute x-irradiation.

Dose (r)	Offspring (No.)	Mutations at 7 loci (No.)	Mean No. of mutations per locus, per gamete ($\times 10^6$)
0	42,833	1	0.33
300	40,408	25	8.85
0	106,408	6	0.81
600	119,326	111	13.29
0	33,972	2	0.84
1000	31,815	23	10.33

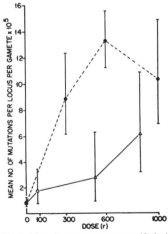

Fig. 1. Mutation rates at seven specific loci in the mouse, with 90-percent confidence intervals. Solid circles represent results with acute x-rays (80 to 90 r/min). Open points represent chronic gamma-ray results (triangles, 90 r/wk); (circle, 10 r/wk). The point for zero dose represents the sum of all controls.

rays. A test of the significance of the difference between the mutation rates per roentgen at these two points gave $P = 0.0008$ for a one-tailed test (8).

In the statistical tests, the fitted curves show no evidence of departure from linearity. The question of whether or not the data may be expected to be truly linear is discussed below. The actual ratio of effectiveness of chronic gamma and acute x-irradiation found may, of course, be valid only for the particular combination of doses and intensities tested. The important point is that the data now available adequately confirm the earlier report (1–3) that chronic gamma radiation is significantly less effective than acute x-radiation in inducing specific locus mutations in spermatogonia.

The conclusions of the preceding paragraphs—that chronic gamma irradiation of mouse spermatogonia is mutagenically less effective than acute x-irradiation—is in sharp contrast to the findings for *Drosophila* spermatozoa, reviewed by Muller (9), which have heretofore been considered to have general applicability and have entered into the basic concepts of radiation genetics.

It is, therefore, of great importance to attempt to determine what factors are responsible for the present result. For this reason, a number of experiments, designed to throw light on this question, have been initiated. In Table 3, preliminary results of these new experiments, as well as older findings already reported elsewhere, are compared with the present data.

Intensity versus Quality

The difference in mutation rate between spermatogonia subjected to chronic gamma irradiation and those subjected to acute x-irradiation could be due to differences either in quality or in intensity of radiation. In order to differentiate between these two factors, the effect of a change in quality alone has been investigated in three separate comparisons (see Table 3). No appreciable differences were found in the effectiveness of acute gamma rays (from Co[60]), on the one hand, and acute x-rays, on the other, in inducing dominant lethal mutations in spermatozoa, specific locus mutations in spermatozoa and other postspermatogonial stages, or specific locus mutations in spermatogonia. (It appears safe to assume the same result also for oocytes, for which no direct quality com-

Table 3. Semiquantitative comparison of mutation rates presented in this article with those obtained earlier and with preliminary results from experiments in progress. Each plus symbol in the table stands for a mutation rate of approximately 5×10^{-8} per roentgen, per locus. The check marks represent arbitrary values that are valid for comparative purposes among the dominant lethal results. They cannot be quantitatively compared with the specific locus mutation rates.

Gametogenic stage irradiated	Genetic effect measured	Type of irradiation		
		Chronic Gamma (Cs^{137})	Acute	
			Gamma (Co^{60})	X-ray
Postspermatogonia	Dominant lethals*†	√ √ √	√ √ √	√ √ √
Postspermatogonia	Specific locus mutations	(++++++)‡	++++++++++§	+++++++++++++‖
Spermatogonia	Specific locus mutations	+	++++§	++++
Oocytes	Specific locus mutations	+¶		+++++++**

* Paper in preparation.
† Chronic and acute gamma rays give approximately equal rates, although comparison is not exact because of difficulty in matching particular postspermatogonial stages irradiated. In the comparison of acute gamma with acute x-rays, the former were found slightly less effective.
‡ Value is based on only 1 mutation in 1613 young, so the mutation rate is not yet reliable.
§ Value based on 4 mutations.
‖ From Russell *et al.* (*17*).
¶ From Russell *et al.* (*12*); see also Carter (*13*) for chronic Co^{60} gamma data.
** From Russell *et al.* (*18*).

parison was made.) These results show that difference in the quality (linear energy transfer) of the gamma rays and x-rays tested, while it may account for a small part, cannot account for the bulk of the difference between the chronic gamma and acute x-ray mutation rate results. It can be concluded that most of the difference must be due to intensity of radiation.

Intensity and Gametogenic Stage

The results summarized in Table 3 show that radiation intensity effects were found only for spermatogonia and oocytes. In the experiments with postspermatogonial stages, radiation intensity had no appreciable effect on the yield of genetic changes. This conclusion can be drawn with near certainty for dominant lethals. The specific locus data, from experiments still in progress, are not yet extensive, but, as far as they go, they are not in disagreement with the dominant lethal result. In both cases, the stages irradiated were spermatozoa and spermatids, with the bulk of the data from the former. It may thus be concluded that dose rate does not influence the frequency of genetic changes produced by irradiation in mouse spermatozoa, but conclusions regarding spermatids and spermatocytes will have to await further work. The spermatozoa results are in agreement with the findings for *Drosophila* spermatozoa. Thus, the classic

finding of intensity independence is supported for spermatozoa (*10*). The explanation for the new phenomenon of intensity dependence resides in gametogenic stage.

Mutation Process versus Cell Selection

The intensity effect in spermatogonia might have been due to secondary causes —that is, selection as a result of cell killing or other interference with the dynamics of the cycle of the seminiferous epithelium, as stated above. This was put forward as one plausible, but not favored, hypothesis in the first detailed publication of the data (*2*). This hypothesis has now been deliberately tested by new experiments on females. Since oogonia are not present in the adult ovary (*11*), and since the completion of the first meiotic division only just precedes ovulation, radiation genetic experiments on adult females deal exclusively with primary oocytes, and the bulk of these are in the uniform dictyate state. Results already reported (*2, 12*) showed that chronic gamma irradiation of oocytes gave mutation rates lower than those from acute x-irradiation of spermatogonia. The new results (Table 3) indicate that acute irradiation of oocytes is at least as effective as acute irradiation of spermatogonia.

In the light of this finding of a dose-rate effect for oocytes as well as for spermatogonia, the hypothesis that the in-

tensity effect on mutation rate is due to cell selection appears to be less tenable. Since oocytes are nonmitotic, since the stages irradiated show no obvious variability, and since, in our chronic irradiation experiment, the continued fertility of the females provides no evidence of extensive killing, selective or otherwise, of the oocytes, it seems highly unlikely that the difference between the mutation rates following chronic and acute irradiation of oocytes can be attributed to any secondary mechanism similar to that put forward as a possible one for spermatogonia. Of course, this mechanism might still be postulated as playing a role in the spermatogonia results, but it is simpler to assume that the explanation for the results in oocytes—namely, that the intensity effect is on the mutation process itself—also applies to spermatogonia.

It should be noted that, at each dose rate tested, there is at present no evidence of marked difference between oocytes and spermatogonia in sensitivity to mutation induction. Therefore, the interpretation by Carter (*13*), who also found a low mutation rate with chronic gamma irradiation of oocytes, and who thought it most likely that this was attributable to sex, is not upheld. His emphasis on the consequence of his interpretation—namely, that only a small part of the genetic hazard from medical irradiation would come from exposure of females—must now be discounted.

Relation to the Linearity Concept

The various results discussed in the three preceding sections and summarized in Table 3 have determined which among the possible factors are the ones responsible for the lower mutation rate from chronic gamma irradiation. It turns out that these are also the more interesting factors. Two of these are radiation intensity, rather than quality; and the mutation process itself, rather than cell selection. Since the finding of an intensity effect on the mutation process was unexpected, the field is now open for new hypotheses about the nature of this process. Such hypotheses are aided, or at least delimited, by the finding of a third factor—namely, that the intensity effect occurs in spermatogonia and oocytes, but apparently not in spermatozoa. Thus, the mechanism for this effect may be found among the characteristics by which the highly specialized spermatozoa differ from spermatogonia and oocytes.

Speculation concerning the nature of the mutation process has a direct bearing on the fundamental problem of what the mutation rates are now likely to be at other doses and intensities. One specific question is already being debated—namely, whether or not the finding of an intensity effect in spermatogonia and oocytes is strong indication that a threshold dose will be found for mutation induction in these cells. This possibility has obvious and vital importance to the problem of genetic hazards.

A strong argument that has long been advanced against the threshold concept is the likelihood that a single direct hit (ion or ion cluster) on such a small target as a gene must sometimes be adequate to cause mutation. This hypothesis has not only seemed plausible on physical grounds but has also been supported by the mutation rate data for *Drosophila* spermatozoa and for other material where an intensity independence or a linear relation with dose has been found (9). The new data from mouse spermatozoa provide additional support. If the intensity effect reported here for mouse spermatogonia and oocytes is taken as evidence for a threshold effect for all mutations induced in these cells, then this necessarily implies that all mutations in spermatogonia and oocytes are induced by a process different from that which has long been, and still can be, assumed for spermatozoa. This may be true, but it would certainly be incautious to jump to this conclusion. In fact, it seems quite plausible to assume that spermatogonia and oocytes may not be completely different from spermatozoa—in other words, that at least a portion of the mutations in them may be induced by a single-hit process.

To make the consequences of this hypothesis easily understandable, they will be presented in terms of a specific model. Thus, it can be postulated that there are two kinds of mutation which, for simplicity in the following discussion, will be called "reparable" and "irreparable." (They could, alternatively, and perhaps more realistically, be looked upon as "preventable" and "not preventable.") It can be further assumed that in spermatogonia and oocytes there is repair of the reparable mutations at the low radiation-intensity (chronic) level so far tested. Such repair is assumed to be impossible, or less probable, because of radiation damage to the repair process, at high radiation intensities (acute) in spermatogonia and oocytes. Repair is also assumed to be impossible at all in-

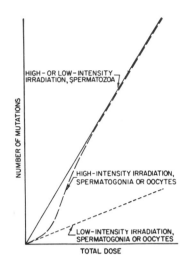

Fig. 2. Theoretical dose curves constructed on the basis of the hypothesis (see text) that "repair" of some mutations is possible in spermatogonia and oocytes but not in spermatozoa.

tensities in spermatozoa, perhaps because of some property—for example, a metabolic activity—lacking in them that is present in spermatogonia and oocytes.

Such a hypothesis could lead to a set of curves something like that shown in Fig. 2. The straight line for chronic irradiation of spermatogonia and oocytes is assumed to be the single-hit curve for irreparable mutation, all reparable ones having been repaired. A steeper straight line is shown for spermatozoa, where it is assumed that none of the reparable mutations are repaired and that both these and the irreparable ones follow a single-hit relation with dose, regardless of intensity. It follows logically that, as is shown in the third curve, acute irradiation of spermatogonia and oocytes would, at total doses low enough to permit repair, duplicate the curve for chronic irradiation, but that, at higher doses, when repair fails, the curve would shift over to a new position approaching that for spermatozoa. (Actually, the curve for observed mutations in spermatozoa is much steeper than the curve for acute irradiation of spermatogonia. The reasons for this, one of which is probably a large chromosomal aberration component of the mutations in spermatozoa (14), are assumed to be irrelevant to the present argument. In Fig. 2, the curve for spermatozoa, as well as the curve for oocytes, may be looked upon as being appropri-

ately adjusted to eliminate the irrelevant factors and to provide an uncomplicated comparison for radiation intensity only.)

No importance is attached to the particular details chosen to make this type of model easily understandable. Thus, "reparable" and "irreparable" need not imply qualitatively different mutational sites. Only one kind of site is necessary if, for example, it is assumed that there is a time lag for the completion of the mutation process and (even with the repair process intact) a probability of less than unity that repair could occur before this completion. Also, the term *repair* is not necessarily restricted to mean the reversal of a damaged gene to normal. In fact, as was mentioned earlier, the term *preventable* might be substituted in place of *reparable*. Prevention could occur at any stage in the mutation process, even at its initiation when there might be diversion, by a "lightning-rod" effect, of ions that might otherwise have caused mutation.

Whether or not the proposed hypothesis is favored, it demonstrates clearly that the discovery of an intensity effect does not necessarily imply that all induced mutations in spermatogonia and oocytes must follow a threshold response. Of course the hypothesis does involve a threshold concept, but it applies to only a portion of the mutations. As demonstrated, the theoretical consequence for chronic irradiation of spermatogonia and oocytes, in this particular model, is a linear relation between mutation rate and dose, even down to the lowest doses, in spite of a lower mutation rate than with acute irradiation.

Other plausible models can, of course, be constructed. Experiments now under way with various intensities of radiation and with fractionated doses will undoubtedly narrow down the possibilities. It should be noted, however, that the range of intensities already tested is tremendous—namely, 10,000-fold (100,000-fold at one point). The fact that this has yielded only a fourfold difference in mutation rate certainly raises the question of whether a further decrease in intensity would be likely to give a further drop in mutation rate. The mutation rate at the lowest intensity tested—10 roentgens per week—and the rate reported by Carter *et al.* (15) for a similar intensity still have such wide confidence intervals that they are not particularly informative in a comparison with the results from the 90-roentgen-per-week intensity.

48

Human Hazards

Caution must be exercised against reaching dangerous conclusions from the present results. Thus, as has been emphasized, it is not safe to conclude that the data imply a threshold dose for all mutations in spermatogonia and oocytes. There might not even be any further reduction in mutation rate with further decrease in intensity. Furthermore, it should not be forgotten that even the lower mutation rates obtained with the present intensity levels are still appreciable and at least as high as *Drosophila* rates for acute irradiation. However, from the results as they stand—results that apply to the germ-cell stages (spermatogonia and oocytes) that are important in appraising human hazards—it does seem safe to conclude that, with at least some intensities of radiation, the genetic damage would not be as great as that estimated from the mutation rates obtained with acute irradiation.

Summary

New data have clearly confirmed the earlier finding that specific locus mutation rates obtained with chronic gamma irradiation of spermatogonia are lower than those obtained with acute x-rays. Since this result is in contrast to classical findings for *Drosophila* spermatozoa, and apparently contradicts one of the basic tenets of radiation genetics, it was important to determine what factors were responsible for it.

Experiments undertaken for this purpose reveal the following: (i) the lower mutation frequency is due mainly to difference in dose rate of radiation, rather than quality; (ii) a dose-rate effect is not obtained in experiments with mouse spermatozoa, confirming classical findings for spermatozoa, and indicating that the explanation for intensity dependence in spermatogonia resides in some characteristic of gametogenic stage; and (iii) a dose-rate effect is found not only in spermatogonia but also in oocytes, where cell selection is improbable, indicating that the radiation intensity effect is on the mutation process itself.

A threshold response for all mutations in spermatogonia and oocytes is not a necessary consequence of the findings. Plausible hypotheses consistent with the present results can lead to other predictions.

From a practical point of view, the results indicate that the genetic hazards, at least under some radiation conditions, may not be as great as those estimated from the mutation rates obtained with acute irradiation. However, it should not be forgotten that even the lower mutation rates obtained with the present intensity levels are still appreciable (16).

References and Notes

1. W. L. Russell and E. M. Kelly, *Science* 127, 1062 (1958).
2. W. L. Russell and L. B. Russell, in *Proc. 2nd Intern. Conf. Peaceful Uses Atomic Energy, Geneva, 1958*, in press.
3. W. L. Russell, L. B. Russell, E. F. Oakberg, in *Radiation Biology and Medicine*, W. D. Claus, Ed. (Addison-Wesley, Reading, Mass., 1958), pp. 189–205.
4. We are greatly indebted to Dr. M. L. Randolph and Mr. D. L. Parrish for the dosimetry, which will be described elsewhere. It should be noted that recent refinements in dosimetry have resulted in some changes subsequent to publication of our preliminary abstract on this subject [W. L. Russell and E. M. Kelly, *Science* 127, 1062 (1958)]. Thus, doses listed in that abstract as 600 r at 100 r/wk and 100 r at 10 r/wk correspond, respectively, to 516 r at 90 r/wk and 86 r at 10 r/wk in later publications (2, 3), including the present one.
5. W. L. Russell, *Cold Spring Harbor Symposia Quant. Biol.* 16, 327 (1951).
6. W. L. Russell, *Genetics* 41, 658 (1956).
7. We are indebted to Dr. A. W. Kimball for statistical advice and computations.
8. A. Birnbaum, *J. Am. Statist. Assoc.* 49, 261 (1954).
9. H. J. Muller, in *Radiation Biology*, A. Hollaender, Ed. (McGraw-Hill, New York, 1954), vol. 1, pp. 475–479.
10. This conclusion is not meant to exclude the possibility of special types of intensity effect such as that reported by A. M. Clark [*Nature* 177, 787 (1956)] for sex-linked lethals in *Drosophila* spermatozoa at high dose rates in the presence of sodium azide.
11. E. F. Oakberg, *Proc. 10th Intern. Genet. Congr.* (1958), vol. 2, p. 207.
12. W. L. Russell, L. B. Russell, J. S. Gower, S. C. Maddux, *Proc. Natl. Acad. Sci. U.S.* 44, 901 (1958).
13. T. C. Carter, *Brit. J. Radiol.* 31, 407 (1958).
14. W. L. Russell and L. B. Russell, *Radiation Research*, in press.
15. T. C. Carter, M. F. Lyon, R. J. S. Phillips, *Nature* 182, 409 (1958).
16. We are grateful to Mrs. M. B. Cupp, Miss J. W. Bangham, and the other members of the Mammalian Genetics and Development Section who assisted with the laboratory work. The Oak Ridge National Laboratory is operated by Union Carbide Nuclear Company for the U.S. Atomic Energy Commission.
17. W. L. Russell, J. W. Bangham, J. S. Gower, *Proc. 10th Intern. Genet. Congr.* (1958), vol. 2, p. 245.
18. W. L. Russell, L. B. Russell, M. B. Cupp, *Proc. Natl. Acad. Sci. U.S.*, in press.

4

Reprinted from *Proc. Natl. Acad. Sci.*, **53**(2), 430–437 (1965)

INDUCTION KINETICS AND GENETIC ANALYSIS OF X-RAY-INDUCED MUTATIONS IN THE AD-3 REGION OF NEUROSPORA CRASSA*

BY B. B. WEBBER AND F. J. DE SERRES

BIOLOGY DIVISION, OAK RIDGE NATIONAL LABORATORY

Communicated by Alexander Hollaender, November 30, 1964

Forward-mutation experiments with a genetically marked balanced heterokaryon of Neurospora have shown that two different classes of mutations are induced in the *ad-3* region by X irradiation.[1] The firstclass consists of reparable mutants (*ad-3^R*) that will grow as homokaryons on adenine-supplemented medium, and the second consists of irreparable mutants (*ad-3^{IR}*) that will not grow as homokaryons either on adenine-supplemented or complete medium.[2] Genetic analyses by means of homology tests[2, 3] indicate that the *ad-3^R* mutants have only the *ad-3A* or *ad-3B* locus inactivated, whereas in the *ad-3^{IR}* mutants the inactivation covers other loci in the immediately adjacent regions. The most reasonable explanation for these data is that the two classes of mutants result from different types of genetic alteration. These differences can be explained if each *ad-3* mutation is a point mutation resulting from a single alteration within the *ad-3A* or *ad-3B* locus, and each *ad-3*[IR] mutation is produced by a chromosome deletion resulting from two alterations—with one or both *outside of* the *ad-3A* and *ad-3B* loci. If this is true, then the difference between *ad-3*[R] and *ad-3*[IR] mutations should be reflected in their induction kinetics.

The development of a direct forward-mutation method for use with wild-type strains of Neurospora[4] has made it possible to obtain precise forward-mutation frequencies for the induction of *ad-3* mutants after mutagenic treatment. In the present experiment, the method has been used to study the kinetics of induction of forward mutations in the *ad-3* region after exposure to X rays. The experiment was designed to investigate these genetic effects over the widest possible range of survival levels. Data derived from genetic analysis of random samples of mutants induced by 1, 2, 5, 10, 20, and 40 kr exposures demonstrate a difference in the induction kinetics of *ad-3*[R] mutants and *ad-3*[IR] mutants and describe the dose dependence of the spectrum of X-ray-induced mutation in the *ad-3* region.

Materials and Methods.—Strains: The strain numbers and genetic markers in the two-component heterokaryon (referred to as a dikaryon) are given in Table 1. The *hist-2* marker is located

TABLE 1

GENETIC COMPOSITION OF EACH COMPONENT OF THE DIKARYON USED IN THE
FORWARD-MUTATION EXPERIMENT

Linkage group	Component I (74-OR60-29A)	Component II (74-OR31-16A)
IR	*hist-2 ad-3A ad-3B nic-2*	*al-2*
III	*ad-2*	—
IV	—	*cot*
V	*inos*	—
VI	—	*pan-2*

in the right arm of linkage group I (LG IR) near the centromere and shows 0.154% crossing over.[5] The *ad-3* loci are about 2.0 map units to the right of *hist-2*, and the *nic-2* locus is about 5.0 map units to the right of *hist-2*.[6] Conidia in the dikaryotic culture are of three different types, namely, (1) homokaryotic for component I; (2) homokaryotic for component II; and (3) heterokaryotic, containing at least one nucleus of each type. Only the heterokaryotic conidia grow on minimal medium, and those heterokaryotic conidia containing mutations in the *ad-3* region in component II can be recovered by supplementing this medium with adenine. The average nuclear number per heterokaryotic conidium has been minimized by growing the culture on minimal medium[7] and by filtering the conidial suspension through thick cotton pads.

The present dikaryon differs from the one used in earlier experiments[1] by the presence of an *ad-2* marker in linkage group III (LG III) in component I. The *ad-2* block precedes the *ad-3* block (or blocks) in purine biosynthesis, so that the presence of this marker prevents the formation and accumulation of purple pigment in heterokaryotic colonies with a skewed nuclear ratio favoring component I. The presence of the *ad-2* marker ensures that purple colonies develop only from those heterokaryotic conidia in which the *ad-3A* and/or the *ad-3B* locus in component II has been inactivated.

Isolation of mutants: The general procedure and media for forward-mutation experiments with homokaryotic wild-type strains of Neurospora have been described previously.[4] Modifications in procedure for experiments with a dikaryon are as follows. To minimize the recovery of spontaneous *ad-3* mutants, single colony isolates (from conidia of the dikaryon plated at 35°C in minimal medium) were made, and these were grown on minimal medium for 2 days at 35°C and then 5–8 days at 24°C. Cotton-filtered suspensions of conidia in water from several of these cultures were adjusted to a concentration of 10^7 per ml by means of hemocytometer counts, and aliquots were irradiated (in Erlenmeyer flasks stirred with a magnetic stirrer) at ice-water temperature with a General Electric Maxitron 250 X-ray source (operated at 250 kv, 30 mA, with 3-mm Al filter) at an exposure rate of 1000 r/min. After the desired exposures had been given, aliquots of suspension containing about 10^6 surviving heterokaryotic conidia were each inoculated into 10 liters of medium containing 12.5 mg/l adenine sulfate, 250 mg/l L-arginine HCl, 10 mg/l nicotinamide, 1.5% sorbose, and 0.1% sucrose in 12-liter Florence flasks and incubated at 30°C for 6 days with aeration in the dark. During this time, each surviving conidium grows into a colony about 2 mm in diameter. The contents of each flask were determined volumetrically and examined for the presence of purple colonies in thin layers in white photographic developing trays. The total number of purple *ad-3* mutant colonies per flask was determined, and each one was isolated and subcultured; the total number of unpigmented colonies was estimated from counts on three samples of 2 or 6 ml taken from each flask, in addition to measurements of the total volume. Heterokaryotic survival was estimated from counts of unpigmented colonies and hemocytometer counts of the original conidial suspension. Estimates of forward-mutation frequency are based on the colony counts of pigmented and unpigmented colonies from each of the incubation flasks and are expressed as *ad-3* mutants (purple colonies) per 10^6 heterokaryotic survivors (total colonies).

Genetic analysis: Conidia from random samples of subcultures of the purple colonies obtained from each exposure were plated, and single-colony isolates that were heterokaryotic (with regard to the two different components of the dikaryon) and adenine-requiring were obtained for each of the purple colonies originally isolated. These single-colony isolates were then analyzed as indicated below to determine (1) genotype (whether *ad-3A*, *ad-3B*, *nic-2*, or any combination of these) and (2) whether the *ad-3* mutation was reparable or irreparable on adenine-supplemented medium. Since histidine supplementation inhibits the general recovery of *ad-3* mutations as purple colonies, *ad-3* mutations affecting the *hist-2* locus were not sought.

Genotype: An *ad-3* dikaryon representing each initial colony was analyzed in heterokaryon tests with four different tester strains on minimal medium to determine which of the known loci in the vicinity of the *ad-3* region had been inactivated. The genotype and number of each tester strain is as follows: (1) *ad-3A*—1-112-13; (2) *ad-3B*—1-112-2; (3) *hist-2 nic-2 al-2*—74-OR33-3A; and (4) *ad-2 inos*—74-OR60-44A. Tester 4 serves as a control, and the heterokaryon test responses with tester strains 1, 2, and 3 indicate whether the *ad-3A* locus, the *ad-3B* locus, and/or the *nic-2* locus, respectively, is inactivated. Each of the four testers bears at least one biochemical marker which is also present in component I; therefore, any growth in minimal medium must reflect complementation between the tester and component II of the dikaryon.

Tests to distinguish between ad-3^R and ad-3^{IR} mutations: Dikaryon test: Conidia (typically 500–2000) from each dikaryon were plated in medium supplemented with 100 mg/l adenine sulfate and 2 mg/l calcium pantothenate and incubated at 35°C for 60–72 hr. Reparability of the *ad-3* mutation in component II is indicated by the presence of *cot* colonies, which are also homokaryotic for *al-2* and *pan-2*. In the dikaryon test it is not possible to distinguish *ad-3^{IR}* mutations from those that are *ad-3^R* + *RL*, where *RL* designates recessive lethal damage elsewhere in the genome.[2] Either type of mutation results in homokaryotic lethality of component II and the absence of *cot* colonies in the dikaryon test.

Trikaryon tests: The *ad-3^{IR}* and *ad-3^R* + *RL* mutations can be distinguished by a trikaryon test employing the dikaryotic tester strain 12-1-18, which was obtained by X irradiation of the dikaryon described in Table 1 and which carries a deletion for *ad-3A*, *ad-3B*, and *nic-2* in component II. The dikaryon to be tested is combined with dikaryon 12-1-18, and conidia from the resulting trikaryon (not a tetrakaryon, since component I in each dikaryon is identical) are plated to determine whether dikaryotic *cot* colonies are formed. If *cot* colonies are present in the trikaryon test with tester 12-1-18 but absent from the dikaryon test, then component II bears irreparable damage outside of the region covered by the 12-1-18 deletion. The absence of *cot* colonies in both the dikaryon test and the trikaryon test indicates that the irreparable damage is within the region covered by the 12-1-18 deletion, and the strain being tested is most probably a true *ad-3^{IR}* mutation.

Dikaryons showing negative results in the dikaryon test and the trikaryon test with tester 12-1-18 were subjected to additional trikaryon tests with an *ad-3A^{IR}* tester strain (12-7-215) and an *ad-3B^{IR}* tester strain (12-5-182), each of which has been shown in extensive homology tests (F. J. de Serres, unpublished) to have irreparable damage closely associated with the inactivated *ad-3* locus. Such tests permit a distinction between *ad-3^{IR}* mutations and *ad-3^{R}* + *RL* mutations, where *RL* designates a separate site of recessive lethal damage closely linked with the *ad-3A* and *ad-3B* loci. The use of trikaryon tests is discussed more fully elsewhere.[2]

Results.—Survival of heterokaryotic fraction of conidia: The survival of hetero-karyotic conidia after various X-ray exposures from four separate experiments, as well as averages from the four experiments combined, are plotted in Figure 1. The decrease in heterokaryotic survival with increasing dose appears to be exponential. Atwood and Mukai[3] have previously shown that after X irradiation the survival of the total conidial population of a heterokaryon (both heterokaryotic and homo-karyotic conidia) exhibits multi-hit kinetics, while the survival of the heterokaryotic fraction is exponential or one-hit in nature. The more rapid inactivation of the heterokaryotic conidia is attributed to nuclear inactivation that results in the rapid conversion of binucleate heterokaryotic conidia into uninucleate (and homo-karyotic) conidia which can no longer grow on minimal medium.

Relation between ad-3 mutant frequency and dose: In Figure 2 the *ad-3* mutant frequencies at 1, 2, 5, 10, 20, and 40 kr exposures are presented. The mutant frequencies increase in proportion to the 1.36 power of the dose. This exponent is significantly different from 1.0 (P < 0.001).

Genetic characterization of ad-3 mutants from each dose: The results of the tests described in *Materials and Methods* for genotype and for reparability of a random sample of mutants from each X-ray dose are presented in Table 2. The numbers of mutants in each of the three classes thus defined (*ad-3^{R}*, *ad-3^{IR}*, and *ad-3^{R}* + *RL*) are included in Table 2. All mutants tested are also classified with regard to geno-type as *ad-3A* mutants, *ad-3B* mutants, or mutants inactivated in two or more loci (*ad-3A ad-3B*, *ad-3A ad-3B nic-2*, or *ad-3B nic-2*, all of which are irreparable).

Estimates of the frequency (expressed as a function of survivors) of all reparable mutants (including both *ad-3^{R}* and *ad-3^{R}* + *RL*), of irreparable mutants, and of reparable mutants with irreparable damage elsewhere in the genome (*ad-3^{R}* + *RL*) may be obtained by multiplying the proportion of tested mutants which falls into each of the categories at any dose by the *ad-3* mutant recovery at that dose. Such estimates are listed in Table 2 for *ad-3^{R}*, *ad-3^{IR}*, and *ad-3^{R}* + *RL* mutants, and in Figure 3 these estimates are plotted as a function of dose. *Ad-3^{R}* mutants increase in proportion to the dose, while *ad-3^{IR}* and *ad-3^{R}* + *RL* mutants increase as the square of the dose. In other words, *ad-3* mutants, which increase as the 1.36 power of the dose, consist of two classes of mutants, namely, *ad-3^{R}* mutants, which exhibit single-hit kinetics, and *ad-3^{IR}* mutants, which exhibit two-hit kinetics. Some *ad-3^{R}* mutants have acquired a second independent alteration elsewhere in the genome which causes irreparable damage unrelated to the mutation affecting the *ad-3* locus. Mutants of this type (*ad-3^{R}* + *RL*) also exhibit two-hit kinetics, as was expected. The actual estimates of the exponent *b* in the equation $y = ax^b$ (where y = mutant yield and x = X-ray exposure) for all *ad-3^{R}* mutants and for *ad-3^{IR}* mutants are 1.08 and 2.09, respectively; these estimates are not significantly different from 1.0 and 2.0 (P = 0.2–0.3 and P = 0.4–0.5, respectively). The point estimate of this parameter for *ad-3^{R}* + *RL* mutants is 1.84 which approximates 2.0; the difference is barely statistically significant (with P slightly less than 0.05), but the regression analysis of *ad-3^{R}* + *RL* mutants is based on only five dose-response points, each of which represents a much smaller number of mutants than the other analyses.

TABLE 2

CLASSIFICATION OF *ad-3* MUTANTS BY GENOTYPE AND REPARABILITY

Dose (kr)	Genotype: *ad-3A* *R*	*ad-3A* *R + RL*	*ad-3A* *IR*	*ad-3A* Total	*ad-3B* *R*	*ad-3B* *R + RL*	*ad-3B* *IR*	*ad-3B* Total	*ad-3A* *ad-3B**	*ad-3A* *ad-3B* *nic-2**	*ad-3B* *nic-2**	Total	Incidence per 10^6 Survivors: *ad-3R*†	*ad-3R + RL*	*ad-3IR*
0	1	0	0	1	3	0	0	3	0	0	0	4	0.3	0	0
1	27	2	0	29	67	0	1	68	2	0	0	99	4.2	0.1	0.1
2	24	0	3	27	68	0	2	70	2	0	0	99	10.6	0	0.8
5	62	2	2	66	98	6	10	114	6	2	0	188	33.3	1.6	4.0
10	44	4	16	64	97	10	22	129	29	5	2	229	64.8	5.9	30.9
20	20	3	3	26	34	8	14	56	24	3	0	109	148.3	25.1	100.5
40	8	6	10	24	21	9	19	49	24	0	2	99	204.6	69.8	255.8

* Mutants in the three classes, *ad-3A ad-3B*, *ad-3A ad-3B nic-2*, and *ad-3B nic-2*, are all irreparable.
Mutants scored as *ad-3R* and mutants scored as *ad-3R + RL* are included.

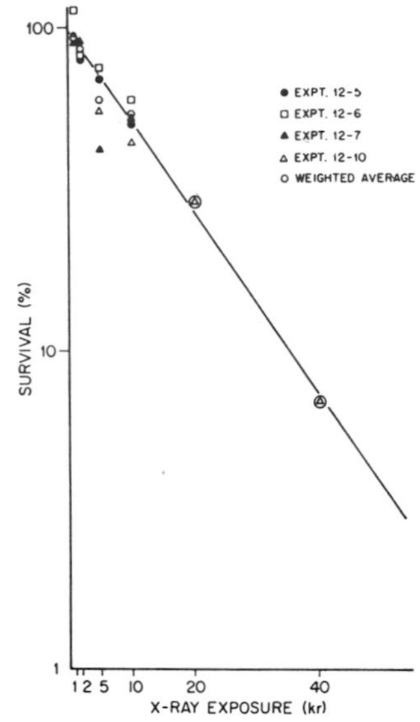

SURVIVAL (%)
100
10
1
1 2 5 10 20 40
X-RAY EXPOSURE (kr)
● EXPT. 12-5
□ EXPT. 12-6
▲ EXPT. 12-7
△ EXPT. 12-10
○ WEIGHTED AVERAGE

FIG. 1.—Survival of heterokaryotic conidia after various exposures to 250-kv X rays.

FORWARD-MUTATION FREQUENCY IN THE ad-3 REGION
10⁻³
10⁻⁴
10⁻⁵
10⁻⁶
1 2 5 10 20 40
X-RAY EXPOSURE (kr)
● EXPT. 12-5
□ EXPT. 12-6
▲ EXPT. 12-7
△ EXPT. 12-10
○ WEIGHTED AVERAGE

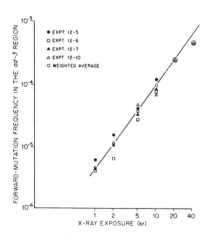

FIG. 2.—Frequency of forward mutations in the *ad-3* region after various exposures to 250-kv X rays.

Discussion.—Two classes of mutants: Two different types of *ad-3* mutants are recovered after X irradiation. One type (*ad-3^R*) is the predominant type at low doses (and low mutant recovery frequencies), and this type increases in proportion to the dose. Mutations of this type result from intragenic alterations. The second type of mutant (*ad-3^{IR}*) is almost negligible in its incidence at low doses, but this increases as the square of the dose, so that at 40 kr, where mutant recovery frequencies are high, about $^2/_3$ of the mutants are *ad-3^{IR}*. The dose-squared kinetics suggest that each *ad-3^{IR}* mutant is derived from the cooperative effect of two alterations which are independent in their probabilities of occurrence.

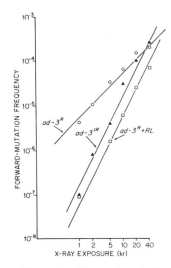

Fig. 3.—Forward-mutation frequencies of various classes of *ad-3* mutations after exposure to 250-kv X rays.

The mode of origin of ad-3^R and ad-3^{IR} mutations: The genetic analysis shows that *ad-3^R* and *ad-3^{IR}* mutants have different modes of origin. One might have expected that *ad-3^R* mutations would be converted to *ad-3^{IR}* mutations by additional hits in one of the immediately adjacent genetic regions (thus causing interstitial deletions rather than point mutations). However, if this were true, the frequency of *ad-3^R* mutants should increase as some power of the dose less than 1.0. Neither the combined induction curve for all *ad-3^R* mutants (Fig. 3) nor the individual induction frequencies for *ad-3A^R* and *ad-3B^R* mutants show a power of the dose significantly different from 1.0, even at the highest doses. This must mean that each *ad-3^{IR}* mutation results from a pair of mutually independent events which are separately not detectable as *ad-3^R* mutations. These independent events are tentatively viewed as alterations or breaks on opposite sides of one or both of the *ad-3* loci which interact to delete the genetic region between them.

Forward mutation rates in the ad-3 region: The yield y of *ad-3* mutants at any X-ray dose x may be described by the equation $y = k + ax + bx^2$. Here k equals the spontaneous mutant frequency (negligible in this case; ax equals the yield of *ad-3^R* mutants; and bx^2 equals the yield of *ad-3^{IR}* mutants. Since ax and bx^2 have been determined experimentally for each dose in the present paper, one may obtain values for a and for b by fitting the data derived from the genetic analysis with ax and bx^2, respectively. The term a then represents the mutation rate for *ad-3^R* mutants and is found to be 5.39×10^{-9} *ad-3^R* mutants·r^{-1}·survivors^{-1}. The rate is 1.75×10^{-9} for *ad-3A^R* mutants and 3.64×10^{-9} for *ad-3B^R* mutants. On the other hand, the term b is a compound term and represents the product of two independent probabilities, namely, the probability per unit dose that the first of the two hits will occur and cooperate to produce an *ad-3^{IR}* mutation, times the probability that the second hit will do so. It seems reasonable to assume that these two probabilities are equal, because the *hist-2–ad-3* and the *ad-3–nic-2* regions are approximately equal in length. Accordingly, one may use $\sqrt{b}$ (4.01×10^{-7} hits·r^{-1}·survivors^{-1}) as an approximation for the frequency of the *occurrence and recovery* of each of the two types of hits. This is necessarily a minimal estimate for frequency of the *occurrence* of these types of hits, since some will probably occur in the *ad-3* region, but their location will be such that neither the *ad-3A* nor the *ad-3B* locus is deleted. Furthermore, the scarcity of *ad-3A ad-3B* double mutants and of *ad-3^{IR}* mutants which have lost the *nic-2* locus suggests that the majority of *ad-3^{IR}* mutants result from pairs of breaks which are close together. For these reasons, 4.01×10^{-7} hits·r^{-1}·survivors^{-1} (about 74 times the frequency of events leading to *ad-3^R* mutants) considerably underestimates the incidence of hits which are capable of interacting in pairs to produce irreparable mutants.

Implication for the comparison of X-ray-induced mutations in haploid and diploid organisms: The finding that the spectrum of X-ray-induced mutations in the *ad-3* region includes both intragenic and extragenic alterations makes it possible to reconcile the differences observed in the analysis of X-ray-induced mutations in haploid and diploid organisms. Whereas in Neurospora there is evidence that X rays cause intragenic alterations,[6, 9, 10] the evidence from similar studies with maize shows that all X-ray-induced mutations have the characteristics of extragenic alterations.[11-13] The problems of the nature of X-ray-induced mutations in Drosophila is one of long standing, with two distinct schools of thought that have only recently been reconciled (see discussions by Muller and Oster,[14] Lefevre and Green,[15] and Green[16]). Some of these workers found only stable X-ray-induced mutations while others found revertible X-ray-induced mutations as well.

Since X-ray-induced mutations in the *ad-3* region are a mixture of one-hit and two-hit alterations, one would expect this to be generally true for other loci in other organisms as well. In general, high exposures are frequently used in forward-mutation experiments to provide a high mutant yield. However, high doses favor extragenic (two-hit) alterations, and in diploid organisms these would constitute the majority class of mutants recovered. However, such irreparable mutations are lethal in forward-mutation studies with haploid organisms, and so in typical experiments with haploid organisms only the intragenic (one-hit) alterations are recovered and analyzed. Whereas the two-hit alterations leading to chromosome deletions should be stable in reversion tests, de Serres[6] has found that 25 per cent of a sample of X-ray-induced *ad-3^R* mutants were revertible spontaneously or after X-ray treatment. Thus, in experiments involving high doses of X rays with diploid organisms, the revertible mutants might constitute a very small fraction of the total number, and a large number of mutants would be required to demonstrate revertibility of X-ray-induced mutants. Data for revertibility of X-ray-induced mutants would be comparable in haploid and diploid organisms only if low X-ray exposures were used in the forward-mutation experiments, so that the majority of mutants would be one-hit in origin. Our evidence indicates that a series of X-ray-induced allelic mutants should be composed of both stable and revertible alleles, and the percentages of each class should be dependent upon the dose and upon the efficiency with which all types of induced mutants can be recovered.

Summary.—(1) An analysis of the induction of *ad-3^R* and *ad-3^{IR}* mutations in a balanced dikaryon of *Neurospora crassa* with X irradiation shows that they have different induction kinetics. *Ad-3^R* mutants show single-hit kinetics, while *ad-3^{IR}* mutants show two-hit kinetics. (2) The spectrum of X-ray-induced forward mutation in the *ad-3* region is dose-dependent; whereas at 1 kr, 97 per cent of the mutants are *ad-3^R* (and affect only the *ad-3A* locus or *ad-3B* locus), at 40 kr, 56 per cent are *ad-3^{IR}* and 26 per cent inactivate both the *ad-3A* and *ad-3B* loci. (3) Estimates of the forward-mutation rates for the two loci in the *ad-3* region are 1.75×10^{-9} mutants$\cdot$r$^{-1}\cdot$survivors^{-1} for *ad-3A^R* mutants and 3.64×10^{-9} mutants$\cdot$r$^{-1}\cdot$survivors^{-1} for *ad-3B^R*$\cdot$mutants. A minimal estimate for the incidence of the hits that cooperate to give *ad-3^{IR}* mutants is 4.01×10^{-7} hits$\cdot$r$^{-1}\cdot$survivors^{-1}. (4) The finding that the spectrum of X-ray-induced mutations in the *ad-3* region includes both intragenic and extragenic alterations makes it possible to reconcile the differences observed in previous analyses of X-ray-induced mutations in haploid and diploid organisms.

We wish to acknowledge gratefully the aid of Dr. Marvin Kastenbaum in the statistical analysis, and the invaluable advice, criticism, and discussion of Dr. Heinrich V. Malling, in addition to the technical assistance of Mrs. Arlee Teasley, Miss Anita Meley, and Mrs. Ida Ruth Miller.

* Research sponsored by the U.S. Atomic Energy Commission under contract with the Union Carbide Corporation.

[1] de Serres, F. J., and R. S. Osterbind, "Estimation of the relative frequencies of X-ray-induced viable and recessive lethal mutations in the *ad-3* region of *Neurospora crassa*," *Genetics*, **47**, 793–796 (1962).

[2] de Serres, F. J., "Genetic analysis of the structure of the *ad-3* region of *Neurospora crassa* by means of irreparable recessive lethal mutations," *Genetics*, **50**, 21–30 (1964).

[3] de Serres, F. J., and B. B. Webber, "Evidence for recessive lethal mutations resulting from gross chromosome deletions in balanced heterokaryons of *Neurospora crassa*," *Neurospora Newsletter*, **3**, 3–5 (1963).

[4] Brockman, H. E., and F. J. de Serres, "Induction of *ad-3* mutants of *Neurospora crassa* by 2-aminopurine," *Genetics*, **48**, 597–604 (1963).

[5] Giles, N. H., F. J. de Serres, Jr., and E. Barbour, "Studies with purple adenine mutants in *Neurospora crassa*. II. Tetrad analysis from a cross of an *ad-3A* mutant with an *ad-3B* mutant," *Genetics*, **42**, 608–617 (1957).

[6] de Serres, F. J., "Studies with purple adenine mutants in *Neurospora crassa*. III. Reversion of X-ray-induced mutants," *Genetics*, **43**, 187–206 (1958).

[7] Huebschman, C., "A method for varying the average number of nuclei in the conidia of *Neurospora crassa*," *Mycologia*, **44**, 599–604 (1952).

[8] Atwood, K. C., and F. Mukai, "Survival and mutation in *Neurospora* exposed at nuclear detonations," *Am. Naturalist*, **88**, 295–314 (1954).

[9] Giles, N. H., "Studies on the mechanism of reversion in biochemical mutants of *Neurospora*," in *Genes and Mutations*, Cold Spring Harbor Symposia on Quantitative Biology, vol. 16 (1951), pp. 283–313.

[10] Giles, N. H., "Forward and back mutation at specific loci in *Neurospora*," in *Mutation*, Brookhaven Symposia in Biology, No. 8 (1956), pp. 103–125.

[11] Stadler, L. J., "The effect of X-rays upon dominant mutation in maize," these PROCEEDINGS, **30**, 123–128 (1944).

[12] Stadler, L. J., and H. Roman, "The effect of X-rays upon mutation of the gene *A* in maize," *Genetics*, **33**, 273–303 (1948).

[13] Nuffer, M. G., "Additional evidence on the effect of X-ray and ultraviolet radiation on mutation in maize," *Genetics*, **42**, 273–282 (1957).

[14] Muller, H. J., and I. I. Oster, "Principles of back mutation as observed in *Drosophila* and other organisms," in *Advances in Radiobiology* (Edinburgh: Oliver and Boyd, 1957), pp. 407–413.

[15] Lefevre, G., Jr., and M. M. Green, "Reverse mutation studies on the forked locus in *Drosophila melanogaster*," *Genetics*, **44**, 769–776 (1959).

[16] Green, M. M., "Back mutation in *Drosophila melanogaster*. I. X-ray-induced back mutations at the yellow, scute and white loci," *Genetics*, **46**, 671–682 (1961).

5

Reprinted from *Nature*, **245**(5426), 460–462 (1973)

Uniformity of Radiation-induced Mutation Rates among Different Species

ONE of the major difficulties in estimating the genetic hazards of ionising radiation to human populations has been our inability to extrapolate with confidence from mutation rate data in lower organisms to man[1,2]. Experimentally observed mutation rates per locus per rad extend over an enormous range of three orders of magnitude. For instance, the forward mutation rate per locus per rad is 1×10^{-9} in *Escherichia coli* B/r at two loci for phage T1 resistance, 2.7×10^{-9} in *Neurospora* at the *ad3* locus, 1.4×10^{-8} in *Drosophila* for eight specific loci (averaged), 1.7×10^{-7} in the mouse for twelve specific loci (averaged), and 1×10^{-6} in barley for three loci.

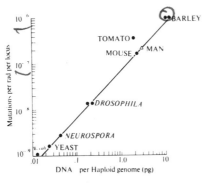

Fig. 1 Relation between forward mutation rate per locus per rad and the DNA content per haploid genome. Line drawn with slope of one through the mouse point. Point for man estimated from DNA content.

The development of target theory in radiobiology and the recognition that the genetic material is the target for many radiobiological endpoints led to the demonstration that the mean lethal dose for survival of irradiated single cells is related to the nuclear DNA content[3,4]. Sparrow *et al.*[5] have noted that in plants, radiosensitivity is closely related to several nuclear parameters such as nuclear volume and interphase chromosome volume and thus to the DNA content per nucleus. In amphibians, too, it has been shown that the mean lethal dose for survival of whole-body irradiated animals is directly related to the amount of DNA per nucleus and per chromosome[6]. The work of Sparrow and his coworkers[7-9] has demonstrated that the relation between nuclear parameters and radiosensitivity is not entirely simple, however, for although the organisms studied do fall into a series of 'radiotaxa' within which a simple linear relation holds, the sensitivities of the radiotaxa differ from one another.

We re-examined the data for radiation-induced forward mutation rates from organisms as disparate as bacteria, fungi, higher plants, insects, and mammals for which we also know the amounts of DNA per nucleus. As seen in Table 1, and as noted above, the mutation rates per locus per rad vary over three orders of magnitude. Nonetheless, when these rates are adjusted for the amount of DNA per nucleus and thus normalised to a common biological baseline, the mutation rates obtained are all essentially the same, varying by a mere factor of three instead of 1,000.

This remarkable consistency in normalised mutation rates in such a wide variety of organisms strongly suggests that extrapolation directly from experimental organisms to man can be done with confidence. For all the species listed in the table, the unweighted average of the mutation rates normalised to the human DNA value is 2.6×10^{-7} per locus per rad. (These rates are all for forward mutations induced by acute irradiation. With chronic radiation or with germ cells irradiated during different cell stages, the mutation rates often vary by a factor of three to four. These differences are conventionally attributed to differential repair. Back mutations, which are often the result of base changes

Table 1 Specific Locus Mutation Rates Normalized for Amount of DNA per Nucleus

Organism	DNA per haploid genome (pg)	Relative amount of DNA as compared with human A	Specific locus mutations per locus per rad B	Normalised specific locus mutations per locus per rad B/A	References
E. coli B/r (Bacterium)	0.013 *	4.5×10^{-3}	1×10^{-9}†	2.2×10^{-7}	*15, †16
Saccharomyces cerevisiae (Yeast)	0.024 *	8.3×10^{-3}	1.6×10^{-9}†	1.9×10^{-7}	*17 †R. Mortimer, personal communication
Neurospora crassa (Fungus)	0.042 *	1.4×10^{-2}	2.7×10^{-9}†	1.9×10^{-7}	*17, †18
Drosophila melanogaster (Fruit fly)	0.17 * 0.22	0.058 0.076	1.4×10^{-8}† (Spermatogonia)	2.4×10^{-7} 1.8×10^{-7}	*19 †20,2.
Lycopersicon esculentum (Tomato)	1.95 *	0.67	3.75×10^{-7}†	5.6×10^{-7}	*22 †23
Mus musculus (Mouse)	2.26 *	0.78	1.7×10^{-7} * (Spermatogonia)	2.2×10^{-7}	*2
Hordeum vulgare (Barley)	10 * 11.5†	3.45 3.97	1×10^{-6}‡	2.9×10^{-7} 2.5×10^{-7}	*24, †17, ‡25
Homo sapiens (Man)	2.9	1	—	2.6×10^{-7} (unweighted average)	

in suppressor loci, are usually lower by a factor of ten.)

The consistency that obtains when the data are adjusted for the amount of DNA per nucleus for each species indicates either that it is the nucleus, and not the locus, that determines target size, or that, on the average, the size of a (radiation-mutable) locus is proportional to the total genome size (DNA content) for the species.

Although the latter conclusion is contrary to the usual genetic expectation that structural genes should be the same size in bacteria as in man, we note that the functional genetic unit in *Drosophila* corresponds to a band in the salivary gland polytene chromosomes[10]. This indicates that complementation groups in higher forms contain much more DNA than is necessary for specification of a protein. It has been postulated that the extra DNA has a regulatory function[11–13] and that mutations in any of the components of the functional genetic unit would fail to complement mutations in any of the other components[14].

Models can also be postulated in which the repair processes vary quantitatively in the different species.

Whatever the basic explanation might be, we conclude that, empirically, the radiation-induced mutation rate is proportional to the total genome size (DNA content) for the species. This correspondence, which is shown graphically in Fig. 1, allows us to extrapolate from mutation rates obtained in experimental organisms to man with greater confidence.

This work was performed under the auspices of the US Atomic Energy Commission and supported in part by a Public Health research grant, and a Research Career award, to A. D. C.

SEYMOUR ABRAHAMSON

Departments of Zoology and Genetics,
University of Wisconsin,
Madison, Wisconsin 43713

MICHAEL A. BENDER

Department of Radiology,
The Johns Hopkins University School of Medicine,
Baltimore, Maryland 21205

ALAN D. CONGER

Department of Radiobiology,
Temple University School of Medicine,
Philadelphia, Pennsylvania 19140

SHELDON WOLFF

Laboratory of Radiobiology and Department of Anatomy,
University of California,
San Francisco, California 94143

Received August 30, 1973.

1 Biological Effects of Ionizing Radiation Advisory Committee, *The Effects on Populations of Exposure to Low Levels of Ionizing Radiation* (National Academy of Sciences, National Research Council, Washington, DC, 1972).
2 United Nations Scientific Committee on the Effects of Atomic Radiation, *Ionizing Radiation: Levels and Effects, Vol. II, Effects. A/8725: General Assembly Official Records, 27th Session, Supplement No. 25* (United Nations, New York, 1972).
3 Terzi, M., *Nature*, **191**, 461 (1961).
4 Terzi, M., *J. theor. Biol.*, **8**, 233 (1965).
5 Sparrow, A. H., Sparrow, R. C., Thompson, K. H., and Schairer, L. A., *Radiat. Bot. Suppl.*, **5**, 101 (1965).
6 Conger, A. D., and Clinton, J. H., *Radiat. Res.*, **54**, 69 (1973).
7 Sparrow, A. H., Underbrink, A. G., and Sparrow, R. C., *Radiat. Res.*, **32**, 915 (1967).
8 Sparrow, A. H., Rogers, A. F., and Schwemmer, S. S., *Radiat. Bot.*, **8**, 149 (1968).
9 Underbrink, A. G., Sparrow, A. H., and Pond, V., *Radiat. Bot.*, **8**, 205 (1968).
10 Judd, B. H., Shen, M. W., and Kaufman, T. C., *Genetics*, **71**, 139 (1972).
11 Britten, R. J., and Davidson, E. H., *Science, N.Y.*, **165**, 349 (1969).
12 Georgiev, G. P., *J. theor. Biol.*, **25**, 473 (1969).
13 Crick, F., *Nature*, **234**, 25 (1971).
14 Shannon, M. P., Kaufman, T. C., Shen, M. W., and Judd, B. H., *Genetics*, **72**, 615 (1972).
15 Gillies, N. E., and Alper, T., *Biochim. biophys. Acta*, **43**, 182 (1960).
16 Demerec, M., and Latarjet, R., *Cold Spring Harb. Symp. quant. Biol.*, **11**, 38 (1946).
17 Sparrow, A. H., Price, H. J., and Underbrink, A. G., in *Evolution of Genetic Systems, Brookhaven Symp. Biol.*, **23**, 451 (Gordon and Breach, New York, 1972).
18 Webber, B. B., and deSerres, F. J., *Proc. natn. Acad. Sci., U.S.A.*, **53**, 430 (1965).
19 Tates, A. D., thesis. Univ. Leiden.
20 Alexander, M. L., *Genetics*, **39**, 409 (1954).
21 Alexander, M. L., *Genetics*, **45**, 1019 (1960).
22 Rees, H., and Jones, R. N., *Int. Rev. Cytol.*, **32**, 53 (1972).
23 Brock, R. D., and Franklin, I. R., *Radiat. Bot.*, **6**, 171 (1966).
24 Bennett, M. D., *Proc. R. Soc., B.*, **181**, 109 (1972).
25 Ehrenberg, L., and Eriksson, G., *Mutat. Res.*, **1**, 139 (1964).

Part II
CHEMICAL MUTAGENESIS

Editors' Comments
on Papers 6 and 7

6   AUERBACH and ROBSON
     The Production of Mutations by Chemical Substances

7   AUERBACH
     The Induction   by Mustard Gas of Chromosomal Instabilities in
     Drosophila melanogaster

THE DISCOVERY OF CHEMICAL MUTAGENESIS

Among the refugees to flee Nazi Germany in the 1930s was Charlotte Auerbach, geneticist. She settled in Edinburgh and became naturalized just in time to avoid detention when the war enveloped Great Britain. Already situated in Edinburgh was the great American geneticist H. J. Muller, who had spent several years in Russia but who had fled that country upon Lysenko's rise to power and the concomitant political suppression of genetics. Muller encouraged Auerbach in her work, particularly in her attempts to produce mutation by chemical treatments. As World War II developed momentum, Muller returned to the United States, and at the same time a pharmacologist, J. M. Robson, began work on a top-secret project concerned with war gases. Robson noticed that mustard gas, $S(CH_2CH_2Cl)_2$, now known to be a bifunctional alkylating agent capable, among other things, of cross-linking the complementary chains of the DNA double helix (see Loveless, 1966), produced cytotoxic effects that were rather similar to those produced by X-irradiation. He accordingly approached Auerbach with the suggestion that she test mustard gas for mutagenicity; the results were spectacularly successful. Wartime censorship prevented the formal publication of their results until 1947 (Paper 6), but a carefully worded telegram to Muller, by then back in the States, ensured the early informal dissemination of this vital information.

Auerbach's initial success was followed very quickly by her description, reprinted in Paper 7, of a phenomenon that turned out to be particularly and quite generally characteristic of chemical mutagenesis—*mosaicism*. Mosaics are organisms (or, in the case of microbes, clones) composed of both mutant and nonmutant cells, and their appearance in, for instance, the progeny of treated *Drosophila* sperm or bacteriophage T4 has been subjected to considerable analysis (see Papers 12, 17, 18, and 19; Green and Krieg, 1961; Lee et al., 1970). In a general sense, the phenomenon of mosaicism is explicable on the basis of two fundamental chemical processes, both of which were quite unknown at the time Paper 7 was written. First, a chemically altered DNA base may mispair only occasionally during replication, so that the primary mutational event may occur only several generations (DNA replications) after the time of treatment. Second, the mispaired configuration must often segregate itself out before the mutant phenotype becomes expressed; this process often reflects organismal differences in chromosome replication and segregation. All would be quite simple if this were the whole story. Unfortunately it is not, for as Auerbach noted in Paper 7, and as had been observed frequently since then (see, for instance, Nasim and James, 1971; Nasim and Grant,1973; and the numerous references therein), a sharply increased tendency to mutate at a specific gene often persists indefinitely in the line of descent from a treated ancestor. Such "induced chromosomal instabilities" (*replicating in stabilities* as they are now called) are very poorly understood, and we shall not speculate here on their nature.

REFERENCES

Green, D. M., and D. R. Krieg. 1961. The delayed origin of mutants induced by exposure of extracellular phage T4 to ethyl methane sulfonate. *Proc. Natl. Acad. Sci., 47:* 65.

Lee, W. R., G. A. Sega, and J. B. Bishop. 1970. Chemically induced mutations observed as mosaics in *Drosophila melanogaster. Mutation Res., 9:* 323.

Loveless, A. 1966. *Genetic and Allied Effects of Alkylating Agents.* Butterworth, London.

Nasim, A., and C. Grant. 1973. Genetic analysis of replicating instabilities in yeast. *Mutation Res., 17:* 185.

——, and A. P. James. 1971. Replicating instabilities in yeast: evidence from single cell isolations. *Genetics, 69:* 513.

6

Reprinted from *Proc. Roy. Soc. Edin.*, **B62**(32) Pt. III, 271–283 (1947)

The Production of Mutations by Chemical Substances.

By **C. Auerbach,** Ph.D., and **J. M. Robson,** M.D., B.Sc. From the Institute of Animal Genetics and Department of Pharmacology, University of Edinburgh. (With Three Text-figures.)

(MS. received March 2, 1946. Read June 3, 1946)

INTRODUCTION

THE production of mutations by the action of chemical substances on germ cells has often been reported. However, the variability of the spontaneous mutation rate and its dependence not only on environmental conditions and physiological factors, but also on the genotype, make it extremely difficult to assess the value of tests in which only small increases over the spontaneous mutation rate have been found. For this reason, Muller was still able to conclude in 1941 that there was no definite proof that chemical substances could exert an effect on the mutation rate. Since then, Thomas and Chevais (1943) have reported results with sulphonamides which, if they can be confirmed, would indicate a real, though slight, action of these substances on the chromosomes, at least as far as gene mutations are concerned. Stubbe (1940), working on plant material, observed a significant increase in mutation rate with phenol and potassium thiocyanate. It is of interest that Auerbach and Robson (1943) independently observed a similar effect with allyl isothiocyanate in experiments on *Drosophila*. Even these definite effects are, however, very slight.

During the last four years we have been testing a number of chemical substances. Among these a certain group has been found which increases the rate of occurrence of mutations and chromosome rearrangements to a similar extent as that brought about by X-rays and similar physical agencies. The best known representative of this group is mustard gas, and the present report deals only with the effects produced by this substance. Results obtained with other effective substances will be published later.

The use of mustard gas as a mutagenic substance was suggested by the observations (1) that the substance produces a prolonged inhibition in the mitotic activity of the vaginal epithelium of the mouse without any appreciable histological effect, *i.e.* that mustard gas is capable of producing a prolonged effect on nuclear activity; and (2) that there are resemblances between burns produced by X-rays and by mustard gas. Both types of lesion are slow in healing and show a tendency to break down again after apparent healing. Since X-rays are known to act on the chromosomes, it was thought possible that mustard gas might have a similar action.

MATERIAL AND METHODS

Drosophila melanogaster was used in all experiments.

In the first experiments on the effect of mustard gas, the flies were exposed to the gas in a large glass chamber (capacity 1 litre); fixed amounts of liquid mustard gas were volatilized by heating in the presence of the flies. It was found that, using this method, the results were variable.

Fig. 1 illustrates the method finally adopted. The flies were contained in a glass tube (6″ × 1″), enclosed at both ends with porcelain filter discs. One end of the chamber was connected through a **T**-tube to an atomizer spray containing a solution of mustard gas in cyclohexane, and to a constant air flow (2 litres of air per minute). The mustard gas was sprayed at regular intervals for fixed periods of time and was swept through the exposure chamber by the stream of air. In a typical experiment a 1 : 10 mixture of mustard gas and cyclohexane was sprayed at 10-second intervals for periods up to 15 minutes. The amount of mustard gas thus administered could be varied by altering either the concentration of the solution, the interval between sprays, or the period of spraying.

The concentration of mustard gas produced in the exposure chamber by the direct method described above was too high for *Drosophila* eggs and larvæ, for which the indirect method illustrated in fig. 2 was used. The eggs were placed in the exposure chamber lying on pieces of agar food on a glass slide, while the larvæ were contained in a small glass tube ($3'' \times \frac{1}{2}''$) enclosed at both ends with fine bolting silk. The chamber was connected to a 10-litre

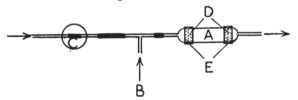

FIG. 1.—Showing the apparatus used for exposing the flies to mustard gas.
A = Exposure chamber.   C = Flowmeter.
B = Spray entry.   D = Porcelain perforated discs.
E = Rubber tubing.

aspirator bottle into which the mustard gas solutions were sprayed at constant intervals for fixed periods, and where the mustard gas spray was mixed, diluted, and swept into the exposure chamber by a constant air flow of 2 litres per minute. After the spraying was

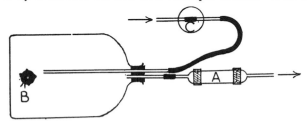

FIG. 2.—Showing the apparatus used for exposing eggs and larvæ to mustard gas.
A, B, and C as in fig. 1.

finished, the mustard gas accumulated in the mixing and diluting chamber was swept through the exposure chamber for a further 90 seconds before it was disconnected. In these experiments with both the direct and indirect methods the temperature was not controlled.

Certain difficulties which are inherent in the use of biological material will be discussed in the appropriate sections.

RESULTS

The Production of Recessive Lethals

The first test for the production of sex-linked lethals by mustard gas was carried out in April 1941. ♂♂ of wild-type stock (Oregon-K) were exposed and then subjected to the *ClB* test. The result left no doubt as to the potency of mustard gas as a mutation-producing agency. Later tests, carried out in connection with other experiments, brought further confirmation. In Table I the data for three different experiments are summarized.

TABLE I.—THE FREQUENCY OF SEX-LINKED LETHALS AFTER TREATMENT OF ♂♂
WITH MUSTARD GAS (*ClB* METHOD)

Serial No. of Experiment	Strain	No. of Tested X-Chromosomes	X-Chromosomes in which a Lethal developed		No. of X-Chromosomes in which a Semi-Lethal developed	Lethals + Semi-Lethals
			No.	Per cent.		Per cent.
H13	Oregon-K	1231	90	7·3	11	8·2
H32	,,	790	68	8·6	4	9·1
H60	*dp*; *e*	115	28	24·2	0	24·2
Control for H13	Oregon-K	1216	3	0·25	0	0·25

In the F_1 of the treated series in the first experiment there were 94 sterile cultures (= 7 per cent.). The difference from the controls, which contained 44 (= 3·5 per cent.) sterile cultures, is statistically significant, and suggests the induction of dominant sterility mutations in addition to recessive sex-linked lethals.

Although no special attempt was made to detect visible mutations in these first tests, a number of them were found incidentally. Some of them occurred among the semi-lethals, others had normal viability. A few of the lethals were found to be associated with chromosome rearrangements which were large enough to be detected by rough location tests. A cytological analysis of about 100 lethals obtained in the first experiments has been carried out by Slizynski and Slizynska (1946).

The Production of Dominant Lethals

Hatchability of eggs laid by treated ♀♀, or by untreated ♀♀ mated to treated ♂♂, is reduced to a degree which depends on the dose. Even with high doses which have a complete sterilizing effect, willingness to copulate persists, and the sperm remains motile as seen by inspection of the seminal receptacles of ♀♀ which have been treated themselves or have been inseminated by treated ♂♂. The possibility has to be considered that treatment either of the sperm or of the ova might create conditions which make fertilization impossible, even though motile spermatozoa are present in the female genital tract. This could only be decided by cytological examination of a sample of eggs, such as has been carried out for X-rays by Sonnenblick (1940), and Demerec and Kaufmann (1941). However, the fact that with medium doses a certain percentage of eggs develop into larvæ shows that a proportion, at least, of the ova become fertilized. Moreover, the ability of mustard gas to break the chromosomes, which had been indicated by the first experiments and subsequently confirmed, makes it probable *a priori* that some of the induced chromosome rearrangements will be of the kind which cause death of the zygote. It seems therefore likely that the reduction in hatchability after treatment of either father or mother with mustard gas is at least partly due to dominant lethality, which has also been shown to be responsible for the low hatchability observed in X-ray experiments on *Drosophila* (Sonnenblick, 1940; Demerec and Kaufmann, 1941).

Inhibition of Gametogenesis

Impaired hatchability of eggs is not the only reason for the reduced fertility of treated flies. Both oogenesis and spermatogenesis are interrupted by the treatment as shown by the following observations:—

(*a*) *Oogenesis.*—When ♀♀ are exposed to sub-lethal doses of mustard gas and subsequently kept on food with ♂♂, the number of eggs produced varies greatly in different individuals. This variation was found to be due, to some extent at least, to the different stages of development at which the ovaries were exposed. In virgins which are treated soon after hatching, or which are kept on a starvation diet from the period of hatching to that of exposure, the ovaries remain tiny and contain mostly young and degenerating follicles without properly formed walls. When such ♀♀ grow older, their abdomen expands and becomes filled with fatty tissue. Sometimes a few eggs are laid, and then the basal portion of the ovariole which produced the egg remains an empty sac. On the other hand, ♀♀ whose ovaries at the time of treatment are already fully developed may lay fairly well for a number of days. The most striking feature about the ovaries of such ♀♀ is the discontinuity of the developmental stages within the egg strings. Eggs ready to be laid are often in immediate proximity to quite small and undeveloped follicles, the intermediate stages being missing. Towards the end of the fertile period this peculiarity becomes very pronounced. Finally, the ovaries become very thin, with empty basal portions and degenerating young follicles in the apical parts. These observations show that only ova which have, at the time of treatment, passed a certain critical stage are able to reach maturity after exposure of the ♀♀ to mustard gas.

(*b*) *Spermatogenesis.*—No histological study was made of the effects of mustard gas on the testes. Since, however, ♂♂ treated even with high doses do not lose their willingness to copulate, the length of the period following treatment during which they retain their capacity of inseminating ♀♀ will give a rough indication of the stage at which spermatogenesis becomes interrupted. The following experiment was carried out to determine this period:—

A sample of some fifty wild-type ♂♂ was divided into 4 groups, namely:

VT kept without ♀♀ up to the time of exposure.

VC controls for VT.

MT mated *en masse* to about twice their own number of virgin ♀♀ on the day preceding exposure.

MC controls for MT.

After exposure of the T-groups all ♂♂ were mated individually to two virgins each. Every 2–3 days these ♀♀ were exchanged for fresh ones. On the 13th day following treatment, the last batch of ♀♀, which had been put with the ♂♂ on the 10th day, was dissected, and the seminal receptacles were examined for the presence of spermatozoa. The result is shown in Table II.

TABLE II.—INSEMINATING CAPACITY OF ♂♂ TREATED WITH MUSTARD GAS ON THE 10TH AND 13TH DAY FOLLOWING EXPOSURE

Series	No. of ♂♂	No. of ♂♂ which, out of two ♀♀, had inseminated:		
		Both	One	Neither
VC	10	10	..	..
MC	9	9	..	..
VT	10	1	4	5
MT	17	3	2	12

VT = virgin ♂♂.
MT = previously mated ♂♂.
VC and MC = controls to VT and MT.

Thus 10 days after exposure, insemination of at least one ♀ was still possible for half of the ♂♂ which, at the beginning of the experiment, had a full supply of sperm, but for less than one-third of those whose store of sperm had been depleted by previous matings. Dissection of a few of the treated ♂♂ showed that the testes were almost empty of spermatozoa. The test was repeated 23 days after exposure on some of the ♂♂. Sperm was found in 10 out of 14 ♀♀ in the control series, but only in one out of 15 in the treated series. If this one exception was not due to some experimental error (*e.g.* non-virginity of the ♀) it may be a case of recuperation. From these data it appears that the critical stage for inhibition of spermatogenesis by mustard gas coincides with the stage at which, after X-ray treatment, a drop occurs in the frequency of induced recessive lethals (Harris, 1929; Hanson and Heys, 1929). Demerec and Kaufmann (1941) observed a drop in X-ray induced dominant lethality as late as 19 days after treatment, but since they did not keep the ♂♂ continuously with the ♀♀, their data are not quite comparable to ours.

Difference in Sensitivity to Mustard Gas

Very early in the course of the work it was found that one of the main difficulties consists in finding a method of standardizing the dose so that the results of different tests are comparable with one another and with results of similar tests carried out with X-rays. Unfortunately, the apparatus used in these experiments did not allow control of temperature and humidity during exposure. But even where it is possible to administer the same amount of gas under strictly comparable conditions, the actual amount which can penetrate to the chromosomes under test—whether they be in the germ cells or in somatic cells—depends on anatomical and physiological factors which vary not only in different lines and individuals of the same line, but also at different stages of development. Thus the same dose, as measured by physical constants, corresponds to quite different effective doses when administered to eggs, larvæ, or imagines; to ♂♂ or ♀♀; to large or small flies, etc. Bodily activity during exposure is also of some importance. This is illustrated by the results of two tests in which the ♂♂ to be exposed were divided into two groups, one group being deprived of their wings previous to exposure. Altogether 807 eggs were tested for hatchability. Of 361 eggs fertilized by wingless ♂♂ 137 hatched, while of 446 eggs fertilized by winged ♂♂ only 29 hatched.

The simplest way of assessing the genetical effectiveness (*i.e.* the amount of gas reaching the chromosomes) of doses applied to adult ♂♂ consists in measuring the degree of dominant lethality induced in their progeny. Using this criterion, it was found that susceptibility is but slightly influenced by age and previous rearing conditions, unless the latter are so extreme that they result in marked reduction of body size with a consequent relative increase in body surface. The most important source of variation in sensitivity to a given dose is undoubtedly the existence of genetical differences between different strains and different individuals. When ♂♂ of various strains are exposed together and in the same container, the percentages of dominant lethality in their progeny show marked and consistent differences, a certain range of variation being characteristic for a given strain. By using tested strains as standards of comparison for new stocks to be tested, it was possible to grade the various strains according to their sensitivity; the least susceptible stock gave a hatchability of 70 per cent. with a dose which completely sterilized the most susceptible one (Table III).

TABLE III.—VARIATION IN SUSCEPTIBILITY TO MUSTARD GAS IN DIFFERENT STRAINS, AS MEASURED BY THE PERCENTAGE OF DOMINANT LETHALITY AMONG THE PROGENY OF TREATED ♂♂

Strain	Dose	No. of Independent Tests	No. of ♂♂ in all Tests	No. of Eggs Laid	Percentage of Eggs Hatched	Percentage Hatched in Separate Tests
Florida-4 . .	I	3	49	916	73·2	70·0 70·8 80·7
sc^{S1} *In S* w^a sc^8 .	I	1	5	96	54·2	..
,, .	II	1	3	60	63·3	..
w sn^3 *B* .	I	1	10	153	31·4	..
,, .	II	1	4	80	62·5	..
Oregon-K .	I	3	32	805	12·5	6·9 10·9 17·3
B . Y^S/sc . Y^L	I	1	10	364	0·0	
f . Y^S/sc . Y^L .	I	3	32	613	0·0	} All 0·0
,, .	II	1	8	160	0·0	

Note.—Dose II, as measured by physical constants, was about 25 per cent. lower than Dose I.

Although variability is high both between individuals used in the same test and between repeated tests carried out with the same dose on the same stock, the data clearly show genetical differences in sensitivity. Thus of the two wild-type stocks, Florida-4 has a much higher tolerance than Oregon-K. The sensitivity of the two stocks carrying the *sc* . Y^L translocation is so exceptionally high that it seemed to suggest an influence of this chromosomal rearrangement on the susceptibility of the spermatozoa themselves. To test this possibility, *f* . Y^S/sc . Y^L ♂♂ and Florida-4 ♂♂ were mated to ♀♀ of the same unrelated line (*y*) and the inseminated ♀♀ were then exposed together to mustard gas. Virgin ♀♀ of the same line were exposed simultaneously and mated subsequently to untreated brothers in order to supply a correction for the effect of the treatment on the maternal chromosome set. The result is shown in Table IV. The fact that the striking difference in dominant lethality among the progeny of treated ♂♂ of the two strains disappeared completely when the spermatozoa were exposed in identical environment, shows that this difference is determined not by the chromosomal constitution of the spermatozoa themselves, but by genetically determined differences. in the anatomy and the physiology of flies of the two stocks.

TABLE IV.—SUSCEPTIBILITY OF SPERM OF TWO DIFFERENT STRAINS TREATED IN THE BODY OF ♀♀ OF THE SAME UNRELATED STRAIN (y)

Type of Sperm	No. of Treated ♀♀	No. of Eggs	Percentage Hatchability *
Untreated . . .	8	251	41
Florida-4, treated . .	7	206	7
$f. Y^S/sc. Y^L$, treated .	11	347	7

* Since attached-X ♀♀ were used for the test, the maximum possible hatchability of eggs from untreated ♀♀ was 50 per cent.

In view of the wide variation in the actual dose necessary to produce a genetical effect, it seemed desirable to use a biological method for assessing the effectiveness of a given dose on the chromosome material. The *ClB* test was chosen for this purpose and was always carried out as a routine on an aliquot of the treated flies. This method makes it possible not only to compare quantitatively the results of different experiments carried out with mustard gas, but also allows of a comparison with X-ray experiments. In the latter case, the dose in a given mustard-gas test is taken to be equivalent to an X-ray dose in r-units which produces the same percentage of sex-linked lethals. We are well aware of the limitation of such a comparison, which is based on merely one of the effects which follow an action of these agents on the nucleus. At present, however, this seems the only method available; but its limitations have to be borne in mind in discussions based on such a comparison.

The Production of Visible Mutations

It has already been mentioned that in the tests for lethals some visible mutations were obtained. In order to detect any specific effects which mustard gas might exert, a special experiment was carried out. Treated wild-type (Oregon-K) ♂♂ were mated to attached-X ♀♀, and their sons were examined for visible abnormalities. An equivalent number of sons from untreated Oregon-K ♂♂ and attached-X ♀♀ was examined in order to determine what hereditary

TABLE V.—THE PRODUCTION OF VISIBLE MUTATIONS BY MUSTARD GAS (TREATED OREGON-K ♂♂ MATED TO ATTACHED-X ♀♀)

Group	Total No. of ♂♂ Examined	Total No. of Aberrations	Genetic Aberrations		Non-genetic Aberrations *
			Proved by Breeding Test †	Presumed by Phenotype	
Treated . . .	2750	148	51	28	69
Control . . .	2750	34	4	2	28

* A breeding test was performed only on a fraction of the ♂♂ listed in the last column. The remainder were judged to be non-genetic because of their phenotype (see Appendix).
† Of these, 10 were sex-linked recessives and 41 autosomal dominants in the treated group. All those in the control group were autosomal dominants.

and non-hereditary aberrations occur normally in these stocks. A small *ClB* test carried out on some of the ♂♂ yielded 3 lethals in 68 chromosomes, enough to show that the dose had been effective. Breeding tests to determine the nature of the aberrations were successfully carried out on about 50 per cent. of the aberrant ♂♂ in both control and treated groups. In Table V the results are summarized; details are given in the Appendix.

Among 2750 sons of treated ♂♂ there occurred 79 mutations or presumed mutations, as compared with 6 among the same number of sons of untreated ♂♂. Since, however, it seems unlikely that the frequency of non-inherited modifications should be significantly greater in the treated than in the control group, it is quite possible that a number of the abnormalities in the treated group classified by phenotype as non-genetic were, in reality, genetic, and that too much caution was exercised in this classification. If this is true, it would make the difference between treated and control group even larger.

The proportion of sex-linked visible mutations was 11 in 2750 = 0·4 per cent., that of sex-linked lethals (in the *ClB* test) was 3 in 68 = 4·4 per cent. This gives a ratio of 1 : 11 between the two frequencies. Although the error attached to this value is considerable on account of the small size of the *ClB* sample, it is clear that the real value lies somewhere within the range observed for irradiated material (*cf.* Muller, 1941).

The data do not provide evidence that mustard gas singles out specific genes or groups of genes for its action. Autosomal *Minutes* formed the majority of the inherited abnormalities. Among the sex-linked recessives, *rough eyes*, *rudimentary wings*, and *forked* were the only ones to occur more than once. In addition, there were two somatic mutations to *lozenge*. With one exception all the mutations observed more than once affected loci, which also show a fairly high degree of mutability after treatment with X-rays. The exception was a new sex-linked mutation *bashed* affecting the shape of the eye (Auerbach, *D.I.S.*, xviii and xix), which occurred three times after treatment with mustard gas and once after a combined treatment with mustard gas and X-rays, in all four cases in the inbred Florida-4 stock. It is not possible, however, to assume that the same mutation cannot be produced by X-rays, since only small samples of this stock have been tested with X-rays for the occurrence of visible sex-linked mutations. The repeated occurrence of *bashed* in these experiments may therefore be due, not to a specific action of mustard gas, but to the presence in the Florida-4 stock of a relatively unstable normal allelomorph at this locus.

The most striking feature in all experiments on visible mutations after mustard-gas treatment was the high percentage of mosaics. This will be dealt with in a separate publication (Auerbach, 1946).

The Production of Translocations

As reported above, a few of the sex-linked lethals produced in the first experiment (H13, Table I) were found to be associated with large chromosome rearrangements. A special test to determine whether mustard gas actually produces translocations was carried out in December 1941. Treated wild-type (Oregon-K) ♂♂ were mated in the same culture bottles to *ClB/scar* and to *y; bw; e* ♀♀. Their Bar daughters were used in a test for sex-linked lethals (see Experiment H32, Table I). Their phenotypically wild-type sons $\left(\dfrac{dp}{+} ; \dfrac{e}{+} \right)$ were back-crossed individually to virgins of the *y; bw; e* stock. This method allows of the detection in F_2 of translocations between X and II (no *dp* sons), X and III (no *e* sons), and II and III (all flies either *dp e* or wild-type). Since inspection of the cultures was carried out without etherization, only those translocations were detected in which the aneuploid classes (*e.g.* the *dp* ♂♂ in an X–II translocation) were completely or almost completely non-viable. The result left no doubt about the ability of mustard gas to produce translocations, but at the same time it indicated that, compared with a dose of X-rays producing the same percentage of sex-linked lethals, mustard gas is considerably less effective in the production of translocations. This conclusion was confirmed in two subsequent experiments in which similar genetical techniques were used for detecting the presence of translocations in the two long autosomes alone (Experiment H89), or in the Y-chromosome and the two long autosomes (Experiment H60). The data from all three experiments are collected in Table VI.

In this table, the last column represents the percentage of translocations which would be expected with a dose of X-rays producing the same frequency of sex-linked lethals as the dose of mustard gas used. The expected frequencies have been calculated from the data of Muller (1940, Tables II and V), and using Muller's rule that the frequency of translocations varies as the 3/2th power of the dose. As Muller's data refer only to translocations between autosomes II and III they are only directly comparable with Experiment H89. In Experiment H32 we also recorded exchanges involving the X-chromosome, and this increases the expected number of translocations by at least one-third (Muller and Altenburg, 1930; Patterson, Stone, Bedichek and Suche, 1934). In Experiment H60 the heterochromatic Y-chromosome was included in the test instead of the X-chromosome. Since, as Neuhaus (1939) has shown,

TABLE VI.—FREQUENCY OF TRANSLOCATIONS PRODUCED BY MUSTARD GAS

Experi- ment	Percentage Sex-linked Lethals *	No. of Fertilized Back-cross Cultures	Chromosomes Tested for Translocations	Translocations		
				No.	Per cent.	Expected Percentage (approx.) †
H32	8·6	816	X, II, III	7 ‡	0·9	7
H89	14·5	981	II, III	21	2·1	10
H60	24·2	33	Y, II, III	1	3·0	More than 10

* In a parallel test on ♂♂ of the same treated batch.
† With a dose of X-rays giving the same percentage of lethals.
‡ 4 II–III, 1 I–II, 2 I–III, 1 I–II–III.

♂♂ carrying a translocation of the Y-chromosome are usually sterile, this entailed the use in F_1 of ♀♀ instead of ♂♂, and the introduction of large inversions to prevent crossing-over. In detail the scheme was as follows:—

$$P_1 \quad \text{♂♂ } dp; e \text{ treated} \times \text{♀♀ } \underline{y}; \frac{Cy}{Bl\ L^2}; \frac{D}{LVM} \qquad \text{mass.}$$

$$F_1 \quad \text{♀ } \underline{y}; \frac{Cy}{dp}; \frac{LVM}{e} \text{ (phenotype } Cy, \text{ non-}D) \times \text{♂ } dp; e \quad \text{pairs.}$$

F_2 The segregation for sex, dp and e was used to determine whether a translocation was present.

Unfortunately, sterility in this experiment was so high that only 33 chromosome sets could be tested; but, since only one translocation occurred with a dose producing 24 per cent. lethals, the result is in agreement with the other two experiments.

The Production of Large Deletions

In two experiments, treated ♂♂ were mated to $y\ v\ f$ ♀♀, and their daughters were examined for the presence of hyperploid ♀♀ carrying a paternal X-chromosome with a large deletion. These ♀♀ are recognizable by the fact that one or two of the three marker genes are covered

TABLE VII.—FREQUENCY OF LARGE DELETIONS IN THE PROGENY OF TREATED
WILD-TYPE ♂♂ AND $y\ v\ f$ ♀♀

Experi- ment	Mutagenic Agent	Percentage Sex-linked Lethals *	No. of F_1 ♀♀ Examined	Large Deletions		
				Observed No.	Expected No.†	Difference
H92	Mustard gas	9	6635	16	41	25 ±6·4
HX1	,,	6	5052	9	18	9 ±4·2
HX1	X-rays, 2000 r	7	4368	14	15	· ·

* In a parallel test on males of the same treated batch.
† With a dose of X-rays giving the same percentage of lethals.

up by the presence of their normal allelomorphs in the deleted fragment. The data, which are summarized in Table VII, prove that mustard gas can induce large deletions in treated sperm.

The "expected number of deletions" shown in Table VII has been calculated on the assumptions (1) that an X-ray dose of 4000 *r* produces about 1 per cent. deletions of the types detectable in our experiments (Bishop, 1938; Pontecorvo, 1940), and (2) that the frequency of deletions varies with the 3/2 power of the dose (Muller, 1940). The result of the X-ray experiment carried out by us (last row of Table VII) confirms that these assumptions are justifiable.

In both experiments with mustard gas the observed number of deletions was significantly smaller than expected, but the divergence was not so marked as in the case of translocations.

Effect of Treated Cytoplasm on Untreated Chromosomes

The possibility has to be considered that mustard gas exercises its effect not directly on the chromosomes, but indirectly by a chain of reactions which starts in the cytoplasm and finishes in the chromosomes. If this were so, then one would expect mutations to develop in untreated chromosomes which have been introduced into treated cytoplasm and have repeatedly divided

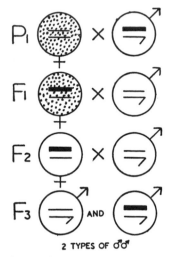

2 TYPES OF ♂♂

FIG. 3.—Scheme for detecting the possible mutagenic action of cytoplasm treated with mustard gas on an untreated chromosome introduced into it.

Stippling indicates treated cytoplasm.   ▬ = X-chromosome under test.
Absence in the F_3 of ♂♂ with the tested chromosome indicates that a lethal has been produced in this chromosome during its sojourn in the treated cytoplasm of the F_1.

in this environment. Such conditions can be created by mating untreated ♂♂ with treated ♀♀ and recording the frequency of lethals which develop in the paternal chromosomes. A positive result, *i.e.* an increased frequency of lethals in the tested chromosomes as compared with similar chromosomes introduced into untreated ♀♀, would indicate either that mustard gas persists in the cytoplasm for a sufficiently long time, and in a sufficiently high concentration, to affect the chromosomes during cleavage and gonad formation, or that a mutagenic substance is formed in the treated cytoplasm. A negative result, on the other hand, would mean that no mutagenic dose of either mustard gas or of a reaction product of mustard gas with the cytoplasm was present during development.

The method was a modification of that used by Muller (1930) and Timoféeff-Ressovsky (1937) for analysing the mutagenic action of X-rays. The essential features of our method are represented in fig. 3, in which the treated cytoplasm is indicated by stippling, and the untreated chromosome under test by a thick line. The ova from which the F_1 developed were treated in the body of the P_1 ♀♀. A few hours after treatment the P_1 ♀♀ were mated to the untreated P_1 ♂♂. A lethal which arises in the paternal X-chromosome during embryonic development of an F_1 ♀ will result in a ♀ which carries the new lethal in part of her body and presumably in most cases only in part of her gonads. A proportion of her daughters, therefore, will be heterozygous for the lethal and will have no sons with the tested chromosome in F_3. Hence a mutagenic action of the treated cytoplasm is detected by an increased mutation rate in the F_3 as compared with controls.

The main difference between our method and that used by the above-mentioned workers consists in the use of ♀♀ instead of ♂♂ in F_2. Since a ♀ is protected from the harmful effects of a recessive lethal on one X-chromosome through the presence in the other of the normal allelomorph, this method safeguards against the possibility that a lethal which arises early during development of an F_1 individual may either kill their carrier or else prevent the germ cells carrying it from developing into mature gametes. The use of ♀♀ in F_1 entailed the introduction of inversions to prevent crossing-over. In detail the scheme was as follows:—

P_1 ♀♀ $w\ sn^3\ B$ (treated) $\times$ ♂♂ $sc^{S1}\ In\ S\ w^a sc^8$ (untreated)   mass.

F_1 ♀ $\dfrac{w\ sn^3\ B}{sc^{S1}\ In\ S\ w^a sc^8}$ (virgin) $\times$ ♂ $Or\ K$ (wild-type)   pairs.

The F_2 was examined for the presence of $w\ sn\ B$ and w^a ♂♂.

F_2 ♀ $\dfrac{Or\ K}{sc^{S1}\ In\ S\ w^a sc^8}$ phenotypically (non-Bar) $\times$ brother   pairs.

The F_3 was examined for the presence of w^a ♂♂.

To guard against the possibility that secondary non-disjunction in an F_2 ♀ mated to a w^a brother might mask the presence of a lethal on the w^a chromosome, the F_2 ♀♀ were collected from cultures which did not produce wild-type sons. An identical scheme of matings was carried through with controls.

The efficiency of the dose of mustard gas to which the cytoplasm was exposed was deduced from the incidence of lethals in the treated *maternal* X-chromosome. These lethals are detected by the absence of $w\ sn\ B$ ♂♂ in F_2. Three lethals were found in 71 chromosomes from treated ♀♀, as compared with none in 76 controls. In addition, one of the 71 F_1 ♀♀ was found to be heterozygous for a lethal in the untreated *paternal* X-chromosome. If this lethal had not arisen spontaneously in the ♂, it may conceivably have been induced in the genital tract of the ♀ or in the ovum between copulation and the first cleavage division.

The results of the experiment are shown in Table VIII. It is clear that no mutagenic effect was produced in untreated chromosomes which underwent all the mitoses of embryogenesis in treated cytoplasm. It would appear therefore that, at least with the dose used,

TABLE VIII.—Frequency of Sex-linked Lethals in Untreated X-Chromosomes which have been Introduced into Treated Cytoplasm

Series	No. of Chromosomes Tested	No. of Lethals
Treated cytoplasm . .	749	1
Controls	1234	1

the chromosomes react directly to mustard gas as they do to radiation, and not secondarily to modifications of the cytoplasm. It is of interest that the dose was high enough to produce 13 per cent. lethals in ♂♂ of the $sc^{S1}\ In\ S\ w^a\ sc^8$ stock which were exposed simultaneously with the P_1 ♀♀.

DISCUSSION

The results reported here prove beyond doubt that mustard gas can produce effects on chromosomes and genes which are qualitatively and quantitatively comparable to those produced by X-rays and γ-rays. Similar results have been obtained with three other vesicants chemically related to mustard gas. These will be described in the future. It should be noted, however, that not all vesicants with a high power of penetration are necessarily mutagenic. A remarkable exception is lewisite. With a dose of lewisite which killed the majority of ♂♂

exposed to it, only 7 lethals were obtained in 1271 chromosomes from treated sperm, as compared with 8 lethals in 891 control chromosomes. It is, of course, possible that the dose of lewisite which is necessary to produce a mutagenic effect is so toxic that all the individuals which receive it die. If this were so, lewisite might still be shown to be mutagenic in an organism less susceptible to its toxic action.

Whether or not physical and chemical agencies exert the same primary actions on the chromosomes is one of the many problems which are opened up by these results. One of the attractions of chemical substances as mutagens lies in the possibility that there might exist affinities between specific chemical groups and certain parts of the genic material which, by the occurrence of selective effects, might provide evidence concerning the chemical structure of the chromosomes.

Specific reactions might occur at two levels. There might, in the first place, exist an affinity of the active substance for certain essential components of the chromosome, *e.g.* certain proteins or amino acids, nucleic acid or one of its constituents. Affinities of this kind would lead to the selective occurrence of certain types of effects, while other types with which this particular component is not concerned would not appear after treatment. If, for example, certain chemical substances could be shown to produce frequent mutations, but no small deficiencies, or *vice versa*, this would strongly suggest that different chemical processes are concerned in these two types of changes. In the second place, the specificity of the substance might be inferred if mutations of specific loci were selectively induced. It would suggest that the substance had a chemical affinity for these genes.

The present data give little, if any, indication of the existence of either type of specificity. Mustard gas and similar active substances have been shown to produce visible mutations, lethals, and gross rearrangements. Furthermore, Slizynski and Slizynska (1946) have shown that small deficiencies are also produced. The visible mutations are comparable to those obtained by radiation both in the frequency of their occurrence and in their type. All the chromosomes, and—as location tests of the sex-linked lethals have shown (Slizynski and Slizynska, 1946)—all euchromatic regions of at least one chromosome are susceptible. Two differences only have so far been found between the effects of X-rays and mustard gas, both of them quantitative, viz.:

(1) A dose of mustard gas produces fewer gross rearrangements (translocations and deletions) than a dose of X-rays which produces the same percentage of sex-linked lethals. This need not necessarily imply that mustard gas is less efficient than X-rays in the production of chromosome breaks. A gross rearrangement occurs when chromosomes are broken in one or more places and join up into a different arrangement. It is conceivable that the treatment with mustard gas favours restitution of broken chromosomes in preference to the formation of new reunions, especially if the breaks involve two different chromosomes. This would be in agreement with the findings that translocations, which, of course, involve two chromosomes, are relatively less frequent than the intrachromosomal deletions. This assumption would gain further support if it were found that the efficiency of mustard gas approached that of X-rays in the production of single breaks, and it is planned to test this.

(2) In the progeny of mustard gas treated ♂♂, mosaics form a significantly higher percentage of all visible mutants than in the progeny of X-rayed ♂♂. This is dealt with in a separate publication (Auerbach, 1946).

It is of interest to note that in respect of both these peculiarities, *i.e.* shortage of gross rearrangements and high percentage of visible mosaics in F_1, mustard gas appears to occupy a curiously intermediate position between X-rays on the one hand and ultra-violet radiation on the other. For it is less efficient than X-rays, and more efficient than ultra-violet rays, in the production of gross rearrangements, and, if Stadler's (1939) data on maize can be compared with our findings on *Drosophila*, the gas produces more mosaics than X-rays, but not quite as many as ultra-violet radiation.

In considering the essential similarity between the nuclear effects of mustard gas and related substances on the one hand, and those of X-rays and other ionizing radiations on the other hand, it has to be borne in mind that the same molecular reaction may be initiated both by physical and chemical stimuli. The ensuing changes may then be due merely to details of gene and chromosome structure which condition the response to the primary process. Such considerations do not, of course, exclude the possibility that certain substances may be capable of producing specific effects; but they do suggest that large numbers of substances may have to be tried before one is found which possesses such specific effects.

SUMMARY

Data are presented which show that mustard gas is capable of producing lethals and visible mutations at rates which are comparable with the effects of X-rays. Up to 24 per cent. sex-linked lethals have been produced in experiments in which mature spermatozoa were treated. Other substances, related in chemical composition to mustard gas, have produced similar effects, and the results will be published later. Lewisite did not produce any such effects.

The induction of dominant lethals by these chemical substances, and their interference with gametogenesis, result in greatly reduced fertility of treated ♂♂ and ♀♀.

Appreciable differences in sensitivity to the gas have been observed in different individuals, different strains, and at different stages of development.

Deletions, inversions, and translocations are produced, but the proportion of translocations and, to a lesser degree, of deletions to sex-linked lethals is significantly smaller than after X-ray treatment.

The rate of lethal mutations was not increased in untreated chromosomes which had been introduced into cytoplasm treated with mustard gas. This suggests that mustard gas, like X-radiation, acts directly on the chromosomes.

The data obtained so far have given no evidence that mustard gas acts selectively on certain constituents, or at certain loci, of the chromosomes.

APPENDIX

DETAILED DATA OF THE EXPERIMENT ON THE PRODUCTION OF VISIBLE MUTATIONS.
(See Table V)

Of the 34 aberrants in the controls, 16 were not tested, or died before they had produced progeny. Most of these obviously were developmental modifications of the types always encountered when large numbers of flies are examined, such as blistered wings, crippled legs, abnormal abdomen; they also included one Minute ♂. Fourteen tested aberrants which did not transmit their abnormality to their progeny consisted mainly of the same type of modifications just mentioned. In addition, they included a fly whose right eye was bi-coloured for red and an eosin-like shade; the most likely interpretation of this case is the occurrence of a somatic mutation in one of the eye-forming cells. Three aberrations were inherited as autosomal dominants of low and irregular penetrance (tendency to notched wings, kinked bristles, rough eyes, irregular venation). Finally, there was one autosomal Minute. The Minute which left no progeny, and the somatic mutation to eosin were counted among the genetic changes.

Of the 148 aberrants in the treated series, 71 were not mated, or died before they could be mated, or did not leave progeny. Among them were 10 Minutes, in which the Minute character of the bristles was often accompanied by other abnormalities such as rough eyes, scalloped wings, irregular venation, which are usually associated with chromosome aberrations. In addition, there were 7 fractional Minutes, again most of them with additional signs of chromosomal unbalance in the affected part of the body. Six more ♂♂ with normal bristles showed similar indications of chromosomal unbalance either over their whole body surface or over part of it. One fly exhibited in one of its eyes what was obviously a somatic mutation to one of the members of the lozenge series. Thus of the 71 aberrants of which no progeny was obtained, 24 were strongly suspected to be chromosome rearrangements or losses. They have been counted as genetic aberrations.

Of 25 aberrants which gave a negative result in a breeding test, one was obviously a somatic mutation to lozenge, two were fractional Minutes, and one had one very small eye of irregular outline, as in the eyeless mutant. These four flies, too, were included among the group of aberrations with a genetical basis.

75

The transmitted abnormalities consisted of 10 sex-linked recessives and 41 autosomal dominants. The sex-linked recessives included forked (twice), wavy, rudimentary (? twice). The other five were not subjected to location tests. They consisted of the following phenotypes: slender bristles; notched wings (semi-dominant); rough eyes (twice, one a semi-lethal); spread wings combined with abnormalities of venation (semi-lethal). In addition two ♂♂ which had been subjected to breeding tests on account of very slight phenotypical irregularities turned out to carry sex-linked lethals. Possibly these ♂♂ were gonosomic mosaics for a lethal, and the phenotypical abnormalities had been only incidental.

The autosomal dominants included 21 total and 13 fractional Minutes. Five abnormalities of bristles, wings, and eyes, all with low penetrance and irregular expression, were similar to those found in the control and were presumably carried in one of the stocks used in the experiment. Of the remaining two autosomal dominants, one was phenotypically like Delta and probably allelomorphic with it; the second, which first appeared as a fractional, was an allelomorph of vg^D and phenotypically like vg^D, although cytologically probably not identical with it (Auerbach, *D.I.S.*, XVII).

We should like to thank Dr M. Y. Ansari, Mr M. Ginsburg, and Mr N. Condon for their help in exposing the flies to mustard gas.

REFERENCES TO LITERATURE

AUERBACH, C., 1944. "Report on new mutants", *Drosophila Information Service* (*D.I.S.*), XVIII, 40.
——, 1945. "New Mutants", *Drosophila Information Service* (*D.I.S.*), XIX, p. 45.
——, 1946. "Chemically induced mosaicism in *Drosophila melanogaster*", *Proc. Roy. Soc. Edin.*, B, LXII, 211–222.
BISHOP, M., 1938. "X-chromosome duplications in *Drosophila melanogaster*", *Genetics*, XXIII, 140.
DEMEREC, M., and KAUFMANN, B. P., 1941. "Time required for *Drosophila* males to exhaust the supply of mature sperm", *Amer. Nat.*, LXXV, 366–379.
HANSON, F. B., and HEYS, F., 1929. "Duration of the effects of X-rays on male germ cells in *Drosophila melanogaster*", *Amer. Nat.*, LXIII, 511–516.
HARRIS, B. B., 1929. "The effect of ageing of X-rayed males upon mutation frequency in *Drosophila*", *Journ. Hered.*, XX, 299–302.
MULLER, H. J., 1930. "Radiation and genetics", *Amer. Nat.*, LXIV, 220–251.
——, 1940. "An analysis of the process of structural change in chromosomes of *Drosophila*", *Journ. Gen.*, XL, 1–66.
——, 1941. "Induced mutations in *Drosophila*", *Cold Spring Harbor Symp. on Quant. Biol.*, IX, 151–165.
MULLER, H. J., and ALTENBURG, E., 1930. "The frequency of translocations produced by X-rays in *Drosophila*", *Genetics*, XV, 283–311.
NEUHAUS, M. J., 1939. "A cytogenetic study of the Y-chromosome of *Drosophila melanogaster*", *Journ. Gen.*, XXXVII, 229–254.
PATTERSON, J. T., STONE, W. S., BEDICHEK, S., and SUCHE, M., 1934. "The production of translocations in *Drosophila*", *Amer. Nat.*, LXVIII, 359–369.
PONTECORVO, G., 1940. "Researches on the mechanism of induced chromosome rearrangements in *Drosophila melanogaster*", Ph.D. thesis, *Univ. Edin.*, 1–146.
SLIZYNSKI, B. M., and SLIZYNSKA, H., 1946. "Genetical and cytological studies of lethals induced by chemical treatment in *Drosophila melanogaster*", *Proc. Roy. Soc. Edin.*, B, LXII, 234–242.
SONNENBLICK, B. P., 1940. "Cytology and development of the embryos of X-rayed adult *Drosophila melanogaster*", *Proc. Nat. Acad. Sc. U.S.A.*, XXVI, 373–381.
STADLER, L. J., 1939. "Genetic studies with ultra-violet radiation", *Proc. 7th Internat. Genet. Congr.*, 269–276.
STUBBE, H., 1940. "Neue Forschungen zur experimentellen Erzeugung von Mutationen", *Biol. Zbl.*, LX, 113–129.
THOMAS, J. A., and CHEVAIS, S., 1943. "Production expérimentale de mutations par les trois amino-phenylsulfamides isomères chez la mouche *Drosophile*. Action sur les cellules mâles", *C.R. Soc. Biol.*, CXXXVII, 185–187.
TIMOFÉEFF-RESSOVSKY, N. W., 1937. "Zur Frage über einen 'direkten' oder 'indirekten' Einfluss der Bestrahlung auf den Mutationsvorgang", *Biol. Zbl.*, LVII, 233–248.

ERRATA

Page 277, paragraph 5, lines 5 and 7: "*y;bw;e*" should read "*y;dp;e.*"

7

Copyright © 1947 by the Royal Society of Edinburgh

Reprinted from *Proc. Roy. Soc. Edin.*, **B62**(37), Pt. III, 307–320 (1947)

The Induction by Mustard Gas of Chromosomal Instabilities in *Drosophila melanogaster*. By **C. Auerbach**, Ph.D., Institute of Animal Genetics, University of Edinburgh. *Communicated by* Dr A. W. GREENWOOD. (With Three Text-figures.)

(MS. received April 17, 1946. Read July 1, 1946)

THE INDUCTION BY MUSTARD GAS OF CHROMOSOMAL INSTABILITIES IN *Drosophila melanogaster*

THE discovery (Auerbach, 1943, 1946; Auerbach and Robson, 1946, 1947) that mustard gas is comparable to X-rays and similar physical agencies in its ability to produce mutations and chromosome rearrangements has opened up a new line of approach to the problem of gene mutation. It is to be expected that a comparative study of the mechanism by which chemical substances on the one hand and physical agencies on the other exercise their mutagenic effects, will further our understanding of the process of mutation itself. One of the first questions to be tackled in the early days of radiation genetics was the possibility of a delayed mutagenic action of irradiation (Muller, 1927; Timoféeff-Ressovsky, 1930, 1931; Grüneberg, 1931). The bulk of the evidence (see, however, Bishop, 1942) indicates that X-ray-induced mutations and chromosome breaks arise as an immediate effect of the irradiation, although after treatment of mature spermatozoa new recombinations of broken chromosomes may be delayed until the spermatozoon has entered the egg. Data obtained by Stadler (1939) suggest that after ultra-violet radiation of pollen grains the mutational process often is not completed before the treated chromosome has split into its two daughter chromatids. This results in a high proportion of mosaics. A similarly high proportion of mosaics has been found in the progeny of *Drosophila* $\male\male$ which had been treated with mustard gas (Auerbach, 1946; Auerbach and Robson, 1946). This raises the question of a possible delayed action of the chemical mutagenic treatment.

Two series of experiments were carried out to investigate this point. The first consisted in a study of the frequency of sex-linked lethals in the F_3 from treated $\male\male$; its object was to detect mutations which had arisen, not in the treated chromosome itself, but in chromosomes descended from it. The second series of experiments consisted in an analysis of apparent "semi-lethals" among the progeny of treated $\male\male$; its object was to discover how many of these semi-lethals were only apparent and caused by the presence of a mutated patch in the gonads of an F_1 individual. Both methods agreed in revealing the frequent occurrence, among the F_1 of treated $\male\male$, of individuals in which part of the gonad carried a mutated gene and the remaining part its normal allelomorph. Mosaics of this type (fig. 2, *a*) will be called "gonadic mosaics" as distinct from "gonosomic mosaics" (Sidky, 1940), in which the whole of the gonad carries a mutated gene while all or part of the soma carries its normal allelomorph (fig. 2, *c*). When the progeny of chemically produced gonadic mosaics was again tested by breeding, it was found that in several instances some of the daughters of a gonadic mosaic in their turn were gonadic mosaics for the same mutation. These findings are taken to suggest a delayed action of the treatment in the sense that it may create instabilities at certain loci, *i.e.* a tendency in these loci to repeated identical mutation.

Most of the experiments were carried out already in 1941, 1942, and 1943, but, owing to a security ban, publication had to be deferred until now.

Frequency of Lethals in F_3

With a scheme of crosses like the classical *ClB* method, or any of its more recent modifications, sex-linked lethals which arise in the germ cells of the P_1 $\male\male$ are detected by the absence of $\male\male$ with the treated X-chromosome in the F_2. Similarly, sex-linked lethals which, after treatment of the P_1, arise in the germ track of F_1 ♂♂ or ♀♀ are detected by the absence of ♂♂ with the treated X-chromosome in F_3. Therefore, if care is taken that no lethals are lost from the treated chromosome by crossing-over, an excess of lethals in F_3 over the controls may be taken to indicate a delayed effect of the treatment.

Tests of this kind have been carried out for irradiated material by Muller (1927), Timoféeff-Ressovsky (1930, 1931), and Grüneberg (1931). They mated X-rayed ♂♂ to attached-X ♀♀ ;

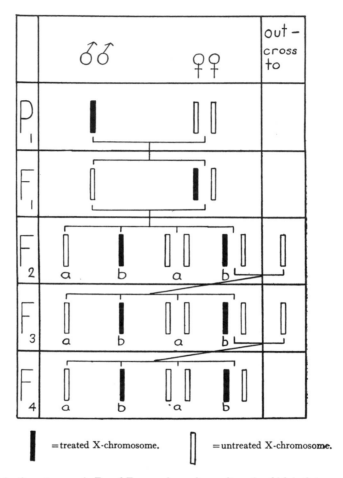

| = treated X-chromosome.   **||** = untreated X-chromosome.

♀♀ used for the out-crosses in F_2 and F_3 were chosen from cultures in which both types of ♂♂ were present.

A lethal which has arisen in the treated X-chromosome becomes manifest as absence of ♂♂
of type *b* in F_2, if the lethal arose in P_1;
in F_3, if the lethal arose in F_1;
in F_4, if the lethal arose in F_2.

Fig. 1.—Scheme of matings for the detection of a delayed effect of chemical treatment.

the F_1 ♂♂, which of course could not carry a sex-linked lethal on their X-chromosome, were subjected to a *ClB*-test. In no instance was a significant difference between experimental series and controls obtained, and this was taken to indicate the absence of an after-effect of the treatment. When this type of cross was used on ♂♂ which had been exposed to mustard-gas vapours (for technique see Auerbach and Robson, 1947), the F_3 of the treated series yielded a slight excess of lethals over the controls, but the difference was not significant.

The negative or inconclusive results obtained with this method are, however, open to the objection that sex-linked lethals, especially small deficiencies, which arise in the germ track of the hemizygous ♂, may prevent the affected cells from developing into mature spermatozoa. This objection holds to a much lesser extent for the ♀, in which the deleterious effect of a lethal on one X-chromosome will in most cases be covered up by the presence of the normal allelomorph in the other. Therefore, tests were carried out in which attention was directed to the rate at which mutations arise in the germ track of daughters of treated ♂♂. The general mating scheme is given in fig. 1, which also shows how the test can be extended to further generations. This method has already been used by Muller (1927) with irradiated flies. He obtained a slight excess of lethals in the F_3 of the treated series, and he considered the possibility that F_1 ♀♀ may be mosaic for a sex-linked lethal, similar to the fractionals for a visible mutation.

Three experiments were carried out in accordance with the general scheme of fig. 1. The detailed mating schemes were slightly different in the three tests; but since in their essential features they were similar, it is sufficient to show one of them in detail.

Genetical Details of one of the Mating Schemes whose General Nature is Illustrated in Fig. 1

P_1 ♂♂ sc^{S1} (*In S*) w^a sc^8 (treated) × ♀♀ $ClB/scar$ mass matings.
F_1 ♀ sc^{S1} (*In S*) w^a sc^8/ClB × brother $scar$ pair matings.
F_2* ♀ sc^{S1} (*In S*) w^a $sc^8/scar$ × ♂ wild-type pair matings.
F_3 Absence of ♂♂ with w^a eyes is due to a sex-linked lethal which arose in the germ track of the F_1 ♀.

The data obtained in three experiments of this type are summarized in Table I.

TABLE I.—FREQUENCY OF SEX-LINKED LETHALS IN THE F_3 OF CHEMICALLY TREATED ♂♂. THE LETHALS OCCURRED FIRST IN THE GERM TRACK OF DAUGHTERS OF THE TREATED ♂♂

Experiment	Percentage of Sex-linked Lethals in F_2 †	No. of Tested Chromosomes	Lethals and Semi-lethals in F_3			
			Lethals		No. of Semi-lethals	Percentage of Lethals + Semi-lethals
			No.	Per cent.		
I . .	9	1161	2	0·2	3 ‡	0·4
Control to I .	..	1165	0	0	0	0
II . .	13	828	21	2·5	7	3·4
Control to II	..	1057	3	0·3	1	0·4
III . .	10·1	1049	45	4·5	not scored	
Total treated	..	3038	68	2·2		
Total controls	..	2222	3	0·13		

It will be seen that the frequency of lethals in the F_3 of treated flies was considerably and significantly in excess of that obtained in the controls; moreover it exceeded markedly the normal range of spontaneous mutability observed in our laboratory stocks of *Drosophila melanogaster*. The same applies to Experiments II and III taken individually. Only in Experiment I, in which the dose of mustard gas was lowest, the difference between treated

* Chosen from F_2 cultures which contained w^a ♂♂.
† This percentage is used to measure the dose of mustard gas which reached the gonads of the treated ♂♂.
‡ All of them almost completely lethal.

series and controls, although in the right direction, did not reach the level of statistical significance (P for lethals and semi-lethals combined between ·05 and ·1).*

Experiment III was carried on for one more generation (fig. 1) so as to obtain the frequency of lethals in F_4. These lethals would have arisen in the germ track of F_2 ♀♀. No lethal was obtained in 1005 tested chromosomes.

It has been said that the lethals observed in F_3 must have arisen in the germ track of F_1 ♀♀. It was of interest to know how many F_1 ♀♀ shared in the production of these lethals. It was conceivable that all lethals were identical and derived from one gonosomic or gonadic mosaic in F_1; on the other hand each lethal might have arisen singly—and therefore presumably late in development—in a different F_1 ♀. These are the two extremes between which the actual distribution of the lethals would be expected to lie. In order to decide this point, which gives an indication of the time at which the lethals arose in F_1, Experiment III was carried through in such a way that every lethal could be traced back to a particular F_1 ♀. Altogether 77 F_1 ♀♀ were used for producing the F_2. The 1049 F_2 ♀♀ were kept in separate groups according to their mother. The distribution of the 45 lethals on these groups ·howed that all of them occurred among the progeny of 3 out of the 77 mothers. One of these 3, F_1 ♀♀, produced only 1 daughter heterozygous for a lethal among 12 tested daughters; this may conceivably have been due to spontaneous mutation. The second ♀ had 33 tested daughters with red eyes (*i.e.* carrying the treated chromosome). Twenty-nine of these were heterozygous for a lethal, the remaining 4 carried a normal wild-type chromosome. The third ♀ had 18 tested daughters with the treated chromosome, 15 of which were heterozygous for a lethal, while 3 carried no lethal on the treated X. In addition, this last ♀ also had one red-eyed daughter which appeared to carry a semi-lethal because she produced only a few wild-type sons among a large progeny. This case turned out to be of special interest, and will be dealt with later. On the whole, then, two, possibly three, among 77 tested daughters of treated ♂♂ were gonadic mosaics for a lethal on the treated X-chromosome.

Gonadic mosaics are rare in the progeny of X-rayed ♂♂. It therefore seemed desirable to compare their relative frequencies after irradiation and mustard-gas treatment. To this purpose an analysis of apparent "semi-lethals" produced by the two types of treatment was undertaken.

Gonadic Mosaicism for a Lethal as a Basis for Spurious "Semi-lethality"

In the F_2 of a *ClB* or equivalent test carried out on ♂♂ which have been treated with an effective mutagen there usually occurs a certain proportion of cultures in which ♂♂ with the treated X-chromosome, although not completely absent as in the case of a lethal, yet form a strikingly low proportion. The usual interpretation of this type of culture is that the F_1 ♀ producing it carried on the treated X-chromosome a recessive mutation which severely impairs the viability of the ♂♂ so that only a small proportion of these reach the imaginal stage. A mutation of this type has been called a "semi-lethal" in contrast to the "full" lethal which impairs viability of the ♂♂ to such an extent that practically none reach the imaginal stage. There is no doubt that in very many cases semi-lethals are responsible for a low proportion of ♂♂ of a particular type in a culture (see particularly Timoféeff-Ressovsky, 1935). However, the same type of culture would also be produced if the F_1 ♀, instead of being heterozygous for a semi-lethal, were heterozygous for a full lethal in part of her ovary and homozygous for its normal allelomorph in the remainder, *i.e.* if she were a gonadic mosaic for a full lethal. A decision between these two alternatives can be made by breeding tests. A ♀ which is heterozygous for a semi-lethal will transmit this gene to all her progeny which receive the treated X-chromosome. Her sons, when mated to attached-X ♀♀, will yield a progeny with low sex-ratio, and her daughters will have few sons with the treated X. If, on the other hand, the F_1 ♀ is a gonadic mosaic for a full lethal, her surviving sons are those which received their X-chromosome from the non-mutated portion of the ovary and thus will yield progenies with normal sex-ratios when mated to attached-X ♀♀. Her daughters will be of two types: those heterozygous for a lethal, and those with no lethal on the treated X-chromosome.

---

* The estimate for P has been based on a contingency table with Yates' correction for small numbers. The author is indebted to Dr O. Kermack for pointing out the advantage of this formula for mutation work.

Table II gives the breeding record of an F_1 ♀ which appeared to be heterozygous for a semi-lethal sex-linked mutation, but which in reality was a gonadic mosaic for a full lethal.

Altogether 35 analyses of this kind were carried out, 15 on semi-lethals produced by X-rays, and 20 on semi-lethals produced by mustard gas. The data are summarized in Table III.

TABLE II.—ANALYSIS OF A SEMI-LETHAL EFFECT IN F_2 CAUSED BY GONADIC MOSAICISM
FOR A LETHAL IN F_1

P_1 treated ♂ *Fo4* × ♀ *ClB/scar*.
F_1 ♀ *ClB*/ + (treated) × ♂ sc^{S1}(*In S*) $w^a sc^8$.
F_2 85 ♀♀, 9 ♂♂.

Test Matings with F_2:

(*a*) 3 ♂♂ mated individually to XX ♀♀ gave 1 : 1 sex-ratios in F_3.
(*b*) 18 ♀♀ sc^{S1} (*In S*) $w^a sc^8$/ + mated individually gave the following progenies:—

♀	daughters	Number of sons w^a	sons +	Treated X	
1	33	10	0		with lethal
2	31	10	11	normal	
3	39	17	0		with lethal
4	27	9	17	normal	
5	26	11	0		with lethal
6	50	15	0		with lethal
7	24	6	0		with lethal
8	22	6	7	normal	
9	26	8	12	normal	
10	36	19	16	normal	
11	38	16	0		with lethal
12	37	7	0		with lethal
13	35	6	11	normal	
14	46	25	27	normal	
15	28	10	7	normal	
16	41	9	0		with lethal
17	42	11	0		with lethal
18	39	13	0		with lethal
				8 normal	10 with lethal

F_3 All the lethals were confirmed in mass-cultures.

TABLE III.—ANALYSIS OF APPARENT SEMI-LETHALS

	No. tested	Viability Mutations	Gonadic Mosaics for a Lethal
X-rays . . .	15	14	1
Mustard gas . .	20	11	9

Whereas about half of the apparent semi-lethals in the chemical series are due to gonadic mosaicism, only one such case was detected among 15 apparent semi-lethals in the X-ray series. If, as these data indicate, gonadic mosaics are frequent after chemical treatment but infrequent after X-radiation of the father, they should tend to swell the number of apparent semi-lethals in mustard-gas tests as compared with X-ray experiments. A valid comparison of this kind would require recording of the sex-ratio in every F_2 culture. This has not been done in any of the present experiments. Nevertheless, it seems worth noting that in one experiment, in which ♂♂ from the same culture bottles were exposed either to X-radiation or to mustard-gas treatment which produced the same percentage of sex-linked lethals as the irradiation, the ratio of striking semi-lethals to full lethals was higher in the progeny of chemically treated ♂♂, the difference being on the border-line of statistical significance.

The question arises: is this difference in the frequencies of gonadic mosaicism in the progeny of chemically treated and of irradiated ♂♂ simply a consequence of the well-established

TABLE IV.—RATIO OF GONADIC MOSAICS TO MOSAICS WITH UNIFORM GONADS

(In the case of lethals, the frequency of mosaics with uniform gonads
has been estimated by calculation—see text.)

	Type of Mutation	Mosaics with Gonads		Ratio A : B	Sample
		Mixed	Uniform		
		A	B		
X-rays (Neuhaus, 1935) (Pontecorvo, 1940)	visible	3	61	I : 2I	
		I	13	I : I3	
	lethal	I	2I	I : 2I	selected *
Mustard gas . . .	lethal	2	4	I : 2	
		2	10	I : 5	selected *
		3	3I	I : 10	selected *

* Only apparent semi-lethals selected for analysis.

fact (Auerbach, 1946; Auerbach and Robson, 1946) that in general mosaicism is very markedly higher after chemical treatment, or is there also a *relative* excess of gonadic mosaics among mosaics of all types in chemical experiments? In order to decide this point, it is

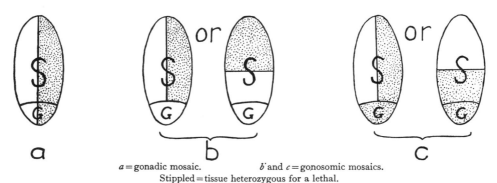

a = gonadic mosaic.   *b* and *c* = gonosomic mosaics.
Stippled = tissue heterozygous for a lethal.

FIG. 2.—Types of mosaics for a lethal.

necessary to establish for each type of treatment the ratio between gonadic mosaics on the one hand and mosaics with uniform gonads on the other. For visible mosaics produced by X-radiation of ♂♂ this has been done repeatedly in the past. The data are included in Table IV. When lethal mutations are studied, it is not possible to detect mosaics with uniform gonads by observation. As fig. 2 shows, they will be scored either as not carrying a lethal at all (type *b*) or as heterozygous for a lethal (type *c*). It is, however, possible to estimate their frequency on the basis of two plausible assumptions: (1) that in ♀♀, in which germinal selection against a sex-linked lethal may be presumed to be absent, the two types *b* and *c* occur with approximately equal frequency; and (2) that the ratio of complete to mosaic mutants in the F_1 of treated ♂♂ is the same for lethals as for visible mutations, *i.e.* approximately 6 : 1 after irradiation (Patterson, 1933) and approximately 1 : 1 after chemical treatment (Auerbach, 1946).

If it be assumed that *l* is the number of apparent heterozygotes for a lethal, including mosaics of type *c*, *2u* the number of mosaics with uniform gonads, and *m* the number of gonadic mosaics of type *a*, the following simple equations obtain for the two kinds of treatment:—

$$\text{X-rays} \qquad \frac{l-u}{2u+m} = 6.$$

$$\text{Mustard gas} \qquad \frac{l-u}{2u+m} = 1.$$

Where *l* and *m* are known from observation, *2u* can be calculated. A source of error is introduced when not all F_1 ♀♀, but only apparent semi-lethals, are analysed by breeding tests, because with this sampling method gonadic mosaics in which the mutated portion of the gonad is small will escape detection. Therefore ratios obtained from such samples (called "selected" in Table IV) will tend to give too low an estimate of the true proportion of gonadic mosaics. Especially in the last experiment listed in the table, selection of semi-lethals for analysis had been carried out in such a way that very probably not all gonadic mosaics could be detected.

Although the data are not suitable for statistical analysis, they suggest a difference between the ratios of the two kinds of mosaics after irradiation and after chemical treatment. In the progeny of irradiated ♂♂, gonadic mosaics always appear to form markedly less than one-tenth of all mosaics. After mustard-gas treatment a ratio as low as 1 : 10 was obtained only in one estimate which for several reasons probably was too low.

Breeding Analysis of the Progeny of Gonadic Mosaics

In general a ♀, who is a gonadic mosaic for a lethal, has two types of daughters (see Table II), those without lethal and those heterozygous for a lethal. In several cases, however, it was observed that a gonadic mosaic in addition to these two expected types produced also one or more daughters which, like their mother, appeared to carry a semi-lethal. One such case has been mentioned in the first part of this paper. When these apparent semi-lethals were analysed by breeding tests they were again found to be due to gonadic mosaicism. The frequency with which this occurred can be seen from Table V.

If ♀♀ D and F are disregarded as not adequately progeny-tested, there remain 9 gonadic mosaics each of which had at least 16 examined daughters. Of these, 4 had again one or two gonadic mosaics among their progeny. For two reasons this figure probably represents an underestimate: (1) Tests on a larger scale might have revealed gonadic mosaics even among the progeny of those ♀♀ which on the present evidence produced only two types of daughters. (2) Only a small proportion of F_2 ♀♀ were tested by breeding from their individual daughters. As in the case of the F_1 ♀♀, mainly those ♀♀ were selected for a thorough progeny testing in which a striking shortage of ♂♂ pointed to the possibility of gonadic mosaicism, and—as pointed out above—this kind of selection will tend to miss gonadic mosaics in which only a relatively small portion of the ovary carries the lethal.

The crucial point for the interpretation of this apparent transmission of mosaicism is the following: were mother and daughter gonadic mosaics for the same mutation, or was mosaicism in the daughter due to spontaneous origin of a new mutation? For ♀ B this point could not be decided on account of inversions in the treated X-chromosome which prevented location of the lethal. For, ♀ I identity of the mutation (a semi-lethal) in the mother and two of her daughters could be established on account of a very characteristic action of the gene on pupal development. The lethals which occurred in ♀ E and one of her daughters were subjected to location tests and found to be situated at practically the same locus; thus they were presumably identical. The case of ♀ J was complicated and requires description.

♀ J was the daughter of a treated wild-type ♂ and a *ClB/scar* ♀. Her genotype was thus *ClB/* treated + . Mated to a sc^{S1} *In S* w^a sc^8 ♂ she produced 124 daughters with either Bar or

TABLE V.—THE FREQUENCY OF GONADIC MOSAICS AMONG DAUGHTERS OF GONADIC MOSAICS. THE MOTHERS WERE ALL DAUGHTERS OF CHEMICALLY TREATED ♂♂

F_1 ♀	No. of Daughters Examined	Genotypes of Daughters (F_2)			Genotypes of Daughters of F_2 Mosaics (F_3)	
		Heterozygous for Lethal	No Lethal	Gonadic Mosaic for Lethal	Heterozygous for Lethal	No Lethal
A	33	29	4	0		
B	18	15	2	1	1	1
C	21	17	4	0	..	..
D	9	4	5	0	..	..
E	18	14	3	1	8	2
F	9	6	3	0	..	..
G	18	10	8	0	..	..
H	16	14	2	0	..	..
I	18	6	10	2	32	4
J	16	10	4	2	8	3
K	16	14	2	0	..	..
	192	139	47	6	49	10

non-Bar eyes, and 11 sons with garnet-coloured eyes. The eye colour of the sons was subsequently shown to be due to an allelomorph of garnet. Sixteen of her non-Bar daughters, which thus were of the constitution w^a */ treated +, were tested by mating them to w^a ♂♂.

The breeding tests carried out on these 16 F_2 ♀♀ showed them to fall into four genotypically distinct classes.

(1) The first class consisted of 3 ♀♀ which produced equal proportions of w^a and g sons. They thus carried on the treated X-chromosome the mutation to garnet which also was apparent in their brothers. Four of their daughters, when tested, proved likewise to carry garnet.

(2) The second class of F_2 ♀♀ consisted of 10 ♀♀ which produced only w^a sons. Thus they carried a lethal on the treated X-chromosome. This is in good agreement with the strikingly small number of their brothers (11 as against 124 ♀♀, see above). The lethal was not allelomorphic with garnet and separated from it by several crossing-over units. The location tests gave no indication of a gross structural rearrangement on the treated chromosome. In the following generation the lethal bred true in 11 pair-matings and several mass-matings. It was not affected by out-crossing and thus not due to complementary action with an autosomal gene.

(3) The third class of F_2 ♀♀ consisted of 1 ♀ only who among her progeny had a normal proportion of wild-type sons. Thus she had received from ♀ J a treated X-chromosome which carried neither the mutation to garnet nor the lethal, each of which were present in some of her sibs. Seven of her daughters were tested and found likewise to carry an unmated wild-type X.

(4) The fourth and most interesting group of F_2 ♀♀ consisted of 2 ♀♀ which bred in a manner resembling that of their mother. Each of these 2 ♀♀ had, in addition to the w^a sons with the untreated X, two types of sons, garnet and wild-type. Of 8 tested daughters 5 were found to be heterozygotes for a lethal, and 3 heterozygotes for a normal wild-type X-chromosome.

Thus ♀ J and two of her daughters transmitted to their offspring three different types of treated X-chromosome: one carrying garnet, one carrying a lethal, and one without mutation. They appear to have been triple gonadic mosaics, one portion of the ovary carrying the unmutated X, another garnet, and a third the lethal. There may of course have been overlapping between the latter two portions, garnet having been present on some of the lethal-

---

* In the following description the symbol w^a will be used to denote the chromosome sc^{S1} *In Ss* w^a sc^8.

bearing chromosomes. This possibility has not been adequately examined: in two cases in which ♀♀ heterozygous for the lethal were mated to garnet ♂♂ it could be shown not to apply. In Table VI the pedigree of ♀ J is summarized on the basis of triple gonadic mosaicism of ♀ J and two of her daughters, a semi-colon between two gene symbols indicating that the chromosomes carrying these genes occur in different portions of the ovary. The absence of $+ ♂♂$ in Class 3 of F_2 and of $\frac{w^a}{g}$ ♀♀ in Class 4 of F_3 may be presumed to be due to sampling error.

TABLE VI.—PEDIGREE OF A TRIPLE GONADIC MOSAIC, ♀ J (TABLE V)

P_1 treated ♂ $+$ × ♀ $ClB/scar$.

F_1 (♀ J) ♀ $\dfrac{ClB}{g;\ l;\ +}$ × ♂ w^a.

F_2	Class 1	Class 2	Class 3	Class 4
♂♂ . . .	11 g	0	0	..
Tested non-Bar ♀♀ (mated to ♂♂ w^a)	$3\,\dfrac{w^a}{g}$	$10\,\dfrac{w^a}{l}$	$1\,\dfrac{w^a}{+}$	$2\,\dfrac{w^a}{g;\ l;\ +}$
F_3				
♂♂ . . .	50% w^a, 50% g	all w^a	50% w^a, 50% $+$	37 w^a, 10 g, 4 $+$
Tested non-w^a ♀♀ .	$4\,\dfrac{w^a}{g}$	$11\,\dfrac{w^a}{l}$	$7\,\dfrac{w^a}{+}$	$5\,\dfrac{w^a}{l}$, $3\,\dfrac{w^a}{+}$

Note.—$w^a = sc^{S1}\ In\ S\ w^a sc^8$, g=garnet, l=lethal, $+$ =X-chromosome with neither g nor l.

Location of the lethal was carried out only once in F_2. Therefore, although it is highly probable that the lethal obtained in F_3 was identical with the original one, this has not been actually proved. However, there definitely was a reoccurrence of an identical visible mutation, garnet, in two daughters of the original gonadic mosaic. This, then, is the third case— out of four—in which identity of the recurring mutation in mother and daughters could be established.

The only gonadic mosaic, which had been obtained in the progeny of irradiated ♂♂, showed a similar history of recurring gonadic mosaicism among her progeny, but with the emphasis on a reversion from complete lethality to viability. Among her 6 daughters there was one which appeared to be heterozygous for a full lethal. However, when her progeny were mass-cultured together, they produced good proportions of sons with the treated X-chromosome. One daughter, tested by pair-mating, was a gonadic mosaic for a lethal. It is not possible to decide whether this case of apparent reversion of a lethal to its normal allelomorph was due to an induced or spontaneous unstable mutation similar to the unstable genes in *Drosophila virilis* (Demerec, 1941), or to some kind of complementary gene interaction.

DISCUSSION

In the first part of this paper it has been shown that in the F_3 of mustard-gas treated ♂♂ the frequency of mutations is significantly increased over that in the controls. Taken on its face value, this finding seems to indicate an after-effect of the treatment leading to a mutation in the generation following the treated one. However, closer analysis revealed that the high mutation frequency in F_3 was due to the presence in F_1 of a few ♀♀ which were gonadic mosaics for a lethal, *i.e.* which carried a lethal in part of their ovaries and transmitted it to part of their progeny. The presence of mosaics in the progeny of treated ♂♂ is by itself no proof of a deferred effect of the treatment.   It is a well-known fact that also after X-radiation of mature spermatozoa chromosome breaks, which have arisen through immediate action of the treatment, may result in flies which are mosaic for an induced aberration (for a discussion of the probable mechanism underlying this type of mosaicism see Muller, 1940).   However, when such X-ray "fractionals" have been analysed by breeding tests (Neuhaus, 1935; Pontecorvo, 1940), it has been found that in the great majority of cases the gonads contained one kind of tissue only, either mutated or unmutated (Table IV, first row).   If, as is generally assumed, X-ray mosaics arise at the first cleavage division, the scarcity of mosaics with mixed gonads among them may be taken to indicate that in the great majority of cases the whole of the future gonad is derived from one of the first two cleavage cells.   If this cell carries the mutation, the resulting mosaic will correspond to type c in fig. 2.   If, on the other hand, it does not carry the mutation, a mosaic of type b will be produced.   The rare cases of X-ray induced gonadic mosaics (type a, fig. 2), if they cannot be attributed to rare occurrences of a delayed action of irradiation, must then be the result of an infrequent type of first cleavage division by which the material of the future gonads is distributed into both first cleavage cells.

If, therefore, it can be shown that among chemically produced mosaics those with mixed gonads are relatively more frequent than after irradiation, it seems plausible to attribute the excess to cases in which mosaicism arose at a later stage than after irradiation, thus later than at least the first cleavage division.   An analysis of apparent semi-lethals revealed that the frequency of gonadic mosaics among them is high, when the fathers had been treated chemically, and negligible when they had been irradiated.   It remained to be seen whether this was simply due to the fact that, in general, mosaics are much more frequent in the progeny of chemically treated than of irradiated ♂♂ (Auerbach, 1946; Auerbach and Robson, 1946), or whether there really was an excess of gonadic mosaics among all mosaics produced by chemical means.   This, owing to difficulties inherent in the material chosen (lethals), could not be established by direct observation; but estimates of the ratio of gonadic mosaics to mosaics with uniform gonads after either kind of treatment strongly suggest that this ratio is markedly higher after chemical treatment.   This, then, was interpreted to indicate a delayed effect of the chemical treatment in the sense that an initially created instability may lead to an effective mutation at a later division than the one following treatment.

This interpretation was confirmed by the results of tests reported in the third part of the paper.   In four cases it was found that among the daughters of a gonadically mosaic ♀ there occurred one or more ♀♀ which in their turn were gonadic mosaics for a mutation.   Quite possibly breeding tests on a larger scale might have revealed a transmission of the mosaic condition also in some of the remaining cases.   In three of the four positive cases the identity of the mutation in mother and daughters could be established.   Here, then, a chemically produced, localized instability on a treated chromosome has been carried through all the cell divisions which separate the fertilized ovum of one generation from that of the next before giving rise to a mutation.

It is noteworthy that gonadic mosaics in the F_2 of treated ♂♂ occurred exclusively among the daughters of F_1 ♀♀ which themselves had been gonadic mosaics.   This is illustrated in fig. 3, in which the various possible female lines originating from treated fathers are represented.   No lethal was obtained in over 1000 chromosomes from F_2 ♀♀ whose mothers had neither been heterozygotes for a lethal nor gonadic mosaics for a lethal (line A).   On the other hand, in no case has reversion of a chemically induced lethal been observed, although ♀♀ heterozygous for such lethals have been mass-cultured very often and sometimes for a great number of generations.   This is true not only for lethals which occurred at first in a heterozygous F_1 ♀ (line B), but equally well for those lethals which at first occurred in a gonadic mosaic (line C) and in certain lines gave repeated origin to mosaics: once such a lethal had established itself in a heterozygous ♀, sister or daughter of a gonadic mosaic for the same lethal, it remained perfectly stable in subsequent generations of this line (line C b).   The facts that stable lines for both the lethal (line C b) and its normal allelomorph (line C a) may be derived from a ♀ which is mosaic for them and hands on the mosaic condition to some of her daughters, suggests that a chemically produced instability which does not lead to at least one mutation during the life cycle of an individual has become stabilized.   It should be pointed out that the tests on F_3 daughters of gonadic mosaics in F_2 (line C c) were carried out on too small a scale (see Table V) to decide whether or not the mosaic condition might not have occurred also in this generation.

The question arises as to the nature of the initially produced instability. Is it a mutation with a tendency to revert to its normal allelomorph, or is it a tendency, induced at a specific locus, to produce a mutation? In the first case, a ♀ which is a gonadic mosaic for a lethal would have started development as a heterozygote for the lethal; but in a portion of her ovary reversion of the lethal to its normal allelomorph would have taken place. In other words, mustard gas would have produced an unstable gene, similar to those found by Demerec (1941) in *Drosophila virilis*. In the second case, a ♀ who started development without a lethal would have become a gonadic mosaic through a subsequent mutation on the affected chromosome.

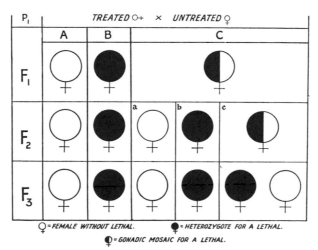

FIG. 3.—The three types of female lines derived from treated males.

The available data do not allow of a definite decision between these two alternatives. Since all 17 gonadic mosaics for a lethal were either direct descendants of treated ♂♂, or daughters of ♀♀, which in their turn had been gonadic mosaics (fig. 3, line C), it is not possible to say whether they started life as heterozygotes for the lethal, or whether the lethal developed in their germ track. At first sight Table V seems to be in favour of the first alternative. Out of 192 tested daughters of gonadic mosaics in F_1, 139 carried the lethal as against 47 which were free of it, and a similar disproportion appears among the daughters of gonadic mosaics in F_2. Indeed, out of altogether 17 gonadic mosaics, there were only 3 which did not transmit the lethal to the majority of their daughters. This is what one would expect if the lethal had disappeared through reversion from parts of an ovary which originally was heterozygous for it. However, it can easily be seen that the same preponderance of lethal-bearing chromosomes is to be expected if the chance for a lethal to develop *de novo* at an affected locus were sufficiently high to result in more than one mutation during the cell divisions which take place in the germ track. This condition will often be fulfilled for instabilities which, as has been shown above, never seem to skip a generation without at least one mutation taking place. Moreover, selection of ♀♀ for breeding tests was such that only those gonadic mosaics could be detected in which a large proportion of the ovary carried the mutation.

In the case of mutations with visible effects a decision as to the origin of a mosaic should be easier to arrive at. An individual which starts with a mutated chromosome is likely to show the effect of the mutation over the greater part of its body, whereas an individual in which the mutation occurs during development will usually exhibit it over less than half of its body, or—if more than one mutation took place—in several small, separate areas. Among visible mosaics in the progeny of mustard-gas treated ♂♂ there never occurred one in which more than one-half of the body showed the aberration; on the contrary, in almost one-third of 75 carefully examined mosaics for Minute less than one-quarter of the body surface showed the aberrant type of bristle (Auerbach, 1946). Considering the irregular cleavage pattern of *Drosophila* (Parks, 1936), such small areas may still have arisen at the first cleavage division and not through delayed action of the treatment, but on the other hand it is highly improbable that this kind of distribution of mutated and unmutated areas would arise through reversion to normal in an originally mutated fly. Thus the data on visible mosaics give no support to the assumption that mustard gas tends to induce unstable mutations which have a tendency to revert to the normal allelomorph.

Also from a purely theoretical point of view, delayed mutation seems a more satisfactory explanation of the observed gonadic mosaics than unstable mutation with a tendency to revert to the normal allelomorph. If the unstable gene is visualized as a gene in which, owing to the chemical treatment, the condition of chemical balance has been altered so that there exists a tendency to revert to the original, stable equilibrium, it is difficult to understand why the same mutation behaves perfectly stable in the majority of closely related lines. It may be objected that in these stable lines the original unstable lethal has mutated once more into a new stable lethal, since a phenotypical distinction between two lethals can usually not be made without embryological investigations. However, in two of the here described cases phenotypical identity of the mutation in stable and unstable lines could be established: one is the garnet mutation in ♀ J, Table V, the second the characteristic semi-lethal in ♀ I, Table V.

The observation that stable lines for both the mutation and its normal allelomorph could be established from gonadic mosaics is best explained by the assumption that the primary instability is somehow poised between two stable equilibria: the original one of the normal allele, and the new one of the mutation. The fact that in the analysed cases stabilization occurred preferentially in the direction of the mutation may be an incidental result of the manner in which gonadic mosaics were selected for analysis. This interpretation would also fit a case of visible mosaicism after chemical treatment which has been more fully discussed elsewhere (Auerbach, 1946). It is the occurrence among the sons of treated wild-type ♂♂ and attached-X ♀♀ of a ♂ who both in his eyes and his testes was a mosaic for two different allelomorphs at the white locus. Both alleles behaved as stable mutations in further breeding tests. Although alternative explanations can be imagined, it seems at least possible that the primary effect of the treatment consisted in the creation at the white locus of an instability which was poised between the two equilibria represented by the two stable allelomorphs. The fact that none of the tested spermatozoa contained an unmutated X-chromosome is easily explained by the assumption that the primary change was too labile to persist through all the cell divisions preceding sperm formation.

Two similar mosaics for different allelomorphs at the same locus have been found among sons of X-radiated ♂♂ (Neuhaus, 1935; Panshin, 1935). If they are interpreted in the same way, they might be taken to suggest that X-rays, too, may induce primary instabilities which later tend to become stabilized through mutation. Apart from these two isolated and not really well understood cases, X-rays have never been reported to induce instabilities of the type described here, nor unstable genes like those described by Demerec, all of which arose spontaneously. This is true in spite of the fact that the number of observed X-ray mutations is infinitely greater than of those which occurred spontaneously or after chemical treatment. This might conceivably be due to the difference in the amounts of energy supplied by chemical reactions on the one hand, hard X-rays such as are generally used for mutation work on the other. Whereas the energy supplied by the impact of X-rays will in the vast majority of cases be sufficient to lift the gene from one stable equilibrium into another—also relatively stable—one, the limited quantity of energy which is supplied by a chemical reaction may sometimes just be sufficient to bring about transformation into an intermediate labile state, the final attainment of a new stable condition or the sliding back into the old one being achieved subsequently under the influence of random temperature oscillations.

If, as appears possible, a proportion at least of the so-called spontaneous mutations arise through the action of naturally occurring mutagenic substances (Auerbach and Robson, 1944), localized chromosome instabilities might be expected to occur also without special treatment. Possibly unstable mutations have arisen in this manner. Another example to the point is Castle's (1929) line of three generations of mosaic rabbits. Perhaps some of the outbreaks of specific mutations in certain stocks (*e.g.* Goldschmidt, 1939; Mampell, 1943) may have been caused in a similar way. It is interesting to speculate whether also in organisms other than *Drosophila* certain types of semi-sterility may be caused by gonadic mosaicism for lethal or semi-lethal genes.

SUMMARY

The frequency of sex-linked lethals in the F_3 of mustard-gas treated ♂♂ was significantly higher than in the controls, when the F_3 was derived from daughters of the treated ♂♂. This was shown to be due to the presence in the F_1 of ♀♀ which were gonadic mosaics for a lethal, *i.e.* which carried a lethal in only a part of their ovaries.

A high proportion of apparent semi-lethals in the progeny of chemically treated, but not of irradiated, ♂♂ was in reality due to gonadic mosaicism of F_1 ♀♀. It was estimated that in the progeny of chemically treated ♂♂ gonadic mosaics form a higher proportion of all mosaics than in the progeny of irradiated ♂♂. This is taken to indicate that the effect of mustard gas may lead to mutation at a division later than the one following treatment.

In several cases daughters of chemically produced gonadic mosaics were again gonadic mosaics for the same mutation: in these cases, an induced chromosomal instability must have been carried through many cell generations before giving rise to a mutation. In the majority of lines derived from gonadic mosaics, both the lethal and its normal allelomorph behaved as perfectly stable.

It is suggested that the primary effect of mustard gas and similar chemicals may sometimes consist in the creation of an instability which may become stabilized in two different ways: either by reverting to the old equilibrium of the normal allelomorph, or by attaining the new equilibrium of a stable mutation.

ACKNOWLEDGMENT

The author is much indebted to Mr N. Condon, Mr M. Y. Ansari, and Dr M. Ginsberg for carrying out the exposures to mustard gas.

Addendum

After this paper had gone to press, H. J. Muller in *DIS* (*Drosophila Information Service*), No. 20, pp. 88–89, reported the occurrence of an unstable mutation in the progeny of X-rayed flies.

REFERENCES TO LITERATURE

AUERBACH, C., 1943. "Chemically induced mutations and re-arrangements", *Drosophila Information Service* (*D.I.S.*), XVII, 48–50.
——, 1946. "Chemically induced mosaicism in *Drosophila melanogaster*", *Proc. Roy. Soc. Edin.*, B, LXII, 211–221.
AUERBACH, C., and ROBSON, J. M., 1944. "Production of mutations by allyl *iso*thiocyanate", *Nature*, CLIV, 81–82.
——, and ——, 1946. "Chemical production of mutations", *Nature*, CLVII, 302.
——, and ——, 1947. "The production of mutations by chemical substances", *Proc. Roy. Soc. Edin.*, B, LXII, 271–283.
BISHOP, D. W., 1942. "Spermatocyte chromosome aberrations in grasshoppers subjected to X-radiation during embryonic stages", *Journ. Morph.*, LXXI, 391–425.
CASTLE, W. E., 1929. "A mosaic (intense-dilute) coat pattern in the rabbit", *Journ. Exp. Zool.*, LII, 471–480.
DEMEREC, M., 1941. "Unstable genes in *Drosophila*", *Cold Spring Harbor Symp. Quant. Biol.*, IX, 145–150.
GOLDSCHMIDT, R., 1939. "Mass mutation in the Florida stock of *Drosophila melanogaster*", *Amer. Nat.*, LXXIII, 547–559.
GRÜNEBERG, H., 1931. "Über die zeitliche Begrenzung genetischer Röntgenwirkungen bei *Drosophila melanogaster*", *Biol. Zbl.*, LI, 219–225.

MAMPELL, K., 1943. "High mutation frequency in *Drosophila pseudo-obscura*, Race B", *Proc. Nat. Acad. Sci.* (*Wash.*), XXIX, 137–143.

MULLER, H. J., 1927. "The problem of genic modification", *Verh. V. intern. Kongr. Vererbungsw.*, 234–260; *Ztschr. indukt. Abst. Vererb. L.*, *Suppl.*, I, 1928.

——, 1940. "An analysis of the process of structural change in chromosomes of *Drosophila*", *Journ. Gen.*, XL, 1–66.

NEUHAUS, M. J., 1935. "Zur Frage der Nachwirkung der Röntgenstrahlen auf den Mutationsprozess", *Ztschr. indukt. Abst. Vererb. L.*, LXX, 257–264.

PANSHIN, I. B., 1935. "The analysis of a bilateral mosaic mutation in *Drosophila melanogaster*", *Trud. Inst. Genet.* (*Mosc.*), X, 227–232. (Russian with English summary.)

PARKS, H. B., 1936. "Cleavage patterns in *Drosophila* and mosaic formation", *Ann. Ent. Soc. Amer.*, XXIX, 350–392.

PATTERSON, J. T., 1933. "The mechanism of mosaic formation in *Drosophila*", *Genetics*, XVII, 32–52.

PONTECORVO, G., 1940. "Researches on the mechanism of induced chromosome re-arrangements in *Drosophila melanogaster*", *Ph.D. Thesis, University of Edinburgh*.

SIDKY, A. R., 1940. "Gonosomic mosaicism involving a lethal", *Journ. Gen.*, XXXIX, 265–271.

STADLER, L. J., 1939. "Genetic studies with ultra-violet radiation", *Proc. 7th Internat. Genet. Congr.*, 269–276.

TIMOFÉEFF-RESSOVSKY, N. W., 1930. "Does there exist an after-effect of X-ray treatment upon the process of genovariability?", *Zh. exp. Biol.*, VI, 79–83. (Russian.)

——, 1931. "Einige Versuche an *Drosophila melanogaster* über die Art der Wirkung der Röntgenstrahlen auf den Mutationsprozess", *Arch. Entw. Mech. Orgn.*, CXXIV, 654–665.

——, 1931. "Does X-ray treatment produce a genetic after-effect?", *Journ. Her.*, XXII, 221–223.

——, 1935. "Auslösung von Vitalitätsmutationen durch Röntgenbestrahlung bei *Drosophila melanogaster*", *Nachr. Ges. Wiss. Göttingen* (*Math.-phys. Kl., Biol.*) N.F., I, 163–180.

Editors' Comments
on Papers 8 and 9

THE STRUCTURE OF DNA

Neither the discovery of X-ray nor of chemical mutagenesis did, after all, immediately reveal the secrets of the gene. This revelation only began when the molecular configuration of the genetic material itself was deduced by Watson and Crick (1953). Their model of the structure of DNA, and their prompt realization of its potential for explaining not only DNA replication itself but also at least certain ways in which infidelities of DNA replication could generate mutations, are described in Papers 8 and 9. As Watson (1968) himself explained, he and Crick were initially frustrated in their model-building efforts by then current ambiguities about the exact structures of the bases. Although deeply influenced by Chargaff's (1951) data showing the base ratio equivalences between adenine and thymine and between guanine and cytosine, they were for some time unaware that the normal forms of thymine and guanine were the keto, and not the enol, configurations. When the basic DNA structure was deduced, it was therefore highly reasonable to suppose that the "incorrect" (e. g., infrequent) enol configurations of these bases could engage in mispairings. Although these mispairings accounted for only certain types of mutations (specifically, A : T $\longleftrightarrow$ G : T $\longleftrightarrow$ G : C), they provided a general conceptual basis for much of the work to follow, which is detailed in numerous of the additional papers in this collection.

There is considerable overlap between Papers 8 and 9; we have abstracted only that portion of the latter paper which deals specifically with the crucial mispairing concept.

REFERENCES

Chargaff, E. 1951. Structure and function of nucleic acids as cell constituents. *Fed. Proc., 10:* 654.

Watson, J. D. 1968. *The Double Helix; a Personal Account of the Discovery of the Structure of DNA.* Atheneum, New York.

——, and F. H. C. Crick. 1953. A structure for desoxyribose nucleic acids. *Nature, 171:* 737.

8

Reprinted from *Nature*, **171**(4361), 964–967 (1953)

GENETICAL IMPLICATIONS OF THE STRUCTURE OF DEOXYRIBONUCLEIC ACID

By J. D. WATSON and F. H. C. CRICK

Medical Research Council Unit for the Study of the Molecular Structure of Biological Systems, Cavendish Laboratory, Cambridge

THE importance of deoxyribonucleic acid (DNA) within living cells is undisputed. It is found in all dividing cells, largely if not entirely in the nucleus, where it is an essential constituent of the chromosomes. Many lines of evidence indicate that it is the carrier of a part of (if not all) the genetic specificity of the chromosomes and thus of the gene itself.

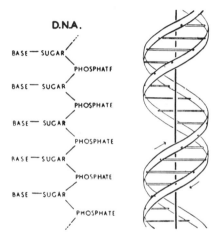

Fig. 1. Chemical formula of a single chain of deoxyribonucleic acid

Fig. 2. This figure is purely diagrammatic. The two ribbons symbolize the two phosphate-sugar chains, and the horizontal rods the pairs of bases holding the chains together. The vertical line marks the fibre axis

Until now, however, no evidence has been presented to show how it might carry out the essential operation required of a genetic material, that of exact self-duplication.

We have recently proposed a structure[1] for the salt of deoxyribonucleic acid which, if correct, immediately suggests a mechanism for its self-duplication. X-ray evidence obtained by the workers at King's College, London[2], and presented at the same time, gives qualitative support to our structure and is incompatible with all previously proposed structures[3]. Though the structure will not be completely proved until a more extensive comparison has been made with the X-ray data, we now feel sufficient confidence in its general correctness to discuss its genetical implications. In doing so we are assuming that fibres of the salt of deoxyribonucleic acid are not artefacts arising in the method of preparation, since it has been shown by Wilkins and his co-workers that similar X-ray patterns are obtained from both the isolated fibres and certain intact biological

materials such as sperm head and bacteriophage particles[2,4].

The chemical formula of deoxyribonucleic acid is now well established. The molecule is a very long chain, the backbone of which consists of a regular alternation of sugar and phosphate groups, as shown in Fig. 1. To each sugar is attached a nitrogenous base, which can be of four different types. (We have considered 5-methyl cytosine to be equivalent to cytosine, since either can fit equally well into our structure.) Two of the possible bases—adenine and guanine—are purines, and the other two—thymine and cytosine—are pyrimidines. So far as is known, the sequence of bases along the chain is irregular. The monomer unit, consisting of phosphate, sugar and base, is known as a nucleotide.

The first feature of our structure which is of biological interest is that it consists not of one chain, but of two. These two chains are both coiled around a common fibre axis, as is shown diagrammatically in Fig. 2. It has often been assumed that since there was only one chain in the chemical formula there would only be one in the structural unit. However, the density, taken with the X-ray evidence[2], suggests very strongly that there are two.

The other biologically important feature is the manner in which the two chains are held together. This is done by hydrogen bonds between the bases, as shown schematically in Fig. 3. The bases are joined together in pairs, a single base from one chain being hydrogen-bonded to a single base from the other. The important point is that only certain pairs of bases will fit into the structure. One member of a pair must be a purine and the other a pyrimidine in order to bridge between the two chains. If a pair consisted of two purines, for example, there would not be room for it.

We believe that the bases will be present almost entirely in their most probable tautomeric forms. If this is true, the conditions for forming hydrogen bonds are more restrictive, and the only pairs of bases possible are :

adenine with thymine ;
guanine with cytosine.

The way in which these are joined together is shown in Figs. 4 and 5. A given pair can be either way round. Adenine, for example, can occur on either

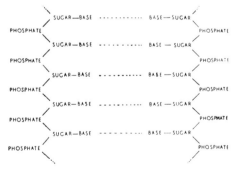

Fig. 3. Chemical formula of a pair of deoxyribonucleic acid chains. The hydrogen bonding is symbolized by dotted lines

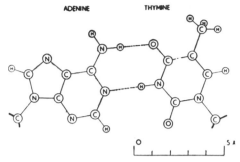

Fig. 4. Pairing of adenine and thymine. Hydrogen bonds are shown dotted. One carbon atom of each sugar is shown

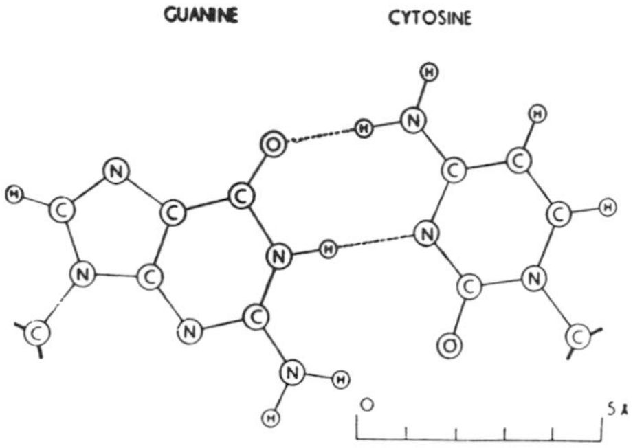

Fig. 5. Pairing of guanine and cytosine. Hydrogen bonds are shown dotted. One carbon atom of each sugar is shown

chain; but when it does, its partner on the other chain must always be thymine.

This pairing is strongly supported by the recent analytical results[5], which show that for all sources of deoxyribonucleic acid examined the amount of adenine is close to the amount of thymine, and the amount of guanine close to the amount of cytosine, although the cross-ratio (the ratio of adenine to guanine) can vary from one source to another. Indeed, if the sequence of bases on one chain is irregular, it is difficult to explain these analytical results except by the sort of pairing we have suggested.

The phosphate-sugar backbone of our model is completely regular, but any sequence of the pairs of bases can fit into the structure. It follows that in a long molecule many different permutations are possible, and it therefore seems likely that the precise sequence of the bases is the code which carries the genetical information. If the actual order of the bases on one of the pair of chains were given, one could write down the exact order of the bases on the other one, because of the specific pairing. Thus one chain is, as it were, the complement of the other, and it is this feature which suggests how the deoxyribonucleic acid molecule might duplicate itself.

Previous discussions of self-duplication have usually involved the concept of a template, or mould. Either the template was supposed to copy itself directly or it was to produce a 'negative', which in its turn was to act as a template and produce the original 'positive' once again. In no case has it been explained in detail how it would do this in terms of atoms and molecules.

Now our model for deoxyribonucleic acid is, in effect, a *pair* of templates, each of which is complementary to the other. We imagine that prior to duplication the hydrogen bonds are broken, and the two chains unwind and separate. Each chain then acts as a template for the formation on to itself of a new companion chain, so that eventually we shall have *two* pairs of chains, where we only had one before. Moreover, the sequence of the pairs of bases will have been duplicated exactly.

A study of our model suggests that this duplication could be done most simply if the single chain (or the relevant portion of it) takes up the helical configuration. We imagine that at this stage in the life of the cell, free nucleotides, strictly polynucleotide precursors, are available in quantity. From time to time the base of a free nucleotide will join up by hydrogen bonds to one of the bases on the chain already formed. We now postulate that the polymerization of these monomers to form a new chain is only possible if the resulting chain can form the proposed structure. This is plausible, because steric reasons would not allow nucleotides 'crystallized' on to the first chain to approach one another in such a way that they could be joined together into a new chain, unless they were those nucleotides which were necessary to form our structure. Whether a special enzyme is required to carry out the polymerization, or whether the single helical chain already formed acts effectively as an enzyme, remains to be seen.

Since the two chains in our model are intertwined, it is essential for them to untwist if they are to separate. As they make one complete turn around each other in 34 A., there will be about 150 turns per million molecular weight, so that whatever the precise structure of the chromosome a considerable amount of uncoiling would be necessary. It is well known from microscopic observation that much coiling and uncoiling occurs during mitosis, and though this is on a much larger scale it probably reflects similar processes on a molecular level. Although it is difficult at the moment to see how these processes occur without everything getting tangled, we do not feel that this objection will be insuperable.

Our structure, as described[1], is an open one. There is room between the pair of polynucleotide chains (see Fig. 2) for a polypeptide chain to wind around the same helical axis. It may be significant that the distance between adjacent phosphorus atoms, 7·1 A., is close to the repeat of a fully extended polypeptide chain. We think it probable that in the sperm head, and in artificial nucleoproteins, the polypeptide chain occupies this position. The relative weakness of the second layer-line in the published X-ray pictures[3a,1] is crudely compatible with such an idea. The function of the protein might well be to control the coiling and uncoiling, to assist in holding a single polynucleotide chain in a helical configuration, or some other non-specific function.

Our model suggests possible explanations for a number of other phenomena. For example, spontaneous mutation may be due to a base occasionally occurring in one of its less likely tautomeric forms. Again, the pairing between homologous chromosomes at meiosis may depend on pairing between specific bases. We shall discuss these ideas in detail elsewhere.

For the moment, the general scheme we have proposed for the reproduction of deoxyribonucleic acid must be regarded as speculative. Even if it is correct, it is clear from what we have said that much remains to be discovered before the picture of genetic duplication can be described in detail. What are the polynucleotide precursors ? What makes the pair of chains unwind and separate ? What is the precise role of the protein ? Is the chromosome one long pair of deoxyribonucleic acid chains, or does it consist of patches of the acid joined together by protein ?

Despite these uncertainties we feel that our proposed structure for deoxyribonucleic acid may help to solve one of the fundamental biological problems—the molecular basis of the template needed for genetic replication. The hypothesis we are suggesting is that the template is the pattern of bases formed by one chain of the deoxyribonucleic acid and that the gene contains a complementary pair of such templates.

One of us (J. D. W.) has been aided by a fellowship from the National Foundation for Infantile Paralysis (U.S.A.).

[1] Watson, J. D., and Crick, F. H. C., *Nature*, **171**, 737 (1953).

[2] Wilkins, M. H. F., Stokes, A. R., and Wilson, H. R., *Nature*, **171**, 738 (1953). Franklin, R. E., and Gosling, R. G., *Nature*, **171**, 740 (1953).

[3] (a) Astbury, W. T., Symp. No. 1 Soc. Exp. Biol., 66 (1947). (b) Furberg , S., *Acta Chem. Scand.*, **6**, 634 (1952). (c) Pauling, L., and Corey, R. B., *Nature*, **171**, 346 (1953) ; *Proc. U.S. Nat. Acad. Sci.*, **39**, 84 (1953). (d) Fraser, R. D. B. (in preparation).

[4] Wilkins, M. H. F., and Randall, J. T., *Biochim. et Biophys. Acta*, **10**, 192 (1953).

[5] Chargaff, E., for references see Zamenhof, S., Brawerman, G., and Chargaff, E., *Biochim. et Biophys. Acta*, **9**, 402 (1952). Wyatt, G. R., *J. Gen. Physiol.*, **36**, 201 (1952).

9

Reprinted from *Cold Spring Harbor Symp. Quant. Biol.*, **18**, 123, 129–130 (1953)

THE STRUCTURE OF DNA

J. D. WATSON AND F. H. C. CRICK

Cavendish Laboratory, Cambridge, England

(Contribution to the Discussion of Provirus.)

It would be superfluous at a Symposium on Viruses to introduce a paper on the structure of DNA with a discussion on its importance to the problem of virus reproduction. Instead we shall not only assume that DNA is important, but in addition that it is the carrier of the genetic specificity of the virus (for argument, see Hershey, this volume) and thus must possess in some sense the capacity for exact self-duplication. In this paper we shall describe a structure for DNA which suggests a mechanism for its self-duplication and allows us to propose, for the first time, a detailed hypothesis on the atomic level for the self-reproduction of genetic material.

[*Editors' Note:* Material has been omitted at this point.]

A POSSIBLE MECHANISM FOR NATURAL MUTATION

In our duplication scheme, the specificity of replication is achieved by means of specific pairing between purine and pyrimidine bases; adenine with thymine, and guanine with one of the cytosines. This specificity results from our assumption that each of the bases possesses one tautomeric form which is very much more stable than any of the other possibilities. The fact that a compound is tautomeric, however, means that the hydrogen atoms can occasionally change their locations. It seems plausible to us that a spontaneous mutation, which as implied earlier we imagine to be a change in the sequence of bases, is due to a base occurring very occasionally in one of the less likely tautomeric forms, at the moment when the complementary chain is being formed. For example, while adenine will normally pair with thymine, if there is a tautomeric shift of one of its hydrogen atoms it can pair with cytosine (Figure 7). The next time pairing occurs, the adenine (having resumed its more usual tautomeric form) will pair with thymine, but

the cytosine will pair with guanine, and so a change in the sequence of bases will have occurred. It would be of interest to know the precise difference in free energy between the various tautomeric forms under physiological conditions.

[*Editors' Note:* Material has been omitted at this point.]

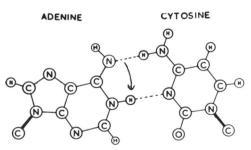

FIGURE. 7. Pairing arrangements of adenine before (above) and after (below) it has undergone a tautomeric shift.

Editors' Comments
on Papers 10 Through 12

NITROUS ACID MUTAGENESIS

Even as early as the 1920s some farsighted geneticists had speculated that the study of viruses might be highly rewarding, mainly on the grounds that viruses might consist of single genes or gene-like entities. This supposition turned out to be qualitatively correct, if quantitatively not very accurate: a median-sized virus actually contains on the order of a few dozen genes. The first papers to demonstrate chemically induced mutagenesis of free virus particles therefore produced a considerable impact; here at last were systems of great genetic simplicity, uncomplicated by surrounding cytoplasm and nucleoplasm, and highly amenable to both genetic and chemical analysis. It was therefore both logical and fitting that the first virus to be employed in studies of chemically induced mutation was tobacco mosaic virus (TMV), the virus that had been most intensively characterized by physical and chemical studies.

In Paper 10, Gierer and Mundry described the induction of mutations in TMV through the use of nitrous acid. Their choice of this agent was clearly inspired by chemical studies conducted by Schuster and Schramm (1958), which indicated that nitrous acid was capable of converting certain nucleic acid bases *in situ* into excellent analogues of other normally occurring nucleic acid bases, just as Watson and Crick had previously realized that the enol form of thymine, for instance, was a good analogue of cytosine. Nitrous acid is a particularly impressive mutagen because of the very simple nature of its main chemical reactions with nucleic acids: the conversion of cytosine to uracil and of adenine to hypoxanthine. It is described in most textbooks on genetics, although these also often incorrectly state that the conversion of guanine to xanthine is innocuous. Xanthine is ionized and probably quite incapable of pairing at all; thus a guanine "hit" is generally lethal (Richardson et al., 1963; Michelson and Grunberg-Manago, 1964). It should also be noted that nitrous acid induces deletions, at least sometimes (compare Tessman, 1962, and Paper 31), but by a still unknown mechanism.

A mutagen with such great chemical specificity was expected to exhibit a corresponding genetic specificity, that is, the exclusive induction of the transitions (see Paper 13) A : T → G : C and G : C → A : T. This mutagenic specificity was first revealed by Vielmetter and Schuster (Paper 11), who demonstrated the quantitative correlation between mutagenesis and the two deamination reactions cytosine → uracil and adenine → hypoxanthine. It is notable that these experiments, as well as many more described in succeeding papers in this collection, were performed with the T-even bacteriophages, T2, T4, and T6. By the late 1940s and early 1950s, the classical T series of bacteriophages had become the systems of choice for genetic and biochemical studies of virus genetics and reproduction. Fortunately for their ultimate relevance to higher genetic systems, these viruses all contain more or less conventional double-stranded DNA. Furthermore, these viruses are sufficiently complicated, containing from about 40 to 200 genes, to allow them to possess at least some "luxury" functions; these functions could be mutated without loss of viral infectivity, but with changes in phenotype that could be employed as useful genetic markers.

Another class of bacteriophages, however, was also about to gain the attention of molecular geneticists. The genomes of viruses such as ϕX174 and S13 are very different from both the genome of TMV and the genomes of the T viruses, for they are composed of single-stranded DNA. It is therefore sometimes possible to identify not merely the base pair, but also the specific base that constitutes the target of a mutagenic treatment. These two phages were employed by Irwin Tessman in an elegant series of studies of induced mutagenesis, of which Paper 12 was the first. This paper demonstrated the localization of nitrous acid induced mutations to a single strand of DNA; thus, although the treatment of a phage like ϕX174 produced only pure mutant clones, treatment of phage T4 produced clones composed, on the average, of equal numbers of mutant and wild-type progeny. [In considering the data of Paper 12, it may be helpful to keep in mind two minor influences that somewhat skew the data. First, nitrous acid mutagenesis of ϕX174 promotes some minor, chemically unexpected pathways (Tessman et al., 1964; Vanderbilt and Tessman, 1970), very possibly by an indirect mechanism such as that described later in Papers 27, 28, and 29; these minor mechanisms may have contributed to the few mixed clones. Second, excision repair induced in the double-stranded genome of phage T4 will sometimes convert a mosaic (strictly speaking, a heteroduplex heterozygote) into a pure-mutant individual (Freese and Freese, 1966).]

REFERENCES

Freese, E. B., and E. Freese. 1966. Induction of pure mutant clones by repair of inactivating DNA alterations in phage T4. *Genetics, 54:* 1055.

Michelson, A. M., and M. Grunberg-Manago. 1964. Polynucleotide analogues, 5. "Nonsense" bases. *Biochim. Biophys. Acta, 91:* 92.

Richardson, C. C., C. L. Schildkraut, and A. Kornberg. 1963. Studies on the replication of DNA by DNA polymerase. *Cold Spring Harbor Symp. Quant. Biol., 28:* 9.

Schuster, H., and G. Schramm. 1958. Bestimmung der biologisch wirksamen Einheit in der Ribosnucleinsäure des Tabakmosaikvirus auf chemischem Wege. *Z. Naturforsch., 13B:* 697.

Tessman, I. 1962. The induction of large deletions by nitrous acid. *J. Mol. Biol., 5:* 442.

——, R. K. Poddar, and S. Kumar. 1964. Identification of the altered bases in mutated single-stranded DNA: I. *In vitro* mutagenesis by hydroxylamine, ethyl methanesulfonate and nitrous acid. *J. Mol. Biol., 9:* 352.

Vanderbilt, A. S., and I. Tessman. 1970. Identification of the altered bases in mutated single-stranded DNA: IV. Nitrous acid induction of the transitions guanine to adenine and thymine to cytosine. *Genetics, 66:* 1.

Production of Mutants of Tobacco Mosaic Virus by Chemical Alteration of its Ribonucleic Acid *in vitro*

WHILE many agents are known to act as mutagens on the living cell, there have been no convincing reports that the genetic character of isolated particles, such as viruses or transforming principles, have been changed *in vitro*. Recent results with tobacco mosaic virus help to overcome this difficulty. The infective unit of the virus is its ribonucleic acid component, which can be isolated in active form[1]. It is made up of 6,000 nucleotides that probably form a single chain[2].

Recently, Schuster and Schramm[3] have used nitrous acid to convert adenine, guanine and cytosine into hypoxanthine, xanthine and uracil, respectively, while the ribonucleic acid strand remains intact. It was shown that the alteration of any one of about 3,000 nucleotides is lethal. Possibly, changes of other bases are mutagenic.

On the basis of this work we studied a simple symptom of mutation, the production of necrotic lesions on a tobacco variety (Java) on which the untreated strain produces chlorotic lesions only. Virus or isolated ribonucleic acid was treated with nitrous acid. After times t, samples were diluted to a constant concentration for assay on Xanthi and Java tobacco plants. The conditions and results are given in Table 1. The total concentration n of infective particles was measured by the number of lesions on Xanthi, and of chlorotic and necrotic lesions on Java tobacco, as compared with standard dilution curves. n decreases exponentially with t:

$$n = n_0 \exp(-t/\tau) \tag{1}$$

where τ is the average time for one lethal conversion of a base per ribonucleic acid molecule.

The necrotic lesions on Java tobacco are due to mutated particles, and their number is expected to be nearly proportional to their concentration m. If the alteration of a single base is mutagenic, and if the average number of such mutations per ribonucleic acid molecule in the time τ is p, the concentration of such mutants is:

Table 1. MUTAGENIC EFFECT OF NITROUS ACID ON RIBONUCLEIC ACID.
2 vol. of 0·19 per cent ribonucleic acid extracted from tobacco mosaic virus *vulgare* by phenol (ref. 1) were mixed with 1 vol. 4 *M* sodium nitrite and 1 vol. 1 *M* acetate *p*H 4·8 at 22° C. After time t, the ribonucleic acid is diluted to 19 μgm./ml. with *M*/15 phosphate, and assayed. Control dilutions of untreated ribonucleic acid contain an equivalent amount of nitrite. Time τ (equation 1) is 18·5 min. Total infections are given as numbers of lesions (per leaf) on Xanthi, and as the sum of chlorotic and necrotic lesions on Java; mutant infections as numbers of necrotic lesions per leaf on Java

t (min.)	t/τ	Total infections		Mutant infections	Mutant infections (per cent)
		Xanthi	Java	Java	Java
1	0·05	72·8	183	1·4	0·8
4	0·22	72·4	130	2·5	1·9
8	0·43	89·8	188	4·5	2·4
16	0·86	59·3	97	5·4	5·6
32	1·73	36·1	63	6·6	10·5
64	3·46	10·4	21	2·1	9·8
96	5·2	2·7	3·5	0·5	15·5
Untreated controls, μgm./ml. ribonucleic acid					
19		155	138	0·3	0·2
1·9		31·2	42	0·1	0·3
0·19		3·1	10	0	—
× 0·019		0·3	0·8	0	—

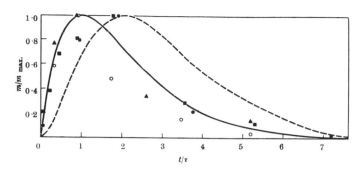

Fig. 1. Dependence of the concentration of mutants (number of necrotic lesions on Java tobacco, relative to maximum value) on the time of incubation with nitrous acid, related to τ. ————, Single-hit curve (equation 3); – – –, double-hit curve; ▲, tobacco mosaic virus *vulgare* (assayed at 0·7 μgm./ml.); ●, ribonucleic acid from tobacco mosaic virus *vulgare* (assayed at 39 μgm./ml.); ■, ribonucleic acid from tobacco mosaic virus *vulgare* (assayed at 19 μgm./ml.); ○, ribonucleic acid from light-green strain *B11* (assayed at 9·5 μgm./ml.)

$$m = n_0 p \frac{t}{\tau} \exp(-t/\tau) = np \frac{t}{\tau} \qquad (2)$$

or in terms of the maximum value of m :

$$\frac{m}{m_{\max.}} = e \frac{t}{\tau} \exp(-t/\tau) \qquad (3)$$

Fig. 1 represents measurements under different conditions with tobacco mosaic virus and with ribonucleic acid derived from two strains. For each case, τ was determined experimentally, using equation (1). The numbers of necrotic lesions on Java tobacco, divided by the maximum value reached in time t, were plotted against t/τ. The experimental results were in agreement with the theoretical relation (3).

The results lead to the following conclusions.

(1) Ribonucleic acid and tobacco mosaic virus treated with nitrous acid give rise to a much larger number of necrotic lesions on Java than untreated material assayed at the same concentration ; it is twenty times the spontaneous level when total infectivity has dropped to only half the original value (Table 1). This appearance of mutants at low rates of inactivation excludes selection, and cannot be due to pre-existing mutants, or to mutants arising in the plant.

(2) The mutagenic effect on tobacco mosaic virus and on its isolated ribonucleic acid is equal when the fraction of bases altered in the ribonucleic acid, measured by t/τ, is the same. Also, tobacco mosaic virus treated with nitrous acid has the same proportion of mutants as ribonucleic acid isolated after the treatment. Thus, the mutants are produced by chemical reactions with ribonucleic acid rather than by aggregation phenomena, or by changes in the protein.

(3) The production of mutants is at first linear with t and follows a single-hit curve, irrespective of varying conditions (Fig. 1). Therefore, alterations of single nucleotides are mutagenic.

At $t = \tau$, where one out of about 3,000 nucleotides is altered[3], 6 per cent of the primary infections on Java plants infected with *vulgare* are found to be necrotic (Table 1). If the plating efficiency is assumed to remain unchanged upon mutation, the alteration of any one out of 180 of the total 6,000 nucleotides in the ribonucleic acid strand would be mutagenic,

leading to necrotic lesions on Java ; this assumption may be subject to later correction.

(4) A variety of other mutants is also produced by nitrous acid. In order to detect them, we isolated individual lesions produced by ribonucleic acid treated with nitrous acid on Xanthi tobacco ($t/\tau = 5\cdot2$, Table 1) and assayed them on Java and Samsun plants. Out of 60 lesions, 20 produced no infection, 33 led to a variety of altered symptoms and only 7 gave apparently unchanged symptoms on Samsun tobacco. The altered symptoms proved genetically stable in a transfer experiment. From a control dilution of untreated ribonucleic acid, 64 out of 65 lesions produced infection and no altered symptoms were detected.

We are led to conclude that the chemical alteration of individual bases of the ribonucleic acid molecule *in vitro* is mutagenic. A variety of different mutants may be produced in this way. It remains undecided which of the three types of chemical change caused by the nitrous acid (see above) are mutagenic, and how the chemically altered ribonucleic acid is related to its progeny. In the case of a conversion of cytosine into uracil, it is conceivable that the altered ribonucleic acid undergoes identical reproduction in the host cell.

We are much indebted to Prof. G. Melchers for generous support and advice, and to Prof. G. Schramm and Dr. H. Schuster for helpful discussions. A detailed account of this work is being published in the *Zeitschrift für Vererbungslehre*.

ALFRED GIERER

Max-Planck-Institut für Virusforschung,
Tübingen.

KARL-WOLFGANG MUNDRY

Max-Planck-Institut für Biologie,
Abt. Melchers,
Tübingen.
Aug. 26.

[1] Gierer, A., and Schramm, G., *Nature*, **177**, 702 (1956) ; *Z. Naturforsch.*, **11b**, 138 (1956).

[2] Gierer, A., *Nature*, **179**, 1297 (1957).

[3] Schuster, H., and Schramm, G., *Z. Naturforsch.* (in the press).

Reprinted from *Biochem. Biophys. Res. Commun.*, **2**(5), 324–328 (1960)

THE BASE SPECIFICITY OF MUTATION INDUCED BY NITROUS ACID IN PHAGE T2

W. Vielmetter and H. Schuster

Max-Planck-Institut für Virusforschung, Tübingen, Germany

Received April 14, 1960

Nitrous acid has been shown to be an effective inactivating and mutagenic agent for the RNA of tobacco mosaic virus (Schuster and Schramm, 1958; Mundry and Gierer, 1958), for the phages T2 (Vielmetter and Wieder, 1959), T4 (Freese, 1959), Ø x 174 (Tessman, 1959) and for transforming DNA (Litman and Ephrussi-Taylor, 1959). In all cases analysed, both inactivation and mutation follow first order kinetics with respect to the time of treatment. Therefore lethal or mutagenic changes are caused by single chemical alterations in the nucleic acid. The chemical nature of these alterations is known to be the deamination of either G,[1] A or C, which is thereby converted into X, HX or U, respectively (Zamenhof, 1953; Schuster and Schramm, 1958; Schuster, 1960).

In DNA, the relative deamination rate ratios, $\alpha A/\alpha G$ and $\alpha C/\alpha G$, are different for different pH values (Schuster, 1960). Therefore, in this paper the deamination rates of G, A and C in phage T2 treated with HNO_2 at different pH values are compared with the corresponding rates of mutation and inactivation. This reveals which types of base deaminations can lead to either mutagenic or lethal changes.

---

[1] G = guanine, A = adenine, C = cytosine, or in phage T2 hydroxy-methyl-cytosine (HMC), X = xanthine, HX = hypoxanthine, U = uracil, or in phage T2 hydroxy-methyl-uracil (HMU).

EXPERIMENTAL

T2 wild type phage were incubated at 20°C in a mixture containing $\sim$2 mg/ml phage, 1 $\underline{M}$ NaNO$_2$, 0.25 $\underline{M}$ acetate buffer and 0.08 $\underline{M}$ NaCl. One experiment was performed at pH 4.20 for 60 hrs and another at pH 5.00 for 550 hrs. At regular time intervals samples were withdrawn and after dialysis the DNA was extracted and purified using the methods of Wyatt and Cohen (1953). After formic acid hydrolysis the bases were chromatographed as previously described (Schuster,1960). The extent of deamination for each sample was measured as the decrease in G, A and HMC and the increase in HMU, assuming T to be constant. Under the same conditions, but at various pH values, the rate of induction of r$^-$ mutation and the inactivation rate were measured as described previously (Vielmetter and Wieder, 1959). The number of plates for each sample was chosen so that at least 50 r$^-$ plaques were observed. The fraction of rII$^-$ mutants was determined by assaying the purified progeny of randomly picked r$^-$ plaques on strain $\underline{E.\ coli}$ K (Benzer, 1957).

RESULTS AND CONCLUSIONS

In Fig.1 the rate of the induction of r$^-$ mutation and the inactivation rate are plotted against the pH. It is found that the log of the mutation rate decreases linearly with increasing pH. Linear regression was used to obtain accurate values for the rate constants at pH 4.20 and pH 5.00. These values can be compared with the deamination rates of the three bases (Table I).The deamination rate of each base has been found to follow first order kinetics for a long period of time. Therefore, it is likely that the same applies also for the beginning of the reaction,at which time mutation and inactivation have been measured.

The results lead to the following conclusions.

(1) The deamination rate, α, of both A and C decreases by a factor of about 90 from pH 4.20 to pH 5.00, whereas the deamination rate of G decreases by a factor of only 35. The corresponding factor for

103

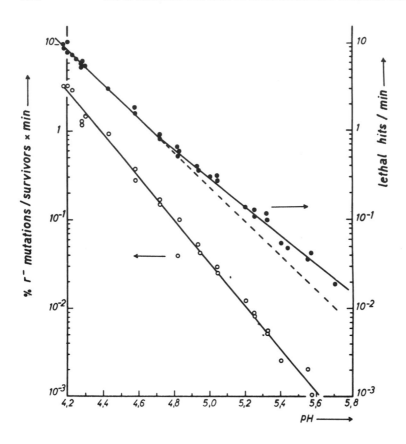

Figure 1.

Dependence of the rate of r⁻ mutation and the inactivation rate on the pH for phage T2 treated with 1 $\underline{M}$ NaNO$_2$ at 20°C

Table I

	Percent of Total DNA	α (Deaminations/min x 10^6)		$\frac{\alpha\ 4.2}{\alpha\ 5.0}$
		pH 4.20	pH 5.00	
C	17	116	1.3	89
A	33	18.5	0.2	90
G	17	297	8.55	35
Average deamination per Total DNA		76.3	1.74	44
% $\frac{r^- \text{mutants/survivors}}{\text{min}}$		2.9	3.3×10^{-2}	88
phage lethal hits/min		9	0.3	30

the decrease of the rate of r^- mutation has been found to be 88. Therefore it is concluded that the deamination of mainly A, C or both, but not of G, is responsible for the induction of mutations.

(2) This result is expected from the base pairing scheme proposed by Watson and Crick (1953). In this scheme A pairs specifically by hydrogen bonding with T and C pairs with G. The pairing specificity is primarily determined by the 6' keto-group in G and T and the 6' amino-group in A and C. The amino-group in G is 2', therefore its removal does not affect the pairing specificity. However, in both A and C deamination leads to analogues (HX and U) which pair like G and T, respectively. Therefore two types of conversions of base pairs (so called "transitions", Freese, 1959), should result from deamination and subsequent replication:

$$CG \longrightarrow TA \text{ (type 1) and } AT \longrightarrow GC \text{ (type 2)}$$

For T2 treated at pH 4.2 it can be calculated, that mutants resulting from conversions of type 1 should be 3-times more frequent as those derived from conversions of type 2.

(3) At either pH 4.20 or pH 5.00 the conversion of any one of about 570 base pairs/phage genome gives rise to a r^- mutation. Since we have found that the proportion of rII mutants/total induced r^- mutants is 65 %, the conversion of any one of about 370 base pairs should lead to a visible rII mutation. This is roughly 1/2 to 1/3 of the total base pairs/rII region, as estimated from data of Benzer (1957) for the closely related phage T4.

(4) The inactivation rate decreases from pH 4.20 to pH 5.00 by a factor of only 30. This indicates that in contrast to the mutagenic inducibility, inactivation can also be caused by the deamination of G (Table I). This conclusion however, remains tentative, because deaminations in the protein might also contribute to the inactivation.

(5) If in T2 the rate of the unspecific inactivation at a low pH of

105

treatment does not exceed 20% of the total inactivation [as indicated from experiments with phage PLT22 (Vielmetter, unpublished) and T4 (Freese, personal communication)] the target size for inactivation due to deamination of A, C and G can be calculated to be about 5×10^4 nucleotide pairs or 25% of the total T2 DNA.

A detailed account and discussion of the results will be published in the Zeitschrift für Naturforschung.

REFERENCES

Benzer, S., in "The Chemical Basis of Heredity", John Hopkins press, Baltimore (1957)

Freese, E., Brookhaven Symp. in Biol. 12, 63 (1959)

Litman, R. and Ephrussi-Taylor, H., Compt.rend.Acad.Sci.,249,838 (1959)

Mundry, K.W. and Gierer, A., Z.f.Vererbl., 89, 614 (1958)

Schuster, H. and Schramm, G., Z.Naturforsch.,13b, 697 (1958)

Schuster, H., Z.Naturforsch. in press (1960)

Tessman, I., Virology, 9, 375 (1959)

Vielmetter, W. and Wieder, C.M., Z.Naturforsch., 14b, 312 (1959)

Watson, J.D. and Crick, F.H.C., Cold Spring Harbor Symp.Quant.Biol., 18, 155 (1953)

Wyatt, G.R. and Cohen, S.S., Biochem.J., 55, 774 (1953)

Zamenhof, S., Alexander, H.E. and Leidy, G., J.exp.Med., 97, 373 (1953)

12

Reprinted from *Virology*, **9**(3), 375–385 (1959)

Mutagenesis in Phages φX174 and T4 and Properties of the Genetic Material[1]

Irwin Tessman[2]

*Biology Department, Massachusetts Institute of Technology,
Cambridge, Massachusetts*

Accepted July 29, 1959

In vitro treatment of phages φX174 and T4 with nitrous acid produces mutants. Mutants of φX174 arise in pure clones, whereas most T4 mutants arise in mixed clones, suggesting that there is only one copy of genetic information in φX174 and at least two copies in T4. The alteration of one of the T4 copies yields a type of heterozygote. The results is consistent with the picture that the deoxyribonucleic acid (DNA) of T4 is in the Watson-Crick double-stranded configuration, both strands carrying the same information, whereas the DNA of φX174 is single-stranded.

The production of mutants in φX174 follows a single-hit kinetics, showing that the alteration of a single nucleotide is sufficient to produce a mutant. The process of nitrous acid mutagenesis in DNA is discussed.

INTRODUCTION

Nitrous acid is an elegant mutagenic agent. In experiments on tobacco mosaic virus (TMV), Schuster and Schramm (1958) showed that nitrous acid deaminates the ribonucleic acid (RNA) of TMV, converting adenine to hypoxanthine, guanine to xanthine, and cytosine to uracil. Their work was complemented by the experiments of Mundry and Gierer (1958), establishing the mutagenic action of nitrous acid on TMV. The virtues of nitrous acid as a mutagen are that a virus or its nucleic acid can be treated *in vitro*, free of the complicating influence of a host cell, and that the effect on nucleic acid is directly on the bases, which are considered to be the coding elements of the nucleic acid. Mutagenic results can be interpreted in terms of molecular properties of the genetic material.

[1] Supported, in part, by research grant E-2028 from the National Institute of Allergy and Infectious Diseases, United States Public Health Service.

[2] Present address: Department of Biological Sciences, Purdue University, Lafayette, Indiana.

The application of this mutagen to a comparative study of phages φX174 and T4 will be presented. The evidence has been growing that the related phages S13 and φX174 are distinguished by having deoxyribonucleic acid (DNA) with unusual properties which are consistent with a single-stranded model of the DNA rather than with the complementary-stranded model of Watson and Crick (1953). These properties have been shown by experiments on the sensitivity of these phages to decay of incorporated P^{32} (Tessman, 1958, 1959) and by physical-chemical studies (Sinsheimer, 1959). It will be seen that the mutagenic effects of nitrous acid add further support to this picture and partially illuminate the genetic properties of single and double-stranded DNA.

MATERIALS AND METHODS

Bacteriophage strains φX174 and T4B were used. The phage were assayed by the agar-layer technique reviewed by Adams (1950). Host-range (h) mutants of φX174 were detected by the clear plaques they give on plating indicator consisting of a mixture of sensitive and resistant cells of *Escherichia coli*, strain C. Rapid-lysing (r) mutants of T4 were detected by their plaque morphology (Hershey, 1946) using *E. coli* B as plating indicator.

Free phage were treated with nitrous acid in an acetic acid-sodium acetate buffer, 0.2 M in acetate ions, at approximately pH 4.0, and at 25°. Nitrous acid was produced by adding KNO_2. Samples of phage were removed at intervals for assay, both for titer and for observation of mutants. Controls were handled in the same way, except that the KNO_2 was omitted.

The population of phage in individual plaques was examined by stabbing the plaques with a fine rod and dispersing the phage picked up by the rod into a liquid medium. The dispersed phage were then suitably diluted and replated for examination.

RESULTS

Kinetics

Figures 1, 2, and 3 show the effect of nitrous acid on φX174 as a function of time. Figure 1 illustrates the survival. The strictly exponential inactivation, confirmed in other experiments down to 10^{-3} survival, makes possible the use of a dimensionless variable, "hits," which is plotted on the abscissa in addition to the time scale. The number of hits h is defined by the equation $s/s_0 = \exp(-h)$, where s/s_0 is the fraction of phage surviving and s_0 is the initial phage titer.

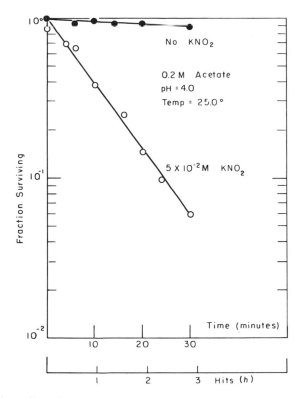

FIG. 1. A semilogarithmic plot of the survival of nitrous acid-treated φX174 as a function of time of treatment. One hit is the time of treatment that reduces the survival to e^{-1}, and equals 10.5 minutes.

Figure 2 shows an absolute increase in the number of host-range mutants that are produced, with a broad maximum occurring at a dose equal to one hit. The same data are shown in Fig. 3, in which instead of the absolute number of mutants, the proportion of mutants among the surviving phage is plotted as a function of time.

The kinetics of mutation production in φX174 is qualitatively the same as was found for TMV by Mundry and Gierer (1958). From Fig. 2 and more evidently from Fig. 3, it is seen that the proportion of mutants is a linear function of the number of hits. Thus, the production of mutants in φX174 is a one-hit process, the deamination of a single base presumably being sufficient to produce a host-range mutant. A number of phenotypically distinct host-range mutants, differing in plaque

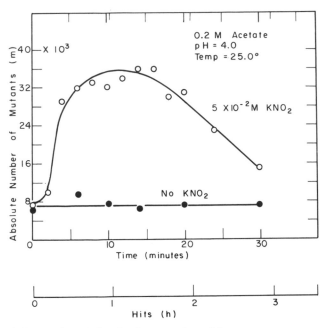

Fig. 2. A linear plot of the absolute number of host range mutants produced by nitrous acid treatment of φX174 as a function of time. The total phage titer at time zero was 1.4×10^9/ml. The phage survival is shown in Fig. 1.

morphology, were observed. It is likely that a base at a different site along the DNA chain was altered in each case.

The mutation frequency as given by the slope of Fig. 3 is 7.0×10^{-5} host-range mutants of φX174/survivor/hit. On one occasion a mutation frequency greater by approximately a factor 2 was observed. A probable reason for the variation is that some of the host-range mutants give only faintly clear plaques on mixed indicator, the efficiency of detection varying from day to day.

Number of Copies of the Genetic Material

The question of the number of copies of the genetic material that are contained in a phage particle was studied by a comparison of the mutagenic effects of nitrous acid on φX174 and T4B to determine whether mutants are produced in pure clones containing only the mutant type, or in mixed clones containing both the original parental type as well as

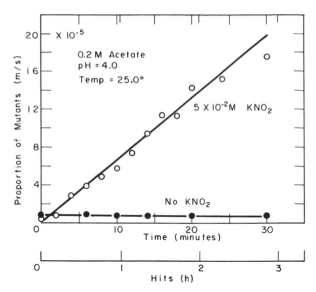

FIG. 3. The proportion of mutants among the survivors of nitrous acid treatment as a function of time. The curves are obtained by combining the data shown in Figs. 1 and 2.

the mutant type. No more than 100 viable nitrous acid-treated phage were seeded on each of many plates to insure that most mutant plaques would be completely isolated and not overlap wild-type plaques. Plaques containing mutant particles were picked and samples of the phage from these plaques were replated to determine the kinds of particles present. A simplified statement of the results is that the mutants of φX174 arise in plaques that contain purely mutant phage, whereas most mutants of T4 arise in plaques containing a mixture of mutant and wild-type phage.

The data for φX174 are presented in Table 1. The clear-cut separation of the results according to the proximity to neighboring plaques indicates that the small percentage of wild-type phage in the 7 "close" plaques can be attributed to contamination from neighboring wild-type plaques. Reconstruction experiments performed by plating mixtures of pure h and h^+ phage verified that such contamination occurs. Plaques 22 and 23, although classified as "distant," were probably marginal cases and similarly contaminated. Only plaque 24 showed an abnormal number of h^+ phage, but this one exception is not significant. It is concluded that the host-range mutants of φX174 arise in pure clones. The possibility

TABLE 1

Sampling of Host-Range Plaques after Plating Nitrous
Acid-treated φX174[a]

Plaque number	Proximity to h^+ plaques	h^+	h	Plaque number	Proximity to h^+ plaques	h^+	h
1	Distant	0	2000	13	Distant	0	42
2	Distant	0	800	14	Distant	0	20
3	Distant	0	700	15	Close	1	800
4	Distant	0	700	16	Close	2	500
5	Distant	0	600	17	Close	2	200
6	Distant	0	500	18	Close	6	250
7	Distant	0	500	19	Close	9	350
8	Distant	0	400	20	Close	2	80
9	Distant	0	350	21	Close	10	200
10	Distant	0	200	22	Distant	2	600
11	Distant	0	150	23	Distant	2	150
12	Distant	0	100	24	Distant	40	250

[a] Phage φX174h^+, treated with 0.5 M KNO$_2$ at pH 4.0 to give a survival of 2×10^{-7}, was plated on mixed indicator. Among 8200 plaques that were examined there were 24 h plaques. Each h plaque was classified according to whether it was separated from all neighboring plaques by more ("distant") or less ("close") than 2 mm between the edges of the plaques. The phage population of the individual h plaques was examined by picking each plaque and replating a sample of the particles. Untreated controls showed no h plaques among 6500 h^+ plaques.

that both h^+ and h particles can descend from a single treated h^+ particle cannot be completely excluded, but the occurrence can only be rare.

A trivial explanation of the pure clones of h particles from nitrous acid-treated φX174 is that h and h^+ particles might not be able to multiply in the same cell. However, such an exclusion does not normally occur. Four h mutants arising from nitrous acid treatment and representing at least three different phenotypic classes were each crossed with the wild type. When plated on mixed indicator 80 % to 90 % of the infected cells produced mottled plaques, which, on being picked and replated, were found to contain similar (± 50 %) numbers of h and h^+ particles. The remaining 10 % to 20 % of the plaques were almost entirely pure h^+ plaques. Single-burst experiments (0.25 infected cells per burst tube) also demonstrated that h and h^+ phage were produced together in comparable numbers in the same cell.

TABLE 2

SAMPLING OF PLAQUES AFTER PLATING NITROUS ACID-TREATED T4B[a]

r	r^+	r	r^+	r	r^+	r	r^+
1000	0	200	0	600	400	300	700
1000	0	50	0	600	400	200	1000
600	0	20	0	300	200	200	1000
400	0	200	1	500	400	200	1000
300	0	500	100	1000	1000	50	300
300	0	500	200	600	800	50	800
200	0	1000	500	500	1000	30	1000
200	0	200	100	200	400		

[a] Phage T4Br^+ was treated with 0.2 M KNO$_2$ at pH 4.0 to give a survival of 2×10^{-1}. Among 6900 plaques 31 contained r mutants. The numbers of r and r^+ phage in samples from each plaque containing an r mutant are shown. Among 15,000 control plaques 13 contained r mutants; 12 of these contained only r particles ($< 1\% r^+$), and one contained roughly equal numbers of r and r^+ particles.

The results for T4 are shown in Table 2. Contamination from neighboring plaques was negligible (possibly due to the low diffusion rate of this large phage). By testing 14 wild-type T4 plaques picked at random, it was found that the background of r phage spontaneously arising in r^+ plaques was about 0.1%; the maximum frequency for any one plaque was 7 r particles among 800 r^+ particles. The nitrous acid sensitivity of one induced r mutant was tested and was found to be the same as the sensitivity of the wild type. It is significant that treatment of T4 with nitrous acid produces mutants in mixed clones. It is appropriate to call the treated parent that yields a mixed clone a heterozygote, similar in behavior to the ones produced naturally in a cross (Hershey and Chase, 1951).

The sampling of the mixed plaques showed that the r and r^+ phage were stable. Very rarely the sampling yielded mottled plaques (less than one in 500), which were probably from naturally occurring heterozygotes or from accidental overlaps of r and r^+ plaques.

T4 treated with nitrous acid to give lower survival values down to 10^{-6} also produced a high proportion of mottled plaques. The high survival was chosen for the experiment shown in Table 2 because it made it certain that mottled plaques arose from single phage particles and not from cross reactivation or multiplicity reactivation on the plate.

DISCUSSION AND CONCLUSIONS

The nitrous acid-induced mutations probably arise from the eventual substitution of a different normal base for the one originally deaminated. In the case of RNA the direct formation of uracil from cytosine can, obviously, account for mutations (Mundry and Gierer, 1958). But the deaminations of adenine and guanine to produce hypoxanthine and xanthine are not so obviously mutagenic, for one or more additional steps would be required in the host cell to replace the abnormal bases by normal ones in the new RNA chain. If, for example, duplication involved specific pairing of nucleotides along the RNA chain, the abnormal bases might provoke a pairing error leading to a new sequence of nucleotides in progeny RNA.

In the case of DNA none of the deamination products would be normal bases. So all the nitrous acid-induced mutations found for φX174 and T4 probably result from some sort of pairing error provoked during duplication by the presence of an abnormal base. The deamination products are essentially base analogs that are entirely contained in the DNA. Base analogs such as 5-bromouracil (Litman and Pardee, 1956) and 2-aminopurine (Freese, 1959) are known to be mutagenic when added to multiplying phage, but it is not known that their effect is due to their incorporation into the DNA.

Deamination of DNA can have at least two sequels:

1. DNA containing an abnormal base might fail to reproduce.

2. The abnormal base might lead, with a probability that may well be considerably less than 1.0, to the substitution of a normal base in a new DNA molecule. If the normal base is the same as the original one, then the original genetic code will have been restored; if different, then a mutant will have been produced.

The two sequels are complementary. The production of a mutant might still be lethal if the new product of the altered genome does not function adequately. This is a matter of whether a missense or a nonsense mutation is obtained (Crick *et al.*, 1957; Delbrück, in Golomb *et al.*, 1958; Levinthal, 1959).

The contrasting results for φX174 and T4 can be explained by the assumption that φX174 contains only one copy of the genetic information while T4 contains two copies (more than two copies cannot be ruled out, but need not be assumed). The results are summarized by the interpretive illustration shown in Fig. 4. The number of copies of genetic information can be directly related to the apparent number of

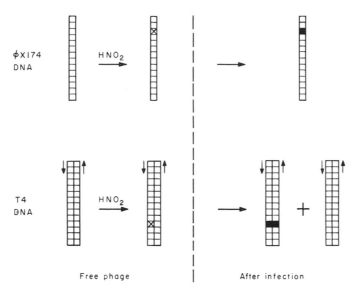

FIG. 4. A schematic drawing of the effect of nitrous acid on φX174 and T4 DNA. The squares represent bases in a DNA chain. ⊠ represents an abnormal base produced by nitrous acid, and ■ represents a normal base in the position corresponding to the ⊠. The arrows on the T4 DNA identify the two complementary strands.

strands in the DNA molecule. It is inferred that heterozygotes induced by nitrous acid can consist of double-stranded DNA with singular regions in which the base pairs are not the normal complementary ones. It still remains to be determined whether the DNA in heterozygotes that normally arise from mixed infection has such a structure, as conjectured by Levinthal (1953).

The results verify what has long been apparent from the normal Watson-Crick structure of DNA, namely, that in the duplex structure each of the two complementary strands carries the genetic information. Experiments by Pratt and Stent (1959) using 5-bromodeoxyuridine to produce T4 heterozygotes also support this conclusion. Recent suggestions about the structure of φX174 DNA (Tessman, 1958, 1959; Sinsheimer, 1959) are supported by the mutation results.

The fact that φX174 mutants arise in pure clones implies that a phage containing a deaminated base cannot have the original nucleotide sequence restored with any appreciable probability in addition to having a new sequence produced. The reason for the one single pathway might be either that the supposed pairing error can lead in only one direction, or that the parental DNA pairs only once.

To obtain a sufficiently large number of isolated φX174 mutants, a high dose of nitrous acid was used. This raises the possibility that if the phage actually contained two copies of the genetic information, heterozygotes might not appear because one of the copies was irremediably damaged by the nitrous acid. For T4 there was no significant decrease in the ratio of heterozygotes to pure r mutants with increasing nitrous acid doses.

In the case of T4 the question may be raised as to why so many of the r mutants arise in pure clones. It is conceivable that individual particles may have hemizygous regions of the genome or that one strand of the DNA molecule may occasionally fail to transmit its genetic information.

Nitrous acid treatment produces a considerably lower mutation frequency in φX174 than was found by Mundry and Gierer' (1958) in TMV. Two possible reasons are that (1) the conversion of cytosine to uracil directly produces a new sequence of normal nucleotides in RNA, but not in DNA; and (2) the number of nucleotide sites which can be transformed by the nitrous acid to produce the observed mutants may be greater in TMV than in φX174.

NOTE ADDED IN PROOF: Vielmetter and Wieder (1959) have studied extensively the production of nitrous acid-induced mutants of T2. The occurrence of T2 heterozygotes is similar to that reported here for T4.

REFERENCES

ADAMS, M. H. (1950). Methods of study of bacterial viruses. In *Methods in Medical Research* (J. H. Comroe, Jr., ed.), Vol. 2, pp. 1–73. Year Book Publishers, Chicago.

CRICK, F. H. C., GRIFFITH, J. S., and ORGEL, L. E. (1957). Codes without commas. *Proc. Natl. Acad. Sci. U. S.* **43**, 416–421.

FREESE, E. (1959). The specific mutagenic effect of base analogues on phage T4. *J. Mol. Biol.* in press.

GOLOMB, S. W., WELCH, L. R., and DELBRÜCK, M. (1958). Construction and properties of comma-free codes. *Kgl. Danske Videnskab. Selskab Biol. Medd.* **23**, 1–34.

HERSHEY, A. D. (1946). Mutation of bacteriophage with respect to type of plaques. *Genetics* **31**, 620–640.

HERSHEY, A. D., and CHASE, M. (1951). Genetic recombination and heterozygosis in bacteriophage. *Cold Spring Harbor Symposia Quant. Biol.* **16**, 471–479.

LEVINTHAL, C. (1953). Recombination in phage: its relationship to heterozygosis and growth. *Cold Spring Harbor Symposia Quant. Biol.* **18**, 13–14.

LEVINTHAL, C. (1959). Coding aspects of protein synthesis. *Revs. Modern Phys.* **31**, 249–255.

LITMAN, R. M., and PARDEE, A. B. (1956). Production of bacteriophage mutants by a disturbance of deoxyribonucleic acid (DNA) metabolism. *Nature* **178**, 529–531.

MUNDRY, K. W., and GIERER, A. (1958). Die Erzeugung von Mutationen des Tabakmosaikvirus durch chemische Behandlung seiner Nucleinsäure *in vitro.* *Z. Vererbungslehre* **89**, 614–630.

PRATT, D., and STENT, G. S. (1959). Mutational heterozygotes in bacteriophages. *Proc. Natl. Acad. Sci. U. S.*, in press.

SCHUSTER, H., and SCHRAMM, G. (1958). Bestimmung der biologisch wirksamen Einheit in der Ribosenucleinsäure des Tabakmosaikvirus auf chemischem Wege. *Z. Naturforsch.* **13b**, 697–704.

SINSHEIMER, R. L. (1959). A single-stranded deoxyribonucleic acid from bacteriophage φX174. *J. Mol. Biol.* **1**, 43–53.

TESSMAN, I. (1958). The unusually high sensitivity of bacteriophages φX174 and S13 to P^{32} decay. *Program and Abstracts. The Biophysical Society*, p. 42.

TESSMAN, I. (1959). Some unusual properties of the nucleic acid in bacteriophages S13 and φX174. *Virology* **7**, 263–275.

VIELMETTER, W., and WIEDER, C. M. (1959). Mutagene und inaktivierende Wirkung salpetriger Säure auf freie Partikel des Phagen T2. *Z. Naturforsch.* **14b**, 312–317.

WATSON, J. D., and CRICK, F. H. C. (1953). A structure for deoxyribose nucleic acid. *Nature* **171**, 737–738.

117

Editors' Comments
on Paper 13

13   FREESE
The Difference Between Spontaneous and Base-Analogue Induced Mutations of Phage T4

BASE-ANALOGUE MUTAGENESIS

In 1956, Litman and Pardee, who had been studying the effects of base analogues on DNA replication, reported that 5-bromouracil was highly mutagenic when present during the replication of bacteriophage T2. Their results suggested a novel mutagenic mechanism, the insertion of a mutagenic base analogue into DNA rather than its creation *in situ,* as exemplified by the action of nitrous acid. Their observation was seized upon by Ernst Freese, who combined it with the pioneering work of Benzer (1957) on the T4*rII* locus to carry out experiments that were to lead to a new conceptual framework for base pair substitution mutagenesis. Freese's key initial observation, reported in Paper 13, was that spontaneous mutants were qualitatively different from mutants induced by base analogues such as 5-bromouracil and 2-aminopurine. He suggested that the latter consisted of the *transitions* $(A:T \longleftrightarrow G:C)$, which were expected on structural grounds to be induced by these base analogues, whereas the former consisted of *transversions* (purine:pyrimidine $\longleftrightarrow$ pyrimidine:purine). That he was wrong about the specific nature of the spontaneous mutants (see Papers 22 and 23) is perhaps less important than that he formulated the basic conceptual distinction, which was to continue to be crucial and to stimulate research in this area for many years.

REFERENCES

Benzer, S. 1957. The elementary units of heredity. In *The Chemical Basis of Heredity* (W. D. McElroy and B. Glass, eds.), p. 70. Johns Hopkins Press, Baltimore.
Litman, R. M., and A. B. Pardee. 1956. Production of bacteriophage mutants by a disturbance of deoxyribonucleic acid metabolism. *Nature, 178:* 529.

13

Reprinted from *Proc. Natl. Acad. Sci.*, 45(4), 622–633 (1959)

THE DIFFERENCE BETWEEN SPONTANEOUS AND BASE-ANALOGUE INDUCED MUTATIONS OF PHAGE T4

By Ernst Freese*

BIOLOGICAL LABORATORIES, HARVARD UNIVERSITY

Communicated by Ernst Mayr, February 24, 1959

This paper describes studies about the induction of reverse mutations by two base analogues, 2-aminopurine (AP)[1] and 5-bromo-deoxyuridine (BD). It will be shown that mutants of phage T4 induced by these base analogues in the forward direction can also be induced to revert. In contrast, proflavine induced and most spontaneous mutants cannot be induced to revert by these base analogues, although they revert spontaneously. Superimposed on this fundamental difference between base analogue inducible and noninducible mutations is a further difference in the relative effect of AP and BD. A further examination of spontaneous and induced revertants indicates that most arise by a back mutation to the standard genotype. A molecular explanation of these results is possible, and provides a better understanding of spontaneous mutations.

MATERIALS

Phages—All mutants examined for the induction of revertants have the *r*II phenotype; i.e., they lack the function necessary for phage growth in bacteria K, and they form *r*-plaques on bacteria B. Their mutations are located in the *r*II region of the phage T4 genome; their isolation and genetic properties have been decribed previously.[2-4]

Bacteria, broth, and *synthetic* medium, see reference 3.

H-Medium, synthetic medium, plus 20 μg/ml histidine, plus 0.25 μg/ml thiamine.

F10 Medium, per liter H-medium 20 mg L-glycine, 20 mg L-methionine, 10 mg L-leucine, 10 mg L-valine, 10 mg L-serine, 10 mg adenine sulfate, 10 mg guanine sulfate, 10 mg uracil, 0.2 mg Ca pantothenate, 0.2 mg pyridoxine, 10 μg Vitamine B12.

Aminopterin (gift of R. B. Angier, American Cyanamid Company).

2-Aminopurine and *5-bromodeoxyuridine*, Sigma Chemical Company. For spot tests, solutions in broth of 10 mg/ml and 5 mg/ml, respectively, were used.

RESULTS

Reversion Rates in Liquid Medium.—Revertants of *r*II-type mutants can be detected, selectively, when the phages of an *r*II-type stock are plated on bacteria K. In most cases plaques are of the wild phenotype; exceptions will be discussed later. For a quantitative measurement of reversion rates, it is best to prepare lysates from a small inoculum of given number (e.g., 100 phages). The introduction of a revertant, already present in the previous stock, is then rare, and can easily be detected by an exceptionally large proportion of revertants in the lysate, when compared to parallel cultures. Fluctuations in the proportion of revertants also arise because the corresponding mutations may occur early or late in phage growth, and the revertants can multiply after their formation. An unusually great fluctuation of this kind will be called a "jackpot." The average ratio of revertants to viable phages (assayed on B), for stocks grown from an inoculum of 100 phages, is called the *reversion index*.

The *induction of revertants* is measured by comparing the reversion index of phage stocks grown in the presence and absence of the mutagen.

The following procedure was used for the measurement of reversion indices with and without induction by AP and BD: 400 phages (in synthetic medium) were added to 20 ml. of the growth medium, containing 10^8/ml bacteria. The content was immediately equidistributed, to 4 test tubes. The tubes were incubated at 37°C for 24 hr, with occasional shaking; chloroform was then added. The lysates were plated on B and K to determine the phage titer and the frequency of revertants.

For the *induction by AP*, bacteria B-97 grown in H medium + 30 μg/ml adenine were added to H medium + 15μg/ml adenine + 500 μg/ml AP.

For the corresponding *control*, the same medium without AP was used.

For the *induction by BD*, bacteria B grown in F 10 + 20 μg/ml thymine were added to F 10 + μg/ml thymine + 50 μg/ml aminopterin + 50 μg/ml BD.

For the corresponding control the same medium was used, containing 50 μg/ml thymidine in the place of BD. The thymidine addition was found necessary since the phage titer otherwise decreased instead of increasing, although the bacteria continued to grow. This shows both the strong inhibition of thymine production by aminopterin and the ability of BD to serve as growth factor in the place of thymidine.

TABLE 1

REVERSION INDEX AFTER PHAGE GROWTH WITH AND WITHOUT BASE ANALOGUES

	Mutant	Reversion Index		Ratio, with/without AP	Reversion Index		Ratio, with/without BD
		With AP	Without AP		With BD	Without BD	
Spontaneous	r111	1,511	265	5.7	408	301	1.4
	r114	0.15	0.11	1.4	0.49	0.29	1.7
	r117	6.6	2.0	3.3	3.0	1.3	2.2
	r131	5.2	0.93	5.6	1.4	1.5	0.9
	r157	1,490	592	2.5	890	933	1.0
	r215	1.04	0.002	500	0.48	0.002	240
BU induced	N11	408	0.53	770	10	0.18	56
	N12	1,705	0.02	95,000	2.3	0.01	23
	N21	98	0.014	7,000	0.14	0.03	4.7
	N24	796	0.63	1,260	2.4	0.29	8.3
	N29	1,705	0.054	32,000	0.5	0.05	10
	N34	{ 270w / 48ti	{ 0.09w / 20 ti	{ 3,000w / 2ti	{ 0.33w / 65 ti	{ 0.04w / 21 ti	{ 8w / 3ti
	N101	559	≤0.01	560,000	2.58	≤0.001	2,600
AP induced	AP12	31	{ 0.14 / +2 ti	220	88	{ 0.13 / +2 ti	680
	AP61	86	0.16	540	4.7	0.05	940
	AP72	74	0.2	370	3,070	0.15	20,000
	AP133	198	5.5	36	5,130	15	340
	AP156	4.2	0.006	700	23.1	0.002	11,500

The table gives the *reversion index*, averaged (arithmetically) over parallel stocks, in *units 10⁻⁶*. The fluctuations from stock to stock are usually in the range of a factor of 5. Excessive fluctuations (jackpots) have not been used for determining the average. The ratio of the reversion index with and without the mutagen is given in the columns "*ratio*." The spontaneous reversion index of mutant N101 is so low that its absolute number could not be determined and only a limiting value can be given, according to the number of phages plated. For the determination of the ratio with/without base analogue the K/B value 0.001 has been used in this case for both AP and BD. The mutants AP12, AP133, and N34 are exceptional in giving rise to two kinds of spontaneous revertants, which on K produce wild (w) and tiny (ti) plaques, respectively.

A typical example for the *results* obtained is given in Table 1. The table includes one mutant for each site of high mutability ("hot spot")[2] and some others of special interest.

The reversion indices of the *controls* obtained in the AP-less and the BD-less media agree quite well, considering the fluctuations in individual stocks.

The increase of the average reversion index, due to the presence of AP or BD in the medium, varied with the mutant and the base analogue used. AP had the largest effect, and increased the reversion index of N12 and N101 by a factor of about 10^5. For some mutants the effect of AP was larger than that of BD; in other cases the reverse was observed.

Five of the spontaneous mutants, belonging to five spontaneous hot spots,[2] showed a small increase of reversion index with AP, and nearly no increase with BD. This effect, however, is so much smaller than the mutagenic effect observed for other mutants that it appears to reflect only the selective action of AP and BD upon phage multiplication. AP has indeed been found[3] to suppress lysis inhibition; it may therefore suppress the slight selection against revertants, which is caused by lysis inhibition during the phage growth in several bacterial cycles. The absence of reversion induction for these spontaneous mutants is furthermore shown by the negative result of spot tests (see later).

The spontaneous mutant r 215 has been included in this list because it constitutes an exception among the spontaneous mutants, as will be seen later. Its spontaneous reversion index is exceptionally small (as compared to other spontaneous mutants), and both AP and BD increase the index by a factor of about 500.

Of special interest is mutant N34 which spontaneously gives rise to some revertants of standard phenotype (w-plaques) and many "partial revertants" of tiny phenotype (ti-plaques). During growth in AP (and less efficiently in BD) the frequency of standard type revertants increases 1000-fold while the frequency of partial revertants does not change significantly.

Spot Test for the Induction of Reversions.—The measurement of reversion rates in liquid medium is accurate but cumbersome, since several parallel cultures must be grown. A qualitative decision, however, for the presence or absence of reversion induction can be obtained for each mutant by means of a simple spot test.[5,3]

In this test different numbers of phages (around 2×10^7) are plated on cold T plates together with a mixture of about 2×10^8 bacteria K and 2×10^7 B. The addition of B enables some of the rII-type phages to multiply and undergo mutations. When all bacteria B are lysed, only the revertants continue to multiply in K and give rise to plaques. The conditions applied seem to be optimal for the detection of small mutagenic effects by AP and BD; high mutagenic effects can also be detected when less bacteria B are added. A small drop containing 10 mg/ml AP was placed on one side of the plate and another drop containing 5 mg/ml BD on the opposite side. The plates were immediately placed in the cold for 2 hr in order to permit a slight diffusion of the base analogues. This cold storage is not necessary for high but useful for small inductions by BD, since the diffusion produces a gradient of BD concentration. Phage growth is inhibited in the center of the BD spot but at the rim mutations are expressed by many plaques. After subsequent incubation in 37°C the plates were inspected. The induction of mutations showed up by a large increase of plaques in the region of the spotted base analogue. In general, AP was more efficiently mutagenic than BD.

Besides being simple the spot test has additional advantages. Instead of comparing the reversion rates of different growth tubes one observes on the same plate the induction of reversions above the spontaneous rate and directly compares the relative effect of AP and BD. The problem of jackpots is eliminated, since each revertant which arises early enough in phage growth gives rise to one and only one plaque. Moreover the number of plaques cannot increase when the base analogue merely affects the degree of lysis inhibition or otherwise preferentially inhibits the development of certain phages; each plated phage revertant can produce only one plaque.

The spot test is quite sensitive, as shown by a comparison to the quantitative measurements in liquid medium. It rarely fails and then only for weak induction (1) when the induced reversion index is very small, (2) when the mutant is rather leaky, (3) when in addition to rare inducible reversions another (tiny) type of frequent noninducible reversions is produced (as for the mutants of site N 34). This rare failure (mostly for BD induction) does not affect our analysis since for all mutants found base analogue inducible, at least one of the two base analogues causes a strong, easily detectable mutagenic effect. Only very rarely can base analogue induction have escaped detection.

Using this spot test the induction of reversions was determined for all available mutants of the rII-type. For each mutant the test was repeated several times, varying also the number of phages plated. The result for spontaneous, AP, and BD induced mutants is given in Figure 1, and summarized in Table 2.

Most striking is the difference between spontaneous and base analogue induced

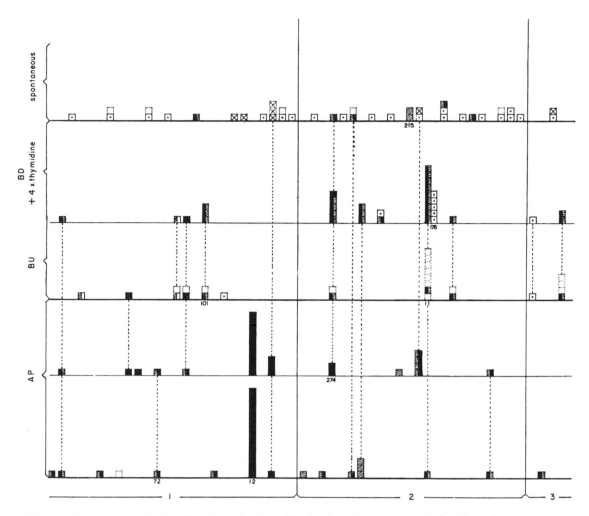

Fig. 1.—Genetic map of the *r*II region of phage T4 showing the approximate location of mutations for independently isolated *r*II-type mutants.   Mutations induced in 5 different experiments are sketched on 5 different horizontal lines.

First horizontal line: spontaneous mutations (*r*) (see Benzer and Freese[2]).

Second horizontal line: mutations (*N*) arising under the action of BD plus 4× as much thymidine (see Freese[3]).

Third horizontal line: mutations (*N*) arising under the action of BU.[2]

Fourth horizontal line: mutations (*AP*) arising under the action of AP in bacteria B-97.[3]

Fifth horizontal line: mutations (*AP*) arising under the action of AP in bacteria B-96.[3]

Mutants for which no recombination has been found are stacked at the same site of the drawing, all others are horizontally displaced.   This map is identical with the one given in an earlier publi-

mutants.   While most of the spontaneous mutants did not show any increase in revertants with either AP or BD, most of the mutants induced by base analogues could also be induced to revert by one or both of the analogues.   This mutual exclusion must correspond to a fundamental difference in the mutagenic mechanism.

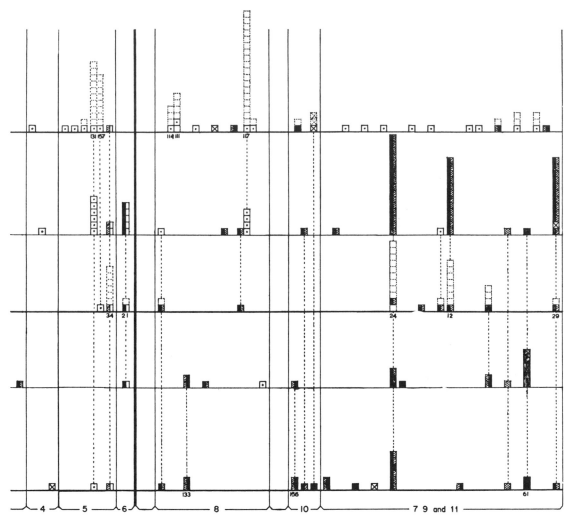

cation[3] but this time the shading of each box indicates the reversion inducibility by AP and BD found in *spot tests:*

⊠ = Nonreverting.

⊡ = Reverts spontaneously, but reversion is not induced by either AP or BD.

▨ = Reverts spontaneously, reversion is strongly inducible by AP and weakly by BD.

▧ = Reverts spontaneously, reversion is strongly inducible by BD and weakly by AP.

■ = Reverts spontaneously, reversion is about equally inducible by AP and BD.

▨] = Reverts spontaneously, reversion is inducible by AP while no BD induction could be observed in spot tests.

⬚ = Reverts spontaneously.   Induction of reversions has not been tested.

Some exceptions to the general rule are found among the spontaneous mutants, showing that some spontaneous mutations are produced by a mechanism akin to that responsible for the induction of mutations by our base analogues.   One of the exceptions is mutant *r* 215 which was included in Table 1 of liquid culture tests.

TABLE 2

Summary of the Spot Tests

Reverting Mutants Which Arose in the Presence of	Total Number	Inducible		Noninducible	
		Number	%	Number	%
BD + 4× thymidine	100	80	80	20	20
BU	64	61	95	3	5
AP	98	96	98	2	2
Spontaneously	110	15	14	95	86
Proflavine	55	1	2	54	98

All untested spontaneous, proflavine, and BU induced mutants are here included which have the same genetic location and spontaneous reversion index as a mutant tested for reversion induction. They are assumed to be reversion inducible or noninducible according to the mutant tested.

A small number of the N- and AP-mutants, which arose in the presence of BU (or BD) and AP, respectively, could not be induced to revert by either AP or BD. All of them may belong to the small background of spontaneous mutants among the induced ones. This becomes especially clear when those N-mutants are inspected which arose under the action of 50 μg/ml BD and 200 μg/ml thymidine (second horizontal line in Fig. 1). In this case the comparison of the mutagenic effect with and without BD had shown[3] that 10 to 20 per cent of these mutants must be of spontaneous origin. Several of these mutants had been found, indeed, to coincide genetically with some of the spontaneous hot spots; this agrees entirely with the induction test for revertants, since all these mutants could not be induced to revert, and therefore behave like spontaneous mutants even in this respect. (This observation also shows that probably most or all nontested mutants of a particular genetic location behave with respect to reversion induction, like the one mutant tested provided they have the same spontaneous reversion index.)

No mutant could be induced to revert which cannot revert spontaneously.

Mutants induced by BU tend to be much more inducible to revert by AP than by BD, and mutants induced by AP tend to be more inducible by BD than by AP.

Table 2 contains also the results of spot tests for proflavine induced mutants (mutants of Brenner, Benzer, and Barnett[4]). Forty genetically different mutants were tested and only one, a single occurrence, was inducible to revert by AP and BD.

Further Examination of Revertants.—A plaque appearing when phages of an rII-type stock are plated with bacteria K can be due to four different reasons:

(1) A "back mutation" to the standard genotype of the rII region.

(2) A "partial reversion" that is a mutation within the rII region to *another*, fully or partially functional, genotype.

(3) A "suppressor mutation" at another than the rII locus.

(4) Leakiness, i.e., ability of the mutants to grow slightly on K so that there is a chance for a plaque to form. Such leaky mutants have not been examined here.

We want to show that most of the rII-type mutants can revert only by back mutations; some can, in addition, undergo a partial reversion, while suppressor mutations do not occur.

As a first examination, several plaques on K are picked and replated on B and K. 55 mutants were examined in this way (21 of them noninducible, 34 inducible; most belonging to different genetic sites). Of the inducible mutants, both spontaneous and induced revertants were analyzed. The results are given in Table 3. It is apparent that *most mutants* give rise to revertants of standard phenotype

TABLE 3

SUMMARY OF THE EXAMINATION OF SPONTANEOUS AND INDUCED REVERTANTS WITH RESPECT TO PLAQUE TYPE ON BACTERIA B AND K

Mutants Tested	Spontaneous "Revertants" on K	Isolated and Replated			Reversions Induced by and Revertants Replated					
					AP			BD		
		No.	On B	On K	No.	On B	On K	No.	On B	On K
21 Base analogue	20 w	16	w	w	..	..	..	..	..	..
noninducible		4	r, w	w						
(17r, 4N)	1 mixed	=	r, w	ti, w	..	..	..	..	..	..
34 Base analogue	26w	25	w	w	24	w	w	20	w	w
inducible		1	r	w	=	r	w	=	r	w
(8r, 14N, 12AP) 5 {w and		=	w	w	5	w	w	3	w	w
{ti		=	r, ti	ti						
3 mixed		=	r, w	ti, w	{1	w	w	..	..	..
					{1	r, w	w, ti	1	r, w	w, ti

The mutants are divided into base analogue noninducible and inducible ones (two horizontal parts). For the inducible mutants the result of spontaneous, AP, and BD induced revertants are given separately; they show that both spontaneous and induced revertants are of the same phenotype. Plaques on K are called "mixed" when their sizes vary between tiny (ti) and wild (w) and a clear distinction of two types was not possible. In contrast, for 5 base analogue inducible mutants such a distinction was possible and only revertants of the wild type were found inducible (this class includes mutants of site N 34).

with respect to plating on both B and K. Induction increases the proportion of these revertants and does not produce any new kind.

Sometimes even a back mutation may not restore phages of the standard phenotype, if the rII-type mutant was in reality a double mutant. These cases are rare, since none of our mutants carry two observable mutations in the rII region or an additional mutation at the rI locus; undetected double mutants may be of the type rII rIII or rII minute.

A small number of induced mutants (e.g., mutants at genetic site N 34) yield some wild type plaques and about 100 times more tiny plaques on K. Isolating and replating the content of the wild type plaques gives again wild type plaques on B and K, while the content of tiny plaques gives r or tiny plaques on B and tiny plaques on K.

Also some other mutants seem to give rise to the two kinds of plaques on K. But they vary more in size between tiny and wild type, and cannot be easily classified. This includes the mutants of hot spot AP 12 and AP 133, which are contained in Table 3; 5 more mutants of site AP 12 were tested with the same result. The revertants of one plaque either produce all wild type plaques on both B and K, or all r-(or large wild) type plaques on B and wild type (or tiny) plaques on K. The partial reversion may in this case be nearly as efficient (on K but not on B) as the standard genotype.

In order to decide whether revertants are due to mutations in the rII region or to suppressor mutations, 20 revertants were backcrossed to standard type phages (all of them had standard phenotype, 5 arose spontaneously, 6 induced by BD, and 9 induced by AP). In all cases, the frequency of r-type phages in the progeny of a cross (mutiplicity 4 each) was not larger than that expected from the frequency of spontaneous r mutants present in the stocks used. The numbers were such that the reverse mutation must have occurred within the rII region, less than 0.2 per cent recombination units distant from the original mutation. (NOTE. The extent of the rII region is about 8 per cent recombination units.)

Suppressor mutations are very unlikely also in all the other cases; for, several rII mutants have never been observed to revert (no revertant in at least 10^{10} phages), spontaneously or by induction; genetic evidence shows that in some of these mutants the mutation cannot extend over more than a very short region of the genome.

A last argument comes from our results about induction of reversion. Suppressor mutations being excluded, we expect that each rII-type mutant which can revert at all can "back mutate," spontaneously or at least under some of the mutagenic treatments.   Whether it can also partially revert by another mutation depends on the particular site and mutation in question.   If these partial reversions at other sites, or at the same site by another mutation, could occur for most, or all, *r*II-type mutants, it would be difficult to understand why these mutations cannot be induced by base analogues for many of the spontaneous and proflavine induced mutants. (Note that the noninducible mutants do not constitute only one class of higher specificity since they comprise spontaneously high and low reverting mutants, hot spot and nonhot-spot mutants.)

We therefore conclude that most *r*II-type mutants revert, in detectable frequency, only by back mutations.

DISCUSSION

The difference between reverting mutants inducible and noninducible by our base analogues is so drastic that it seems to reveal two kinds of fundamentally different mutagenic effects: (1) a mutagenic effect by which our base analogues induce forward mutations and their reversions, and (2) a mutagenic effect which is responsible for the formation and reversion of those mutants which cannot be induced to revert by our base analogues.   Proflavine induces mutations of the second kind.   Since about 10 per cent of the spontaneous mutants can be induced to revert and about 90 per cent cannot, we assume that 10 per cent of the spontaneous mutations are of the first kind, and about 90 per cent of the second.   Both base analogue inducible and noninducible mutants can therefore revert spontaneously. The existence of these two spontaneous mutagenic effects is consistent with the properties of mutant *N* 34 and others of its kind (Table 1).

The two different mutagenic effects must correspond to two fundamentally different changes in DNA.   It is possible to arrive at a simple molecular explanation of these changes if we assume that each *new* chain of replicating DNA is formed along a preexisting complementary chain (e.g., according to the scheme of Watson and Crick[6]) and that all mutations with which we deal in this paper are due to mistakes in DNA replications.

(1)   The *mutagenic effect of the first kind*, i.e., the induction of mutations by BU and AP probably arises by mistakes in base pairing when, or after, a base analogue is incorporated into DNA.[3]   Whatever the exact mechanism, a purine would always be replaced by another purine, and a pyrimidine by another pyrimidine.   Hence the only possible transitions a nucleotide pair can finally undergo are

$$\begin{matrix} A \\ T \end{matrix} \longleftrightarrow \begin{matrix} G \\ H \end{matrix} = \text{mutagenic effect of the first kind} \qquad (1)$$

One would expect that for most base analogue induced mutants, the change of just one nucleotide pair is responsible for the mutant phenotype.   This agrees with the fact that all induced mutants are well localized in the genome, and only 4 out of 300 do not revert.

It has been discussed[3] that each base analogue can induce the transitions (1) in both directions (from A–T into G–H and vice versa).   This explains why most base analogue induced mutants can also be induced to revert.   In most cases the

127

only inducible transition seems to be that of a back mutation to the standard genotype.

A given base analogue may produce changes in one direction (1) more often than
in the other. This may explain why BU induced mutants tend to be more reversion
inducible by AP than by BD, and vice versa for the AP induced mutants. The
preferred mutagenic direction of each base analogue is still unknown.

(2) The *mutagenic effect of the second kind* must involve molecular transitions
different from (1). Even the possibility of two or more neighbouring nucleotide
pairs being changed by the mechanism (1) does not seem to explain the data, since
base analogue noninducible mutants readily revert spontaneously. One would have
to assume di- or poly-nucleotides in the medium inducing both forward mutations
and reversions.

A simple explanation, however, can be obtained when one assumes that the mutagenic effect of the second kind involves the "transversion" of a nucleotide pair, in
which a purine is replaced by a pyrimidine and vice versa. Each (revertable)
mutation would correspond to the change of one nucleotide pair only and the possible final changes are sketched in (2).

$$
\begin{array}{cc}
A & T \\
T \xleftrightarrow{} A \\
\updownarrow \quad \updownarrow \\
H & G \\
G \xleftrightarrow{} H
\end{array}
\quad = \text{mutagenic effect of the second kind} \qquad (2)
$$

Such a transversion may occur (one at a time) in replicating DNA in a number of
ways. Most plausible for spontaneous mutations is the mistake incorporation of a
purine in place of a pyrimidine or vice versa. Figure 2 shows schematically how
such a mistake leads to the transversion of a nucleotide pair which subsequently
replicates as such, indefinitely. The assumed purine-purine or pyrimidine-
pyrimidine base pairs do not exactly fit into the Watson-Crick structure of
DNA, and their presence should cause a minute distortion of the double helix. But
mutational changes are very rare, and both the pairing by one or two hydrogen
bonds and the rare incorporation of noncomplementary bases are structurally
feasible.

It is not possible at present to decide which of the conceivable base pairings are
most frequently incorporated, although the pairs A–G and T–H seem to have a
higher chance, for spontaneous mutations, than pairings of identical bases. In
some cases, incorporation of the wrong base may bend the end of the growing chain
sufficiently off the growing direction so that further replication is difficult or impossible. This in turn may invite further mistakes leading to nonreverting mutants
and larger alterations observed for some spontaneous mutants.

This picture of two mutually exclusive mutagenic effects has a number of testable
consequences.

(1) All mutagenic agents which, by any conceivable mechanism, convert one
purine of DNA into another purine, or one pyrimidine into another pyrimidine,
can only induce mutations of the first kind. These mutations can be induced to
revert by the same or, sometimes more efficiently, by another agent of this class.

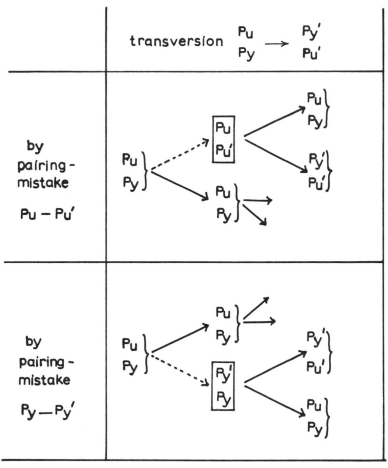

FIG. 2.—The sequence of events which may lead to the spontaneous "transversion" of a nucleotide pair. During DNA replication rarely a purine may pair with a purine- or a pyrimidine with a pyrimidine- (DNA precursor), and the wrong base pair may get incorporated into DNA. In the next DNA replication each base attaches to it its normal complementary base and thereby finalizes the purine-pyrimidine exchange. Pu = purine, Py = pyrimidine.

(2) The incorporation of a base analogue into DNA should not be able to increase the frequency of transversions, since apparently no base analogue would make the necessary pairing mistakes more often than any of the normal bases. Certain pairing mistakes of the second kind may, however, be increased, e.g. by (a) a large unbalance of the nucleotide ratio at the direct DNA precursor level (e.g., by thymine starvation); (b) agents which separate hydrogen bonds of DNA (e.g., higher temperature) or which otherwise tend to pull or keep DNA chains apart; (c) agents which attach (e.g., at the bases) to one nucleotide in DNA and one non-complementary nucleotide in the medium and provide a bridge of the right distance between their sugar-phosphates to enable incorporation.

An agent of kind (b) or (c) may be proflavine (= 3,6- diamino-acridine) which carries two amino groups in about 10 Å distance. One cannot decide, at present, which of the possible attachments to nucleotides are responsible for its mutagenic effect.   But this effect, apparently, is of the second kind, since proflavine induced mutants could not be induced to revert by our base analogues, although they revert spontaneously.

(3)   For other organisms the same two kinds of mutagenic effects should exist. Their relative rate among the "spontaneous" mutations cannot be predicted, since it depends on the natural formation of mutagens in different organisms.

(4)   The results are of obvious importance for the coding of proteins by DNA. It should be possible to group all amino acids into different classes, so that mutations of one kind can only *intra*convert amino acids of the same class, while *inter*conversions of different classes are only possible by mutations of the other kind.

I wish to thank Dr. J. D. Watson for his generous support, and both him and Mr. D. Schlesinger for helpful discussions.

* Research Fellow of the Damon Runyon Memorial Fund for Cancer Research.

[1] Abbreviations: 2-aminopurine = AP, 5-bromouracil = BU, 5-bromodeoxyuridine = BD, adenine = A, guanine = G, thymine = T 5-hydroxymethyl cytosine = H.

[2] Benzer, S., and E. Freese, these PROCEEDINGS, **44,** 112 (1958).

[3] Freese, E., *J. Mol. Biol.,* 1959 (in press).

[4] Brenner, S., S. Benzer, and L. Barnett, *Nature,* **182,** 983 (1958).

[5] Iyer, V. N., and W. Szybalski, *J. Applied Microbiology,* **6,** 24 (1958).

[6] Watson, J. D., and F. H. Crick, *Cold Spring Harbor Symposia Quant. Biol.,* **18,** 23 (1953).

Editors' Comments
on Papers 14 Through 16

14   FREESE et al.
      The Chemical and Mutagenic Specificity of Hydroxylamine

15   PHILLIPS et al.
      The Efficiency of Induction of Mutations by Hydroxylamine

16   TESSMAN et al.
      Mutagenic Effects of Hydroxylamine in vivo

HYDROXYLAMINE MUTAGENESIS

Until the advent of hydroxylamine, even the most specific of the new chemical mutagens, nitrous acid and the base analogues, were not completely specific; they induced *both* transitions. What was needed was a unidirectional mutagen, for it alone would allow one to distinguish between the two possible transitions. A search for such an agent led the Freeses to hydroxylamine, and their initial successes were reported in Paper 14. (Further evidence for the high genetic specificity of hydroxylamine is to be found in Paper 17.) Although numerous other mutagens of good specificity have accumulated in the succeeding years, only hydroxylamine and bisulfite (Summers and Drake, 1971) are completely base specific (at least when applied to bacteriophages or to naked DNA), inducing only $G:C \rightarrow A:T$ transitions (or $C \rightarrow T$ in the case of $\phi X174$). As such, hydroxylamine has remained an indispensable tool in the analysis of mutagenic specificities, and is referred to repeatedly in succeeding papers.

The impressive successes achieved with simple chemical mutagens applied to bacteriophages soon stimulated efforts by biochemists to emulate these systems *in vitro*. The first such study to be reported involved 5-bromouracil-induced misincorporation in a synthetic system employing only a simple DNA template of defined sequence, DNA polymerase, and appropriate nucleotide triphosphates (Trautner et al., 1962). Although the results were less simple than had been anticipated (and were, in fact, largely artifactual), the approach employed was to become quite successful (see, for instance, Hall and Lehman, 1968, and Paper 37). The first application of this type of analysis to the mutagenicity of hydroxylamine is described in Paper 15. (This study actually employed methoxyamine, a close analogue of hydroxylamine, and an RNA-synthesizing rather than a DNA-synthesizing system.) In this paper, Phillips, Brown, and Grossman showed that the chemical interaction of hydroxylamine with cytosine produces an *in situ* analogue of thymine which (mis)pairs with adenine with a very high efficiency.

The third paper on hydroxylamine mutagenicity originated from the Tessman group and nicely illustrates the dangers that can arise when simple mutagens first characterized in viruses are extended to cellular systems. The mutagenic target in a virus is almost necessarily the viral DNA, whereas the primary mutagenic target in a cellular system can be much more indirect; it could, for instance, be a nucleotide triphosphate precursor of DNA, or even an enzyme of DNA synthesis such as the DNA polymerase itself. (Further examples of indirectly acting mutagens are discussed in the Editors' Comments on Papers 27 Through 29.) In Paper 16, Tessman, Ishiwa, and Kumar showed that the high mutagenic specificity of hydroxylamine mutagenesis is largely lost when infected cells, rather than free phage particles, are treated. Furthermore, simply pretreating the cells with hydroxylamine and then removing at least the residual extracellular mutagen produced mutants in the phage particles that later infected the treated cells.

REFERENCES

Hall, Z. W., and I. R. Lehman. 1968. An *in vitro* transversion by a mutationally altered T4-induced DNA polymerase. *J. Mol. Biol., 36:* 321.

Summers, G. A., and J. W. Drake. 1971. Bisulfite mutagenesis in bacteriophage T4. *Genetics, 68:* 603.

Trautner, T. A., M. N. Swartz, and A. Kornberg. 1962. Enzymatic synthesis of deoxyribonucleic acid, 10. Influence of bromouracil substitutions on replication. *Proc. Natl. Acad. Sci., 48:* 449.

14

Reprinted from *Proc. Natl. Acad. Sci.*, 47(6), 845–855 (1961)

THE CHEMICAL AND MUTAGENIC SPECIFICITY OF HYDROXYLAMINE*

By Ernst Freese, Ekkehard Bautz, and Elisabeth Bautz Freese

DEPARTMENT OF GENETICS, UNIVERSITY OF WISCONSIN

Communicated by M. R. Irwin, March 21, 1961

The hereditary changes that convert one base pair (or base) of nucleic acid into another one have been divided into two classes—transitions and transversions.[1] This paper is concerned with the further subdivision of mutagenic transitions. For this purpose, the specific mutagen hydroxylamine[2] (HA) has been employed, which, for the normal DNA bases, reacts with cytosine (C) and 5-hydroxymethyl-cytosine (HMC) but not or only little with 5-methylcytosine (MeC) or thymine (T). The reaction with C has been examined in detail, since it is apparently responsible for the mutagenic effect of HA. Other bases, e.g., 5-bromouracil (BU), also react strongly with HA, but for BU–containing phages this effect is mainly lethal. In contrast to HA, the chemically similar agent hydrazine reacts much more with T than with C and is much less mutagenic.

The induction of reverse mutations by HA shows a clear bipartition into highly and little inducible phage T4 *r*II mutants; this indicates which base pair is present at the mutant site.

Abbreviations used: A = adenine, HMC = 5-hydroxymethylcytosine, C = cytosine, G = guanine, T = thymine. AP = 2-aminopurine, BU = 5-bromo-uracil, EES = ethylethane sulfonate, HA = hydroxylamine. CR = cytosine riboside, CdR = cytosine deoxyriboside, etc. dHMP = 5-hydroxymethylcytosine-deoxyribotide. G-C = guanine-cytosine pair in DNA, where C stands for any one of the cytosines.

Material and Methods.—The phage T4 *r*II mutants, bacteriological media, and genetic methods have been described previously.[2–5]

Chemicals: dHMP and glucose-dHMP (of phage T2) were gifts of Dr. Lehman, Stanford University. 2,4-diaminopyrimidine was a gift of Dr. Bendich, Sloan Kettering Institute. All other agents are commercially available.

Mutation induction by hydroxylamine (HA): Method used throughout this paper.

(1) *Reaction mixture:* Reagents: A. 2.5 M $NH_2OH \cdot HCl$ plus 2.0 M NaCl, kept in cold (up to one week); B. 1 M $MgSO_4$; C. 1.5 M Na_2HPO_4; D. 10 N NaOH. Shortly before use, take 10 ml A, 0.02 ml B, 1 ml C; add D (*ca.* 2 ml) until pH is nearly 7.5; fill to 20 ml volume with distilled water; add more D, cautiously, until pH is 7.5.

(2) *Dilution broth:* Bactotryptone, 5 gm; NaCl, 60 gm; H_2O, 1 liter. The high salt concentration is useful, since without salt even low concentrations of HA inactivate phages very rapidly.[2]

(3) *Stopping mixture:* Dilution broth plus 2 ml acetone per 100 ml medium (gives about 0.4 M acetone). This concentration of acetone has no inactivating effect on phages, but it reacts very fast with HA and thereby annihilates it. Nevertheless, it is advisable to plate the phages as soon as possible, since an additional inactivation slowly takes place.

(4) *Reaction procedure:* Take reaction mixture, ice-cold, 2 ml. Add phages, ice-cold, 0.5 ml. Mix for 30 seconds. Take 0.2 ml aliquots into 10 ml ice-cold stopping mixture (= "zero control"). Place remainder in 37°C (= zero time). Take at various times 0.2 ml aliquots into 10 ml ice-cold stopping mixture.

Measurement of the decrease of UV absorption: Reaction mixtures: (a) 0.5 M hydroxylamine $\times$ HCl; 0.04 M Na_2HPO_4; 5×10^{-4} M $MgSO_4$; pH 7.5. (b) 0.5 M hydrazine $\times$ 2 HCl; 10^{-3} M $MgSO_4$; pH 8.5. In most cases, *ca.* 15 μl of a 20 μM/ml solution of the nucleosides or nucleotides were added to 2 ml of reagent mixture such that an OD_{276} of just below 1 was obtained. The mixture was kept at 37°C in glass-stoppered cuvettes (silica greased) and the OD_{276} read against both H_2O and the reagent mixture alone, which was incubated in the same way. We used 0.5 M HA and hydrazine, since the 1 M concentrations already have an appreciable UV absorption, which significantly decreases during the time of measurement. This self-reaction is accompanied by the production of gas bubbles; more of these bubbles are observed in the presence of one of the bases, indicating their catalytic effect, and therefore, exact measurements of the alteration of the bases themselves become more difficult.

Preparation of BU-containing phages: Both methods of Litman and Pardee[6] and of Freese[4] were used.

Column chromatography: In order to keep the UV absorption of the eluent low, we eluted with dilute H_2SO_4 or 0.05 M $NaHPO_4$, at various pH's, off Dowex 50 and with dilute ammonia or 0.05 M NH_4 chloride off Dowex 1. Length of column was 10 cm.

Reduction by titanium chloride, $TiCl_3$: The eluent was adjusted to pH 5 and excess phosphate (0.2 M). 1/100 volume of a 20% $TiCl_3$ solution was added at room temperature and the precipitate removed. The filtrate was applied to a Dowex 50 column at pH 2.

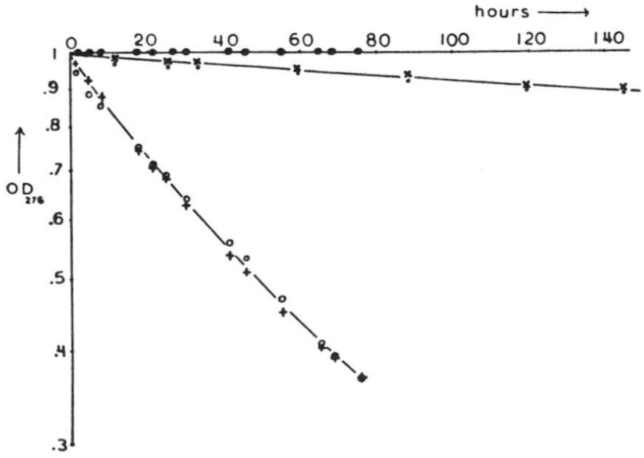

FIG. 1.—Decrease of UV absorption at 276 mμ of various nucleotides in 0.5 M HA at pH 7.5. ● = dTMP, · = dHMP, $\times$ = dHMP − glucose, + = dCMP, O = UMP.

Results.—The reaction of hydroxylamine with the pyrimidine bases: HA reacts with certain pyrimidines[2] and causes the decrease of UV absorption in the range of 260–280 mμ; this is shown in Figure 1 for UMP, dCMP, dHMP, and dHMP-glucose. No reaction has been observed for dTMP, MeC, pseudouridine, or the purines. The thymine analogs BU or BUdR, however, react even faster than the corresponding uracil compounds. Electron-donating (e.g., methyl) groups seem to reduce and electron-attracting groups (e.g., Br) seem to increase the reactivity with HA. This reactivity also increases with the pH.

The HA reaction with U and BU or their ribosides leads to the apparently irreversible loss of the UV absorption peak in the 260–280 mμ range. BU is readily debrominated, as can be observed by the heavy precipitate of AgBr formed when nitric acid and silver nitrate are added to the reaction mixture.

In contrast, the decrease of the UV absorption after HA reaction with C can be almost completely reversed by exposure of the reaction mixture to low pH. Figure 2 illustrates how the UV

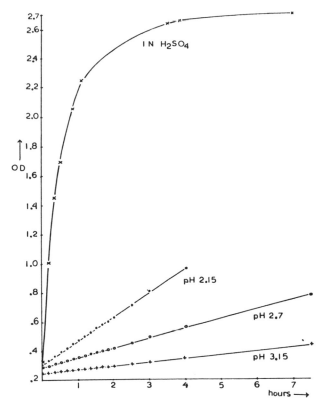

Fig. 2.—Increase of UV absorption at 276 mμ of the pH 7.5 reaction product of C + HA when acidified by H_2SO_4.

absorption returns to nearly the same absorbency that was originally observed at low pH by a first-order kinetics and with a rate depending on the pH. Such a reversal can also be observed for CdR but the absorbency does not return completely, even after long times of treatment at pH's 0–1.

When the reaction mixture of C (compound I) with HA is carefully acidified to pH 3 and immediately applied to a Dowex 50 column, the main reaction product (II) elutes at once and can be isolated when care is taken that the pH never drops much lower and that the eluted material is immediately neutralized. Figure 3a shows the spectrum of this compound II. Upon acidification, this material recovers a UV absorption peak at about 276 mμ with a first-order kinetics similar to that shown in Figure 2. When compound II is reacted for several hours with 1 N HCl and then eluted on a Dowex 50 column (with 0.05 M NaH_2PO_4 at pH 3.5), a new compound III comes off. Although its spectrum, shown in Figure 3b, closely resembles that of C at acid pH, it differs from it at neutral and basic pH. Also its pK value must be lower (about 3) than that of C (pK 4.6), which readily elutes only when the pH is above 5. Mild reduction of compound III by $TiCl_3$ at pH 5 and room temperature gives rise to a major product (IV) that elutes off a Dowex 50 column at the same time as C and indistinguishable from it when the two are chromatographed together. Also the spectrum of this compound IV (Fig. 3c) is the same as that of C.

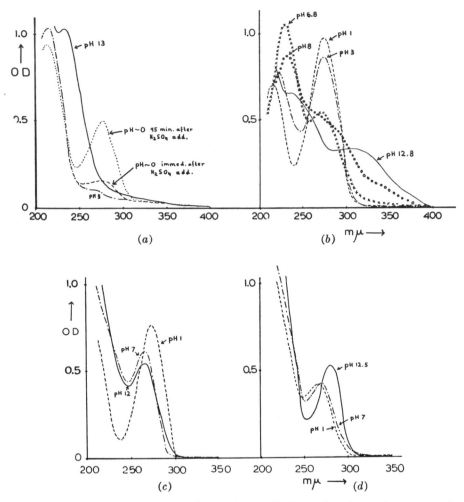

FIG. 3.—UV spectra of various reaction products of C after purification on Dowex 50 cation exchanger.
(a) Main product of reaction of C with 1 M HA at pH 7.5. = Compound II in Figure 4.
(b) Compound (III) obtained (from II) by acidification (pH 0).
(c) Compound (IV = I) obtained from the acidified product (III) by reaction with TiCl₃ at pH 3. Spectrum is identical with that of C.
(d) Spectrum of 2,4-diaminopyrimidine.

One might have expected that HA would predominantly attack the keto group of C and that compound III would be the ketoxime. Our results, however, exclude this possibility as the major reaction, since the reduction of this 2-oximo-4-amino-pyrimidine should result in 2,4-diamino-pyrimidine, which we do not obtain. 2,4-diamino-pyrimidine has absorption spectra shown in Figure 3d and elutes from a Dowex 50 column only at very high pH; it is clearly different from compound IV. Hence, we conclude that compound IV is actually C and that the major reaction products of the chemical reactions analyzed above are those given in Figure 4. At present we cannot exclude the alternative possibility that the initial attack of HA breaks the pyrimidine ring between the 3-N and 4-C position, followed by ring closure upon acidification to yield compound III. The initial attack of HA is probably the same as that of hydrazin, at the 4-carbon of cytosine; for hydrazine, however, the other NH_2 group reacts with the 6-carbon and thus 3-aminopyrazole[7] is produced.

Compound III is quite stable in 2 N H_2SO_4 at 100°C, but treatment with 0.5 N NaOH for $^1/_2$ hour at 80°C diminishes the UV absorption (partial destruction) and alters it towards that of U. The reaction mixture was applied to a Dowex-1-formate column, and a compound was eluted when the same pH (9.0) was reached at which normally U comes off. Its spectra, at pH's 0, 9, and 14, were also those of uracil. We therefore conclude that high pH easily hydrolyzes the =NOH group, at the 4-position of compound III, resulting in a Keto group, i.e. giving uracil.

Similar to C, CdR also produces with HA a compound (II) which is immediately eluted off Dowex 50 at pH 3 and upon acidification is converted into another compound (III) with a spectrum similar to that obtained from C (see Fig. 5). Reaction of III at high pH apparently produces UdR. Prolonged reaction of CdR with HA gives rise to some additional compound which elutes only at very high pH and has again a different spectrum.

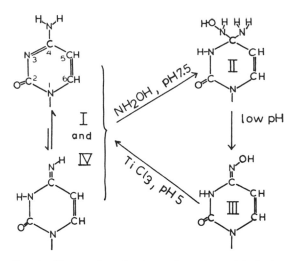

FIG. 4.—Proposed mechanisms of reactions with C, described in the text.

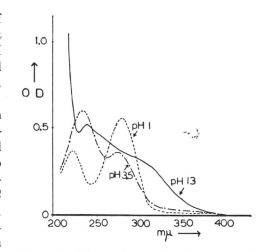

FIG. 5.—UV spectrum of the acidified reaction product (III) of CdR with HA.

The reaction of hydrazin with the pyrimidine bases: Concentrated hydrazin readily reacts with the pyrimidine nucleotides CMP and UMP to give pyrazole derivatives, urea, and ribose phosphate.[7] The reaction with RNA[8] and DNA[9] produces "apyrimidinic acid." We have examined the decrease of the UV absorption of various nucleotides caused by the reaction with hydrazin under milder conditions (0.5 M, pH 8.5) (see Fig. 6). In this case, the reactivity of dCMP is negligible compared to that of dTMP, opposite to the effect of HA.

The mutagenic effect of hydroxylamine: We have shown previously[2] that HA exerts two chemically different effects on phage T4 treated *in vitro*: a rapid action which is only lethal and not mutagenic and a slow but highly mutagenic effect. The nonmutagenic effect can be almost completely suppressed by high concentra-

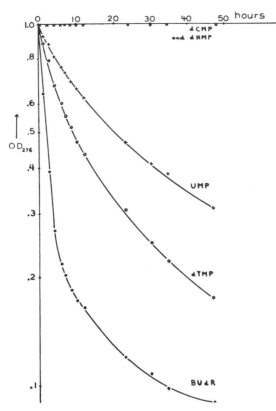

FIG. 6.—Decrease of UV absorption at 276 mμ of various nucleotides in 0.5 M hydrazine at pH 8.5.

tions of both NaCl and HA. The induction of forward mutations under such conditions has been examined already.[2]

The induction of reverse mutations is illustrated in Figure 7 for a highly inducible mutant. Reversion induction with HA apparently is a one-hit phenomenon which follows the equation

$$\frac{K}{B} = \frac{K_0}{B_0} + \alpha t = \frac{K_0}{B_0} + \frac{\alpha}{\beta} n.$$

(K or B are the titers of phages on bacteria K or B. K_0 or B_0 are these titers at zero-time treatment. α is the probability of reversion induction per unit time and phage, while α/β is this probability per lethal hit n ($= \ln B_0/B$) and phage.)

Curves of this kind have been examined in more detail in a previous paper[5] for the case of nitrous acid. The mutagenic effect of HA, however, is more pronounced since, owing to HA treatment, the concentration of phages (per ml) that produce plaques on K may increase by as much as a factor of 100.

HA exhibits such a strong effect on BU, as compared to T, that it seemed worth while to examine its action on T4 phages which had been grown in the presence of BUdR. Figure 8 shows that these phages are inactivated by HA much more rapidly than those not containing BU. This is similar to the effect of UV on such phages.[10] The inactivation slope in Figure 8 later levels off, probably because about 5 per cent of the phages in our stock contained little or no BU. The frequency of r mutants per viable phage also decreased during the initial part of the HA treatment, i.e. r-mutant phages died faster than certain nonmutant ones, and new HA-induced mutants could significantly increase the r-mutant frequency only after longer HA treatment. This proves both that the rapidly killed BU-phages contained more r mutants than the surviving non-BU-phages and that the reaction of HA with BU in DNA is lethal but little or not at all mutagenic. The same was observed when a BU stock of the rII-type mutant N 24 was treated by HA: no increase in the frequency of revertants per viable phage could be observed during the initial rapid inactivation.

The effect of hydrazin: $N_2H_4 \times 2$ HCl acts similarly to $NH_2OH \times$ HCl in that it inactivates phages faster at 0.1 M than at 1 M concentrations and the effect increases with pH. However, the mutagenic effect of hydrazin on phage T4 at pH

8.3 is much smaller than that of HA. The frequency of forward (r) mutations barely increases, and when the rII mutants AP 275, N29, and N17 were tested for reversion induction with 1 M hydrazin, the respective α/β values were 3, 0.01, and 0.005 in units 10^{-6}, i.e. 10 to 100 times smaller than with HA. Although we cannot exclude the possibility that hydrazin inactivates phages mainly by affecting their protein, the more likely explanation of its small mutagenic effect seems to be that it preferentially attacks thymine in DNA with a predominantly lethal effect.

The specificity of reversion induction: The induction of reverse mutations by HA was examined quantitatively for a number of rII-type mutants which had been induced by different means. For each mutant, the inactivation curve and the induction curve (see Fig. 7b) was determined by at least 5 points. The approximate α/β values obtained in this way are summarized in Table 1. It can be seen that certain mutants, induced by 2-aminopurine, display a much larger reversion inducibility by HA than others. (It should be mentioned that for mutants which are highly reversion-inducible by HA the K_0/B_0 values given in Table 1 are slightly higher than the values in the phage stock, since even the short exposure to HA in the cold exerts a slight mutagenic effect.)

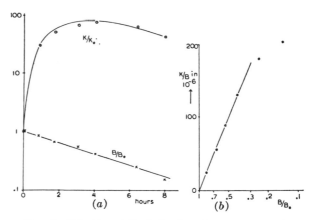

Fig. 7.—Kinetics of the induction of reverse mutations for the highly reversion inducible T4 rII mutant AP 72.
(a) Titer of viable phages (B/B_0) and titer of revertants (K/K_0) plotted logarithmically against the time of HA treatment.
(b) Frequency of revertants per viable phage (K/B) plotted linearly against the logarithm of the phage titer (B/B_0). For longer treatment, the curve deviates from the straight line, since the relative proportion of mutagenic versus lethal hits decreases.[2]

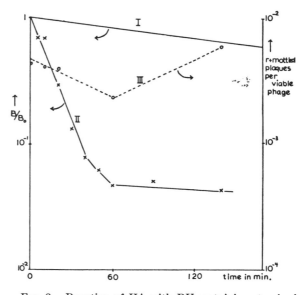

Fig. 8.—Reaction of HA with BU-containing standard type phage T4.
Curve I: Survival of phages not containing BU.
Curve II: Survival of BU-containing phages.
Curve III: Change in the frequency of rII mutants per viable phage for BU-containing standard type phages.

When in previous work the reversion induction by nitrous acid[5] had been analyzed, it was observed that certain mutants could not fully express all of their

TABLE 1
Rates of Reversion Induction by Hydroxylamine*

Mutant	K_0/B_0†	α/β	K_0'/B_0	α'/β
AP 114	4.4	450	70	1040
AP 275	1.2	280	6	240
AP 72	0.7	134	13	158
AP 12	0.7	78	13	194
AP 41	0.3	62	5	265
N 24	0.3	0.2	30	10
N 29	0.05	0.1	5	1
HA 54	0.08	0.1	3	3
AP 61	0.13	0.08	21	18
N 19	0.04	0.06	5	6
HA 45	0.01	0.06	1	4
N 7	0.02	0.04	7	2
HA 11	0.02	0.04	7	8
N 12	0.04	0.03	6	2
N 17	0.02	0.03	—	—
N 31	0.02	0.03	3	6
HA 25	0.003	0.001	0.5	0.1
AP 156	0.008	0.002	5	5
AP 70	0.04	0.01	10	6
AP 83	0.006	0.005	—	—

* All values are given in units 10^{-6}.
† K_0/B_0 = Frequency of revertants per viable phage at zero time treatment. K_0'/B_0 = Same, but with preadsorption on bacteria B before plating on K. α/β = Frequency of induced revertants per viable phage and per lethal hit. α'/β = Same, but with preadsorption.

induced revertants when plated directly on bacteria K, but they could do so when they were first allowed to multiply in (UV-killed) bacteria B, which then were plated on K. Therefore, again the HA reversion inducibility of all mutants was also examined by preadsorbing the HA-treated phages on bacteria B. The results are given again in Table 1, in the last two columns. It is apparent that for the five highly reversion-inducible mutants the α'/β values obtained with preadsorption are not much larger than the α/β values without preadsorption. For the mutants of low inducibility, the background of spontaneous revertants (K_0'), after preadsorption on B, is so large that we could not measure accurately the α'/β value but can give only an upper limit. It is clear, however, that even with preadsorption none of the other mutants has a high enough α'/β value to be a candidate for the class of highly reversion-inducible mutants.

Some further examination of induced revertants: For a chemical interpretation of the observed mutagenic specificity, it is important to know whether induced revertants arise by genuine backmutations to the standard genotype or whether they are due to suppressor mutations at another site. It has been mentioned previously[1] that no *unspecific* suppressor locus exists such that *each* rII mutant could revert by a mutation at this locus. *Specific* suppressor mutations, however, do exist for certain rII mutants. They may give rise to phages with different phenotypic properties from the standard type, or they may be detectable only by more refined techniques, i.e., by recombination or mutation tests. E.g., several rII mutants produce revertants giving tiny plaques on bacteria K in addition to the normal revertants with wild type plaques. These tiny revertants have not been counted at all in the preceding analysis.

By purely genetic means we cannot prove unequivocally that a revertant arose by a genuine backmutation and does not have a genotype different from that of the standard type phages. But certain results make it at least likely that most revertants of *r*II mutants that produce wild type plaques on both bacteria K and B correspond to genuine backmutations.

The functional property of revertants obtained after HA treatment of phages was examined by replating isolated revertants on both bacteria B and K.   Ten to 20 revertants were analyzed in this way for most of the *r*II mutants mentioned in Table 1.   Most of them gave wild type plaques on both bacteria.   Exceptions were some revertants of AP 114 and all of AP 70 which produce large wild type plaques on B and all induced revertants of AP 12 (or AP 28, which is at the same AP hot spot as AP 12) which gave *r* plaques on B.   These exceptions may be due either to a suppressor mutation or to a second mutation present already in some or all phages of the stock.   For AP 114, crosses between such revertants and standard type phages gave only the same two types (large wild and wild) and not the original *r* mutant, showing that both explanations are still possible.   These exceptions have not been investigated further.

Twenty-three wild type revertants which originated from HA induction of AP 41, 72, 114, and 275 and showed wild type plaques on both bacteria K and B were backcrossed to standard type phages.   None of the lysates of a cross showed any significant increase in the frequency of *r*-type plaques above the frequency expected from the background in the stocks.   Hence, no evidence for suppressor mutations was observed for these revertants.

In addition to the functional and recombination test, the observed mutagenic specificity argues against the possibility that most *r*II mutants can revert to a wild phenotype by suppressor mutations at several sites, for one would expect the mutagenic specificity of a suppressor mutation in simple base-pair exchanges to be, in general, independent of the specificity of the original mutation.   It would then be hard to understand why none of the 5-bromouracil (N)- and HA-induced mutants was as highly inducible to revert as some of the AP mutants.

We conclude that most induced revertants which produce plaques of standard phenotype on both bacteria B and K probably constitute genuine backmutations to the standard genotype.

Discussion.—Both hydroxylamine (HA) and hydrazine react with pyrimidines and not with purines, inverse to the effect of treatment by low pH or ethylating agents.   They are very specific chemicals, since they differ in their action on U, T, and C, and HA even separates UR and the various derivatives of C from pseudouridine.   HA can be employed to mutate preferentially certain bases or base pairs.

The main mutagenic effect of HA is apparently caused by its reaction with C. Although some reaction with T could also be expected, it must be slow, since so far it has not been detected chemically.   The reaction product with C is apparently compound II in Figure 4*b*.   The attachment of HA to C opens the resonance structure of the pyrimidine ring and destroys the normal pyrimidine absorption spectrum.   Compound II is stable for several days at neutral pH but on acidification is converted into compound III.   We do not know whether or not this conversion occurs already at neutral pH when the HA-treated phage DNA enters the bacterium.   But whichever compound may be present, the HA-reacted C should no longer be able to pair with G since the 3-position hydrogen interferes sterically with the hydrogen atom on G.   However, this compound can make at least one hydrogen bond with A and thus cause its incorporation in the place of G.   HA is small and far enough out of the pyrimidine plane that it should not sterically hinder the pairing between A and compound II. Whatever the detailed chemical events may be it seems clear that HA predominantly induces the base pair transitions

$$\begin{array}{cc} C & T \\ G & A \end{array} \quad (1)$$

Corresponding to this chemical directional effect, we found that transition mutants fall into two classes with respect to the induction of reverse mutations by HA. Five mutants that had been induced by AP are highly inducible by HA to revert. Most revertants of four of these mutants could not be distinguished from standard type phages by functional or recombinational tests. It seems likely, therefore, that these mutants carry a G-HMC pair at their mutant site. The fifth mutant, AP 12, yields only revertants with *r* plaques on bacteria B; it is still open whether these revertants arise by a suppressor mutation or whether they already contain a second *r* mutation at some other site. These five AP mutants are also especially reversion-inducible by nitrous acid (after preadsorption on B),[5] by ethylethane sulfonate,[11] and by low pH treatment.[12] Thus, both the chemical and the mutagenic observations agree that all these agents induce transitions preferentially in the direction shown in equation (1). This directional effect is the same irrespective of whether G or C is attacked by the agent.

The directional effect of HA is much stronger than for any of the other agents, for the difference in induction rates, α/β, between highly and slightly or not inducible mutants is larger than 100. It seems likely, therefore, that these lowly or noninducible mutants have an A-T pair at their mutant site unless few of them are caused by an exceptional DNA change. Three mutants (AP 70, 83, 156) that respond little to HA, nitrous acid, or low pH gave a stronger response to ethylethane sulfonate.[11, 12] This indicates some higher mutagenic specificity of the corresponding DNA sites with respect to the various mutagens and shows that the mutagenic direction can be safely deduced only when mutagens are employed that have an especially high preference for one base.

When HA is compared to other mutagens, its lethal effect is very small in relation to its mutagenic effect. This follows from the large frequency of forward mutations per lethal hit[2] as well as from the large α/β value of about 10^{-4} for the highly reversion-inducible mutants. Since there are about 7×10^4 HMC bases in a T4 phage (assumed to be the same as measured for phage T2[13, 14]), the observed probability, α/β, of reversion induction per lethal hit is about that expected if a reversion is induced by the HA action on a *certain* HMC base (or few of them) while the action on most other HMC bases is lethal. This agrees with our picture that the mutations concerned here are due to the alteration of a single nucleotide pair in DNA. Furthermore, it seems probable that HA exerts its lethal effect on DNA mostly by the same mechanism which at another site is mutagenic. This is in contrast, e.g., to nitrous acid, whose effect on G is not mutagenic but apparently lethal.[15]

It is worth-while to compare the data on reversion induction with those on spontaneous reversion indices,[1] for transition mutants that are highly inducible to revert by HA or the other agents mentioned before also revert spontaneously with especially high frequency. This suggests that, at least in phage T4, spontaneous transitions occur more frequently in the direction of equation (1) than in the opposite direction. Similar observations have been made for spontaneous transversions.[12]

We wish to thank A. Bendich for his valuable criticism.

* Paper No. 826 from the Department of Genetics, University of Wisconsin. This work was supported by grants from the American Cancer Society and the Atomic Energy Commission.

[1] Freese, E., these PROCEEDINGS, **45**, 622 (1959).
[2] Freese, E., E. Bautz Freese, and E. Bautz, *J. Mol. Biol.* (in press).
[3] Benzer, S., and. E. Freese, these PROCEEDINGS, **44**, 112 (1958).
[4] Freese, E., *J. Mol. Biol.*, **1**, 87 (1959).
[5] Freese, E. Bautz, and E. Freese, *Virology*, **13**, 19 (1961).
[6] Litman, R. M., and A. B. Pardee, *Nature*, **178**, 529 (1956).
[7] Baron, F., and D. M. Brown, *J. Chem. Soc.*, Aug., 2855 (1955).
[8] Takemura, S., *J. Biochemistry (Japan)*, **44**, 321 (1957).
[9] Takemura, S., *Bull. Chem. Soc. (Japan)*, **32**, 920 (1959).
[10] Stahl, F. W., J. M. Crasemann, L. O. Kun, E. Fox, and C. Laird, *Virology*, **13**, 98 (1961).
[11] Bautz, E., and E. Freese, these PROCEEDINGS, **46**, 1585 (1960).
[12] Freese, E. Bautz, these PROCEEDINGS, **47**, 540 (1961).
[13] Hershey, A. D., J. Dixon, and M. Chase, *J. Gen. Physiol.*, **36**, 777 (1953).
[14] Wyatt, G. R., and S. S. Cohen, *Biochem. J.*, **55**, 774 (1953).
[15] Vielmetter, W., and H. Schuster, *Biochem. Biophys. Research Commun.*, **2**, 324 (1960).

15

Reprinted from *J. Mol. Biol.*, **21**, 405–419 (1966)

The Efficiency of Induction of Mutations by Hydroxylamine

J. H. Phillips†, D. M. Brown

University Chemical Laboratory, Cambridge, England

AND

L. Grossman

Graduate Department of Biochemistry,
Brandeis University, Waltham, Mass., U.S.A.

(*Received 31 May 1966*)

Treatment of poly C‡ with methoxyamine leads to formation of 5,6-dihydro-6-methoxyaminocytosine residues in the polymer. These direct incorporation of adenylate residues into the poly G synthesized when the modified poly C is used as a template for RNA polymerase. The process is analogous to that of hydroxylamine-induced errors in replication.

Comparison of the uptake of [^{14}C]methoxyamine by the poly C with the percentage of adenylate incorporated in the poly G shows that the induction of errors is a highly efficient process: every modified cytosine residue directs the incorporation of one adenylate residue. A probable explanation for this is that the incorporation is mediated by hydrogen-bonding between the adenine and a predominant imino tautomer of the cytosine adduct.

1. Introduction

We have recently reported experiments which suggest that hydroxylamine-induced mutations in transforming DNA and in cytosine-containing bacteriophages result from base-pairing of an adenylate residue with a 5,6-dihydro-6-hydroxylaminocytosine§ residue in the template nucleic acid (Brown & Phillips, 1965; Phillips, Brown, Adman & Grossman, 1965). We now report further experiments, designed to investigate the probability with which this reaction product is replicated as if it were a uracil or thymine residue.

Lawley & Brookes (1962) suggested that there were two reasonable mechanisms by which unnatural bases, formed in the template by the action of a mutagen, could induce anomalous base-pairing. 7-Alkylguanines and 5-bromouracil can pair with thymine and guanine respectively if, compared with the natural bases, they have a high probability of existing either in an unusual tautomeric form or as anions at the pH of replication. Two equivalent mechanisms may be applied to the case of hydroxylamine mutagenesis: the unnatural base, 5,6-dihydro-6-hydroxylamino-cytosine (I), has a certain probability of existing either as the imino tautomer or as the cation (Fig. 1), and it is conceivable that either of these may form a base-pair

† Present address: Department of Biochemistry, University of Uppsala, Uppsala, Sweden.

‡ Abbreviations used: poly A, polyadenylic acid; poly C, polycytidylic acid; poly G, polyguanylic acid; poly AG, copolymer of adenylic and guanylic acids; TCA, trichloroacetic acid.

§ The pyrimidine ring is numbered so that $N_{(1)}$ is attached to the glycosidic carbon atom.

with adenine (Fig. 2). It is possible that the positively charged cation may interact with a Lewis base, such as adenine, and that this interaction may be sufficiently strong to overcome the steric effect due to proximity of the two amino groups (Fig. 2(b); Ono, Wilson & Grossman (1965) proposed a similar interaction in the case of cytosine hydrate). The uncharged amino tautomer, on the other hand, is analogous to the parent cytosine, and would be expected to pair with guanine, and not adenine.

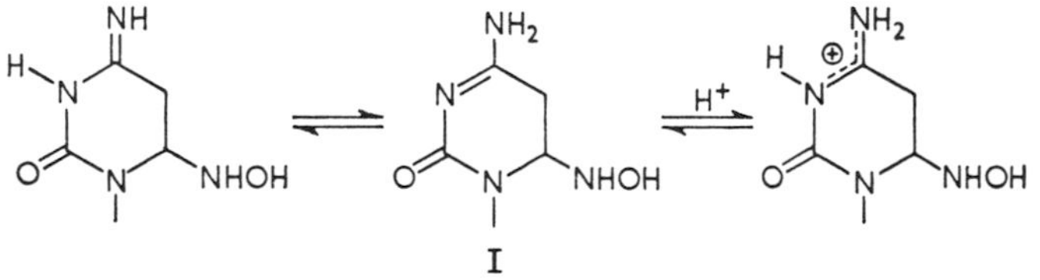

I

FIG. 1

(a)

(b)

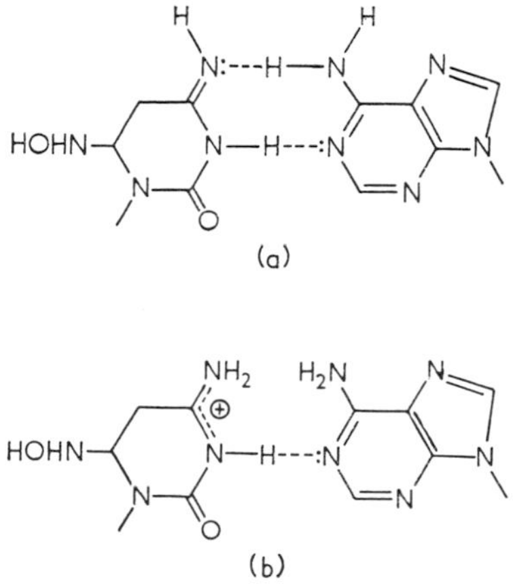

FIG. 2

The probabilities of such interactions are clearly related to the tautomeric constant of the base in the first case, and to its pK and the pH of the medium in the second. One can clearly construct analogous mispairing schemes for the case of cytosine itself. Since the pK_a of the base is about 4·4, approximately one base per thousand will be protonated at pH 7·5; the tautomeric constant of cytosine is more than 10^4 in favour of the amino form (Kenner, Reese & Todd, 1955) and the probability of a cytosine residue existing as the imino tautomer is therefore less than 10^{-4}. In the case of 5,6-dihydro-6-hydroxylaminocytosine, the tautomeric constant is unknown; the pK_a, however, may be roughly estimated from theoretical considerations (see Clark & Perrin, 1964); it is presumably intermediate between the pK_a of dihydrocytosine (pK_a 6·3, Janion & Shugar, 1960) and that of cytosine hydrate (5,6-dihydro-6-hydroxycytosine, pK_a 5·6, Johns, LeBlanc & Freeman, 1965). In other words, less than one 5,6-dihydro-6-hydroxylaminocytosine residue in ten exists as a cation at pH 7·5.

The experiments described in this paper were designed to determine how many adenylate residues are incorporated into polyguanylic acid when polycytidylic acid containing a known number of 5,6-dihydro-6-hydroxylaminocytosine residues is used as a template for RNA polymerase. It was found that adenylate is incorporated with very high efficiency. The experiments also provide further evidence that the species responsible for hydroxylamine-induced mutations is 5,6-dihydro-6-hydroxyl-aminocytosine and not its N^4-hydroxy derivative, nor N^4-hydroxycytosine.

2. Experimental Approach

The conversion of cytosine in poly C to residues with saturated 5,6 bonds can be measured directly by observing the decrease in absorbance of the polymer with time. For very small extents of reaction, however, use of this method involves large extrapolations from measured rates. [¹⁴C]Methoxyamine was therefore used instead of hydroxylamine itself, and its uptake by poly C measured directly. Parallel incubations of poly C with unlabelled methoxyamine were used for the preparation of modified poly C as a template for poly G synthesis in the presence of RNA polymerase; from a series of such experiments the percentage of adenylate incorporated could be compared with the percentage of cytidylate residues modified.

Methoxyamine reacts with cytosine analogously to hydroxylamine (Fig. 3). This has been shown by the isolation and characterization of products of types IV and V (Brown & Schell, 1965) and is also demonstrated in the present experiments.

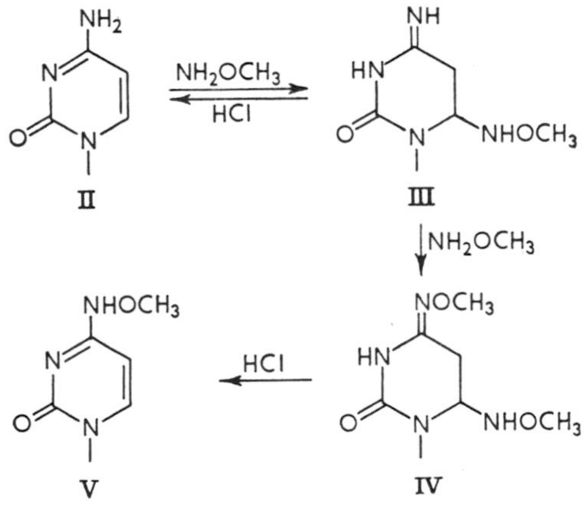

Fig. 3

The reagents are chemically very similar (Jencks & Carriuolo, 1960), the main difference being in their pK_a values which are 4·60 and 5·97, respectively (Bissot, Parry & Campbell, 1957). Further, methoxyamine is a mutagen for transforming DNA (Freese & Freese, 1964) and for T-even bacteriophages (Chubukov & Tatarinova, 1965); it is less powerful than hydroxylamine and its action has not been studied in detail.

3. Experimental Procedure

(a) Materials

RNA polymerase was isolated and purified from dried *Micrococcus lysodeikticus* cells, according to procedure A of Nakamoto, Fox & Weiss (1964). Following DEAE fractionation, the enzyme was stored in 50% glycerol at −15°C and used as such.

Poly C was purchased from Miles Chemical Co. It had a sedimentation coefficient of 6·05 s. [14C]Poly C was prepared from [14C]CDP by a viscometric method using purified polynucleotide phosphorylase, as described previously (Grossman, 1963).

14C-labelled and unlabelled nucleoside triphosphates and [14C]CDP were purchased from Schwarz BioResearch Inc. and [14C]methoxyamine hydrochloride was purchased from New England Nuclear Corp. The pyruvate methoxime derived from it had electrophoretic and chromatographic properties identical with those of authentic material.

(b) *Methods*

(i) *Treatment of poly C with methoxyamine*

2·5 M-Methoxyamine hydrochloride, adjusted to pH 5·5 with aqueous sodium hydroxide, was stored frozen. An appropriate volume of the cold solution was added to an aqueous solution of the polymer; this was mixed and then incubated in a stoppered conical tube at 37°C for the desired length of time. Reaction was halted by cooling in ice and adding a volume of cold 0·1 M-sodium pyruvate equivalent to the methoxyamine present. Portions of the resulting solution were used directly for assay with the polymerase.

(ii) *RNA polymerase assay*

Incubation conditions and method of assay have been described previously (Phillips *et al.*, 1965). Contents of incubation mixtures are indicated in the legends to the Figures. Routine assays were performed at room temperature for 60 min.

(iii) *Kinetics of reaction between poly C and methoxyamine*

1·60 ml. of 2·5 M-methoxyamine, pH 5·5, was added to 0·40 ml. aqueous poly C solution (1·25 mg/ml.) and the mixture incubated in a stoppered conical tube at 37°C. A similar solution without poly C was incubated in the same way, as a blank. The reaction was followed by withdrawing portions at intervals and measuring their absorbance at 269 mμ in cuvettes of 1 mm path-length in a Zeiss spectrophotometer (PMQII). At similar times 0·1-ml. portions were added to 0·9 ml. ethylene glycol in cuvettes of 1 cm path-length. The solution was mixed and left to stand at room temperature for 20 min before the absorbance at 269 mμ was recorded.

When [14C]methoxyamine was used as the reagent a more concentrated poly C solution was used (final concentration 2 mg/ml.), and a blank solution was prepared with unlabelled material and water. Decrease of optical density was followed by adding 0·1 ml. to 2·9 ml. water or ethylene glycol in 3-ml. cuvettes of 1 cm path-length. In addition, 0·1-ml. portions were withdrawn at intervals and added to 0·5 ml. ice-cold 10% TCA; the mixtures were shaken and filtered immediately through Millipore filters (type HA) and the precipitates washed 5 times with 4 ml. of cold 5% TCA. The filters were dried in open scintillation vials for 5 min at 110°C. Twenty ml. of toluene scintillation fluid was added and the radioactivity determined in an Ansitron scintillation counter, Wallingford, Conn.

The percentage of cytidylate residues that had undergone reaction was derived from the decrease in absorbance of the polymer in ethylene glycol; the concentration of cytidylate residues present at the beginning of reaction was determined from the absorbance in water. Uptake of [14C]methoxyamine in mμmoles was computed as follows: a known volume of the reaction mixture was added to 20 ml. of a dioxan scintillation fluid (Bray, 1960) and its radioactivity determined. A known volume of an aqueous solution of [14C]poly C was counted similarly, and the same volume was precipitated, filtered and its radioactivity determined in toluene scintillation fluid, as above. From these measurements a ratio could be obtained for counts in aqueous solution added to Bray's counting fluid to counts precipitated on filters and counted in toluene scintillation fluid. From this ratio, the cts/min of [14C]methoxyamine in the reaction mixture could be related to the cts/min precipitated. (Methoxyamine hydrochloride is volatile, and so cannot be conveniently counted on planchets.)

(iv) *Comparison of uptake of methoxyamine by poly C and incorporation of adenylate into poly AG*

0·8 ml. 2·5 M-[14C]methoxyamine (pH 5·5, 70 μc/m-mole) was added to 0·8 ml. water and 0·4 ml. poly C solution (10 mg/ml. in 0·1 M-NaCl) and the mixture incubated at

37°C in a stoppered conical tube. At intervals, 2 samples of 0·1 ml. were withdrawn and added to 0·5 ml. of ice-cold 10% TCA containing 2 M-hydroxylamine hydrochloride. The first sample was filtered immediately through a Millipore filter and the precipitate washed 3 times with 4 ml. of cold 5% TCA; with three 10-ml. portions of 5% TCA containing 2 M-hydroxylamine hydrochloride; and finally with another 10 ml. of 5% TCA. The filter was then dried for exactly 5 min in an open scintillation vial at 110°C before the radioactivity was determined, using toluene scintillation fluid. The second sample was centrifuged in the cold and the pellet washed 3 times with cold 5% TCA containing 2 M-hydroxylamine hydrochloride. 0·2 ml. of 1 N-hydrochloric acid at 60°C was then added to the pellet; after solution had occurred, the sample was maintained overnight at room temperature. It was then taken to dryness *in vacuo*, dissolved in 0·05 ml. of 0·05 M-ammonium formate buffer at pH 3·0, and applied quantitatively to Whatman no. 3 paper. Carrier N^4-methoxycytidylic acid (prepared by acid hydrolysis of a solution of cytidylic acid in methoxyamine that had stood at room temperature for 24 hr) was added and the paper subjected to electrophoresis in the formate buffer at 15 v/cm for 2 hr. The paper was then dried, the ultraviolet-absorbing area cut out and its radioactivity determined directly in toluene scintillation fluid.

The 2 samples were compared quantitatively, using data from the following control experiment: the radioactivity of a solution of [^{14}C]poly C was determined directly by counting in Bray's dioxan scintillation fluid and also, after application to Whatman no. 3 paper, by counting the paper in toluene scintillation fluid. Other samples were subjected to the acid hydrolysis procedure and then applied to the paper. Radioactivity in toluene scintillation fluid was then determined before and after electrophoresis. Less than 0·4% of the recovered radioactivity was found at the origin; 1·1% was recovered between the origin and the ultraviolet-absorbing area of cytidylic acid.

At the same time as the reaction with [^{14}C]methoxyamine, a series of samples of poly C were incubated with unlabelled reagent. Each reaction mixture contained 0·06 ml. 2·5 M-methoxyamine, pH 5·5, 0·03 ml. poly C solution (10 mg/ml. in 0·1 M-NaCl) and 0·06 ml. water in a stoppered conical tube. It was incubated at 37°C for the desired length of time, after which 3·0 ml. of cold 0·05 M-sodium pyruvate were added. The solutions were stored frozen before assay with the RNA polymerase. Four series of assays were performed. Each contained 0·75 ml. of the pre-treated poly C, the nucleoside triphosphates added being as listed in the legend to Fig. 8; in addition, one series contained [^{14}C]GTP as the only nucleoside triphosphate. 0·2-ml. portions were removed at intervals, added to 0·5 ml. of ice-cold 10% TCA and filtered through Millipore filters (type HA). After washing with cold 5% TCA, the filters were dried on planchets and the radioactivity of the precipitates determined using a Nuclear–Chicago thin-window Geiger counter (model 8703). Percentages of adenylate incorporated were determined as described in the text, correction being made for zero-time values and for adenylate incorporation directed by untreated poly C.

4. Results

(a) *The effect of methoxyamine on the template properties of poly C*

Use of poly C as a template for poly G synthesis in the presence of RNA polymerase from *M. lysodeikticus* leads to the formation of a complex of the two homopolymers. This has recently been described in detail by Haselkorn & Fox (1965). Use of the system for investigating the effect of the reagent methoxyamine on template properties was essentially as described previously for hydroxylamine (Phillips *et al.*, 1965).

Initial experiments were carried out to investigate the effect of incubation of poly C with different concentrations of methoxyamine on the rate of poly G synthesis. It was found that the reagent had less inactivating effect than hydroxylamine at the same concentration, a result which was expected in view of its lower rate of reaction with cytosine derivatives (Brown & Schell, 1965). The pH optimum for the reaction is lower, and incubations were carried out at pH 5·5; inactivations with hydroxylamine were always performed at pH 6·5. The lower pH probably has an effect on the secondary structure of the poly C, since this polynucleotide has a pK for transition from a double-stranded hemiprotonated structure to a single-stranded form at pH 5·7 (Akinrimisi, Sander & Ts'o, 1963; Fasman, Lindblow & Grossman, 1964). Reaction with methoxyamine was halted by adding sodium pyruvate and then chilling, and portions of the treated poly C were incubated at pH 7·5 with purified RNA polymerase and nucleoside triphosphates.

Figure 4 shows the result of an experiment on the inactivation of the poly C template. It may be compared with a similar experiment on inactivation by hydroxylamine (Fig. 5 of Phillips et al., 1965). Conditions were chosen such that the inactivation

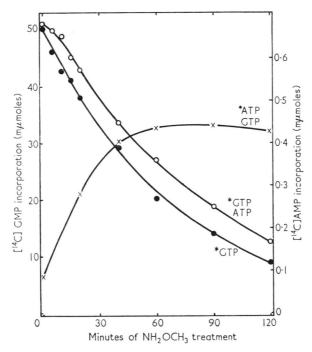

Fig. 4. Incorporation of GMP in the presence (—○—○—) and absence (—●—●—) of ATP, and incorporation of AMP in the presence of GTP (—×—×—), directed by poly C pre-treated with methoxyamine (2·0 M, pH 5·5, 37°C). Incubation mixtures (0·30 ml. each) contained 30 μ moles Tris buffer, pH 7·5, 1·0 μmole MnCl₂, 14 μg of pre-treated poly C, 23 μg of RNA polymerase and the following nucleoside triphosphates: (i) 0·25 μmole [¹⁴C]GTP (8·9 × 10⁴ cts/min/μmole); (ii) 0·25 μmole [¹⁴C]GTP + 0·04 μmole ATP; and (iii) 0·25 μmole GTP + 0·04 μmole [¹⁴C]ATP (2·1 × 10⁶ cts/min/μmole). The asterisk denotes the nucleotide labelled with ¹⁴C.

produced by the methoxyamine (2·0 M, pH 5·5, 37°C) was approximately the same as that produced by hydroxylamine (0·5 M, pH 6·5, 37°C). The Figure shows three series of incubations. They contain (i) [¹⁴C]GTP as the only nucleoside triphosphate: this curve for inactivation of the template is exponential; (ii) [¹⁴C]GTP + [¹²C]ATP: this shows that addition of ATP leads to some recovery of poly G synthesis, especially with low extents of modification of the template, as in the case of inactivation by hydroxylamine; (iii) [¹⁴C]ATP + [¹²C]GTP, demonstrating increasing incorporation of adenylate into acid-precipitable form as the template becomes more modified. It may be noted that the recovery of guanylate incorporation which can be effected by adding ATP is rather small; complete recovery of the poly G synthesis is only found after very short treatments of the poly C with methoxyamine. This suggests that, under the conditions used, the steady-state concentration of the species in the polymer directing adenylate incorporation (probably III, Fig. 3) must be very low. Comparison with results obtained using hydroxylamine suggests that when the poly C template has been extensively modified considerable poly A synthesis may occur; however, this was not verified by a nearest neighbour analysis.

This experiment shows that methoxyamine has an effect similar to that of hydroxylamine on the template properties of poly C, and that it may be used as a model for hydroxylamine action. However, to eliminate the risk of homopolymer synthesis concomitant with the poly AG synthesis directed by the modified template, it was desirable to work with low concentrations of reagent (0·5 M and 1·0 M) and short reaction times (less than one hour) in subsequent experiments.

(b) *The reaction between methoxyamine and poly C*

Before measuring uptake of [^{14}C]methoxyamine by poly C it was necessary to follow the reaction on a large scale and to demonstrate its over-all stoichiometry. A typical reaction curve, as followed by change of optical density at maximum absorbance, is shown in Fig. 5, curve A. The reaction proceeds with an apparent induction

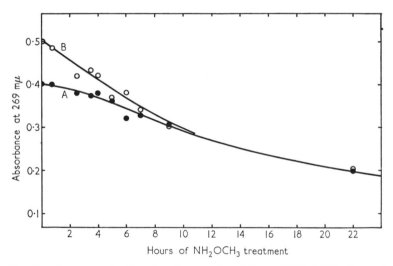

FIG. 5. Reaction between poly C and methoxyamine (2·0 M, pH 5·5, 37°C). Curve A: reaction followed by recording optical density at 269 mμ in cuvettes of 1 mm path-length. Curve B: reaction followed by diluting portions with 9 vol. of ethylene glycol and recording optical density at 269 mμ in cuvettes of 1 cm path-length.

period, analogous to the profile found using hydroxylamine as reagent (Brown & Phillips, 1965). The curve is in fact compound, resulting from the decrease in absorbance due to addition of the reagent to the cytosine residues, and the increase in absorbance from loss of the secondary structure of the polymer.

The true reaction curve was measured by removing portions of the reaction mixture and diluting them with ethylene glycol (final concentration of glycol, 90%) before measuring their optical density. This effects destruction of the helical content of the polymer (Fasman *et al.*, 1964); the loss of optical density is thus due entirely to the saturation of the 5,6-double bond of the cytosine residues (Fig. 5, curve B). The reaction curve is exponential over the first ten hours under the conditions used (2·0 M, pH 5·5, 37°C). After this time all the secondary structure of the polymer is lost, as shown by the coincidence of the two curves.

These curves were then followed again, using [^{14}C]methoxyamine as reagent, and the uptake of radioactive reagent followed as a function of time. The total number of cytidylate residues present at the beginning of reaction was calculated from the optical density at 269 mμ in water, using $\lambda_{max.} = 6\cdot3 \times 10^3$ (Warner, 1957); the percentage reacted at any given time was calculated from the decrease in absorbance at 269 mμ in 90% ethylene glycol, assuming that the reaction product does not absorb at this wavelength (Brown & Schell, 1965). From these figures the number of mμmoles of cytidylate which had reacted was derived. The uptake of labelled reagent by the poly C is plotted as a function of cytidylate residues that have undergone reaction (up to 25% of the total number of residues) in Fig. 6. It is shown by the slope of the curve that two moles of methoxyamine are equivalent to one mole of cytidylate, confirming that the final product is the 5,6-dihydro-N^4-methoxy-6-methoxyaminocytosine residue, IV (Fig. 3). It was shown by a control experiment that there was no measurable loss of methoxyamine from the reaction product owing to acid-catalysed elimination under the precipitation conditions used.

The high background adsorption of reagent to the polynucleotide on the filter was reduced considerably in subsequent experiments by precipitating and washing in

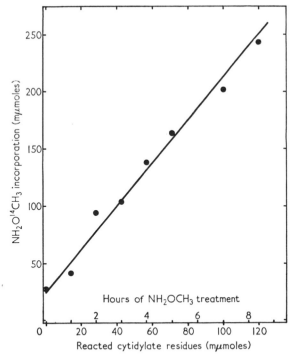

FIG. 6. Uptake of [^{14}C]methoxyamine by poly C. Reaction mixture contained poly C (2 mg/ml.) in 2·5 ml. [^{14}C]methoxyamine (2·0 M, pH 5·5, 37°C; 9·8 × 10^6 cts/min/m-mole). Portions of 100 μl. were removed for (i) dilution with 29 vol. water and recording of optical density at 269 mμ; (ii) dilution with 29 vol. ethylene glycol and recording of optical density at 269 mμ; and (iii) precipitation with cold TCA and counting of radioactivity.

TCA containing 2 M-hydroxylamine hydrochloride, but could never be removed entirely; measurements of extent of reaction by following uptake of radioactive reagent were thus accompanied by a high background "noise" level which made interpretation difficult.

(c) Uptake of [^{14}C]methoxyamine by poly C over short reaction periods

Although the over-all stoichiometry of the reaction is NH_2OCH_3:cytidylate equals 2:1, in the initial stages of the reaction the stoichiometry must be 1:1. A plot of uptake of reagent as a function of time should thus reveal a change of slope from an initial value of one to a final value of two. This is not apparent in Fig. 6, presumably because the concentration of III is very low under these conditions (cf. Fig. 4). However, even using lower concentrations of methoxyamine, it was not possible to demonstrate this change of slope directly because, in spite of extensive washing of the filter, it was impossible to eliminate the ^{14}C-background which led to scatter of the experimental points and masking of the change of slope (see Fig. 7).

An indirect method was therefore used to confirm the nature of the initial product, and to estimate it quantitatively. This utilized the technique of acid hydrolysis of portions of the poly C reaction mixture, as used previously to investigate the reaction with hydroxylamine (Brown & Phillips, 1965). Poly C treated with [^{14}C]methoxyamine was degraded to mononucleotides by acid treatment after various extents of reaction. This led to simultaneous reversion of residues of 5,6-dihydro-6-methoxyaminocytosine (III) in the polymer to non-radioactive cytosine residues (II), and conversion of residues of 5,6-dihydro-N^4-methoxy-6-methoxyaminocytosine (IV) to radioactive N^4-methoxycytosine residues (V, Fig. 3). Cytidylic and N^4-methoxycytidylic acids were then separated by electrophoresis and the uptake of [^{14}C]methoxyl groups into the latter was measured.

The result of such an experiment is given in Fig. 7 (curve A). This shows that, under these conditions (1·0 M-methoxyamine, pH 5·5, 37°C; molar ratio NH_2OCH_3: cytidylate $=213$:1), there is little formation of the N^4-exchange product (IV) during the first hour of reaction, although both optical density measurements after dilution by ethylene glycol and measurements of uptake of [^{14}C]methoxyamine by the poly C (filled circles in Fig. 7) indicate that reaction is initiated immediately. This confirms that the product formed initially, during the first hour of reaction, must be III, the monoadduct, as is the case when hydroxylamine is used as the reagent (Brown & Phillips, 1965).

In another experiment, 2·4 M-[^{14}C]methoxyamine was used as the reagent under otherwise similar conditions, and N^4-methoxycytidylic acid again separated by electrophoresis after acid hydrolysis of portions of the reaction mixture. In this case the curve for the production of the N^4-exchange product reached its maximum slope after approximately 20 minutes, compared with a value of about two hours under the milder conditions of Fig. 7.

The uptake of methoxyamine by poly C may be calculated from the data of such acid-hydrolysis experiments, since it is known that the hydrolysis procedure results in elimination of exactly one of two molecules of reagent taken up by a cytosine residue to yield IV. The reaction curve for cytidylate residues is linear during the period under consideration (up to about 8% of reaction) and the slope of the initial uptake of reagent by the polymer is therefore equal to the final slope of curve A in Fig. 7. This curve, its slope representing the equivalence of one mole of reagent to

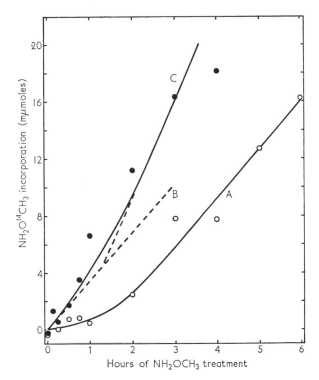

FIG. 7. Uptake of [^{14}C]methoxyamine by poly C. Reaction mixture contained poly C (2 mg/ml.) in 2·0 ml. [^{14}C]methoxyamine (1·0 M, pH 5·5, 37°C; 1·82 × 10^8 cts/min/m-mole). Optical density measurements showed that 100 μl. of reaction mixture contained 469 mμmoles of cytidylate residues. Curve A: 100-μl. portions subjected to acid degradation and hydrolysis; —○—○— represents incorporation of [^{14}C]methoxyamine into N^4-methoxycytidylic acid. (A background equivalent to 0·73 mμmole has been subtracted from all values.) Curves B and C: total incorporation of [^{14}C]methoxyamine into poly C in 100-μl. portions of reaction mixture, calculated from curve A (see text). —●—●—: experimentally determined total incorporation of [^{14}C]methoxyamine into poly C in 100-μl. portions. (A background equivalent to 3·8 mμmoles has been subtracted from all values.)

one mole of cytidylate, is plotted in Fig. 7 as the straight line B. As the reaction proceeds, it changes to a straight line with twice the slope when the initial product III is converted to the final product IV. This *calculated* curve is plotted on Fig. 7 (C) where the total uptake of methoxyamine by poly C calculated from curve A is also shown (line connecting B and C). It can be seen that the experimentally determined uptake of methoxyamine by the poly C (determined in the same experiment; filled circles in Fig. 7) is in reasonable agreement with this calculated curve, but the scatter of the points does not permit deduction of the curve from the experimental data.

(d) *Determination of the percentage of adenylate incorporated in poly G as a function of the percentage of cytidylate residues reacted in the template*

Poly C was treated with [^{12}C]methoxyamine in an experiment parallel to that described in the previous section, under identical conditions. The two experiments were, in fact, performed at the same time, as small differences in temperature may exert a marked effect on the percentage of monoadduct III in the polymer. Portions of the incubation mixture were removed and the reaction halted at known time intervals. These polymer samples were then incubated with appropriate nucleoside triphosphates and RNA polymerase, and the synthesis of poly AG complementary to the modified poly C template was followed kinetically. A series of results for polymers treated for different times with methoxyamine was thus obtained: a typical result is shown in Fig. 8 (50 minutes treatment).

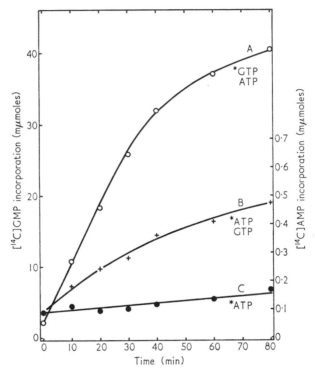

FIG. 8. Kinetics of poly AG synthesis directed by poly C treated with methoxyamine (1·0 M, pH 5·5, 37°C) for 50 min. Incubation mixtures (1·5 ml. each) contained 150 μmoles Tris buffer, pH 7·5, 5 μmoles MnCl₂, 71 μg of pre-treated poly C, 114 μg of RNA polymerase and the following nucleoside triphosphates: curve A, 1·25 μmoles [¹⁴C]GTP (9·4 × 10⁴ cts/min/μmole) + 1·25 μmoles ATP; curve B, 1·25 μmoles [¹⁴C]ATP (8·0 × 10⁵ cts/min/μmole) + 1·25 μmoles GTP; curve C, 1·25 μmoles [¹⁴C]ATP as the only nucleoside triphosphate. Portions of 0·2 ml. were removed for assay of incorporated radioactivity. The asterisk denotes the nucleotide labelled with ¹⁴C.

The percentage of adenylate in the poly AG was calculated as follows: the incorporation of adenylate in the absence of GTP (incubation mixtures containing [¹⁴C]ATP as the only nucleoside triphosphate) was found to be almost constant for each polymer; values taken at ten-minute intervals along the curve (Fig. 8, curve C) were subtracted from appropriate values of adenylate incorporation in the presence of GTP (curve B). The percentage of adenylate incorporated at each interval was then obtained from this figure and from the value for the total number of nucleotides incorporated (from curve A + curve B — curve C). Control experiments were performed, in which the cold pyruvate solution used for halting the reaction with methoxyamine was added before the polyC had been incubated at 37°C. Significant incorporations of adenylate were found in the case of each of these zero-time controls (0·33%); this may possibly be attributable to a corresponding percentage of uridylate residues in the polymer, to some initiation effect, or to a small extent of poly A synthesis. The control percentage at each time interval was subtracted from the value found for the poly AG synthesized from the treated polymer.

It was found that in all cases the percentage of adenylate incorporated was nearly constant after more than 30 minutes of synthesis, and a percentage could thus be derived for each polymer. In the first 30 minutes the total extent of synthesis was not great enough for accurate values to be obtained.

In Fig. 7, curve B indicates the total number of cytidylate residues that have undergone reaction; curve A indicates those that have been further modified to yield N^4-methoxy derivatives, IV. Subtraction of A from B thus gives a value for those residues which have taken up one molecule of reagent to yield III; they have been expressed as a percentage of the total number of residues in the polymer and plotted against the percentage of adenylate incorporated into the poly AG in Fig. 9.

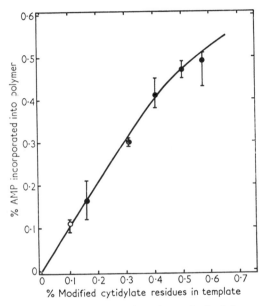

FIG. 9. Percentage of adenylate found in poly AG synthesized from poly C templates pre-treated with methoxyamine. Filled circles: poly C treated with 1·0 M-methoxyamine, pH 5·5, 37°C, for 15, 30, 40, 50 and 60 min. Open circle: poly C treated with 0·5 M-methoxyamine, pH 5·5, 37°C, for 30 min. Vertical lines indicate range of percentages found at different times of polymer synthesis.

The points indicate mean values and the vertical lines indicate the range of adenylate percentages found at ten-minute intervals from 40 to 80 minutes of polymer synthesis. It can be seen that the points are fitted by a curve with a slope of unity for small extents of treatment of the poly C; for more extensive treatment the slope decreases. The minimal and maximal possible values of the slope for the initial part of the curve (from a consideration of different curves through the open circles of Fig. 7) are 0·8 and 1·1. We conclude that every modified cytosine residue directs the incorporation of one adenylate residue.

5. Discussion

Before discussing the significance of these results, we must consider whether the data obtained have real meaning. There are several indications that measurement of adenylate incorporation may be difficult to interpret due to "end-effects" (see Bremer, Konrad, Gaines & Stent, 1965; Maitra & Hurwitz, 1965) or to poly A synthesis (Krug, Gomatos & Tamm, 1965). Clearly, the adenylate incorporation must not be over-estimated. First, subtraction of the zero-treatment control eliminates random adenylate incorporation, affecting all samples similarly. Second, very mild conditions were chosen for inactivation: incorporation of [^{14}C]GMP in the absence of ATP after 60 minutes of treatment with 1·0 M-methoxyamine was still 83% of the value for untreated poly C (cf. Fig. 4). Poly A synthesis is minimized as a result (Phillips *et al.*, 1965). Third, the kinetic analyses of each polymer were designed to reveal poly A synthesis. Since poly A synthesis is probably linear with time in the presence of excess ATP (Gomatos, Krug & Tamm, 1964) the percentage of adenylate incorporated should also increase linearly. It was found, in fact, that the percentage incorporated did not generally vary much with time over the last 40 or 50 minutes of synthesis (the variation is shown by the vertical lines in Fig. 9); after only 10 or 20 minutes the calculated values were somewhat lower, except in one case. As there was no regular variation, however, no correction was made, apart from the zero-treatment control (which probably includes a small proportion of poly A synthesis). It is possible that some error has been introduced into our results by undetected poly A synthesis, although this should not be great enough to affect the conclusions.

Is it possible to reconcile adenylate incorporation directed by some template species other than III, or in addition to III, with the results we have presented? First, we may dismiss deamination products as relevant contributors to the effect. The data of Johns *et al.* (1965) on the deamination of the related species, cytosine hydrate, suggest that deamination will be very slight under our conditions, in agreement with our previous conclusions (Phillips *et al.*, 1965). Second, N^4-methoxycytosine (V, Fig. 3) residues in the template may be excluded since under our conditions and during the reaction times used their formation is negligible. (The total number of N^4-methoxycytosine residues is in all cases less than the number of IV in the template, and the following reasoning, which excludes IV as a sole contributor to the mutagenic effect, applies with greater force to V.) The possibility that residues of IV in the template are alone responsible for adenylate incorporation may be investigated as follows: their proportion in the template is given by curve A (Fig. 7) and may be plotted against adenylate incorporation in a graph analogous to Fig. 9. Such a plot has been made, using the steepest possible curve that may be drawn through the open circles of Fig. 7: the minimum initial slope that may be drawn through the points gives an equivalence of 2·2 adenylate residues per residue of IV, and values taken from curve A of Fig. 7 give an equivalence of five adenylate residues per residue of IV. This would indicate that one altered template residue leads to more than one erroneous base in the product. This seems very unlikely, since experiments using mixed copolymers show that there is a high fidelity of replication in this system (Adman & Grossman, manuscript in preparation).

Values for modified residues taken from curve B (Fig. 7) alone give values for the template content of III + IV. If these values are plotted against adenylate incorporation, a curve is obtained which, although it has a slope of unity at first, rapidly decreases in slope after about 30 to 40 minutes of polymer treatment (more than 0·4% of cytidylate residues modified). A rapid decrease in this case is difficult to explain since, if both III and IV direct adenylate incorporation, the curve should be strictly linear. On the other hand, if only III directs incorporation, and IV inhibits replication, as suggested by experiments with hydroxylamine (Phillips et al., 1965), the curve of Fig. 9 is expected to be linear at first, but its slope is expected to decrease until it becomes zero, when the steady-state concentration of III is reached. Fig. 7 indicates that this concentration is about 0·9% of the residues. This seems to explain the curve of Fig. 9 well. Our data are not sufficiently precise, however, to enable the possibility of mis-coding by IV to be eliminated completely unequivocally, although, taken in conjunction with our earlier experiments, the results do suggest that IV cannot be replicated by the enzyme.

After one hour of treatment of the poly C with methoxyamine (1·0 M, pH 5·5, 37°C), conversion of monoadduct residues III to the final product IV becomes extensive (Fig. 7); after approximately 160 minutes the concentrations of the two species become equal. The extent of synthesis of poly AG should decrease with increasing treatment of the poly C if IV cannot be replicated, but the percentage of adenylate incorporated should reach a constant value, unless it decreases due to an acceleration of the second reaction by changes in the secondary structure of the polymer. It was not possible to investigate this, however, since the formation of N^4-methoxy residues (IV) may lead to initiation of poly A synthesis, as in the case of N^4-hydroxy residues.

The experiments therefore provide direct evidence that the monoadduct formed by the action of hydroxylamine on cytosine, namely 5,6-dihydro-6-hydroxylamino-cytosine (I, Fig. 1), can direct incorporation of adenylate residues in nucleic acid replication, a result which had been inferred from indirect evidence (Phillips et al., 1965); it is not necessary for this to be converted to the N^4-hydroxy derivative by reaction with a second molecule of reagent before the species gives rise to a replication error. Indeed, we consider it unlikely that in a bacteriophage genome, where the hydroxylamine-mutable sites are widely distributed (Champe & Benzer, 1962), such a second reaction will occur readily, except possibly at a site where the polynucleotide has a less rigid secondary structure. Although it seems likely that the mechanism elucidated applies not only to cytosine-containing genomes but also to T-even bacteriophages, it must be admitted that the stability of the monoadduct of 5-hydroxy-methylcytosine is unknown. Janion & Shugar (1965) provide evidence that there is little, if any, formation of such an adduct with 5-alkylcytosines in aqueous solution and suggest that N^4-hydroxycytosine derivatives must therefore be implicated as the relevant "unnatural" bases. This clearly points to the need for further chemical experiments, using biological systems to supplement the information derived from model systems such as the one we have used, and work is in progress to investigate the biological significance of these chemical results.

We may further conclude from our experiments that methoxyamine and, presumably, hydroxylamine are mutagens with high efficiency. Essentially every residue of 5,6-dihydro-6-methoxyaminocytosine in the poly C directs incorporation of an adenylate residue (Fig. 9). This must be explained by a highly efficient "all-or-none" process. From a consideration of pK values (p. 406), it seems unlikely that interactions of the cation of the monoadduct could account for the results. We conclude, tentatively, that the replication error is to be ascribed to specific hydrogen-bonding between the monoadduct and the adenine residue, and that the former is present predominantly as the imino tautomer under the conditions pertaining at the point of replication. Experiments on tautomeric equilibria in dihydrocytosine derivatives will be reported later (Brown & Hewlins, unpublished work).

This work was supported in part by grants from the National Institutes of Health (HD–1258), the National Science Foundation (GB 3490) and the Atomic Energy Commission (AT–30–1–3449). A travel grant from the Wellcome Trust (to J. H. P.) is gratefully acknowledged.

This is publication no. 452 from the Graduate Department of Biochemistry, Brandeis University.

REFERENCES

Akinrimisi, E. O., Sander, C. & Ts'o, P. O. P. (1963) *Biochemistry*, **2**, 340.
Bissot, T. C., Parry, R. W. & Campbell, D. H. (1957). *J. Amer. Chem. Soc.* **79**, 796.
Bray, G. A. (1960). *Analyt. Biochem.* **1**, 279.
Bremer, H., Konrad, M. W., Gaines, K. & Stent, G. S. (1965). *J. Mol. Biol.* **13**, 540.
Brown, D. M. & Phillips, J. H. (1965). *J. Mol. Biol.* **11**, 663.
Brown, D. M. & Schell, P. (1965). *J. Chem. Soc.* 208.
Champe, S. P. & Benzer, S. (1962). *Proc. Nat. Acad. Sci., Wash.* **48**, 532.
Chubukov, V. F. & Tatarinova, S. G. (1965). *Zh. Mikrobiol., Epidemiol., Immunobiol.* **42**, 80.
Clark, J. & Perrin, D. D. (1964). *Quart. Rev.* **18**, 295.
Fasman, G. D., Lindblow, C. & Grossman, L. (1964). *Biochemistry*, **3**, 1015.
Freese, E. F. & Freese, E. (1964). *Proc. Nat. Acad. Sci., Wash.* **52**, 1289.
Gomatos, P. J., Krug, R. M. & Tamm, I. (1964). *J. Mol. Biol.* **9**, 193.
Grossman, L. (1963). *Proc. Nat. Acad. Sci., Wash.* **50**, 657.
Haselkorn, R. & Fox, C. F. (1965). *J. Mol. Biol.* **13**, 780.
Janion, C. & Shugar, 'D. (1960). *Acta Biochim. Polonica*, **7**, 309.
Janion, C. & Shugar, D. (1965). *Acta Biochim. Polonica*, **12**, 337.
Jencks, W. P. & Carriuolo, J. (1960). *J. Amer. Chem. Soc.* **82**, 1778.
Johns, H. E., LeBlanc, J. C. & Freeman, K. B. (1965). *J. Mol. Biol.* **13**, 849.
Kenner, G. W., Reese, C. B. & Todd, A. R. (1955). *J. Chem. Soc.* 855.
Krug, R. M., Gomatos, P. J. & Tamm, I. (1965). *J. Mol. Biol.* **12**, 872.
Lawley, P. D. & Brookes, P. (1962). *J. Mol. Biol.* **4**, 216.
Maitra, U. & Hurwitz, J. (1965). *Proc. Nat. Acad. Sci., Wash.* **54**, 815.
Nakamoto, T., Fox, C. F. & Weiss, S. B. (1964). *J. Biol. Chem.* **239**, 167.
Ono, J., Wilson, R. & Grossman, L. (1965). *J. Mol. Biol.* **11**, 600.
Phillips, J. H., Brown, D. M., Adman, R. & Grossman, L. (1965). *J. Mol. Biol.* **12**, 816.
Warner, R. C. (1957). *J. Biol. Chem.* **229**, 711.

16

Reprinted from *Science*, **148**(3669), 507–508 (1965)

Mutagenic Effects of Hydroxylamine in vivo

Abstract. *Hydroxylamine can induce any one of the four types of transition mutations in bacteriophage S13 if the mutagen is added directly to agar plates seeded with phage and bacterial indicators. The phage can also be mutated by treating the host cell with hydroxylamine before infecting it with phage. These effects of hydroxylamine in vivo contrast with the mutagenic effect in vitro, which seems to be exclusively on cytosine.*

Hydroxylamine has a highly specific mutagenic effect on bacteriophage treated in vitro. From a study of the induction of forward and reverse mutations, it has been inferred that in the single-stranded DNA of phage S13, only one of the four bases is altered in mutations induced by treatment of the free phage with hydroxylamine (*1*). By comparing that inference with data on the chemical reactivity of hydroxylamine (*2, 3*), it was concluded that it is cytosine that is altered, being replaced after replication by thymine. This result is consistent with mutation experiments on double-stranded DNA (*2, 4*), although in double-stranded DNA the mutagenic effect cannot be restricted to cytosine, but must ultimately involve both members of the guanine-cytosine base pair. In S13, it was concluded that the mutagenic specificity of hydroxylamine for cytosine was at least 500 times greater than for adenine, guanine, or thymine, in sharp contrast with the effects of two other frequently used mutagens, nitrous acid and the alkylating agent, ethyl methanesulfonate, each of which appeared to mutate all four bases (*1*).

In this report we discuss the mutagenic effect of hydroxylamine on S13 when the treatment is in vivo—that is, when the host cell is treated either during or before infection. Under these conditions, hydroxylamine causes the mutation of all four bases almost indiscriminately.

A set of reference mutations that includes all four types of transitions was studied (Table 1). The base change for each mutation was previously inferred from the effects of a variety of chemical mutagens (*1, 5*), and reasons have been given (*1*) for the belief that each mutation involves only the one particular base change indicated. The mutants have been described (*1*) except *su*HT9, which is a hydroxylamine-induced mutant that is able to infect both *Escherichia coli* C and *Shigella dysenteriae* Y6R, but which can produce progeny only in the *Shigella* strain. It is shown in Table 1 that all the mutations were induced when hydroxylamine was added directly to the overlay agar together with the parental phage and bacterial indicators. The identities of the mutants were verified by picking at least three mutant plaques of each type and replating on hydroxylamine-free plates. Of all the mutations shown in Table 1, only $h_i1 \longrightarrow h+$ can be induced in vitro; therefore it is clear that hydroxylamine is less specific in its mutagenic action in vivo, regardless of what base changes are assumed in Table 1.

The initial concentration of hydroxylamine was $2 \times 10^{-2}M$, but the rate of diffusion into the 35 ml (approximately) of bottom-layer agar is not known. The mutations must have occurred within 1 hour after plating because mutant plaques were already clearly visible after only 4 hours of incubation at 37°C. The effect of adding from 10 to 80 μmole of hydroxylamine was tested on the mutation $h_i1 \longrightarrow h^+$. The optimum amount of hydroxylamine was between 40 and 60 μmole. Above 60 μmole, the bacterial lawn would not develop well.

The plates contained free phage, infected cells, and uninfected cells, and any of the three could have participated in the mutagenic process. It seems that a mutagenic effect can occur without a direct interaction of the hydroxylamine with the phage because we were able to induce phage mutations by infecting cells that had previously been treated with hydroxylamine but which had been separated from it by centrifugation before infection. This is shown in Fig. 1 for an experiment in which the cells were both grown and treated in broth. The hydroxylamine concentration ($2.5 \times 10^{-2}M$) used for the treatment was nearly the same as the initial concentration in the overlay agar in the plate experiments. We do not know whether some hydroxylamine might become bound to the cells and thus not be removed by centrifugation. We also do not know why there was a large decrease in the number of mutants after 30 minutes of the treatment.

The host, *E. coli* C, used in this experiment is normally nonpermissive for growth of the parent phage, *su*HT9. Two other mutations, $h^+ \longrightarrow h_i1$ and $hH43 \longrightarrow h^+$, were also tested for inducibility by prior treatment of *E. coli* C with hydroxylamine, and in these

Table 1. Induction of the four types of transition mutations by addition of hydroxylamine (HA) directly to the agar plates. The plates (*1*) were incubated at 37°C within 10 minutes after the overlay agar containing phage and indicator bacteria were added. Hydroxylamine hydrochloride (Fisher) was made up as a $1M$ solution in $0.2M$ sodium phosphate buffer, pH 6.0, and was stored frozen at -10°C. Similar stored stocks were used in previous in vitro experiments (*1*). Mixed bacterial indicators had to be used for the host-range mutants (*1*). The pH of the plates was measured with pHydrion papers (Micro Essential Laboratory). With *E. coli* C in the overlay agar, the pH changed from 6.3 at the time the plates were seeded, to approximately 6.0 after 6 hours of incubation at 37°C, and to approximately 5.5 after overnight growth.

Mutation			Amount of parental phage plated	No. of mutant phage plaques appearing on a plate	
Phage		Base change*		No HA	50 μmole HA†
Parent	Mutant				
$h^+ \longrightarrow h_i1$		$G \longrightarrow A$	4×10^7	20	450
$h_i1 \longrightarrow h^+$		$A \longrightarrow G$	1×10^8	8	100
$h_i1 \longrightarrow h'$		$C \longrightarrow T$	1×10^8	8	90
$hH43 \longrightarrow h^+$		$T \longrightarrow C$	1×10^8	1	110
$hH76 \longrightarrow h^+$		$T \longrightarrow C$	1×10^8	2	80
suHT9 $\longrightarrow su^+$		$T \longrightarrow C$	1×10^8	6	$\geqq$300
suHT9 $\longrightarrow su^+$		$T \longrightarrow C$	2×10^7	1	100

* The abbreviations are: G, guanine; A, adenine; C, cytosine; T, thymine. † Added to 2.5 ml of overlay agar.

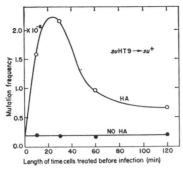

Fig. 1. Mutation frequency of *su*HT9 ⟶ *su*⁺ relative to the number of input phage as a function of the length of time the host cells were treated with hydroxylamine (HA) before infection. *Escherichia coli* C was grown to 1×10^8/ml in tryptone broth (13 g of Bacto tryptone per liter of H_2O), pH 6.4. To 20 ml of the bacterial suspension was added 0.5 ml of $1M$ HA at pH 6.0. This was shaken at 37°C. At the indicated times, 5.0 ml was removed and centrifuged, the pellet then being resuspended in 5.0 ml tryptone broth, centrifuged again and resuspended to give a cell concentration between 5 and 10×10^8/ml in 1.0 ml tryptone broth-salt (13 g of Bacto tryptone plus 7 g of NaCl per liter of H_2O) containing $1 \times 10^{-2}M$ $CaCl_2$. To 0.2 ml of this bacterial suspension was added 0.05 ml of phage at 1×10^9/ml. Ten minutes at 37°C was allowed for adsorption, and then the entire phage and cell mixture was plated with indicator bacteria for phage assay. A control culture of cells was handled in the same way but with no HA added. Before treatment with HA, approximately 100 percent of the bacteria observable in the microscope could form colonies, but after 120 minutes of treatment, approximately 50 percent could not.

cases the host normally permits growth of the parent phage types. Both mutations were induced.

A similar effect could be obtained in an inorganic buffer. Cells grown in broth were treated with hydroxylamine in $0.1M$ sodium phosphate buffer, pH 6.8, containing $10^{-3}M$ $MgSO_4$. The one mutation studied, *su*HT9⟶*su*⁺, was

induced. At 10 minutes the effect was about half as great as the value shown in Fig. 1 for broth. But no mutagenic effect was observed at 30 and 60 minutes. The reason could well be the extremely rapid inactivation of cells treated in the buffer; the survival decreased roughly exponentially, reaching 10^{-2} at 60 minutes.

When cells were treated with hydroxylamine, centrifuged, and then aerated in tryptone broth at 37°C, the mutagenic effect of the hydroxylamine was reduced, but even after 80 minutes of aeration there was still a threefold increase in the mutation *su*HT9⟶*su*⁺ compared with control cells not treated with hydroxylamine.

Because hydroxylamine is such a highly specific mutagen in vitro, it might at first seem surprising that it should be so unspecific in vivo. But the mutagenic effects in vivo are actually not unreasonable considering the widespread effects that HA and its derivatives have on *E. coli*. At $10^{-3}M$ concentrations, hydroxylamine is known to seriously affect many distinct cellular processes, immediately stopping DNA, RNA, and protein synthesis (6); and inhibitory effects have been observed even at $10^{-4}M$ (7). It should not be surprising if such damages were to impair the fidelity of the replication process and thus increase the frequency of many types of mutations.

It is also conceivable that hydroxylamine can react with cytosine to produce an analogue for both cytosine and thymine that is mutagenic. Furthermore, mutagens might be produced by the reactions of hydroxylamine with other substances besides cytosine. For example, a large class of hydroxamates could be formed (8), particularly with the aid of intracellular enzymes; and some of the hydroxamates would be expected to react with bases other than cytosine (9). Perhaps in that way each base could be chemically modified. But

actually, the formation of just one base analogue could be enough to explain the induction of all the mutations observed (5). In any case, it is clear that for the in vivo mutagenic effects there are many conceivable explanations that require further study.

The S13 experiments suggest possible limitations of hydroxylamine as a specific cellular mutagen. Hydroxylamine appears to be highly specific for just cytosine in vitro, but cellular mutations are usually induced in vivo. It is possible to mutate some cells by treatment in vitro of transforming DNA or transducing viruses, in which cases we would expect only cytosine to be affected. But if cellular mutations were produced by hydroxylamine treatment of the cells directly, then the in vivo results for phage S13 should alert us to the possibility of a mutagenic effect on any one of the four bases. Other mutagens might also behave quite differently in vivo than they do in vitro.

IRWIN TESSMAN
HIROMI ISHIWA
SANTOSH KUMAR
*Department of Biological Sciences,
Purdue University, Lafayette, Indiana*

References and Notes

1. I. Tessman, R. Poddar, S. Kumar, *J. Mol. Biol.* **9**, 352 (1964).
2. E. Freese, E. Bautz-Freese, E. Bautz, *ibid.* **3**, 133 (1961); E. Freese, E. Bautz, E. B. Freese, *Proc. Nat. Acad. Sci. U.S.* **47**, 845 (1961).
3. H. Schuster, *J. Mol. Biol.* **3**, 447 (1961); D. M. Brown and P. Schell, *ibid.*, p. 709; D. W. Verwoerd, W. Zillig, H. Kohlhage, *Z. Physiol. Chemie* **332**, 184 (1963).
4. S. P. Champe and S. Benzer, *Proc. Nat. Acad. Sci. U.S.* **48**, 532 (1962).
5. B. Howard and I. Tessman, *J. Mol. Biol.* **9**, 364 (1964).
6. H. S. Rosenkranz and A. J. Bendich, *Biochim. Biophys. Acta* **87**, 40 (1964).
7. S. Béguin and A. Kepes, *Compt. Rend.* **258**, 2427, 2690 (1964).
8. W. P. Jencks, M. Caplow, M. Gilchrist, R. G. Kallen, *Biochemistry* **2**, 1313 (1963).
9. N. K. Kochetkov, E. I. Budowsky, R. P. Shibaeva, *Biochim. Biophys. Acta* **87**, 515 (1964).
10. This research was supported, in part, by NSF research grant GB-2748.

1 February 1965

Editors' Comments
on Papers 17 and 18

17   CHAMPE and BENZER
     *Reversal of Mutant Phenotypes by 5-Fluorouracil: An Approach
     to Nucleotide Sequences in Messenger-RNA*

18   BRENNER et al.
     *Genetic Code: The "Nonsense" Triplets for Chain Termination
     and Their Suppression*

COMBINATIONS AND PERMUTATIONS

The mutagens described in Papers 10 through 18 are all characterized by their simplicity and specificity of action. Before we go on to consider more complex mutagenic mechanisms, it is instructive to consider two papers which illustrate the analytical power that can be achieved by the imaginative application of these simple mutagens, plus a curious extension of the concept of base-analogue mutagenesis.

In Paper 17, Champe and Benzer first classify a collection of T4*rII* mutants as to the DNA base pair at a mutated site. (Note that their Table 1 lists the deduced *wild type*, not *mutant*, base pair. For instance, a mutant reverted by 2-aminopurine but not by hydroxylamine would contain an A:T base pair at the mutant site, and, presuming that it arose by a transition in the first place, a G:C base pair at that site in the wild type.) To strengthen their assignments, which depend heavily upon the responses of the mutants to hydroxylamine, the authors show in passing that the very minor chemical reaction between hydroxylamine and thymine does *not* correlate (in its pH dependency) with the mutagenicity of hydroxylamine. They then go on to study the behavior of a base analogue, 5-fluorouracil, which only enters RNA and not DNA. As Paper 17 illustrates, a kind of nonheritable phenotypic mutagenesis results from 5-fluorouracil-induced mispairings during gene expression; the pattern of such "phenotypic suppression" allowed Champe and Benzer to deduce the *orientation* of base pairs at sites that generated *rII* mutations by G : C $\rightarrow$ A : T transitions. (Note that the genetic maps of Figures 5 and 6 of Paper 17 are extensions of the grander genetic map of Paper 30.)

This novel application of 5-fluorouracil appears to have provided three research groups with the same idea simultaneously, for just two years later each reported the induction of true mutations in RNA viruses using this base analogue (Cooper, 1964; Davern, 1964; Kramer et al., 1964). Furthermore, the development of a very different method for deducing the orientation of DNA base pairs was also probably stimulated by this study, for Tessman reported in 1962 and Levisohn in 1967 that an *rII* mutant just induced by nitrous acid, or just reverted by hydroxylamine, can grow in the nonpermissive host only when it is the strand encoding the messenger RNA, which is wild type in a heteroduplex heterozygote. The same relationship was to be crucial to the work reported in Paper 18.

Paper 18, from the Brenner group, is one of the few in this collection that stresses the applications of mutagen specificity for the analysis of gene function more than it reveals aspects of the mutagenic process itself. Paper 18 announced the formal deduction of the composition of the two most important chain termination codons, amber (UAG) and ochre (UAA). Previous work (Benzer and Champe, 1962; Sarabhai et al., 1964) had shown that amber mutants terminate polypeptide chain elongation, and the UAG and UAA codons were good candidates for the signals responsible for chain termination on the basis of the codon catalogue as it was then known. This very tightly reasoned genetic analysis was required, however, to prove the point unequivocally. Paper 18 also postulated a mechanism of suppression mediated by alterations in the anticodons of tRNA molecules; the first biochemical proof of this supposition was given by Goodman et al. in 1968

REFERENCES

Benzer, S., and S. P. Champe. 1962. A change from nonsense to sense in the genetic code. *Proc. Natl. Acad. Sci., 48:* 114.

Cooper, P. D. 1964. The mutation of poliovirus by 5-fluorouracil. *Virology, 22:* 186.

Davern, C. I. 1964. The inhibition and mutagenesis of an RNA bacteriophage by 5-fluorouracil. *Australian J. Biol. Sci., 17:* 726.

Goodman, H. M., J. Abelson, A. Landy, S. Brenner, and J. D. Smith. 1968. Amber suppression: a nucleotide change in the anticodon of a tyrosine transfer RNA. *Nature, 217:* 1019.

Kramer, G., H. G. Wittmann, and H. Schuster. 1964. Die Eizeugung von Mutaten des Tabakmosaikvirus durch den Einbau von Fluoruracil in die Virusnucleinsäure. *Z. Naturforsch., 19B:* 46.

Levisohn, R. 1967. Genotypic reversion by hydroxylamine: the orientation of guanine-hydroxymethylcytosine at mutated sites in phage T4rII. *Genetics, 55:* 345.

Sarabhai, A. S., A. O. W. Stretton, S. Brenner, and A. Bolle. 1964. Co-linearity of the gene with the polypeptide chain. *Nature, 200:* 13.

Tessman, I. 1962. Mutagenesis and the functioning of genetic material in phage. In *The Molecular Basis of Neoplasia*, p. 172. University of Texas Press, Austin, Tex.

17

Reprinted from *Proc. Natl. Acad. Sci.*, **48**(4), 532–546 (1962)

REVERSAL OF MUTANT PHENOTYPES BY 5-FLUOROURACIL: AN APPROACH TO NUCLEOTIDE SEQUENCES IN MESSENGER-RNA

By Sewell P. Champe and Seymour Benzer

DEPARTMENT OF BIOLOGICAL SCIENCES, PURDUE UNIVERSITY

Communicated February 27, 1962

Genetic information in DNA is apparently expressed via transcription into RNA messengers[1-5, 10] which in turn act as the templates for protein synthesis. Thus, incorporation of base analogues into messenger-RNA could lead to errors in the reading of the message into an amino acid sequence. The effect would be an alteration of phenotype without a permanent change in the DNA genotype. A promising analogue for this purpose is 5-fluorouracil (5FU), which is readily incorporated into RNA, mostly in place of uracil.[6] Modification of proteins by 5FU has been reported by Naono and Gros[7] and by Bussard *et al.*[8] who found that the enzymes alkaline phosphatase and β-galactosidase synthesized in the presence of the analogue are abnormal.

A very sensitive method for the detection of induced errors in the translation of genetic information is to begin with a mutant that is defective in some function so that errors causing the appearance of a small amount of activity can be easily observed. This approach has been successful[9] using *r*II mutants of phage T4, which are ordinarily unable to grow on strain K of *E. coli*. Addition of 5FU after infection of the cell can partially reverse the phage mutant phenotype, leading to active development of the phage. The response is highly specific, occurring only with certain *r*II mutants and not others within the same cistron. Since, after phage infection, synthesis of ribosomal RNA and S-RNA is almost totally arrested,[10] it is plausible that the effect of 5FU on phage mutants is due to its incorporation into messenger-RNA.

Introduction of a fluorine atom at the 5-position in uracil would be expected to induce a positive charge elsewhere in the molecule, increasing the probability of loss of the proton from the 1-position. This would make it possible for 5FU to pair with guanine. Thus, 5FU might produce an effect by either (or both) of two mechanisms: by entering the messenger in place of U and behaving sometimes like C, or by going in (occasionally) like C and later pairing like U. In either case, assuming that the single-stranded messenger-RNA copies the DNA according to the Watson-Crick rules of base pairing, 5FU could reverse the effect of a mutation at a particular DNA site only if the corresponding base in the messenger is a pyrimidine.

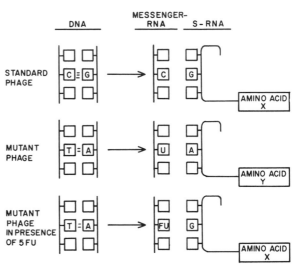

FIG. 1.—Proposed mechanism for the specific action of 5FU. The GC pair in the DNA of the standard (normal) phage is translated as C in the messenger-RNA, and a complementary S-RNA adaptor pairs with the latter, thus specifying amino acid X. For a mutant arising by a GC → AT transition, the base in the messenger-RNA becomes U, specifying an incorrect amino acid Y. In the presence of FU, however, U may be replaced by the analogue, which, when occasionally pairing like C, produces a normal message.

An additional requirement, if the altered message is to correspond to the original nonmutant one, is that the mutant DNA must have been derived from the standard type by a "transition" mutation,[11] i.e., a substitution in which the orientation of purine and pyrimidine between the two DNA chains remains unchanged.

Figure 1 illustrates the proposed mechanism for a mutant that has arisen by substitution, in the phage DNA, of an adenine-thymine (AT) pair for a guanine(hydroxymethyl) cytosine (GC) pair. At the corresponding site in the messenger-RNA, the mutant has U instead of C, so that the message is incorrect. In the presence of 5FU, however, U may be replaced by the analogue, which, when functioning as C, produces a normal message.

Among mutants containing an AT base pair, the A can be in either strand of the DNA. If each strand of the DNA is transcribed into a functional messenger, all transition mutants containing AT as the mutant base pair should respond to 5FU. On the other hand, if only one strand of the DNA is transcribed into a useful messenger, about one half of the transition mutants containing AT as the mutant base pair should respond to 5FU.

In this paper, rII mutants of phage T4 are analyzed to characterize the mutations in their DNA (by induction of reverse mutations with DNA base analogues) and also for phenotypic response to 5FU. The results indicate that only one functional messenger is produced and suggest the assignment of bases to the various sites in the messenger-RNA.

Materials and Methods.—Bacterial strains: For crosses and for nonselective plating, *E. coli* B was used. *E. coli* BB, which does not discriminate between standard type (r^+) phage and rII mutants, was used for growing all phage stocks and as the host for measuring induction of reverse mutations by base analogues. *E. coli* K (the specific variety used here being the KB strain) supports the growth of r^+ but not rII mutants, and was used as selective host in detecting reversion from rII to r^+. K was also used as the host in testing the 5FU effect on rII mutants. *E. coli* K10 and its phosphatase negative derivatives[12, 13] were obtained from Dr. Alan Garen.

rII mutants:[14] All were derived from phage T4B, with the exception of those designated by ED, which are spontaneous mutants derived from T4D. *Spontaneous mutant* numbers have either no prefix, or the prefixes SN or SD. Single-letter prefixes from A through J designate spontaneous mutants derived from revertants of spontaneous rII mutants. *Mutagen-induced mutants* are prefixed as follows: NA, NB, or NT induced by nitrous acid; EM by ethyl methane sulfonate; HB by hydroxylamine; N or M by 5-bromouracil or 5-bromodeoxyuridine; AP by 2-aminopurine; DAP by 2,6-diaminopurine; BC by 5-bromodeoxycytidine; P by proflavine; PT, PB by heat at low pH; UV by ultraviolet light. Many of the mutants were contributed by J. Drake, R. Edgar, E. Freese, M. Meselson, and I. Tessman.

Genetic mapping of rII mutants was done by techniques previously described.[14]

Media: Unless otherwise noted, the medium was broth (1% Difco bacto-tryptone plus 0.5% NaCl). For plates, 1.2% agar was added for the bottom layer and 0.7% for the top layer. In experiments involving 5-fluorouracil, a supplemented synthetic medium (M9S) was used containing, per liter of solution, 5.8 gm Na_2HPO_4, 3.0 gm KH_2PO_4, 0.5 gm NaCl, 1.0 gm NH_4Cl, 0.25 gm $MgSO_4 \cdot 7H_2O$, 2.7 mg $FeCl_3 \cdot 6H_2O$, 4.0 gm glucose, 20 mg L-tryptophan, and 2.5 gm Difco vitamin-free casamino acids. M9 is the same medium minus the casamino acids. M9 buffer is the same medium minus the carbon compounds. For experiments involving induction of alkaline phosphatase, Tris-glucose medium was used which contained, per liter, 12.1 gm Tris (Sigma Chemical Co.), 3.0 gm $MgSO_4 \cdot 7H_2O$, 1.0 gm $(NH_4)_2SO_4$, 0.5 gm sodium citrate, 50 mg methionine, and 1.0 gm glucose, pH adjusted to 7.4 with HCl. This medium was used either with 10 gm per liter KH_2PO_4 ("high phosphate") to repress phosphatase synthesis, or with 5 mg per liter ("low phosphate") to induce phosphatase synthesis.[15] *5-fluorouracil* was kindly donated by Dr. R. Duschinsky of the Hoffmann-LaRoche Company.

Reversion induction by base analogues: *E. coli* BB was grown in M9 to 10^8 cells/ml and the phage in question was added to give about 200 particles per ml. One ml of this mixture was then added to each of three tubes containing, respectively, 1 ml of (*a*) plain M9, (*b*) M9 plus 5-bromodeoxyuridine (0.1 mg/ml) and (*c*) M9 plus 2-aminopurine (1 mg/ml). The tubes were incubated at 37°C for 20 hr, then shaken with a few drops of chloroform to complete lysis. Each lysate was assayed on strain B for total plaque-forming particles and on strain K for revertants; the ratio of the titer on K to the titer on B is the *reversion index*. For all cases in which the reversion index appeared to be raised by a mutagen, the test was repeated at least once to assure that the effect was not due to a "jackpot"[16] (i.e., an abnormally large clone due to early appearance and subsequent replication of a revertant).

Reversion induction with hydroxylamine was tested by the procedure of Freese, Bautz, and Bautz-Freese.[17] One volume of phage stock was added to four volumes of a freshly made reaction mixture containing 1.25 M $NH_2OH \cdot HCl$, 1.0 M NaCl, 0.001 M $MgSO_4$, 0.075 M Na_2HPO_4, the pH being adjusted to 7.5 with NaOH. Samples were taken before and after four hours of treatment at 37°C, diluting into a cold reaction-stopping mixture consisting of 0.5% bacto-tryptone,

6% NaCl, and 2% by volume of acetone. Under the conditions used, inactivation of the phage particles by hydroxylamine amounted to roughly 50 per cent. The results are given directly as the reversion index (i.e., ratio of titer on K to titer on B) before and after treatment. This was preferred to expressing them in terms of mutations per lethal hit, the latter quantity being difficult to measure when the killing is so small.

Verification of revertants: For any mutant that was induced to revert a number of plaques (from 4 to 30) was picked from the K plates and replated on strain B as a check on plaque type. In some cases, the plaques were r type on B, showing that the "revertants" were false, i.e., did not represent a return to the original standard type. Such false reversions may be due to suppressor mutations at some site other than the original mutation.[18, 19] Therefore, data are reported only for mutants most of whose induced revertants looked genuine on B. This is, of course, a necessary but still not sufficient criterion for genuineness of the revertants.

Assay of phosphatase activity:[20] p-nitrophenyl phosphate was dissolved at 5mg/ml in 1 M Tris buffer, pH 8.0, and extracted with ether until colorless. This removes traces of p-nitrophenol and increases the sensitivity of the assay at low levels of enzyme. The reagent was added to an equal volume of appropriately diluted bacterial culture (previously shaken with a few drops of chloroform) and incubated at 37°C. The rate of color development measured at 410 mμ is a measure of the phosphatase activity.

Results.—Effect of 5FU on rII mutants: Figure 2 shows the effect of 5FU on the activity of various rII mutants in *E. coli* strain K. The procedure was to infect K cells with the phage mutant, and dilute into media with or without 5FU. At various times thereafter, the infected cells were further diluted into broth and allowed to lyse. Reversal of the phenotype of an rII mutant is reflected in appearance of phage progeny, the yield per infected cell, as compared with that obtained for r^+ phage under the same conditions, being a measure of the degree of activity. While two of the mutants in Figure 2 produce a sizable fraction of the r^+ yield, the other three mutants show practically no response. This striking specificity, which applies to mutants of both cistrons, shows that the effect of 5FU cannot be a general removal of the block against rII mutants. Although 5FU is in general somewhat inhibitory, depressing the yield of the standard type phage by about twofold, the increase shown by responsive mutants is over and above this general inhibition.

Stimulation by 5FU was almost completely prevented by adding uracil at a concentration of 20 γ/ml. On the other hand, inclusion or omission of thymidine at the same concentration made no detectable difference. This indicates that the stimulating effect of 5FU is not due to its interference[21] with DNA synthesis.

It is essential to stress that the effect of 5FU is a physiological one and not due to mutagenesis. The progeny phage yielded by K in the presence of 5FU were no more active (in the absence of 5FU) than their parents.

Kinetics of the 5FU effect: To localize the time of action of 5FU in phage-infected cells, 3-minute pulse exposures to the analogue were given at various times before and after infection. The results (Fig. 3) show that pre-infection exposure causes little effect and that 5FU is most effective if present during the first 18 minutes at 26°. This is the first third of the eclipse period, a time during which there is active synthesis[1, 2] of messenger-RNA but before DNA synthesis has begun.[38]

That 5FU is rapidly incorporated into messenger-RNA has been shown by Gros *et al.*[22] Their studies were on uninfected cells, but, as indicated by the work of Nomura, Hall, and Spiegelman,[10] messenger-RNA may in fact be the only kind of RNA made after phage infection.

As Garen[23] and Nomura[24] have shown, the metabolism of an rII-infected K cell is

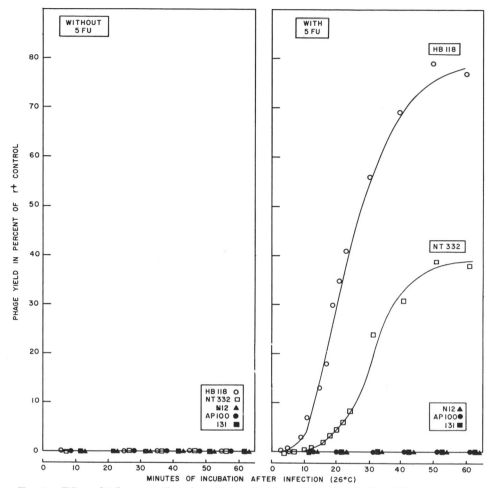

Fig. 2.—Effect of 5-fluorouracil on the activity of various *r*II mutants in *E. coli* K.

To K growing exponentially in M9S at 2.3 × 10⁸ cells/ml, 0.01 *M* NaCN was added followed by phage at 0.3 particle per bacterium. After 10 min adsorption at 37°C, anti-T4 serum was added to eliminate unadsorbed phage. Ten min later, phage development was initiated (time zero on the graph) by diluting 500-fold into M9S + 20 γ/ml thymidine at 26°C either without or with 10 γ/ml 5FU. At various times thereafter, samples were diluted 50-fold into broth supplemented with 20 γ/ml of uracil and 20 γ/ml of thymidine. After one hour at 37°C, the broth tubes were shaken with a few drops of chloroform to complete lysis and assayed for phage on *E. coli* B. As a control, *r*⁺ phage (not shown) was measured in precisely the same way. The *r*⁺ yield decreased somewhat with progressively later shifts from synthetic medium to broth medium so that the data for the mutants are expressed as per cent of the yield for *r*⁺.

at first normal and even proceeds as far as the production of DNA replicas before the cell ceases to function. Nevertheless, standard type recombinants arising during replication are unable to remove the block.[25] This decision is presumably dictated by messenger-RNA made directly from the DNA of the infecting phage particles.

Generality of the 5FU effect: If the mechanism postulated for the action of 5FU on *r*II mutants is correct, the same phenomenon should be observable with mutants affected in other cistrons. This has indeed been found for mutants of *E. coli* strain K10 that are defective in the enzyme alkaline phosphatase. Twenty-five phosphatase-negative mutants,[12, 13] generously supplied by Dr. Alan Garen, were tested. Each strain was grown in high phosphate medium (to repress formation of any enzyme) and then shifted to low phosphate medium, in which condition formation of the enzyme (if any) is induced.[15] When induction was done in the presence of 5FU, two of the mutants formed active enzyme. Figure 4 compares a responsive mutant (U8) with a nonresponsive mutant (U12). Thus, the 5FU effect, although highly site-specific within a given cistron, is of general applicability to other cistrons.

The increase in enzyme activity of U8 induced in the presence of 5FU, while some 30-fold over the control, was small (about 0.5%) relative to the phosphatase-positive strain under the same conditions. In the case of the *r*II mutants shown in Figure 2, the phage yield may, of course, be far out of proportion to the amount of r^+ activity.

Identification of DNA base pairs by mutagens: When DNA replicates in the presence of analogues of the normal DNA bases, mutations are readily induced,[26] apparently due to errors in pairing which lead to the permanent substitution of one base pair for another. As argued by Freese,[27] such errors in base pairing should lead to the substitution of a purine for a purine and a pyrimidine for a pyrimidine

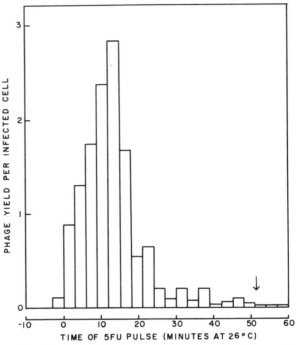

FIG. 3.—Effect of a three-minute pulse of 5-fluorouracil, applied at various times, on *r*HB118 infecting *E. coli* K. The procedure was as in Fig. 2 except that the infected cells were allowed to develop in M9S medium for various times before exposure to 5FU. The pulse was applied by diluting into an equal volume of the same medium plus 20 γ-ml 5FU. Three minutes later, the pulse was ended by dilution into uracil-containing broth and the experiment continued as in Fig. 2. Due to the short exposure to the analogue the phage yield per infected cell was much smaller than in Fig. 2. The arrow at 52 minutes indicates the end of the "eclipse" period, i.e., the time at which the number of mature phage particles reaches, on the average, one per cell for cells infected with r^+ in the absence of 5FU.

without changing their orientation with respect to the two DNA chains (transitions). Two base analogues that have been studied extensively are 2-aminopurine (2AP) and 5-bromodeoxyuridine (BDU). These induce mutations specifically at certain genetic sites,[28, 27, 14] and the mutants are, as a rule, also inducible by base analogues to revert to standard type.[11]

From the finding that mutants induced by BDU tend to be strongly revertible by 2AP, and vice versa, Freese[11] suggested that one of the mutagens favored the transition GC → AT while the other induced AT → GC. An indication of which is which came from experiments with hydroxylamine.[17, 29] Cytosine and its derivatives were shown to react more readily than the other bases that occur in DNA.

It is difficult, however, to extrapolate with assurance from the gross chemical effects of mutagens to the actual mutational events, which could be caused by a minority reaction. Schuster[30] and Brown and Schell[31] have shown that hydroxylamine indeed reacts to a slight extent with thymine in DNA. However, Schuster also showed that the rate of reaction of hydroxylamine with cytosine (in RNA) decreases with higher pH, whereas the reaction with uracil increases. The pH dependences of the reactions with thymine and hydroxymethylcytosine in phage DNA, although not yet determined, might be expected to be analogous to those of uracil and cytosine, respectively. To obtain further evidence as to which reaction is responsible for the mutagenic effect, we have examined the pH dependence of the mutagenic effect of hydroxylamine on the rII mutant $rAP275$. The observed mutation rates (induced revertants per survivor per unit time) at pH values of 6, 7.5, and 9 were in the ratio $28:12:1$. The rates found by Schuster for the reaction with cytosine at similar pH values were in the ratio of $32:13:<4$, whereas uracil reacted in the ratio $<1:13:30$. Six other mutants were tested and all

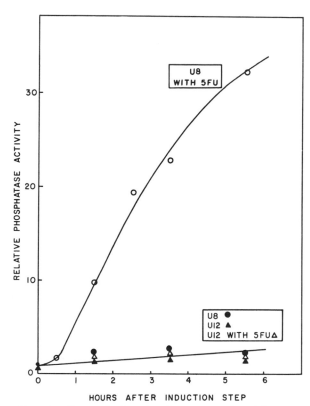

FIG. 4.—Reversal of a phosphatase-negative defect by 5-fluorouracil. Overnight cultures of the bacteria grown in repressing medium ("high phosphate" Tris-glucose) were diluted 20-fold into the same medium and aerated at 37°C for 3 hours. The cells were centrifuged, resuspended in chilled medium (less glucose and phosphate), centrifuged again, and resuspended in inducing medium ("low phosphate" Tris-glucose). The cultures were then divided into 2 parts, to one of which was added 20 γ/ml of 5FU, and both were aerated at 37°C (starting at time zero on the graph). Samples were taken at various times and assayed for alkaline phosphatase activity as described in *Methods*. The ordinate is in arbitrary units relative to the activity of mutant U8 at zero time.

showed the same pH dependence as r AP275.

This result reinforces the conclusion that hydroxylamine predominantly induces the transition GC → AT, and the correlation between hydroxylamine and BDU (Table 1, below) supports the conclusion that BDU acts in the same way. The action of 2AP, on the other hand, seems less specific than originally supposed. In the data of Freese et al.,[17] as well as those given below, 2AP-induced mutants are about equally divided with respect to their revertibility by hydroxylamine.   2AP apparently induces both transitions with comparable ease.

Therefore, the following rules may be used for identification of base pairs. Given a mutant (of any origin), if it is inducible to revert by 2AP, it can be classed as a transition mutant.   If it reverts strongly in response to BDU and hydroxylamine as well, this suggests that the mutant base pair is GC; if it does not, the mutant base pair indicated is AT.   These criteria identify the base pair at the corresponding site in the standard type phage as AT in the former case and GC in the latter case.

*r*II mutations have been located at some hundreds of distinct sites in the two *r*II cistrons.   Most of the sites have many recurrences, some spontaneous, some induced with one mutagen or another.[14]   One representative of each site in the *r*II region was chosen (usually the first mutant to indicate the site) and tested quantitatively for induction of reversions by 2AP and BDU.   Where an increase over the spontaneous rate was observed, the revertants were examined for plaque type on strain B (see *Methods*).

Of the 339 mutants tested, 69 were unambiguously inducible to revert by 2AP (and/or BDU) to a form which was judged to be phenotypically r^+ on strain B. (Some of these also showed false revertants in smaller numbers.)   The mutants are listed in Table 1 (Groups I, II, and III) along with 40 mutants (Group IV) chosen at random from the larger group which showed no reversion induction.   All 109 mutants were further tested for reversion by hydroxylamine.

Mutants of Groups I and II are those that are reverted strongly by 2AP but weakly, or not at all, by BDU.   According to the rules based upon the mutagenic specificity of the analogues, these mutants must have arisen by GC → AT transitions.   The base pairs deduced to be in the standard type DNA at the corresponding sites are listed in the table as GC.

Group III consists of mutants that respond not only to 2AP but also to BDU. It is seen that there is a good correlation between hydroxylamine and BDU induction, consistent with the idea that both of these agents act predominantly on GC base pairs.   For mutants responding to *both* BDU and hydroxylamine, the base pair indicated in the standard type is AT.   In the few cases where a mutant responds to BDU and not hydroxylamine, or vice versa, identification of the base pair cannot be made with assurance.   The mutants of Group IV, not revertible by base analogues, could be transversions,[11] i.e., substitutions in which the orientation of purine and pyrimidine has been reversed, or tiny deletions or insertions.[32]   For these mutants, no conclusion as to the original base pair can be reached.

Figure 5 shows the positions of the various mutations within the two cistrons of the *r*II region, as determined by genetic mapping.   Mutations which showed no induced reversions by 2AP or BDU are represented as open circles.   Those represented as solid circles are inducible to revert (either truly or falsely) by one or more of the mutagens.   Where the base pair can be labeled unambiguously according to the data of Table 1, it is shown in the figure.

The remainder of the mutations indicated as responsive to mutagens showed either feeble induction or a large component of manifestly false revertants, and therefore no base pair designation can be given without further study.   Of course, a revertant can be demonstrated to be false, but cannot be proved, by genetic analysis alone, to be identical with the original standard type, so that the possibility is not

TABLE 1

RESPONSE OF rII MUTANTS TO MUTAGENS AND TO 5FU

	Reverse Mutation Induction by Mutagens Reversion Index (K/B) in Units of 10^{-7}					Response to 5FU Progeny per cell		Summary				Indicated Bases r^+	r^+
	Spont.	2AP	BDU	NH_2OH Control	NH_2OH Treated	Control	5FU	2AP	BDU	NH_2OH	5FU	Phage DNA	Messenger-RNA
r^+						126.	79.						
GROUP I													
rN55	1.4	1700	2.8	0.	0.5	0.00	0.9	+	0	0	+	GC	C
rHB118	0.5	2300	0.	0.5	0.	0.00	56.	+	0	0	+	GC	C
rC204	0.0	730	0.2	0.	0.	0.00	1.3	+	0	0	+	GC	C
rN11	1.3	2800	5.3	0.	0.	0.00	1.5	+	0	0	+	GC	C
rEM64	0.4	2500	1.3	0.9	0.5	0.00	33.	+	0	0	+	GC	C
rHB35	0.1	1400	0.2	2.2	2.0	0.00	9.6	+	0	0	+	GC	C
rHB129	0.2	1000	0.3	1.6	0.	0.00	24.	+	0	0	+	GC	C
rHB84	0.0	750	0.1	0.	0.	0.00	14.	+	0	0	+	GC	C
rN21	0.0	450	0.0	0.	0.	0.01	29.	+	0	0	+	GC	C
rEM84	0.2	1600	0.6	0.4	3.1	0.00	5.2	+	0	0	+	GC	C
rN24	3.0	4600	5.9	8.6	20.	0.02	8.2	+	0	0	+	GC	C
rHB74	0.3	3000	0.5	0.8	3.	0.00	0.6	+	0	0	+	GC	C
rNT332	0.2	1400	1.6	1.9	1.6	0.00	22.	+	0	0	+	GC	C
rB94	0.0	2800	0.3	2.	5.	0.00	7.4	+	0	0	+	GC	C
rSD160	0.5	1500	3.6	0.	3.	0.01	2.8	+	0	0	+	GC	C
rHB232	0.1	300	0.5	0.9	2.5	0.02	11.	+	0	0	+	GC	C
rAP53	0.6	890	0.5	0.3	2.5	0.04	48.	+	0	0	+	GC	C
GROUP II													
rNA27	0.1	1600	0.5	1.	3.	0.00	0.05	+	0	0	0	GC	G
rN74	0.0	390	0.5	0.5	0.	0.04	0.2	+	0	0	0	GC	G
r1274	0.0	330	7.0	2.8	3.5	0.00	0.00	+	0	0	0	GC	G
r1249	0.1	1600	20.	1.8	1.7	0.00	0.01	+	0	0	0	GC	G
rNB4777	0.6	630	58.	2.4	4.8	0.00	0.05	+	0?	0	0	GC	G
rHB122	0.1	14000	4.6	0.7	0.5	0.00	0.08	+	0	0	0	GC	G
rBC11	0.0	15000	0.	1.	0.5	0.00	0.06	+	0	0	0	GC	G
rUV47	0.2	280	9.	0.9	0.8	0.00	0.05	+	0	0	0	GC	G
rAP211	0.0	200	0.5	0.	1.2	0.00	0.06	+	0	0	0	GC	G
rUV1	0.0	350	0.1	0.	0.	0.00	0.03	+	0	0	0	GC	G
r425	0.3	3300	1.4	0.	0.	0.00	0.00	+	0	0	0	GC	G
rEM20	0.0	3000	0.8	1.	0.5	0.00	0.01	+	0	0	0	GC	G
rUV122	0.5	1600	18.	2.2	3.7	0.16	0.06	+	0	0	0	GC	G
rHB32	0.1	4800	1.8	1.8	2.7	0.00	0.09	+	0	0	0	GC	G
r585	0.0	330	0.7	0.	5.	0.00	0.01	+	0	0	0	GC	G
r1310	0.0	1000	0.0	0.	0.5	0.00	0.00	+	0	0	0	GC	G
rHB80	0.0	3500	12.	0.4	0.8	0.00	0.07	+	0	0	0	GC	G
rAP126	0.0	2900	0.7	0.	0.8	0.00	0.06	+	0	0	0	GC	G
rUV375	0.6	6700	24.	0.9	1.7	0.00	0.04	+	0	0	0	GC	G
r360	0.0	530	0.6	0.	0.	0.00	0.03	+	0	0	0	GC	G
r375	0.3	2200	2.0	0.8	1.1	0.00	0.00	+	0	0	0	GC	G
rN17	0.0	1500	0.4	0.	0.	0.00	0.09	+	0	0	0	GC	G
rN90	0.4	8600	4.5	1.	2.	0.01	0.06	+	0	0	0	GC	G
rUV199	2.4	5900	7.1	0.7	0.3	0.00	0.2	+	0	0	0	GC	G
rN12	0.2	8500	1.3	0.6	1.5	0.02	0.05	+	0	0	0	GC	G
rN29	0.0	7600	0.	0.	4.	0.00	0.00	+	0	0	0	GC	G
rAP61	1.2	500	2.9	1.9	4.1	0.02	0.1	+	0	0	0	GC	G
r979	1.5	3200	24.	5.7	7.1	0.01	0.08	+	0	0	0	GC	G
rEM7	0.1	2000	0.6	0.2	0.0	0.00	0.04	+	0	0	0	GC	G
r1814	1.	8200	10.	0.9	72.	0.04	0.2	+	0	+	0	..	..
r287	2.1	2800	1.0	6.4	150.	0.00	0.00	+	0	+	0	..	..

The mutants are divided into groups according to their responses to mutagens and to 5FU. Within each group, they are listed according to their order in the recombination map (Fig. 5). In cases of positive response to 2AP or BDU, the reversion test was done two or more times and the value listed is the average reversion index, excluding extreme values occasionally observed due to "jackpots." A zero in the last decimal place indicates that the value in that place is less than one.

The control for hydroxylamine reversion is in some cases different from the spontaneous reversion index due to fluctuations of the r^+ background from one stock to another. Procedures for the mutagen tests are described in *Methods*. The measurements of response to 5FU were made as described in Figure 2, the dilution from the test medium being made about 45 minutes after the initiation of phage development.

ruled out that the apparent specificity is incorrect in certain cases. A more stringent test of genuineness is to cross the presumed revertant to standard type, in which case a suppressor mutation at another site may be revealed by the appearance of the unsuppressed mutant as a recombinant. A still more stringent test is to show that the revertant has the same forward mutability, at the same site, as

TABLE 1 (Continued)

RESPONSE OF rII MUTANTS TO MUTAGENS AND TO 5FU

	Reverse Mutation Induction by Mutagens Reversion Index (K/B) in Units of 10^{-7}					Response to 5FU Progeny per cell		—Summary—				Indicated Bases r^+ r^+	
	Spont.	2AP	BDU	NH₂OH Control	NH₂OH Treated	Control	5FU	2AP	BDU	NH₂OH	5FU	Phage DNA	Messenger-RNA
GROUP III													
rAP129	4.1	6200	2100.	4.	680.	0.1	0.4	+	+	+	0	AT	..
rAP218	2.6	1200	250.	1.	730.	0.01	0.2	+	+	+	0	AT	..
rH221	0.4	460	220.	0.	26.	0.00	0.04	+	+	+?	0	AT	..
rAP100	2.0	80	1800.	3.	2600.	0.00	0.01	+	+	+	0	AT	..
r607	0.9	340	7100.	1.9	1500.	0.03	0.02	+	+	+	0	AT	..
rEM114	30.	3200	1200.	100.	890.	0.5	0.8	+	+	+	0	AT	..
rNT88	0.5	230	500.	0.1	160.	0.00	0.04	+	+	+	0	AT	..
rDAP56	3.6	5500	2000.	16.	3100.	0.00	0.01	+	+	+	0	AT	..
r380	0.5	6600	2400.	0.	220.	0.4	1.8	+	+	+	+?	AT	..
rSN103	0.9	580	1800.	2.2	370.	0.01	0.07	+	+	+	0	AT	..
r263	7.8	3700	8900.	27.	3600.	0.6	3.3	+	+	+	+?	AT	..
rF72	0.7	120	680.	2.6	770.	0.00	0.01	+	+	+	0	AT	..
rJ33	1.5	2000	890.	4.	230.	0.00	0.04	+	+	+	0	AT	..
rBC35	2.1	46	760.	1.2	660.	0.00	0.01	+	+	+	0	AT	..
rAP275	1.9	580	4500.	4.1	1600.	0.00	0.04	+	+	+	0	AT	..
rUV363	0.7	26	3000.	3.5	600.	0.00	0.03	+	+	+	0	AT	..
rUV181	0.07	940	470.	0.8	1.4	0.00	0.06	+	+	0	0	..	..
rG178	0.07	970	86.	0.	0.5	0.00	0.02	+	+	0	0	..	..
r2074	4.7	25000	390.	19.	16.	0.00	0.01	+	+	0	0	..	..
r609	3.3	8800	450.	3.0	1.0	0.02	0.04	+	+	0	0	..	..
r1221	1.9	1900	170.	9.2	14.	0.2	5.6	+	+	0	+	..	..
GROUP IV													
r681	3.1	1.7	3.4	8.	16.	0.00	0.00	0	0	0	0	..	..
rUV11	0.1	0.1	0.4	0.	0.	0.02	0.02	0	0	0	0	..	..
r569	0.9	1.3	4.7	0.5	0.6	0.02	0.1	0	0	0	0	..	..
r1176	0.2	0.0	0.2	0.	0.	0.00	0.00	0	0	0	0	..	..
rUV68	1.2	1.1	3.0	2.6	1.4	0.00	0.01	0	0	0	0	..	..
rNT311	0.0	0.1	0.0	0.9	0.8	0.00	0.01	0	0	0	0	..	..
rBC81	0.0	0.0	0.1	0.0	0.0	0.00	0.01	0	0	0	0	..	..
r227	0.1	0.0	0.	0.	0.	0.00	0.00	0	0	0	0	..	..
r577	0.1	2.2	0.0	0.	0.	0.00	0.00	0	0	0	0	..	..
r1084	0.4	2.5	2.2	1.	0.	0.00	0.00	0	0	0	0	..	..
r205	0.0	0.0	0.1	0.5	0.5	0.00	0.01	0	0	0	0	..	..
rH201	3.5	0.8	3.5	0.8	2.9	0.00	0.00	0	0	0	0	..	..
r465	1.7	3.0	4.2	7.6	6.5	0.15	0.04	0	0	0	0	..	..
r1470	0.2	0.2	0.2	0.	0.	0.00	0.01	0	0	0	0	..	..
r447	0.4	0.3	0.2	0.8	1.4	0.00	0.01	0	0	0	0	..	..
rDAP66	1.0	2.0	1.3	0.7	1.	0.00	0.00	0	0	0	0	..	..
rH51	0.2	0.3	0.5	0.5	0.3	0.00	0.02	0	0	0	0	..	..
r2232	1.8	1.9	1.9	4.	2.	0.00	0.02	0	0	0	0	..	..
rP5	2.2	8.1	3.7	6.	5.	0.00	0.05	0	0	0	0	..	..
rF27	4.1	1.0	5.3	1.2	27.	0.00	0.00	0	0	+?	0	..	..
r285	0.6	0.3	0.4	0.6	3.3	0.00	0.00	0	0	0	0	..	..
rC135	0.2	0.2	0.2	0.	0.	0.00	0.01	0	0	0	0	..	..
rSN181	2.4	0.7	2.2	10.	8.	0.01	0.02	0	0	+?	0	..	..
rD2	3.9	4.7	4.0	3.5	43.	0.00	0.02	0	0	+?	0	..	..
r173	0.0	0.0	0.0	0.	0.	0.01	0.02	0	0	0	0	..	..
rUV118	0.3	0.2	0.4	0.7	0.	0.00	0.03	0	0	0	0	..	..
r240	0.1	0.0	0.1	3.6	3.7	0.00	0.01	0	0	0	0	..	..
r131	2.0	3.7	6.6	4.1	3.0	0.01	0.03	0	0	0	0	..	..
r244	8.0	6.1	22.	13.	14.	0.00	0.01	0	0	0	0	..	..
r326	0.3	0.6	1.2	0.	0.	0.00	0.01	0	0	0	0	..	..
rP4	0.3	0.0	0.	0.1	0.	0.00	0.00	0	0	0	0	..	..
rAP176	0.1	0.2	0.0	1.8	1.7	0.00	0.02	0	0	0	0	..	..
r117	16.	2.7	5.0	27.	37.	0.00	0.00	0	0	0	0	..	..
r1467	6.3	5.7	3.4	8.9	7.5	0.00	0.01	0	0	0	0	..	..
rB13	0.0	0.1	0.2	0.3	0.0	0.00	0.00	0	0	0	0	..	..
rJ241	0.1	0.1	0.7	0.	0.7	0.00	0.01	0	0	0	0	..	..
rEM113	0.0	0.1	0.1	0.	0.	0.00	0.01	0	0	0	0	..	..
rEM29	0.0	0.3	0.0	0.	0.	0.00	0.01	0	0	0	0	..	..
rUV124	0.0	0.0	0.0	0.5	0.0	0.00	0.02	0	0	0	0	..	..
rEM87	0.0	0.0	0.	0.5	0.6	0.00	0.01	0	0	0	0	..	..

does the standard type.   While both of these criteria are satisfied by some revertants that have been studied, the tests have not yet been applied to the various mutants of Table 1.   It is possible also that the 5FU effect could, in certain instances, act at a site distant from the one in question, producing the phenotypic equivalent of a suppressor mutation.

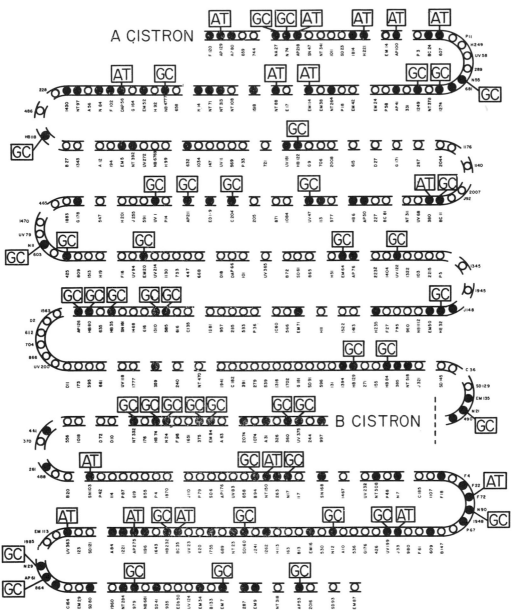

Fig. 5.—Genetic map of the *r*II region showing the base pairs in the standard type as deduced from the data of Table 1. Each circle represents a distinct mutational site. Breaks in the map indicate segments as defined by the ends of deletions. While the order of the segments is known, the arrangement of sites within any one segment has not been determined.

● Indicates mutants which are inducible to revert by one or more of the mutagens tested (2AP, BDU, or hydroxylamine). Where reversion was weak or false, no base pair is assigned.

○ Indicates mutants for which no induced reversion was detected with 2AP or BDU.

Specificity of response to 5FU: All of the mutants in Table 1 have been tested for reversal of phenotype by 5FU and the results are listed.

The first expectation from the proposed mechanism, namely, that only transition

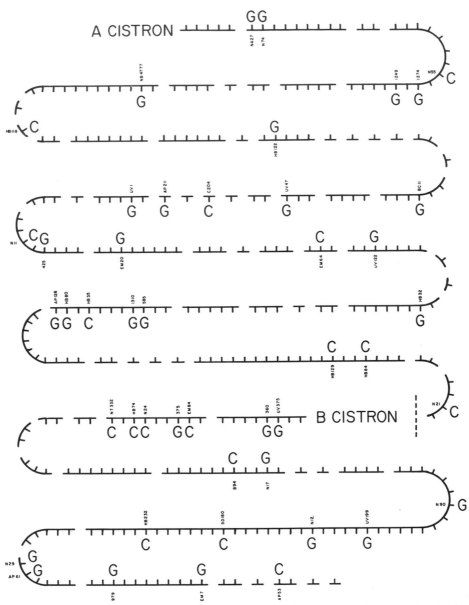

FIG. 6.—Map of the messenger-RNA for the *r*II region showing the nucleotides (for the standard type phage) suggested by the data of Table 1 (Groups I and II).

mutations (Groups I, II, and III) should respond to 5FU, is seen to be rigorously fulfilled.   Out of the 20 mutants judged to have a positive response to 5FU, 17 are among those induced to revert only by 2AP.   The other three (having rather high control values and weak response) are induced to revert also by BDU.   This distribution implies that 5FU acts predominantly by being incorporated into messenger-RNA at sites corresponding to AT in the DNA.   That is, it enters in place of U and acts partially like C.   The weak but reproducible response observed with several

BDU-reverting mutants is not surprising if 5FU can also (but more rarely) be incorporated in place of C.

Unfortunately there is not an absolutely sharp separation between 5FU responders and nonresponders. A slight though variable response to 5FU can be detected for almost every mutant. This small nonspecific effect can be seen for some of the Group IV nontransition mutants, and occurs even with deletion mutants, so it would seem to be due to a partial undoing of the general block against *r*II mutants. Even for mutants that definitely respond, the degree of response varies. This is perhaps ascribable to the influence of neighboring bases on the efficiency of incorporating the analogue or on its properties once it is in.

From Table 1 (Groups I and II) it is evident, nevertheless, that the mutants revertible solely by 2AP include some (17) that are responsive to 5FU and others (29) that are not. This is close to the expectation if only one DNA strand is copied as a useful messenger. Response to 5FU would thus specify the relative orientation of the DNA base pairs at the sites in question. Given a GC base pair in the standard type, which becomes, in the mutant, AT, the latter will respond to 5FU only if the A is in the "major" strand of the DNA, i.e., the one which is transcribed into messenger-RNA. Thus, a 5FU-responsive mutant would be one which has G at the corresponding site in the major strand of the DNA of the standard type. It follows that the base in the standard type messenger-RNA is C. Figure 6 summarizes the various nucleotides in normal *r*II messenger-RNA deduced according to this scheme.

Discussion.—Unlike the analogues BDU and 2AP, which cause heritable changes in the DNA, 5FU causes a temporary change in phenotype. It is effective on phage only if added in the period shortly after infection during a time when messenger-RNA is actively synthesized but soluble RNA and ribosomal RNA are not. Mutants that are not induced to revert by base analogues do not respond to 5FU. Of those mutants identified by chemical mutagenesis as having an AT pair at the mutant site, somewhat less than half respond more or less strongly to 5FU. Conversely, among mutants identified as containing GC pairs, response to 5FU is uncommon.

These observations are consistent with the scheme presented in Figure 1. The fact that roughly half of the AT mutants respond to 5FU would appear to suggest that only one strand of the DNA is transcribed as functional messenger-RNA. The 5FU effect makes it possible to define the orientation of the base pair in the two DNA strands and thus provides the key needed to reduce the genetic map to nucleotide sequences along one of the DNA strands. The messenger-RNA would then have the complementary sequence. 5FU permits the orientation of only one of the two kinds of DNA base pairs. It would obviously be desirable to find an analogue of cytosine or guanine, which would serve for the other.

The results suggest that only one *physiologically active* messenger is made from the two-stranded DNA, as also suggested by certain genetic experiments of Tessman.[33] However, it is known from the work of Chamberlin and Berg[34] that, *in vitro*, both strands of DNA *can* be copied as RNA by ribonucleotide polymerase. It may be, nevertheless, that for an intact chromosome, *in vitro*, there is a mechanism fixing the direction in which the messenger copies the DNA. Suggestive evidence for one-strand copying comes from the experiments of Bautz and Hall,[35] who have isolated T4 messenger-RNA and measured its base composition. Their results reveal a deviation from identity between the base compositions of T4 DNA and the messenger-RNA. This indicates that messenger-RNA copies only one DNA strand, and further that the bases are not equally distributed between the two strands. They found 21 per cent G and 15 per cent C in the purified T4 messenger-RNA. Although the number of G and C sites so far indicated for the rII messenger (Fig. 6) is not large enough to give a statistically reliable sampling and may also be a random sample of all possible G and C sites, it is interesting to note that they deviate from equality in the same direction, as observed by Bautz and Hall.

In the set of mutants studied so far, ones arising from changes at GC sites are about twice as numerous as those arising at AT sites. This might appear to be in contradiction with the known composition of T4 DNA, in which the relative abundance of the base pairs is just the reverse. However, this may very easily be due to the fact that most of the mutagens so far used in producing the mutants turn out to act preferentially on GC pairs. If this is so, it should prove possible to detect many more sites by exhaustive mapping of mutants induced by 2AP.

Regarding the behavior of 5FU in base pairing, Lengyel *et al.*[36] found polyfluorouridylic acid to be rather ineffective in stimulation of polyphenylalanine synthesis (as compared to polyuridylic acid, which is extremely effective).[37] This is evidence that the fluorine atom does cause an alteration in the behavior of U. By studying the properties of a mixed polymer made of U and 5FU, it should be feasible to determine directly whether 5FU indeed acts partially like C in coding for amino acids.

Summary.—5-Flourouracil partially reverses the defective phenotypes of certain rII mutants of phage T4 (as well as certain phosphatase-negative mutants of *E. coli*). From the kinetics of the effect and the specificity of action on various rII mutants, taken together with the responses of the mutants to specific mutagens, it is concluded that 5FU acts mainly by incorporation into messenger-RNA in place of uracil, there acting partially like cytosine. The results are consistent with the idea that only one strand of the DNA duplex is copied into useful messenger-RNA. This provides a means for determining the orientation of the nucleotide pairs with respect to the two DNA chains, as well as identification of the nucleotides at the corresponding sites in the rII messenger-RNA.

We wish to express our gratitude to Mmes. Daine Auzins, Judith Berry, and Barbara Williams for their tireless and able assistance with the experiments. This research has been supported by grants from the National Science Foundation and the National Institutes of Health.

[1] Volkin, E., and L. Astrachan, *Virology*, **2**, 149 (1956).

[2] Astrachan, L., and E. Volkin, *Biochim. et Biophys. Acta*, **29**, 536 (1958).

[3] Jacob, F., and J. Monod, *J. Mol. Biol.*, **3**, 318 (1961).

[4] Brenner, S., F. Jacob, and M. Meselson, *Nature*, **190**, 576 (1961).

[5] Gros, F., H. Hiatt, W. Gilbert, C. G. Kurland, R. W. Risebrough, and J. D. Watson, *Nature*, **190**, 581 (1961).

[6] Horowitz, J., and E. Chargaff, *Nature*, **184**, 1213 (1959).

[7] Naono, S., and F. Gros, *C.R. Acad. Sci.* (*Paris*), **250**, 3889 (1960).

[8] Bussard, A., S. Naono, F. Gros, and J. Monod, *C.R. Acad. Sci.* (*Paris*), **250**, 4049 (1960).

[9] Benzer, S., and S. P. Champe, these PROCEEDINGS, **47**, 1025 (1961).

[10] Nomura, M., B. D. Hall, and S. Spiegelman, *J. Mol. Biol.*, **2**, 306 (1960).

[11] Freese, E., *J. Mol. Biol.*, **1**, 87 (1959).

[12] Levinthal, C., in *Structure and Function of Genetic Elements*, Brookhaven Symposia in Biology, **12**, 76 (1959).

[13] Garen, A., in *Microbial Genetics*, ed. W. Hayes and R. C. Clowes (Cambridge University Press, 1960).

[14] Benzer, S., these PROCEEDINGS, **47**, 403 (1961).

[15] Torriani, A., *Biochim. et Biophys. Acta*, **38**, 460 (1960).

[16] Luria, S. E., and M. Delbrück, *Genetics*, **28**, 491 (1943).

[17] Freese, E., E. Bautz, and E. Bautz-Freese, these PROCEEDINGS, **47**, 845 (1961).

[18] Crick, F. H. C., L. Barnett, S. Brenner, and R. J. Watts-Tobin, *Nature*, **192**, 1227 (1961).

[19] Feynman, R., personal communication.

[20] Garen, A., and C. Levinthal, *Biochim. et Biophys. Acta*, **38**, 470 (1960).

[21] Cohen, S. S., J. G. Flaks, H. D. Barner, M. R. Loeb, and J. Lichtenstein, these PROCEEDINGS, **44**, 1004 (1958).

[22] Gros, F., W. Gilbert, H. H. Hiatt, G. Attardi, P. F. Spahr, and J. D. Watson, in *Cellular Regulatory Mechanisms*, Cold Spring Harbor Symposia on Quantitative Biology, vol. 26 (in press).

[23] Garen, A., *Virology*, **14**, 151 (1961).

[24] Nomura, M., *Virology*, **14**, 164 (1961).

[25] If K is infected with two *r*II mutants containing mutations far apart in the A cistron, r^+ recombinants should occur in many of the cells since DNA replicas are known to be formed. However, in such an experiment there is still no progeny yield such as would be expected if the recombinants could express their phenotype.

[26] Litman, R. M., and A. B. Pardee, *Nature*, **178**, 529 (1956).

[27] Freese, E., these PROCEEDINGS, **45**, 622 (1959).

[28] Benzer S., and E. Freese, these PROCEEDINGS, **44**, 112 (1958).

[29] Freese, E., E. Bautz-Freese, and E. Bautz, *J. Mol. Biol.*, **3**, 133 (1961).

[30] Schuster, H., *J. Mol. Biol.*, **3**, 447 (1961).

[31] Brown, D. M., and P. Schell, *J. Mol. Biol.*, **3**, 709 (1961).

[32] Brenner, S., L. Barnett, F. H. C. Crick, and A. Orgel, *J. Mol. Biol.*, **3**, 121 (1961).

[33] Tessman, I., in *The Molecular Basis of Neoplasia*, 15th Annual Symposium on Fundamental Cancer Research, Houston, in press.

[34] Chamberlin, M., and P. Berg, these PROCEEDINGS, **48**, 81 (1962).

[35] Bautz, E., and B. Hall, these PROCEEDINGS, **48**, 400 (1962).

[36] Lengyel, P., J. F. Speyer, and S. Ochoa, these PROCEEDINGS, **47**, 1936 (1961).

[37] Nirenberg, N. M., and J. H. Matthaei, these PROCEEDINGS, **47**, 1558 (1961).

[38] Cohen, S. S., *J. Biol. Chem.*, **174**, 218 (1948).

ERRATUM

Page 545 line 8 should read: "may also be a non-random sample. . . ."

18

Reprinted from *Nature,* **206**(4988), 994–998 (1965)

GENETIC CODE: THE 'NONSENSE' TRIPLETS FOR CHAIN TERMINATION AND THEIR SUPPRESSION

By Dr. S. BRENNER, F.R.S., Dr. A. O. W. STRETTON and Dr. S. KAPLAN

Medical Research Council, Laboratory of Molecular Biology, Cambridge, England

THE nucleotide sequence of messenger RNA is a code determining the amino-acid sequences of proteins. Although the biochemical apparatus which translates the code is elaborate, it is likely that the code itself is simple and consists of non-overlapping nucleotide triplets. In general, each amino-acid has more than one triplet corresponding to it, but it is not known how many of the sixty-four triplets are used to code for the twenty amino-acids. Triplets which do not correspond to amino-acids have been loosely referred to as 'nonsense' triplets, but it is not known whether these triplets have an information content which is strictly null or whether they serve some special function in information transfer.

The evidence that there are nonsense triplets is mainly genetic and will be reviewed later in this article. A remarkable property of nonsense mutants in bacteria and bacteriophages is that they are suppressible; wild-type function can be restored to such mutants by certain strains of bacteria carrying suppressor genes. It was realized early that this implies an ambiguity in the genetic code in the sense that a codon which is nonsense in one strain can be recognized as sense in another. The problem of nonsense triplets has become inextricably connected with the problem of suppression and, in particular, it has proved difficult to construct a theory of suppression without knowing the function of the nonsense triplet.

In this article we report experiments which allow us to deduce the structure of two nonsense codons as UAG and UAA. We suggest that these codons are the normal recognition signals in messenger RNA for chain termination and, on this basis, propose a theory of their suppression.

Nonsense mutants and their suppressors. One class of suppressible nonsense mutants which has been widely examined includes the subset I ambivalent *r*II mutants[1], the suppressible mutants of alkaline phosphatase[2], the *hd* or *sus* mutants of phage λ[3], and many of the *amber* mutants of bacteriophage *T*4 (ref. 4). These mutants have been isolated in various ways and the permissive (*su+*) and non-permissive (*su–*) strains used have been different. When isogenic bacterial strains, differing only in the *su* locus, are constructed, it can be shown that all these mutants respond to the same set of suppressors. They are therefore of the same class and we propose that all these mutants should be called *amber* mutants.

We may now consider the evidence that these mutants contain nonsense codons. Garen and Siddiqi[2] originally noted that *amber* mutants of the alkaline phosphatase of *E. coli* contained no protein related immunologically to the enzyme. Benzer and Champe[1] also showed that the mutants exert drastic effects and suggested that *amber* mutants of the *r*II gene interrupt the reading of the genetic message. In the *r*II genes, a deletion, *r*1589, joins part of the A cistron to part of the B cistron; complementation tests show that this mutant still possesses B activity although it lacks the A function. It may therefore be used to test the effects of mutants in the A cistron on the activity of the B. Double mutants, composed of an A *amber* mutant together with *r*1589 did not have B activity on the *su–* strain; this effect is suppressed by the *su+* strain which restores B activity. This result is explained by our finding that each *amber* mutant of the head protein produces a characteristic fragment of the polypeptide chain in *su–* bacteria[5]. More recently we have shown that the polarity of the fragment is such that it can only be produced by the termination of the growing polypeptide chain at the site of the mutation[6]. In *su+* strains, both the fragment and a completed chain are produced, and the efficiency of propagation depends on which of the

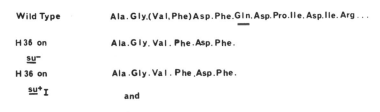

Fig. 1. Amino-acid sequences of the relevant region of the head protein in wild-type *T4D* and in the *amber* mutant *H36* on *su⁻* and *su⁺₁* strains

suppressors is carried by the strain. In $su^+{}_I$, the efficiency of propagation is about 65 per cent[7], and the completed chain contains a serine at a position occupied by a glutamine in the wild type[6] as shown in Fig. 1.

In addition to the *amber* mutants, there are other mutants, called *ochre* mutants, which are suppressed by a different set of suppressors[8]. *Ochre* mutants of the A cistron of *r*II abolish the B activity of *r*1589. This effect is not suppressed by *amber* suppressors, which shows that the *ochre* mutants are intrinsically different from the *amber* mutants.

Thus, nonsense mutants may be divided into two types, *amber* and *ochre* mutants, depending on their pattern of suppression. In Table 1, which is abstracted from a larger set of results[8], it can be seen that strains, carrying the *amber* suppressors $su^+{}_I$, $su^+{}_{II}$, $su^+{}_{III}$ and $su^+{}_{IV}$, suppress different but overlapping sets of *amber* mutants, but do not suppress any *ochre* mutants. This is the feature which distinguished the two classes of mutants from each other. Table 1 also shows that *ochre* mutants are suppressed by one or more of the strains carrying the *ochre* suppressors $su^+{}_B$, $su^+{}_C$, $su^+{}_D$ and $su^+{}_E$. These suppressors are also active on various *amber* mutants, and we have not yet been able to isolate suppressors specific for *ochre* mutants. Table 1 also shows that the suppressor strains can be differentiated by the set of mutants they suppress.

Table 1. SUPPRESSION OF *r*II MUTANTS BY *su*⁺ STRAINS OF *E. coli* Hfr H(λ)

	Amber suppressors				*Ochre* suppressors			
	$su^+{}_I$	$su^+{}_{II}$	$su^+{}_{III}$	$su^+{}_{IV}$	$su^+{}_B$	$su^+{}_C$	$su^+{}_D$	$su^+{}_E$
Amber mutants								
*r*IIA HD120	+	+	+	+	0	+	0	0
N97	+	poor	+	0	poor	poor	+	0
S116	0	0	+	0	0	+	+	0
N19	+	poor	+	0	poor	0	+	0
N34	+	+	+	+	+	+	+	0
*r*IIB HB122	+	+	+	+	poor	+	+	0
HB74	+	+	+	+	+	+	+	+
X237	+	poor	+	0	0	+	+	0
HB232	+	+	+	+	0	+	+	+
X417	+	+	+	+	0	poor	+	0
Ochre mutants								
*r*IIA HD147	0	0	0	0	+	+	+	+
N55	0	0	0	0	+	+	+	0
X20	0	0	0	0	+	0	+	0
N21	0	0	0	0	+	0	0	+
*r*IIB UV375	0	0	0	0	+	+	+	+
360	0	0	0	0	+	0	+	+
375	0	0	0	0	+	+	+	+
HF208	0	0	0	0	+	+	0	+
N29	0	0	0	0	0	0	+	0

Unlike some of the *amber* suppressors, all the *ochre* suppressors are weak[7,8]. This has made the isolation of *ochre* mutants of the head protein impossible. We therefore do not know the molecular consequences of *ochre* mutants, but we shall assume that, like the *amber* mutants, they too result in chain termination.

If we accept that both types of nonsense mutants result in termination of the polypeptide chain, we have to ask: at which level of information transfer is this effect exerted ? We have recently shown that it is likely that chain termination occurs as part of protein synthesis, since both types of mutants vanish when the phase of reading of the genetic message is altered[9]. This leads us to conclude that *amber* and *ochre* mutants produce different triplets

which have to be read in the correct phase and which are recognized as signals for chain termination.

Decoding of amber and ochre triplets. We have done two types of experiments which allow us to deduce the structures of the *amber* and *ochre* triplets. First, we studied the production and reversion of *r*II *amber* and *ochre* mutants using chemical mutagens. We show that the two triplets are connected to each other and that we can define their possible nucleotide compositions. Next, we investigated head protein *amber* mutants, to define the amino-acids connected to the *amber* triplet. Comparison of these results with known amino-acid codons allows us to deduce the structures of the *amber* and *ochre* triplets.

The experiments with *r*II mutants depend on the specificity of the mutagenic agent, hydroxylamine. This reacts only with cytosine in DNA, and although the exact structure of the product has not yet been defined, the altered base (called U′) appears to act like T with high efficiency, producing base-pair transitions of the G—C → A—T type[10]. Hence, response of any particular site of the DNA to hydroxylamine is evidence for the existence of a G—C pair at that site. Usually, phage particles are treated with hydroxylamine and, since these contain double-stranded DNA, the alteration of C occurs on only one of the strands. In any given gene, only one of the strands is transcribed into messenger RNA (ref. 11). This is the *sense* strand; it carries the genetic information proper and contains a nucleotide sequence which is the inverse complement of the sequence of the messenger RNA. The other strand, the *antisense* strand, has the same sequence as the messenger. In a phage, treated with hydroxylamine, the altered base, U′, could be on either the sense or the antisense strand of the DNA. Since the *r*II genes express their functions before the onset of DNA replication[12], only sense strand changes will register a phenotypic effect in the first cycle of growth; changes on the antisense strand, while still yielding altered DNA progeny, will go unexpressed (Fig. 2).

We can now examine the reversion properties of *amber* and *ochre* mutants. Champe and Benzer[12] studied reversion of a large number of *r*II mutants using different mutagens. They noted that no *amber* mutant was induced to revert by hydroxylamine. Some of the mutants they studied can now be identified as *ochre* mutants, and their results show that *ochre* mutants are equally insensitive to hydroxylamine. However, in their experiments, the treated mutant phages were plated directly on the bacterial strain which restricts the growth of *r*II mutants. They could therefore detect *sense* strand changes only, and any mutation on the antisense strand would not have been expressed. Strictly speaking, then, their results tell us that neither the *amber* nor the *ochre* triplet contains a C on the sense strand of the DNA, or, if any one does, it is connected by a C → U change to another nonsense codon. To extend this result and to recover all possible mutational changes, we grew the mutagenized phages in *E. coli* B, in which the *r*II functions are unnecessary, and then plated the progeny on strain *K* to measure reversion frequency. Table 2 shows that *amber* and

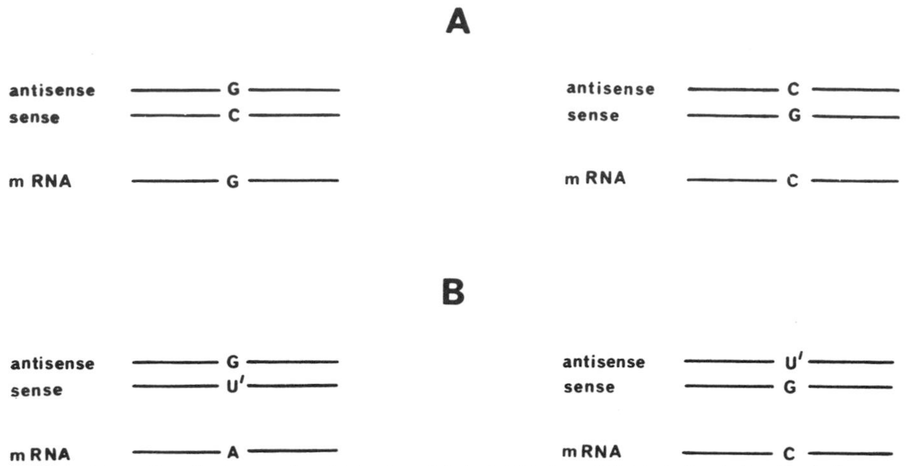

A

antisense	——————— G ———————		antisense	——————— C ———————
sense	——————— C ———————		sense	——————— G ———————
m RNA	——————— G ———————		m RNA	——————— C ———————

B

antisense	——————— G ———————		antisense	——————— U' ———————
sense	——————— U' ———————		sense	——————— G ———————
m RNA	——————— A ———————		m RNA	——————— C ———————

Fig. 2. Diagram illustrating the expression of the two types of G–C pairs (*A*) before and (*B*) after treatment with hydroxylamine

Table 2. REVERSION OF AMBER AND OCHRE MUTANTS AFTER ALLOWING DNA REPLICATION

| | | Reversion index × 10⁻⁷ | |
		Control	NH₂OH
amber mutants	S116	0·04	0·03
	HD26	0·1	0·2
	S24	0·4	0·9
	S99	0·1	0·1
	N19	0·15	0·13
	HD59	0·0	0·05
	HB232	0·1	0·4
ochre mutants	UV375	0·8	2·0
	360	0·6	0·8
	X27	0·3	0·5
	375	0·8	0·9
	X511	0·2	0·3
	UV256	1·0	**280**

Phages were incubated in a solution of M NH₂OH in 2 M NaCl and 0·05 M sodium phosphate (pH 7·5) for 2 h at 37°. The reaction was terminated by dilution into acetone broth. About 10⁴ phage particles were used to infect a culture of *E. coli* B which was grown to lysis, and the progeny assayed on *E. coli* B and *E. coli* K12(λ). The reversion index is the *K/B* ratio. The control was treated in the same way except that hydroxylamine was omitted. The mutant *UV*256, which is not an *ochre* or an *amber* mutant, was used to check the efficacy of the mutagenic treatment.

ochre mutants are not induced to revert by hydroxylamine, and we conclude that in neither mutant, does the triplet in the DNA contain G—C pairs, or, if a G—C pair is present, that triplet is connected by a G—C → A—T transition to another nonsense codon. In other words, subject to the last important reservation, we can conclude that the codons on the messenger RNA contain neither G nor C.

However, we next discovered that *ochre* mutants can be converted into *amber* mutants by mutation. Since *ochre* mutants are not suppressed by *amber* suppressors, plating on strains carrying such suppressors selects for *amber* revertants. Wild-type revertants also grow, but the two can be distinguished by testing revertant plaques on the *su⁻* strain. Twenty-six rII *ochre* mutants have been studied and, of these, 25 have been converted into *amber* mutants. A sample of the results is given in Table 3, which shows that the mutation is strongly induced by 2-aminopurine, as strongly as the reversion of the *ochre* mutant to wild type. Other experiments, not reported here, show that the mutations of *ochre* mutants both to the *amber* and to the wild type are also induced by 5-bromouracil, but the induction is weaker than with 2-aminopurine. These results prove that the *amber* and *ochre* triplets differ from each other by only one nucleotide base, and must have the other two bases in common. 2-Aminopurine is a base analogue mutagen inducing the transition A—T ⇌ G—C in both directions[13].

This tells us that one of the triplets has a G—C pair in the DNA. The experiment reported in Table 4 shows that *ochre* mutants cannot be induced to mutate to *amber* mutants with hydroxylamine, even after the treated phages have been grown in *E. coli* B. This shows that it is the *amber* triplet which has the G—C pair and the *ochre* which contains the A—T pair.

Although the insensitivity of the mutants to reversion induction by hydroxylamine might suggest that they contain A—T base pairs only, the conversion of *ochre* mutants to *amber* mutants shows that, in one position, the *amber* mutant contains a G—C pair. The other two bases must be common to both triplets, but we cannot conclude that both are A—T pairs. In fact, both could be G—C pairs and the triplets may be connected to other nonsense triplets by G—C → A—T changes. However, we know that *amber* and *ochre* mutants can be induced by hydroxyl-

Table 3. MUTATION OF *ochre* MUTANTS TO *amber* MUTANTS

| | | Reversion index × 10⁻⁷ | |
| | Spontaneous | 2-Aminopurine | |
rIIA cistron	(wild type + *ambers*)	wild type	*amber*
N55	0·5	830	280
X20	0·05	100	2,100
X372	0·1	300	2,100
X352	0·06	340	1,500
HD147	0·3	370	50
rIIB cistron			
X511	0·2	710	65
N17	0·2	610	80
SD160	0·6	380	390
N29	1·0	3,900	330
AP53	0·7	350	15

Cultures of *E. coli* B in minimal medium with and without 2-aminopurine (600 μg/ml.) were inoculated with about 100 phages and grown to lysis. These were plated on *E. coli* B and on *E. coli* K12(λ) *su⁺*₁. About 50 induced revertants were tested on *E. coli* K12(λ) *su⁻* to measure the relative frequencies of *amber* revertants.

Table 4. INDUCTION OF THE *ochre→amber* MUTATION

| | | Reversion index × 10⁻⁷ | |
		r⁺	*amber*
360	Control	0·6	0·2
	Hydroxylamine	0·8	0·3
	2AP	200	1,200
UV375	Control	0·8	0·4
	Hydroxylamine	2·0	1·0
	2AP	660	140
X27	Control	0·3	<0·1
	Hydroxylamine	1·0	0·5
	2AP	7·0	73
375	Control	0·8	0·4
	Hydroxylamine	2·0	1·0
	2AP	1,400	1,700

Hydroxylamine treatment and growth of the mutagenized phages, and 2-aminopurine induction, were carried out as described in Tables 2 and 3.

amine from wild type. This proves that both triplets have at least one common A—T pair.

We now present an experiment which shows that the *amber* triplet has two A—T base pairs, and which also establishes the orientation of the pairs with respect to the two strands of DNA. Let us suppose that the *amber* triplet in the messenger RNA contains a U. This corresponds to an A in the sense strand of DNA of the *amber* mutant, implying that the wild-type DNA contains a G in this strand and a C in the antisense strand. When the wild-type DNA is treated with hydroxylamine to alter this C the change is not effective and normal messenger is still made (Fig. 2, right). On the other hand, if the *amber* triplet in the messenger contains an A, the mutant will be induced by the action of hydroxylamine on a C in the sense strand of the wild-type phage DNA, and provided that the U' produced acts identically to U in messenger synthesis, mutant messenger will be made. This argument has been tested by the following experiment. Wild-type *T4r+* phages were treated with hydroxylamine to induce *r* mutants to a frequency of 1 per cent. In set B, the phages were then grown on *E. coli B*, in which the *rII* functions are not required, to recover all mutants. In set K, the phages were grown through *E. coli K12(λ) su-*, to eliminate from the population all phages with an immediate mutant expression. *Amber* and *ochre* mutants were then selected and mapped. Table 5 summarizes the results. About the same number of *rI* mutants were recorded in each set, and since these mutants show no difference in growth on the two bacterial strains, this shows that the results may be compared directly. It will be seen that *amber* mutants at the sites, *N97, S24, N34, X237* and *HB232*, recur many times in set B, but are absent or rarely found in set K. At other *amber* sites, such as *HB118, HB129, EM84* and *AP164*, mutants occur with approximately equal frequency in both sets. The first class fulfils the expectation for a C → U change on the sense strand, while the second class must arise by C → U changes on the antisense strand. This shows that

Table 5. HYDROXYLAMINE INDUCTION OF *amber* AND *ochre* MUTANTS

A. No. of mutants isolated

	Set B	Set K
rI	2,010	1,823
Leaky or high reverting *rII*	720	508
non-suppressible *rII*	1,144	433
amber	319	121
ochre	83	82
Total	4,276	2,967

B. Recurrences found at different sites

Amber mutants No. found at each site Site	Set B	Set K	*Ochre* mutants No. found at each site Site	Set B	Set K
A cistron			A cistron		
HB118	27	15	*HD147*	2	0
C204	1	0	*HF220*	1	0
N97	44	1	*HF240*	1	0
S116	31	2	*N55*	19	19
N11	3	3	*X20*	9	8
S172	9	5	*HF219*	1	0
S24	44	3	*HF245*	1	0
HB129	14	25	*N31*	3	5
S99	12	16	*HM127*	0	1
N19	15	9	*N21*	2	2
N34	8	0			
B cistron			B cistron		
HE122	1	0	360	11	10
EM84	29	21	*UV375*	2	0
HB74	16	5	*N24*	6	4
X237	14	2	375	2	5
AP164	28	12	*N17*	5	3
HB232	21	1	*HF208*	1	2
X417	1	0	*N7*	4	12
HD231	1	1	*N12*	5	7
			X234	0	2
			X191	1	0
			HE267	5	0
			AP53	2	2

T4Br+ was treated with M hydroxylamine (see Table 2) for 2 h at 37° C. Survival was 50 per cent, and the frequency of *r* and mottled plaques, 1 per cent. $1 \cdot 2 \times 10^9$ phage particles were added to 10^8 cells of *E. coli B* (set B) and to *E. coli K12(λ) su-* (set K). After 8 min, the infected bacteria were diluted a thousand-fold into 2 litres of broth, incubated for 35' and lysed with CHCl₃. The burst sizes in both sets were about 60. *r* mutants were isolated from each set using less than 2 ml. to ensure that the mutants selected had mostly arisen from independent events. These were picked and stabbed into B and K, and *rI* mutants and leaky mutants discarded. The *rII* mutants were then screened on *su+III* and *su+B* to select for *amber* and *ochre* mutants which were then located by genetic mapping.

the *amber* triplet in the messenger contains both an A and a U. The same should be true of the *ochre* mutants. However, as shown in Table 5, *ochre* mutants are not as strongly induced by hydroxylamine as are *amber* mutants, and we cannot separate the two classes with the same degree of confidence. Nevertheless, since we have already shown that the mutants are connected, it follows that the *ochre* triplet must also contain an A and a U. We conclude that the *amber* and *ochre* triplets are, respectively, either (UAG) and (UAA), or (UAC) and (UAU). If we had a strain which suppressed *ochre* mutants only, we could specify the third base by studying the induction of the *amber* → *ochre* change with hydroxylamine.

Fortunately, we can resolve the ambiguity by determining the amino-acids to which the *amber* triplet is connected by mutation. In particular, we note that it should be connected to two and only two amino-acid codons by transitions, corresponding, in fact, to the two types of origin of the mutants described here. The third codon to which it is connected by a transition is the *ochre* triplet. As mentioned earlier, the head *amber* mutant *H36* has arisen from glutamine (Fig. 1). This mutant was induced with hydroxylamine. We have evidence that two other mutants, *E161* and *B278* induced by 2-aminopurine and 5-bromouracil respectively, have arisen from tryptophan. In a recent study of two *amber* mutants of the alkaline phosphatase, Garen and Weigert[14] found one mutant to arise from glutamine and the other from tryptophan; and Notani *et al.*[15] have found an *amber* mutant to arise from glutamine in the RNA phage *f2*. In addition, we have examined 2-aminopurine induced revertants of 10 different head *amber* mutants. Ten to 12 independently induced revertants of each of the mutants have been screened for tryptophan containing peptides by examining the ¹⁴C-tryptophan labelled protein. Among a total of 115 revertants, 62 are to tryptophan. Determination of glutamine involves sequence analysis and takes more time. So far, among the remaining 53 revertants, glutamine has been identified in one revertant of *H36*. These results suggest that the two amino-acids connected to the *amber* triplet are glutamine and tryptophan. If the *amber* triplet is (UAC), then one of these must be (CAC) and the other (UGC); if it is (UAG), then the corresponding codons are (CAG) and (UGG). Nirenberg *et al.*[16] have shown that poly AC does not code for tryptophan, but does for glutamine. However, they find that the triplet for glutamine clearly has the composition (CAA) and is definitely not (CAC). Since this latter triplet corresponds neither to glutamine nor to tryptophan we can eliminate the first alternative. We note with satisfaction that (UGG) is the composition of a codon assigned to tryptophan[16,17], and this assignment of (UAG) to the *amber* triplet suggests that glutamine is (CAG).

We can also make a reasonable assignment of the order of the bases in the triplet. Our original argument was based on deductions from a few known triplets and from amino-acid replacement data; it will not be given here. The order of the bases follows directly from a recent demonstration by Nirenberg *et al.*[18] that the triplet CAG does, in fact, correspond to glutamine. The *amber* triplet is therefore UAG and the *ochre* triplet UAA. This assignment is supported by the following additional evidence. We have found a tyrosine replacement in 21 independent spontaneous mutants of the head *amber* mutant, *H36* (ref. 19). This change must be due to a transversion because we have already accounted for all the transitions of the *amber* triplet. In support of this, we find that the change is not induced by 2-aminopurine. There are six possible transversions of the *amber* triplet, namely, AAG, GAG, UUG, UCG, UAU and UAC. It has recently been shown that both UAU and UAC correspond to tyrosine[20] which confirms the order. The spontaneous revertants of the *amber* mutants to leucine, serine and glutamic acid found by Weigert and Garen[21] are further evidence for the assignment. UUG, a transversion of the *amber* triplet, does in fact code for leucine[22], and reasonable allocations for serine and glutamic acid are UCG and GAG, respectively. Weigert and Garen[21] also find revertants of an *amber* mutant to either lysine or arginine. This may be the final transversion expected since AAG is a codeword for lysine[20].

It should be noted that in the foregoing discussion it has been tacitly assumed that the *amber* and *ochre* signals are triplets. Examination of revertants of *amber* mutants has supported this assumption, since in 41 independent revertants of *H*36 the amino-acid replaced is always at the site of mutation, and never in adjacent positions. The 21 revertants that Weigert and Garen[21] isolated reinforce this conclusion.

Function of amber and ochre triplets and the mechanism of suppression. According to present-day ideas of protein synthesis, it is expected that the termination of the growth of the polypeptide chain should involve a special mechanism. Since the terminal carboxyl group of the growing peptide chain is esterified to an *s*RNA (ref. 23), chain termination must involve not only the cessation of growth, but also the cleavage of this bond. Since the *amber* mutants have been shown to result in efficient termination of polypeptide chain synthesis, it is reasonable to suppose that this special mechanism may be provided by the *amber* and *ochre* triplets.

We postulate that the chain-terminating triplets UAA and UAG are recognized by specific *s*RNAs, just like other codons. These *s*RNAs do not carry amino-acids but a special compound which results in termination of the growing polypeptide chain. There are many possible ways of formulating the mechanism in detail, but all are speculative and will not be considered here. The essential feature of this hypothesis is to make the process of chain termination exactly congruent with that of chain extension.

In suppressing strains, a mechanism is provided for competing with chain termination; it is easy to visualize this process as being due to two ways of recognizing the nonsense codon—one by the chain-terminating *s*RNA, and the other by an *s*RNA carrying an amino-acid. Mechanisms of suppression can be classified according to which *s*RNA carries the amino-acid to the nonsense codon.

Alteration in the recognition of the chain-terminating *s*RNA might allow the attachment of an amino-acid to this *s*RNA. This could be brought about either by modifying normal activating enzymes so as to widen their specificity, or by changing the chain terminating *s*RNAs to allow them to be recognized by activating enzymes.

Another possibility is that the region of an amino-acyl *s*RNA used for triplet recognition is modified so that it can recognize the nonsense triplet. Clearly, this alteration must not affect the normal recognition of its own codon by the amino acyl *s*RNA because such a change would be lethal. Either there must be more than one gene for the given *s*RNA, or else the change must produce an ambiguity in the recognition site so that it can read both its own codon and the nonsense codon. Such ambiguity could result not only from mutation in the *s*RNA gene but also by enzymatic modification of one of the bases in the recognition site. The ambiguity, however, must be narrowly restricted to prevent the suppression from affecting codons other than the *amber* and *ochre* triplets. Moreover, the amino-acids which are inserted by the *amber* suppressors must be those the codons of which are connected to UAG. It should be noted that this condition is fulfilled by *su*+ı which inserts serine, since serine has been found as a reversion of an *amber* mutant. This theory does not easily explain the *ochre* suppressors. Since these recognize both *amber* and *ochre* mutants the *s*RNA must possess this ambiguity as well.

Another quite different possibility for suppression that has been considered is that the suppressors alter a component of the ribosomes to permit errors to occur in the reading of the messenger RNA (ref. 24). This is probably

the explanation of streptomycin suppression[25], but suppression of *amber* and *ochre* mutants cannot be readily explained by this theory. It is scarcely likely that such a mechanism could be specific for only one or two triplets, and for this reason it might be expected to give us suppression of mutants which are not nonsense, but missense, and this has not been found[1,8]. Moreover, the efficiency of *amber* suppression argues strongly against such a mechanism. It is unlikely that a generalized error in reading nucleotides could produce the 60 per cent efficiency of suppression found for *su*+ı without seriously affecting the viability of the cell.

It is a consequence of our theory that normal chain termination could also be suppressed in these strains. Since the *amber* suppressors are efficient we have to introduce the *ad hoc* hypothesis that the UAG codon is rarely used for chain termination in *Escherichia coli* and bacteriophage *T*4 and that UAA is the common codon. This is supported by the fact that all *ochre* suppressors thus far isolated are weak[7,8]. Another possibility is that neither is the common chain terminating triplet. We cannot exclude the existence of other chain terminating triplets which are not suppressible.

To summarize: we show that the triplets of the *amber* and *ochre* mutants are UAG and UAA, respectively. We suggest that the 'nonsense' codons should be more properly considered to be the codons for chain termination. In essence, this means that the number of elements to be coded for is not 20 but more likely 21. We propose that the recognition of the chain-terminating codons is carried out by two special *s*RNAs.

We thank our colleagues for their advice, and Dr. M. Nirenberg for allowing us to quote his unpublished results.

[1] Benzer, S., and Champe, S. P., *Proc. U.S. Nat. Acad. Sci.*, **47**, 1025 (1961); **48**, 1114 (1962).

[2] Garen, A., and Siddiqi, O., *Proc. U.S. Nat. Acad. Sci.*, **48**, 1121 (1962).

[3] Campbell, A., *Virology*, **14**, 22 (1961).

[4] Epstein, R. H., Bolle, A., Steinberg, C. M., Kellenberger, E., Boy de la Tour, E., Chevalley, R., Edgar, R. S., Susman, M., Denhardt, G. H., and Lielausis, A., *Cold Spring Harbour Symp. Quant. Biol.*, **28**, 375 (1963).

[5] Sarabhai, A., Stretton, A. O. W., Brenner, S., and Bolle, A., *Nature*, **201**, 13 (1964).

[6] Stretton, A. O. W., and Brenner, S., *J. Mol. Biol.* (in the press).

[7] Kaplan, S., Stretton A. O. W., and Brenner, S. (in preparation).

[8] Brenner, S., and Beckwith, J. R. (in preparation).

[9] Brenner, S., and Stretton, A. O. W. (in preparation).

[10] Brown, D. M., and Schell, P., *J. Mol. Biol.*, **3**, 709 (1961). Freese, E., Bautz-Freese, E., and Bautz, E., *J. Mol. Biol.*, **3**, 133 (1961). Schuster, H., *J. Mol. Biol.*, **3**, 447 (1961). Freese, E., Bautz, E., and Freese, E. B., *Proc. U.S. Nat. Acad. Sci.*, **47**, 845 (1961).

[11] Tocchini-Valentini, G. P., Stodolsky, M., Aurisicchio, A., Sarnat, M., Graziosi, F., Weiss, S. B., and Geiduschek, E. P., *Proc. U.S. Nat. Acad. Sci.*, **50**, 935 (1963). Hayashi, M., Hayashi, M. N., and Spiegelman, S., *Proc. U.S. Nat. Acad. Sci.*, **50**, 664 (1963). Marmur, J., Greenspan, C. M., Palacek, E., Kahan, F. M., Levene, J., and Mandel, M., *Cold Spring Harbor Symp. Quant. Biol.*, **28**, 191 (1963). Bautz, E. K. F., *Cold Spring Harbor Symp. Quant. Biol.*, **28**, 205 (1963). Hall, B. D., Green, M., Nygaard, A. P., and Boezi, J., *Cold Spring Harbor Symp. Quant. Biol.*, **28**, 201 (1963).

[12] Champe, S. P., and Benzer, S., *Proc. U.S. Nat. Acad. Sci.*, **48**, 532 (1962). Tessman, I., Poddar, R. K., and Kumar, S., *J. Mol. Biol.*, **9**, 352 (1964).

[13] Freese, E., *J. Mol. Biol.*, **1**, 87 (1959). Freese, E., *Proc. U.S. Nat. Acad. Sci.*, **45**, 622 (1959). Howard, B. D., and Tessman, I., *J. Mol. Biol.*, **9**, 372 (1964).

[14] Garen, A., and Weigert, M. G., *J. Mol. Biol.* (in the press).

[15] Notani, G. W., Engelhardt, D. L., Konigsberg, W., and Zinder, N., *J. Mol. Biol.* (in the press).

[16] Nirenberg, M., Jones, O. W., Leder, P., Clark, B. F. C., Sly, W. S., and Pestka, S., *Cold Spring Harbor Symp. Quant. Biol.*, **28**, 549 (1963).

[17] Speyer, J. F., Lengyel, P., Basilio, C., Wahba, A. J., Gardner, R. S., and Ochoa, S., *Cold Spring Harbor Symp. Quant. Biol.*, **28**, 559 (1963).

[18] Nirenberg, M., Leder, P., Bernfield, M., Brimacombe, R., Trupin, J., and Rottman, F., *Proc. U.S. Nat. Acad. Sci.* (in the press).

[19] Stretton, A. O. W., and Brenner, S. (in preparation).

[20] Trupin, J., Rottman, F., Brimacombe, R., Leder, P., Bernfield, M., and Nirenberg, M., *Proc. U.S. Nat. Acad. Sci.* (in the press). Clark, B. F. C., presented before the French Biochemical Society, February, 1965.

[21] Weigert, M. G., and Garen, A., *Nature* (preceding communication).

[22] Leder, P., and Nirenberg, M. W., *Proc. U.S. Nat. Acad. Sci.*, **52**, 1521 (1964).

[23] Gilbert, W., *J. Mol. Biol.*, **6**, 389 (1963). Bretscher, M. S., *J. Mol. Biol.*, **7**, 446 (1963).

[24] Davies, J., Gilbert, W., and Gorini, L., *Proc. U.S. Nat. Acad. Sci.*, **51**, 883 (1964).

[25] Gorini, L., and Kataja, E., *Proc. U.S. Nat. Acad. Sci.*, **51**, 487 (1964).

Editors' Comments
on Papers 19 Through 21

19   LOVELESS
      *The Influence of Radiomimetic Substances on Deoxyribonucleic
      Acid Synthesis and Function Studied* in Escherichia coli/*Phage
      Systems: III. Mutation of T2 Bacteriophage as a Consequence of
      Alkylation* in vitro: *The Uniqueness of Ethylation*

20   KRIEG
      *Ethyl Methanesulfonate-Induced Reversion of Bacteriophage T4rII
      Mutants*

21   LOVELESS
      *Possible Relevance of O-6 Alkylation of Deoxyguanosine to the
      Mutagenicity and Carcinogenicity of Nitrosamines and Nitrosa-
      mides*

ALKYLATION MUTAGENESIS

Alkylating agents comprise the most diverse group of mutagens that
interact with DNA by covalent bonding. They act by two, or perhaps
three, quite different mechanisms. Those to be considered here are the
directly acting alkylating agents, which react with a nucleotide so as to
generate a directly mispairing analogue. Others react with DNA in ways
that inhibit DNA replication, but do not directly trigger mispairing; some
of these are considered in Papers 6, 7, and 29. Still others act by promo-
ting the mutagenic efficiency of intercalating frameshift mutagens; they
are described in Paper 26.

The first report of alkylation mutagenesis in a phage system was by
Loveless (Paper 19), who observed that ethylation by ethyl methanesulfonate (EMS) was highly mutagenic, but that methylation by methyl
methanesulfonate (MMS) was apparently nonmutagenic. This curious difference prompted many investigations of the chemistry of the interactions of simple alkylating agents with DNA (see, for instance, Lawley
and Brookes, 1963; Brookes and Lawley, 1963; Paper 21). The development of improved techniques for scoring mutagenesis in phage T4, where
mosaicism may occur so extensively as to obscure the mutant phenotype (Drake, 1966), later revealed that MMS was, in fact, quite muta-

genic in this system (Green and Drake, 1974). Whereas EMS acts primarily by a direct mechanism, MMS acts primarily by triggering misrepair (see Papers 27 through 29). It should also be noted that this difference tends to become obscured at least in certain higher systems. The direct mechanism of EMS mutagenesis can be substantially overcome by more sophisticated DNA polymerases (Drake and Greening, 1970), while EMS mutagenesis in yeast seems to occur almost exclusively by misrepair (Prakash, 1975). Readers who are interested in more experimental details than given in Paper 19 should consult the first paper in this series (Loveless and Stock, 1959).

The mutagenic specificity of EMS was first examined by Bautz and Freese (1960), but was most thoroughly explored by Krieg (Paper 20). Unlike the alkylating agents that act via misrepair, and which may induce virtually all types of mutations (Green and Drake, 1974), EMS turned out to induce predominantly transitions, with the pathway G :C → A : T being somewhat favored. Both Paper 20 and Green and Krieg (1961) showed that mispairing is not as efficient as was the case with hydroxylamine (Paper 15), presumably because of the intermediate ability of the T4 DNA polymerase to overcome these pairing errors (Drake and Greening, 1970). The reader should also be warned that Krieg's observations about the EMS revertibility of some proflavin-induced mutants (putative frameshift mutants) probably occurs by atypical mechanisms not directly relevant to the general mechanism of EMS mutagenicity.

The first clue as to how alkylating agents such as EMS achieve their mutagenic specificity and are able to induce mutations by directed mispairing was obtained by Loveless (Paper 21). Alkylation of the O-6 position of guanine (or of the O-4 position of thymine; see Lawley, 1974) converts these bases to an enol-like configuration (see Papers 8 and 9). Whereas both EMS and nitrosoguanidine alkylate these sites efficiently, MMS does so very inefficiently, which accounts for its relatively weaker mutagenicity in T phages.

REFERENCES

Bautz, E., and E. Freese. 1960. On the mutagenic effect of alkylating agents. *Proc. Natl. Acad. Sci., 46:* 1585.

Brookes, P., and P. D. Lawley. 1963. Effects of alkylating agents on T2 and T4 bacteriophages. *Biochem. J., 89:* 138.

Drake, J. W. 1966. Ultraviolet mutagenesis in bacteriophage T4: I. Irradiation of extracellular phage particles. *J. Bacteriol., 91:* 1775.

——, and E. O. Greening. 1970. Suppression of chemical mutagenesis in bacteriophage T4 by genetically modified DNA polymerases. *Proc. Natl. Acad. Sci., 66:* 823.

Green, D. M., and D. R. Krieg. 1961. The delayed origin of mutants induced by exposure of extracellular phage T4 to ethyl methane sulfonate. *Proc. Natl. Acad. Sci., 47:* 65.

Green, R. R., and J. W. Drake. 1974. Misrepair mutagenesis in bacteriophage T4. *Genetics, 78:* 81.

Lawley, P. D. 1974. Alkylation of nucleic acids and mutagenesis. In *Molecular and Environmental Aspects of Mutagenesis* (L. Prakash, F. Sherman, M. W. Miller, C. W. Lawrence, and H. W. Taber, eds.), p. 17. Charles C Thomas, Springfield, Ill.

——, and P. Brookes. 1963. Further studies on the alkylation of nucleic acids and their constituent nucleotides. *Biochem. J., 89:* 127.

Loveless, A., and J. C. Stock. 1959. The influence of radiomimetic substances on deoxyribonucleic acid synthesis and function studied in *Escherichia coli*/phage systems: I. The nature of the inactivation of T2 phage *in vitro* by certain alkylating agents. *Proc. Roy. Soc. (Lond.), B 150:* 486.

Prakash, L. 1975. Lack of chemically induced mutation in repair-deficient mutants of yeast. *Genetics, 78:* 1101.

Reprinted from *Proc. Roy. Soc. (Lond.)*, **B150**, 497–508 (1959)

III. Mutation of $T2$ bacteriophage as a consequence of alkylation *in vitro*: the uniqueness of ethylation

By A. Loveless

Chester Beatty Research Institute, Institute of Cancer Research:
Royal Cancer Hospital, London, S.W. 3

(*Communicated by A. Haddow, F.R.S.—Received* 22 *July* 1958—
Revised 11 *November* 1958)

Of a comprehensive set of alkylating agents tested, only two, namely, ethyl methane sulphonate and diethyl sulphate, have been found so to interact with $T2$ bacteriophage that cells of *Escherichia coli*, infected with phage treated extracellularly, manifest a considerably increased likelihood of yielding mutated phage. Since this increase can occur where the infective titre of the phage and the latent period and average burst size of the infected bacteria remain unchanged, it is considered that the increased mutation rate is a direct consequence of the chemical treatment, although the alkylation itself does not constitute the mutation. A study of the manner of inactivation of the phage by these agents has not revealed any characteristic difference between ethylation and other alkylations which could be held to account for its apparent uniqueness.

Introduction

In part I of this series (Loveless & Stock 1959) was reported a study of the inactivation of $T2$ bacteriophage by a series of alkylating agents comprising mono- and bifunctional representatives of the sulphur and nitrogen mustards and epoxides. Although, naturally, alert to the possibility, we saw no evidence of any increase in the mutation rate of the phage as a result of reaction with any of these agents. In order to extend the comparison of mono- and bifunctional analogues reported in part I, attention was directed to the alkyl methane sulphonates, and here we found, through the agency of ethyl methane sulphonate, the first evidence that chemical treatment of $T2$ *in vitro* could result in an enhanced rate of mutation when such virus was subsequently grown in sensitive bacteria (Loveless 1958). The investigation was therefore extended to a study of further alkylating agents chosen for their ability to introduce a graded series of alkyl and substituted alkyl groups into the phage particle. The choice of compounds available or potentially available for such a purpose is limited by their solubility and reactivity and the range now covered appears to be virtually exhaustive. In addition to experiments designed specifically to test the mutagenicity* or otherwise of these compounds, the manner of their inactivation of phage has also been studied for comparative

* What was actually under investigation was the relative number of mutants arising in the progeny of bacteria singly infected with phage previously treated *in vitro*. To avoid cumbersome repetitiveness, the terms mutagenic, induced mutation, etc., will be used loosely in this connexion.

186

purposes according to the characteristics of inactivation by alkylating agents described in part I.

MATERIALS AND METHODS

The phage employed in all the work to be reported was $T2$ (wild-type) of the same stock as used for all the other work reported in this series. (Attempts were made to produce host-range mutants of $T4$ but these did not succeed.) For plaque-type mutants, all platings were with strain B of *Escherichia coli* ('B' hereinafter) and for host-range mutants, with a strain of $B/2$ isolated in this laboratory.

The chemical compounds used were: ethylene oxide, ethylenimine, methyl, ethyl, n-propyl and n-butyl methane sulphonates, 1:5-di(methane sulphonoxy)-pentane, and diethyl sulphate. All the methane sulphonates were made available by the chemistry department of this Institute; 1:5-di(methane sulphonoxy)pentane, originally described by Timmis (1950), was especially prepared for this investigation by Dr W. C. J. Ross. The halogenated ethyl methane sulphonates, used in mutation studies in other systems (Fahmy & Fahmy 1956; Kølmark, private communication), proved of too low solubility and reactivity for the present purpose.

All chemical treatments were carried out in $M9$ medium lacking glucose ('$M9$ buffer') at 37 °C and terminated before changes of pH consequent upon hydrolysis or other reaction of the agent involved any deleterious effect upon the phage.

'Dilution' and 'thiosulphate' survival curves were estimated in the manner described in part I as were the dose- and pH-dependence of inactivation by mono-functional compounds. Tests for mutagenicity were performed according to the following procedure.

Treated phage was added, at the time of termination of treatment, unless otherwise indicated, to an aerated culture of B in $M9$ at *ca.* 5×10^8/ml. to give a multiplicity of infection of *ca.* 0·05. This dilution reduced the concentration of the chemical agent to an innocuous level. Aeration was continued for 10 min to assure virtually complete adsorption and thus eliminate the necessity for the use of antiserum. At the tenth minute samples were taken: (1) for plating undiluted with $B/2$ for host-range mutations, and (2) *via* appropriate dilution into a second $M9$ growth tube in which the concentration of B was reduced to at most 10^5/ml. Samples were plated from this tube prior to burst, i.e. before the twentieth minute, and after burst, i.e. after the fortieth minute from the commencement of adsorption. For assessment of plaque-type, the experiments were designed to give approximately 500 plaques per plate and the number of plates was always ten or more. Samples (0·1 ml.) for plating were mixed with 0·1 ml. of a fully grown broth culture of B or $B/2$ in 2 ml. of 0·6 % broth agar, poured onto 1 % broth agar containing 0·1 % glucose and incubated overnight. Only m (mottled) and r plaques were scored. Plaques were only scored as mottled when showing considerable areas of r-type lysis (cf. Loveless 1958).

These experiments also served to indicate the infectivity of the treated phage and burst size of the infected cells, but some thorough one-step growth experiments (Ellis & Delbrück 1939) were also carried out as a check.

A few 'single burst' experiments (cf. Adams 1950) were carried out with phage treated with ethyl methane sulphonate, although these were limited in extent by the facilities, particularly for incubation, available. For this purpose, all operations, including growth of the host organism but excluding chemical treatment of the phage, were carried out in broth, since it was found that cells grown and infected in $M 9$ gave very low average burst sizes when transferred to broth. Also, since the induced mutation rates were low (about 2 % of infected cells at most yielding mutants), the average number of bacteria (*ca.* 1) added to each burst tube was higher than that normally employed.

RESULTS

(1) *Survival curves*

The purpose in mind in determining the characteristics of phage inactivation by the alkylating agents now studied was threefold: first, to confirm that these compounds behaved qualitatively in a similar manner to those previously examined; secondly, to compare them quantitatively in order to determine equivalent doses, and thirdly, to seek any differences between their modes of action which might underlie their mutagenicity or otherwise.

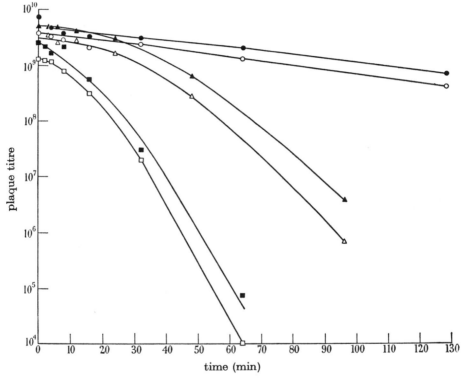

FIGURE 1. Dilution and thiosulphate survival curves for $T 2$ in methyl ($0 \cdot 02$ M), ethyl ($0 \cdot 4$ M) and n-propyl ($0 \cdot 25$ M) methane sulphonates. Squares, methyl; triangles, ethyl; circles, propyl. Open symbols, dilution; solid symbols, thiosulphate.

(a) *Thiosulphate and dilution curves*

These have been obtained for $T 2$ treated with the following compounds: methyl, ethyl and n-propyl methane sulphonates, 1:5-di(methane sulphonoxy)-pentane and ethylenimine. Those for ethylene oxide were published in part I.

Diethyl sulphate and *n*-butyl methane sulphonate were not studied in this respect, since the former hydrolyzes very rapidly and the acid liberated overcomes the buffer in less than 30 min, whilst the latter showed a very low 'toxicity'. None of the other compounds tested undergoes appreciable hydrolysis under the conditions of the experiments. The results are illustrated in figures 1, 2 and 3.

These curves are all of the same type as those given by ethylene oxide-treated $T2$, indicating (1) the absence of a thiosulphate-reversible step in the inactivation and (2) the dose-dependence of the post-treatment decay process which is incomplete at the time of plating (24 h after treatment).

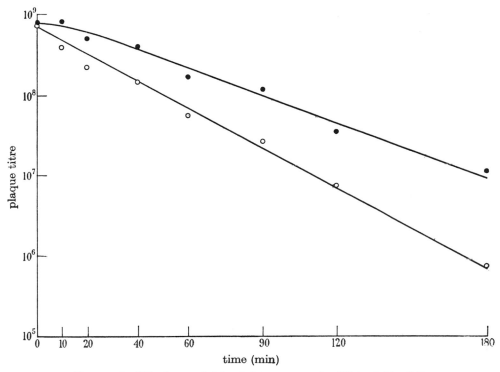

FIGURE 2. Dilution and thiosulphate curves for $T2$ in 0·04M-1:5-
di(methane sulphonoxy)pentane. ○, Dilution; ●, thiosulphate.

These curves resemble those obtained previously with di(2:3-epoxypropyl)ether, showing that there is a thiosulphate-reversible phase of inactivation characteristic of a bifunctional alkylating agent. They may not give an accurate picture of this owing to the relatively low competition factor of thiosulphate for alkyl methane sulphonates (Marshall 1955) as compared with 'mustards' and epoxides. In the case of the mono-alkyl methane sulphonates, reaction with thiosulphate was evident from the formation of alkyl mercaptans in the incubation tubes.

Again, the curves are typically those of a monofunctional compound.

(b) Decay rate: dose-dependence

As noted above, the convexity of the survival curves itself indicates that the post-treatment decay is dose-dependent. This dependence was further confirmed following arbitrary doses of ethylene oxide and ethyl methane sulphonate by following the plaque titres afforded by phage treated for various lengths of time and then diluted into buffer. (There is no need, in the case of these compounds, for a further incubation in thiosulphate prior to plating, since the loss of infectivity between sampling and adsorption on the plate is negligible.) The decay curves, which are given in figures 4 and 5, are not corrected for the decay in untreated controls. Evidently the rate of decay is more influenced by dose in the case of ethylene oxide.

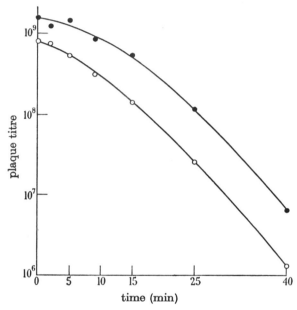

FIGURE 3. Dilution and thiosulphate curves for $T2$ in $0\cdot01$ M-ethylenimine. ○, Dilution; ●, thiosulphate.

(c) *Decay rate: pH-dependence*

This has been ascertained for methyl and ethyl methane sulphonates and for ethylene oxide and ethylenimine. The loss of titre on storage of untreated phage at 37 °C and at the appropriate pH's was concomitantly determined and the decay curves, given in figures 6, 7 and 8, have been corrected for this. The stock used proved to be most stable at pH 6. Thus the stability of alkylated phage on the acid side of neutrality appears to be less where hydroxyethyl or aminoethyl groups have been inserted than where the alkyl groups are unsubstituted.

(2) *Mutagenicity tests*
(a) *Screening experiments*

Table 1 (p. 505) presents the results of tests for induction of mutation at the *r* and *h* loci. It will be seen that so far as possible the compounds have been tested at doses which were comparable in respect of the inactivation which would have occurred had the treated phage been allowed to decay. It had been hoped also to make comparative tests at doses which were equivalent in terms of the amount of alkylation produced, but this was not possible owing to the great variation in toxicity between compounds of similar reactivity. A similar difficulty was experienced by Kølmark (1956) in comparing the mutagenic action of dimethyl and diethyl sulphates upon *Neurospora*. The concentrations of *n*-propyl and *n*-butyl methane sulphonates, 1:5-di(methane sulphonoxy)pentane and diethyl sulphate used corresponded approximately to saturation. The time of treatment with diethyl sulphate could not be extended owing to the amount of acid liberated by its hydrolysis.

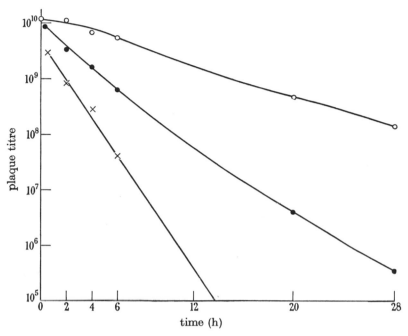

FIGURE 4. Post-treatment decay of $T2$ at pH 7 and 37 °C after treatments of various durations with 0·4M-ethylene oxide. ○, 10 min; ●, 20 min; ×, 30 min.

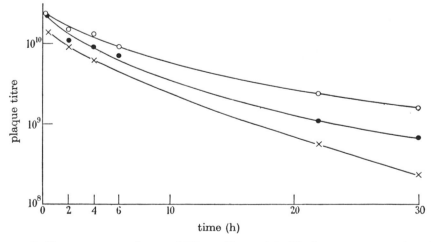

FIGURE 5. Post-treatment decay of $T2$ at pH 7 and 37 °C after treatments of various durations with 0·4M-ethyl methane sulphonate. ○, 10 min; ●, 20 min; ×, 30 min.

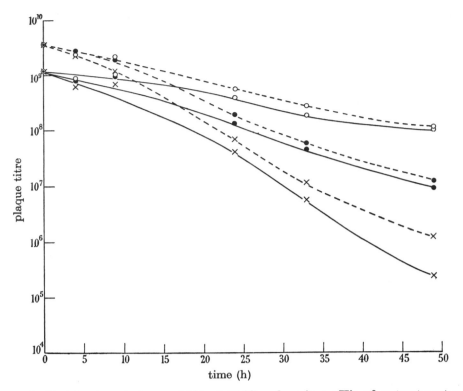

FIGURE 6. Post-treatment decay of $T2$ at 37 °C and various pH's after treatment with 0·01M-methyl methane sulphonate or 0·4M-ethyl methane sulphonate for 40 min. ——, Methyl methane sulphonate; – – –, ethyl methane sulphonate; ○, pH 7; ●, pH 6; ×, pH 5.

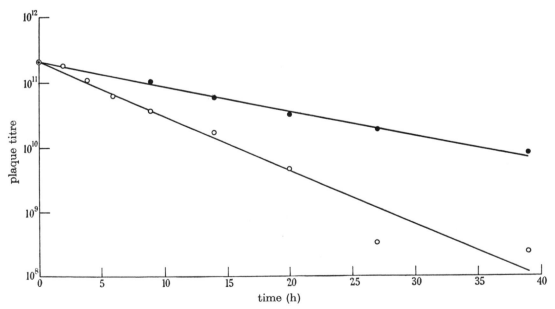

FIGURE 7. Post-treatment decay of $T2$ at 37 °C and various pH's after treatment with 0·4M-ethylene oxide for 10 min. ●, pH 7; ○, pH 5.

Allowing the treated phage to decay up to 24 h after treatment did not effect the proportions of mutant plaques.

(b) Single burst experiments with ethyl methane sulphonate

The dose given was 0·4 M for 40 min which, as can be seen from table 1, reduces the infective titre to about 70 % and results in an average burst size of 27. This same average was obtained in the single burst experiments. Allowing for the reduced infectivity, the cells were diluted following adsorption to give approximately one infected bacterium per burst tube, the actual average achieved being 0·96.

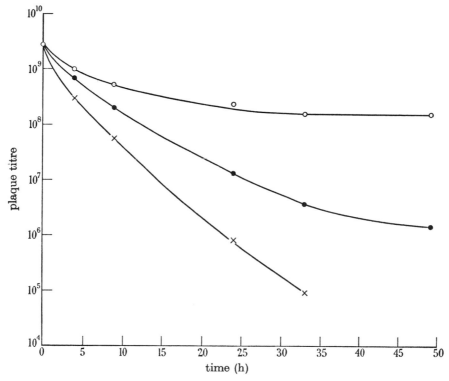

FIGURE 8. Post-treatment decay of $T2$ at 37 °C and various pH's after treatment with 0·01 M-ethylenimine for 40 min. O, pH 7; ●, pH 6; ×, pH 5.

Out of 200 plates examined, five showed r plaques, corresponding very closely to the proportion of mottled plaques found by plating cells infected with $T2$ similarly treated. The expected number for untreated phage would have been 0·2 or less. In addition, two plates had a large proportion of minute plaques and one plate a proportion of plaques in appearance intermediate between r and r^+. On picking and replating the latter, they all yielded, in addition to the parental type, other plaque types though no r. Replating of these again yielded a similar mixture.

Table 2 gives the plaque composition of those plates bearing r mutants. It seems that, although the mutations occurred in the first cycle of infection, there was no constancy in the time during the latent period at which they arose.

TABLE 1. PLAQUE COUNTS OBTAINED IN SCREENING EXPERIMENTS

reagent	group introduced	dose concn. (M)	time (min)	survival (%)	burst size	plaque counts on B pre-burst total	m	r	post-burst total	m	r	plaque counts on B/2 (≡ total pre-burst count on B × 5000)
none	—	—	—	100	40/55	5163	4	7	3631	5	8	300
methyl methane sulphonate	—CH$_3$	0·04	6	65	40	10525	4	9	8060	4	11	436
ethyl methane sulphonate	—CH$_2$.CH$_3$	0·4	20	100	42	10753	126	26	4870	5	45	5688*
		0·4	40	70	27	8775	137	39	2227	6	47	—
diethyl sulphate	—CH$_2$.CH$_3$	0·065	20	60	27	4090	51	12	2280	2	27	2344
ethylene oxide	—CH$_2$.CH$_2$.OH	0·1	10	100	50	15000	6	15	7216	0	7	—
		0·4	10	100	63	4420	—	—	—	—	—	571
ethylenimine	—CH$_2$.CH$_2$.NH$_2$	0·01	10	100	55	6360	7	10	6950	8	10	403
		0·01	20	100	53	6725	7	7	6992	5	14	509
propyl methane sulphonate	—CH$_2$.CH$_2$.CH$_3$	0·25	32	50	45	2835	2	1	2560	1	5	301
butyl methane sulphonate	—CH$_2$.[CH$_2$]$_2$.CH$_3$	0·065	120	100	52	7670	3	11	8000	2	14	513
1:5-di(methane sulphonoxy)pentane	—CH$_2$.[CH$_2$]$_3$.CH$_2$—	0·04	60	100	55	12500	10	8	13800	5	17	605

* Value adjusted for disparity in number of plates counted.

TABLE 2. SINGLE BURST EXPERIMENTS: PLAQUE-TYPE DISTRIBUTION ON
PLATES BEARING r PLAQUES

m	r	all others
2	56	30
0	8	46
0	4	16
0	5	247
0	2	207

DISCUSSION

All previous claims to having induced an increased rate of mutation of bacterial viruses by artificial means have depended upon treatment, either of the host/virus complex, or of the host cells, or host cells and phage separately prior to infection. In the case of λ (Weigle 1953), $T3$ (Weigle & Dulbecco 1953) and $T1$ (Tessman 1956) treated by ultra-violet-irradiation, the increased production of mutants has been shown (Stent 1958) to be very likely due to recombination of virus with host material. This explanation applies also to the high proportion of host-range mutants of $T3$ produced in cells of abnormally extended latent period (Fraser 1957) and to λ mutants arising in heavily irradiated cells (Jacob 1954).

The mutagenicity experiments described herein, particularly those involving host-range mutation, established that the mutants scored arose in the first cycle of growth of the treated phage. The host-range and single burst experiments further ensured that where a positive result was obtained this was due to an increase in the number of cells yielding mutants, and not to an increased yield of mutant phage from cells which would normally have produced mutants. The remaining point to be established, before it can be unambiguously claimed that mutation has been induced, is that the cells producing mutants do not have an abnormally long latent period, nor a consistently high burst size as a result of infection by the treated phage. Of the five plates yielding r mutants in the single burst experiments, four had plaque counts greater than the average burst size, two markedly so. However, at the average multiplicity of bacteria per burst tube employed, some 35 % of plates would be expected to have plaques derived from two or more bursts. It is not possible to say, therefore, on the basis of these limited experiments whether the mixed bursts arose from bacteria giving abnormally high burst sizes. In the screening experiments for plaque-type mutation all post-burst sampling was done at the fortieth minute after commencement of adsorption; hence the phage plated must have been liberated by bacteria having a maximum latent period of 40 min. In any event, a recombinational origin of phage mutants is unlikely in the case of $T2$ where no 'mating' of viral and host material is known to occur.

It remains to consider how alkylation of the phage prior to infection can result in an enhanced mutation rate during subsequent proliferation and why ethylation should apparently be unique in this respect. Since the mutations arise during the vegetative phase it is evident that the alkylation does not *ipso facto* constitute the mutation. In this respect the phenomenon appears similar to that reported by Litman & Pardee (1956), who found that the forced incorporation of 5-bromouracil into phage particles during synthesis led to an increased yield of plaque-type mutants. The proportion of mutants resulting from infection by ethylated phage achieved in the experiments reported above was considerably less than that attained by Litman & Pardee; no attempt, however, has been made to determine the proportional yield at low survival levels, since we have been concerned to demonstrate induction of mutation with the least ambiguity. The situation following ethylation is probably similar to that following bromouracil incorporation in so far as the chemical reaction alters the phage *DNA* without necessarily reducing its 'viability' but increasing the chances of faulty replication.

The inactivation studies give no clue to any difference in the mode of action of ethylating agents compared with other alkylating agents to which their mutagenicity might be ascribed. Whilst having a much lower toxicity than methyl methane sulphonate, ethyl methane sulphonate is no different in this respect from the non-mutagenic propyl and butyl homologues all of which have essentially the same reactivity. It is well known that methylating agents are relatively far more toxic than ethylating agents, and this has been found to apply in other systems studied for mutational purposes (e.g. Kølmark 1956).

Lastly, since insertion of the ethyl radical by either diethyl sulphate or ethyl methane sulphonate led to mutation, it must be concluded that ethylation *qua* ethylation is the important determinant. One is led to suppose that related radicals inserted into the phage *DNA* in similar positions are either innately incapable of determining mutation at the loci studied, or that the metabolic machinery of the infected bacteria is able to eliminate them before this occurs. Induction of mutation of *T* 2 is not the only example of the uniqueness of ethylation in the production of mutations; Kølmark (1956) has found diethyl sulphate to be the only one of a considerable number of otherwise mutagenic substances to cause appreciable reversion at the inositol locus in *Neurospora*.

This work could not have been performed without the skilled assistance of Miss Jocelyn Stock which is gratefully acknowledged.

The work has been supported by grants to the Chester Beatty Research Institute (Institute of Cancer Research: Royal Cancer Hospital) from the British Empire Cancer Campaign, the Jane Coffin Childs Memorial Fund for Medical Research, the Anna Fuller Fund, and the National Cancer Institute of the National Institutes of Health, U.S. Public Health Service.

REFERENCES

Adams, M. 1950 *Meth. Med. Res.* **2**, 1.
Ellis, E. L. & Delbrück, M. 1939 *J. Gen. Physiol.* **22**, 365.
Fahmy, O. G. & Fahmy, M. J. 1956 *Nature, Lond.* **177**, 996.
Fraser, D. 1957 *Virology*, **3**, 527.
Jacob, F. 1954 *C.R. Acad. Sci., Paris*, **238**, 732.
Kølmark, G. 1956 *C.R. Lab. Carlsberg, Sér. physiol.* **26**, 205.
Litman, R. M. & Pardee, A. B. 1956 *Nature, Lond.* **178**, 529.
Loveless, A. 1958 *Nature, Lond.* **181**, 1212.
Loveless, A. & Stock, J. C. 1959 *Proc. Roy. Soc.* B, **150**, 486.

Marshall, R. D. 1955 Thesis for Ph.D., University of London.
Stent, G. S. 1958 *Advanc. Virus Res.* **5**, 95.
Tessman, E. 1956 *Virology*, **2**, 679.
Timmis, G. M. 1950 *Ann. Rep. B.E.C.C.* p. 58.
Weigle, J. J. 1953 *Proc. Nat. Acad. Sci.*, *Wash.* **39**, 628.
Weigle, J. J. & Dulbecco, R. 1953 *Experientia*, **9**, 372.

Addendum to Parts I, II and III

Since completion of these three papers our attention has been drawn to a publication by L. Silvestri (1949 *Boll. Ist. sieroter. Milano*, **28**, 326). This work reports results which are relevant to all three papers now submitted. In particular Silvestri has shown that the inactivation of $T2$ treated with $HN2$ is not complete until 2 h after removal from the agent, although the conditions which he employed did not permit an accurate assessment of this nor reveal its cause. He also claims to have found an increased proportion of host-range mutants in the progeny of cells of *E. coli* infected with $HN2$-treated $T2$ at multiplicities greater than 1; this demonstration also involved incidental evidence of multiplicity reactivation of the treated phage.

We have been unable, however, to obtain convincing evidence of an increase in the proportion of cells yielding host-range mutants following infection with $HN2$-treated $T2$.

We are indebted to Dr David R. Krieg of the Oak Ridge National Laboratory for drawing our attention to the above publication.

CORRIGENDA

A. Loveless, *Proceedings* B, **150**, 497–508

p. 500, l. 6. *For* 'These curves are...' *read* 'The curves in figure 1 are...'

p. 500, l. 4 from bottom. *For* 'Again, the curves are...' *read* 'Again, the curves in figure 3 are...'

Reprinted from *Genetics*, **48**(4), 561–580 (1963)

ETHYL METHANESULFONATE-INDUCED REVERSION OF BACTERIOPHAGE T4rII MUTANTS

DAVID R. KRIEG

Biology Division, Oak Ridge National Laboratory,[1] Oak Ridge, Tennessee

Received December 11, 1962

ETHYL methanesulfonate (EMS[2]) is an alkylating agent that can react with extracellular virus particles to produce many mutations with rather little killing (LOVELESS 1958). The induced mutations may be delayed by several generations (GREEN and KRIEG 1961). The mutagen reacts with three of the four bases naturally occurring in DNA (REINER and ZAMENHOF 1957; BROOKES and LAWLEY 1960, 1961a, 1962; PAL 1962), yet there is reason to suspect that it might be quite specific as to the type of base pair changes it usually induces. The purpose of this investigation was to compare the frequencies of EMS-induced mutations at various sites within the rII region of bacteriophage T4, and to test the notion that predominantly one type of base pair substitution might be induced.

Four years ago in this laboratory, during a search for strongly EMS-revertible point mutants, D. M. GREEN found that some AP mutants (mutants which had been produced from standard r^+ phage by AP) were quite EMS-revertible, but that none of the 18 EMS-produced mutants he examined were induced by EMS to revert to a comparable extent. This prompted the working hypothesis that one of the base pair transitions—either GC to AT or AT to GC—was much more strongly inducible by EMS than was the other transition. It was decided to test this possible specificity by checking a group of EMS mutants for EMS-induced reversion by a more sensitive test than had been previously employed, and to test similarly additional base analog and proflavin mutants.

While this work was in progress, the results of another investigation with similar intent were reported (BAUTZ and FREESE 1960; FREESE 1961). The results differed in several respects for mutants that were included in both investigations; most striking was that the other group of workers found much less induced reversion. Although the other workers used EES, the chemistry of the two ethylating agents would be expected to be very similar. It seemed that the difference must have arisen from the different techniques employed. This possibility led to an expansion of our work, and the chief reason for the conflicting results— which will be discussed in context in this paper—has now been clarified. We shall

---

[1] Operated by Union Carbide Corporation for the U. S. Atomic Energy Commission.

[2] The abbreviations used are: G, guanine; C, cytosine or 5-hydroxymethylcytosine; A, adenine; T, thymine; EMS, ethyl methanesulfonate; EES, ethyl ethanesulfonate; AP, 2-aminopurine; BU, 5-bromouracil or 5-bromodeoxyuridine; P, proflavine; and DNA, deoxyribonucleic acid.

conclude that EMS induces the GC to AT transition at a much higher rate than the AT to GC transition. As previously concluded, transversions—the replacement of a purine-pyrimidine pair by a pyrimidine-purine pair—may also be induced to some extent by EMS (BAUTZ and FREESE 1960; FREESE 1961), but we do not consider the existing evidence to be compelling. We shall describe several hypothetical molecular mechanisms of EMS-induced mutation, including one that was not considered by the other group and that would generate only GC to AT transitions by pairing errors in replicating DNA. BAUTZ and FREESE (1960) concluded that any base could be incorporated opposite a "gap" in the DNA template produced by the hydrolytic removal of 7-ethylguanine from DNA prior to replication. We suggest that, if such gaps are not lethal, they might lead to base pair deletions.

MATERIALS AND METHODS

General: Bacteriophage T4B and derivative *r*II mutants were obtained from SEYMOUR BENZER, ERNST FREESE, and JOHN DRAKE. An *r*II mutant is defined as one that has the distinctive *r*-type plaque morphology on plates with strain B bacteria and that does not normally form plaques when plated with bacteria lysogenic for phage lambda. All such mutants are closely linked and occur in two adjacent cistrons. *Escherichia coli* strain Bb, originally isolated from Berkeley B by SYDNEY BRENNER and obtained from BENZER, was used to prepare phage stocks and as the host for nonselective growth. Strain B was used for nonselective platings, and two lambda-lysogenic strains, both previously designated K and now redesignated KB (K-12 from BENZER) and KT [112–12 (λ h) No. 3, from the California Institute of Technology collection] were used for selective platings (BENZER and CHAMPE 1961). Growing bacteria were centrifuged and resuspended in cold broth for use as plating bacteria. Host bacteria were grown to about 5×10^7 cells/ml, and cyanide was added before centrifuging.

H-broth (8 g Nutrient Broth, 5 g NaCl, 1 g dextrose, 5 g Peptone/liter H_2O) was used for growth of plating and host bacteria. The medium used for production of progeny phage in experiments was identical to H-broth except that NaCl was omitted, and indole was usually added at 0.1 g/liter. M9 buffer contained, per liter of distilled H_2O: 5.8 g Na_2HPO_4, 3 g KH_2PO_4, 0.5 g NaCl, 1 g NH_4Cl, and 0.01 g gelatin. The modified M9 medium used for growth of phage stocks was made by adding, to M9 buffer, separately autoclaved solutions to give 4 g/l dextrose, 1 mM $MgSO_4$, 2 μM $FeCl_3$, 25 mg/l L-tryptophan, and either 100 mg/l histidine or 2.5 g/l vitamin-free casamino acids. (It was found that phage were inactivated in the presence of dextrose unless either histidine or casamino acids were added (KLIGLER and OLENICK 1943); this inactivation is also prevented by materials present in lysates). Plating media were those described by HERSHEY and ROTMAN (1949), except that peptone was used instead of tryptone.

Procedure for standard reversion experiments: A phage lysate was diluted in M9 buffer to 1×10^{10}/ml, and an equal volume of 1 M phosphate buffer at pH 7 was added. Phage were exposed at 45°C by adding this buffered preparation to EMS in the ratio 1 ml to 0.025 ml. Exposures were terminated by chilling and dilution into four volumes of cold M9 buffer containing $Na_2S_2O_3$ at a final concentration of 0.04 M. For controls, the procedure was similar except for addition of EMS, and dilution was made with 19 volumes to give the same titer of viable phage. The dilutions were kept cold until used, then incubated ten to 30 minutes at 37°C to allow residual EMS to be eliminated by the thiosulfate. An equal volume of resuspended host bacteria was added, and the mixture was gently aerated ten minutes for adsorption. The ratio of viable phage to bacteria was about 0.6, and 0.001 M NaCN was present to synchronize infection. Adsorption of EMS-exposed phage is impaired, but about 85 percent of the viable phage are adsorbed under these conditions. Growth bottles were made by adding 0.5 ml aliquots of the adsorption mixtures to 100 ml of salt-free broth. After two hours incubation at 37°C the growth bottles were shaken with chloroform and kept cold. Assays were made with B bacteria for total progeny phage and with KB or KT bacteria for revertant phage. Any plaques found on KB or KT are attributed to revertants, without necessarily implying that they represent true back mutants. Platings of standard r^+ phage were made routinely to determine the efficiency of plating for the bacterial strains. Usually, two growth bottles were made from each adsorption mixture, and the results were averaged. In the earliest experiments a single growth bottle was made, and at 30°C with a slightly longer incubation. The mean burst size was usually more than 100. In a few experiments a much lower progeny titer was found in one or more growth bottles, and the mutants were checked again; since similar revertant frequencies were found, both experiments are reported.

EXPERIMENTAL

Relation between duration of exposure, extent of killing and induced mutation

Figure 1 shows the fraction of phage found to be viable at various times after EMS was added to a phage suspension. The rate of phage killing increases uniformly throughout the entire period studied. Since a constant slope is not achieved, this curve is not of a "multiple hit" or "multiple target" type. The increasing rate of killing is apparently not due simply to a progressive diffusion of the poison into the phage; when EMS was added to a cold phage suspension and the mixture was kept in the refrigerator for one or 24 hours, during which killing was 12 percent or less, a survival curve of the same general shape was generated once the reaction mixture was brought to 45°C. Moreover, T4o5 stocks, having a mutant gene for osmotic shock resistance and thus a higher permeability to ions, give a similar survival curve. Additional experiments (not shown here) established that the progressively steepening survival curve is not due to the gradual production of a toxic substance in the reaction mixture. When additional phage were added to the reaction mixture 80 minutes after EMS was put into solution, the newly added phage were killed no faster than was initially observed. A survival curve of the same general shape was generated, but with a somewhat lower initial rate of killing, a result suggesting that the effective concentration of the poison was decreasing rather than increasing.

200

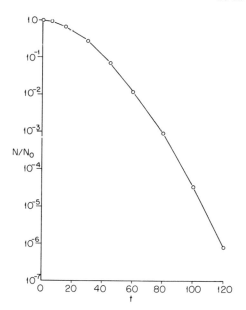

FIGURE 1.—Semilogarithmic plot of the fraction of phage surviving an exposure to EMS for t minutes. N = viable phage titer at time t, N_0 = original phage titer.

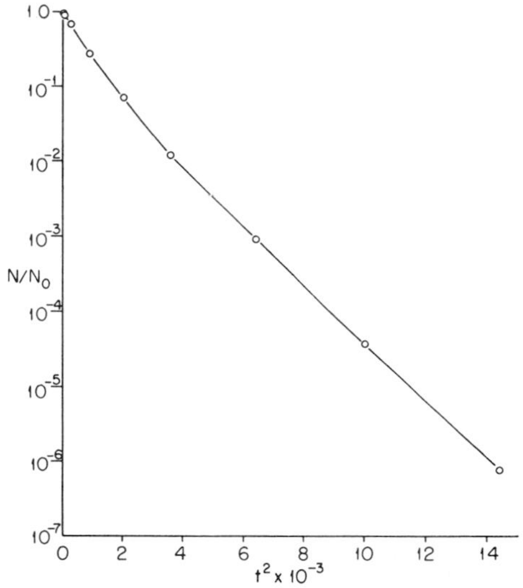

FIGURE 2.—Semilogarithmic plot of the fraction of phage surviving versus the square of the duration of exposure to EMS.

Interpretation of survival curve: The survival curve can be interpreted according to the following hypothesis. We assume that each phage has a large number, S, of sites of potential lethality that are ethylated with pseudo-first order kinetics at a rate a, proportional to the EMS concentration. We further assume that an ethylated site is not in itself lethal, but that it can be converted to a lethal damage by a first order reaction that has a rate constant b. For a duration of exposure, t, that is small relative to $1/a$ and $1/b$, the survival curve predicted by this model is approximately given by $\log (N/N_o) = -(abS/2) \, t^2$, where N_o is the initial titer and N the titer of viable phage at time t.

To test the applicability of this equation, the data presented in Figure 1 are replotted in Figure 2, with t^2 as the abscissa. As predicted, an approximately straight line is found. The gradually decreasing slope may reflect a gradual reduction in the effective concentration of EMS. The compound is unstable in aqueous solution, and Figure 2 suggests a half-life of several hours for EMS. This postulated two-step mechanism of killing accounts for the instability of ethylated phage after an exposure to EMS, which was reported previously (LOVELESS 1959; BAUTZ and FREESE 1960; STRAUSS 1961).

Kinetics of induced mutation: The relation between induced mutation and duration of EMS exposure was determined for reversions of the *r*II mutant AP72 and is shown in Figure 3. It can be seen that although killing is approximately proportional to t^2, the frequency of induced mutation is approximately proportional to t and, therefore, is not proportional to the number of phage-lethal damages.

This suggests that mutations are produced by ethylated bases contained within the phage at the time of infection and that killing is due to a secondary event that also occurs during the exposure interval, such as a breakdown or hydrolysis of ethylated molecules.

It should be noted that the frequency of induced reversions was reported to be proportional to the frequency of lethal damages for two other mutants, AP156 and EES66 (BAUTZ and FREESE 1960; FREESE 1961). It is not known whether the different dose dependence reported is due to the different mutants or to the different techniques employed.

Comparative survey of EMS-induced revertibility of
AP, BU, EMS and P mutants

An exposure of 30 minutes was chosen as the standard condition for a comparative survey of mutants. This exposure allowed convenient production of progeny populations with a titer of about 10^8/ml, and since it induced revertants at a frequency of about 10^{-4} for AP72, it offered a highly sensitive test of low-level revertibility.

The mutants initially chosen for analysis included a set of nine produced by AP and seven produced by 5-bromouracil (indicated here by the prefix BU, instead of their former prefix, N). These were selected to permit comparison with the studies by FREESE and coworkers of their induced revertibility by various agents, including base analogs (FREESE 1959), nitrous acid (FREESE and FREESE 1961), and by DRAKE (personal communication) for UV-induced reversion. A set of 11 EMS-produced mutants (indicated by the prefix EM) and ten proflavin-produced mutants (indicated by the prefix P) were obtained from BENZER, as a random sample of those he mapped in the B cistron (BENZER 1961). Mutants added for various reasons after the survey had been begun are: AP50, BU90, EM126, EES66, and r207.

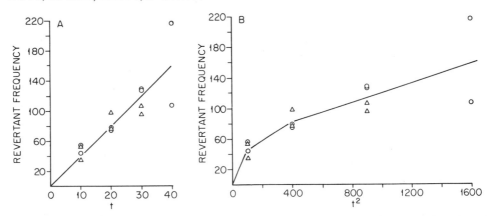

FIGURE 3.—Frequency of induced AP72 revertants per 10^6 progeny phage as a function of duration of exposure to EMS, t. Triangles are for exposure of rAP72 and circles are for exposure of rAP72o5, which carries a gene for osmotic shock resistance. A separate point is indicated for the duplicate growth bottles made for each exposure.

Table 1 presents the results of all standard reversion experiments performed. Induced revertant frequency refers to the difference between the results for the treated and the control populations.

Plaque morphologies of revertants: The plaques produced by revertants on the plates with a selective strain of bacteria were generally indistinguishable from those produced by the standard r^+ phage under those plating conditions. Generally, a small fraction of the plaques had an aberrant plaque morphology, but this could reasonably be expected from separate plaque morphology mutations incidentally produced in the same line of descent as the rII reversion, so no notation is made in the tables if the frequency of aberrant plaque morphologies amounted to less than ten percent of the revertants. Some mutants, however, regularly produced revertants that had distinctively smaller plaques. In these

TABLE 1

EMS-induced reversion

		Frequency of revertants per 10^6 progeny phage[†]						
Mutant	Site[*]	Control			Induced			
AP114	DAP56	2.8	s,	0.9 t	290.	s,	30.	t
AP72	607	1.7			143			
		1.2			129			
		1.4			153			
		1.1			183			
		0.4			144			
AP12	F102	0.9	s,	18. t	81.	s,	37.	t
		0.8	s,	24. t	111.	s,	32.	t
AP41	AP41	2.6	s,	44. t	77.	s,	154.	t
AP275	N38?	1.4			85.			
		1.0			50.			
		1.4			55.4			
		0.4			71.			
		1.5			49.			
AP50	AP50	0.74	s,	16. t	37.	s,	42.	t
AP61	AP61	0.2			0.6			
AP156	EM9	0.02			0.4			
		0.02			0.4			
		0.03			0.36			
AP70	?	0.00			0.5			
		0.00			0.15			
AP83	EM9	0.03			0.2			
BU90	N90	0.8	s,	0.4 t	23.2	s,	5.6	t
BU24	N24	0.7			1.3			
BU7	N7	0.01			0.6			
BU29	N29	0.2			0.6			
		0.15			0.00			
BU19	N19	0.07			0.18			
BU12	N12	0.04			0.15			
BU21	N21	0.00	s,	0.06 t	0.05	s,	0.09	t
BU101	NT71	0.005			0.02			

TABLE 1—Continued

Mutant	Site*	Frequency of revertants per 10^6 progeny phage†					
		Control			Induced		
EM126	607	0.35			147.		
EM34	EM34	0.4	s,	15. t	22.	s,	36. t
EM43	N24	0.3			2.3		
		0.2			0.3		
EM11	BC35	0.17 s,		0.10 t	1.1	s,	2.7 t
EM3	EM3	3.3			0.9		
EM9	EM9	0.02			0.5		
EM30	117	0.0			0.3		
		0.5			0.5		
EM41	163	0.04			0.02		
		0.12			0.55		
EM7	EM7	0.03			0.16		
EM26	EM26	0.96			0.11		
		0.03			0.15		
EM87	EM87	0.01			0.00		
EM15	N7	0.28			—0.05		
P85	326?	1.7	s,	27. t	60.8 s,		76. t
P42	P42	12.	s,	1. t	—1.	s,	35. t
		10.6	s,	1.6 t	0.3	s,	24. t
P28	NT23	0.3			0.7		
P67	P67	0.35			0.60		
P83	P83	0.07			0.10		
P48	P48	0.05			0.1		
P79	P79	0.06			0.08		
P87	P87	0.07			0.07		
P13	1074	0.27 s,		0.1 t	0.04 s,		1.0 t
P31	EM113	0.11			—0.02		
*r*207	?	85.4			32.4		
		85.6			31.3		
EES66	?	8.4			4.2		
		9.1			3.7		

* As designated by BENZER (1961).
† s = standard sized plaques on selective platings; t = tiny plaques on selective platings.

cases, the tabulated results distinguish between apparently standard size plaques (designated *s*) and the "tiny" plaques (designated *t*). It must be emphasized that this criterion refers to the appearance of the plaques on the strain initially used for the detection of revertants. In certain cases, including all those in which EMS-induced revertants had a frequency of more than one per 10^6 progeny, five or more revertant plaques were picked and their contents plated with the bacterial strains B, KB, and KT. Some revertants were indistinguishable from standard r^+ phage with respect to plaque morphology and efficiency of plating under all conditions that were tried. Some revertants that produced full-size plaques with one selective strain produced tiny plaques or had a reduced efficiency of plating with the other strain, and sometimes this detection depended on the physiological state of the plating bacteria. Some revertants were indistinguishable from standard r^+ phage on both selective strains but were found to give distinctively different plaque morphologies with strain B. For instance, the BU90 revertants designated s were found to give nearly r-like plaques with strain B. Most s-type revertants from AP50 resemble standard r^+, as do a fraction of those from AP41 and AP114; but none resembling standard r^+ were observed from AP12 or EM34.

Back-crosses of apparent wild-type revertants: AP72, AP275, and AP156 regularly gave EMS-induced revertants which were indistinguishable from standard r^+ phage. Stocks were made of one such revertant from each, and these were back-crossed to standard-type phage, with UV irradiation to enhance recombination. If the revertants were caused by suppressor mutations, the original r-type phage would be segregated out, and perhaps the isolated suppressor might also be recognized as an r plaque-former (E. B. FREESE 1962; CRICK, BARNETT, BRENNER and WATTS-TOBIN 1961). The frequency of r plaque-formers among the progeny was not greater than 0.1 percent in any of these backcrosses, but as a further test of closely linked suppressor mutations, all r plaques found among 3500 to 8000 progeny were picked and spot-tested (BENZER 1957) with the original r mutant. In each case there was a positive indication of recombination showing that it was a different mutant. Hence, the EMS-induced revertants from AP72, AP275, and AP156 are either true back mutants or due to very closely linked suppressors. In contrast to these results, when a back-cross was made between standard phage and an s-type revertant from BU90, r plaques were found at a frequency of 0.1 percent and four out of the six r plaques found had phage that were indistinguishable from BU90 by the criteria of spontaneous revertibility and standard crosses to BU90. The presumed suppressor was not recovered, but it is not known whether that is because insufficient r plaques were tested or because the isolated suppressor fails to make a plaque type distinguishable from standard phage and the s revertant. (The backcross progeny in this case were preabsorbed on B bacteria and plated on a mixture of B and KB bacteria, on which the r-like plaques of the s revertant could be distinguished by their clear centers from the more turbid plaques of rII mutants.)

Contrast of induced revertant frequencies with previously published data: When the results shown in Table 1 for EMS-induced reversion of base-analog-produced mutants are compared with the results reported for EES-induced reversion of the same mutants (BAUTZ and FREESE 1960; FREESE 1961), two features are especially striking. First, the mutants that we find most strongly revertible were also the most revertible in their data, but the absolute values are much higher in our study. For instance, the induced revertant frequencies we find for AP72 and AP275 are 149 and 62 per 10^6, while they report only 1.0 and 1.5 per 10^6, respectively. Second, mutants they found to give induced revertants at a lower frequency are also revertible in our studies, but the absolute values are not enhanced to the same extent. For instance, AP156 and AP70 give 0.4 and 0.3 per 10^6 in our study, and 0.08 and 0.065 per 10^6 in their study.

The implications of these findings for inferring base-pair changes induced by ethylating agents will be discussed later. The results prompted the next group of experiments to determine the reasons for the differences observed.

A comparison of EMS-induced revertant frequencies obtained from direct selection of exposed phage and selection of progeny

It seemed likely that the direct plating of exposed phage with selective bacteria might give only a partial recovery of the induced revertants, and that this could account for the lower values observed in the experiments of Freese and co-workers. We tested this by directly plating a portion of phage exposed under our standard conditions and subjecting another portion of the same phage preparation to our standard cycle of reproduction in a nonselective host bacterial strain before selective platings for revertants. Table 2 presents the results of such experiments. It also includes the results of a reconstruction experiment performed with a mixture of AP156 and standard r^+ phage. The mixture was subjected to

TABLE 2

Recovery of induced revertants from direct selective platings of exposed phage compared to platings of progeny phage

	Frequency of revertants per 10^6				
	Direct platings		Progeny platings*		Induced revertants (direct/progeny)
Mutant	Control	Induced	Control	Induced	
AP72	0.8	4.6	1.1	183	0.03
	0.4	2.4	0.4	143	0.02
EM126	0.1	7.2	0.4	147	0.05
AP275	1.9	10.2	0.4	71	0.14
	2.7	9.2	1.5	49	0.19
*r*207	39	—11	85.4	32.4	. . .
	53	—14	85.6	31.3	. . .
EES66	3.9	0.8	8.4	4.2	0.2
	4.8	—1.4	9.1	3.7	. . .
AP156	0.005	0.67	0.03	0.36	1.9
Mixture of AP156 and r^+	3.69	0.60	4.60	0.57	. . .
BU29	0.04	0.00	0.15	0.00	. . .
AP114	3.9 s	40. s	2.8 s	290. s	0.14 s
	1.6 t	9. t	0.9 t	30. t	0.30 t
AP12	0.25 s	3.1 s	0.8 s	111. s	0.03 s
	2.6 t	7.9 t	24. t	32. t	0.25 t
AP41	0.3 s	9.3 s	2.6 s	77. s	0.12 s
	15. t	11. t	44. t	154. t	0.07 t
P85	0.5 s	5.2 s	1.7 s	60.8 s	0.09 s
	15.2 t	24.0 t	26.8 t	75.7 t	0.32 t
AP50	0.3 s	30. s	0.7 s	37. s	0.81 s
	9.5 t	6. t	16. t	42. t	0.14 t
EM34	0.36 s	0.3 s	0.4 s	22. s	0.01 s
	3.7 t	1.3 t	15. t	36. t	0.04 t

* These data are also included in Table 1.

206

standard EMS exposure and control procedures, and dilutions for assays, etc., were identical with those used for AP156 alone in a parallel experiment. This experiment was designed to evaluate the reliability of detection of r^+ phage among a large exces of rII phage, since it seemed possible that some plaques of spurious origin might appear on the selective platings of exposed phage. For instance, if enough plating bacteria were infected by more than one phage, plaques might arise from killed r^+ phage by multiplicity reactivation or cross reactivation. The data show that this is negligible under these conditions and the increased frequency of r^+ plaques among exposed phage can be attributed to induced revertants.

The data of Table 2 indicate there is indeed a markedly reduced recovery of potential induced revertants in direct platings of exposed phage on selective bacteria. The ratio of the revertant frequency detected in this way to that observed among progeny differs from mutant to mutant. (This variability does not simply come from daily variations in plating conditions, since it was found for different mutants checked at the same time.) While our most revertible mutants show a ratio as low as a few percent of the potential induced revertants, AP156 regularly gives about the same frequency by the two techniques; indeed, it appears that a slightly higher frequency may be found on direct platings. These results will be discussed later in terms of mechanisms of induced reversion, but it may be noted here that the apparent relative revertibility of different mutants from our direct platings agrees approximately with the results of direct platings of EES-exposed phage (BAUTZ and FREESE 1960; FREESE 1961). The fact that our data still give generally larger rates may arise from other differences of procedure that affect the various mutants to a similar extent; for instance, we did not use the 24 hour incubation at 37°C which the other group used after EES exposure.

Some technical comments seem to be appropriate. There was a wide range of plaque sizes on selective platings of exposed phage, grading from full size plaques like those typically made by standard r^+ phage down to very small plaques. This was true even for AP275, AP72, AP156 and the reconstruction experiment involving AP156, although progeny populations from the same exposed phage preparations gave typically s-type plaques and no "tiny" revertants. The reduced plaque sizes can probably be partly attributed to nonheritable damage from the EMS, for instance, reduced burst size or extended latent period. Delayed revertants might also be expected to give a smaller plaque, since only a fraction of the phage issuing from the first infection—in some cases, a single phage particle—would be revertant and would thus get a relatively "late start" at making a plaque. Counting all the revertant plaques becomes a greater problem for a mutant like AP12, which gives two types of revertants. Generally, the s and t types are fairly distinct in selective platings in progeny. However, the variable reduction in plaque size in direct platings of exposed phage results in a blurring of the distinction: some standard revertants probably make plaques as small as a typical tiny revertant, and plaques from the tiny revertants grade down to a size so small as to be overlooked. Poor reproducibility among experiments for the same mutant suggests that differences in plating conditions from day to day may also be more of a problem in platings of such mutants.

Possible molecular mechanisms of EMS-induced mutations

Alkylating agents like EMS react with guanine, adenine, cytosine, and 5-hydroxymethylcytosine when the free bases or nucleotides are exposed in solution (BROOKES and LAWLEY 1960, 1961a, 1962; PAL 1962). The reaction of this agent with DNA produces 7-ethylguanine (BROOKES and LAWLEY 1961b), and 3-ethyladenine and other products may occur to a lesser extent. We shall consider two distinct mutagenic mechanisms that may be postulated for 7-ethylguanine, and one mechanism for 3-ethyladenine.

1. *GC to AT transitions mediated by 7-ethylguanine paring errors:* Whereas guanosine has a pK of 9.2 for the dissociation of the hydrogen atom at the N_1 position, the corresponding pK for a 7, 9-substituted guanine is shifted closer to 7 (LAWLEY and BROOKES 1961; PFLEIDERER 1961). Thus, ionization of the N_1 may occur while the DNA strand is serving as a template for DNA synthesis, which would permit pairing of 7-ethylguanine with thymine (Figure 4A) as well as with cytosine. This would generate a GC to AT transition. If a template molecule of 7-ethylguanine was not ionized at the moment of DNA replication it would pair like guanine with cytosine and hence make a normal, nonmutated copy. The possibility of making a later copy error would be retained by the DNA

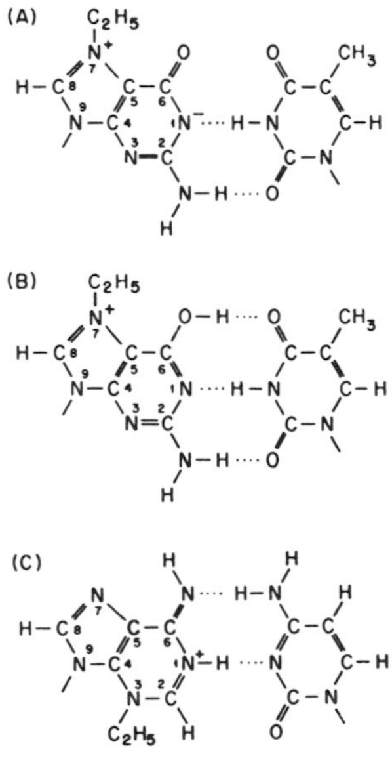

FIGURE 4.—Pairings postulated to occur during replication of DNA containing ethylated purines. Hydrogen bonds are shown by dotted lines. A: 7-ethylguanine and thymine. (The negative charge may be located on C_6O as well as on N_1.) B: Enol tautomer of 7-ethylguanine and thymine. C: 3-ethyladenine and cytosine.

strand containing the analog, regardless of the number of normal copies made first. The probability of a pairing error would be a function of the pH at the time and place of replication and of the effective pK of the analog present in a DNA strand: if the pH were one unit lower than the effective pK, the chance of a pairing error might be about 0.1; if the pH were two units lower, about 0.01. However, we do not know the actual pK of 7-ethylguanine in DNA, nor the effective pH during replication. The effective pK for 7-ethylguanine could well be several units higher in an ordered DNA molecule than for the nucleotide, by analogy with observations on guanine and thymine (CAVALIERI and STONE 1955). Thus, the absolute probability of the pairing error may be quite small. Nevertheless, the effective pK of 7-ethylguanine should still be lower than that of guanine in DNA. Thus, the probability of a pairing error with thymine would be larger for the guanine analog than for guanine.

Another mechanism by which 7-ethylguanine in DNA might make pairing errors during replication can be mentioned as a variant of this hypothesis. At pH values below the pK for ionization, this analog and also guanine may possibly exist to some extent in the enol rather than the keto tautomeric form. If so, the enol tautomer would be expected to pair with thymine by three hydrogen bonds (Figure 4B). If the frequency of the presumably rare enol tautomer were higher in DNA for 7-ethylguanine than for guanine, it would permit another mechanism for pairing errors which would be nearly indistinguishable in its effects from the mechanism outlined above.

2. *Replication errors from gaps produced by 7-ethylguanine hydrolysis:* Alkylation of the N_7 of deoxyriboguanylic acid produces a quaternary ammonium, and the positive charge may be shared by the N_9 atom. This weakens the glycoside linkage, which may lead to hydrolysis of the 7-ethylguanine from the DNA prior to replication (LAWLEY 1957; BAUTZ and FREESE 1960). The biological effect on DNA synthesis of a "gap" from a missing base is not known and it may be lethal. However, if replication can proceed past this point it may result in a base pair deletion or, as has been previously suggested, in a base pair substitution at the site formerly occupied by the GC pair (BAUTZ and FREESE 1960). This hypothesis is supported by the observed mutagenicity of low pH on phage (FREESE 1961). The chance of a 7-ethylguanine producing a replication error by this mechanism will be the product of the incidence of hydrolysis at that site and the probability of the replication error occurring at such a gap. If any of the four types of precursor molecules may be incorporated randomly opposite a gap, there may be a comparable chance of obtaining a normal copy, a GC to AT transition, or a GC to CG or GC to TA transversion. However, when the strands of DNA separate and serve as a template for replication, the forces normally tending to "stack" adjacent bases may draw together the two previously separated bases neighboring the gap produced by 7-ethylguanine hydrolysis. Replication then might result in the production of a single base pair deletion, equivalent to one of the types of "reading frame shift" mutations postulated for acridine mutagenesis (CRICK et al. 1961).

3. *AT to GC transitions mediated by 3-ethyladenine:* 3-Methyladenosine has

been shown to have an imino group in place of the amino group of adenine, and to be positively charged at pH 8 and below (Pal and Horton 1962). The presumed structure indicates that during DNA replication, pairing would be possible between 3-ethyladenine and cytosine by two hydrogen bonds (Figure 4C), and that pairing with thymine would not occur. This would imply that the pairing error from 3-ethyladenine would have a probability near one.

Some implications of different copy error probabilities

The probability of copy error (c) characteristic of a given mutational mechanism will determine whether the frequency of induced mutations is as high as or lower than the fraction of the bases that have been altered at a given site in the DNA during exposure to the ethylating agent, whether the induced mutations are immediate or delayed, and how the frequency M_p of mutants among progeny derived from exposed phage is related to the fraction M_s of survivors that give an induced mutation. We shall illustrate these relations by considering several cases and then consider what this suggests as to the relative importance of the three hypothetical mechanisms.

If c is equal to one, the induced mutation occurs at the first duplication. If half the clone produced by a phage is descendant from the strand of the DNA duplex that contains the altered base, and this subclone is entirely mutant, then $M_p = 0.5 M_s$. This is the case for most nitrous acid-induced mutations (Tessman 1959). As will be discussed in connection with AP156, we infer that this is the case for AT to GC transitions mediated by 3-ethyladenine.

If c is smaller than one, and a normal nonmutated copy can be made, then normal duplications may precede the occurrence of a copy error at the site of an altered base. A population of phage containing altered bases producing mutations by such a mechanism would generate mutant subclones of various fractional sizes distributed as $\frac{1}{2}^n$ of the total clones with n reflecting the duplication at which a particular copy error took place. We refer to this as a pattern of delayed mutation. If c is very small, the altered bases could generate an equal incidence of half-, quarter-, eighth-mutant clones, etc. (Green and Krieg 1961). This distribution implies that half of the total mutants in the progeny population would be contained in half-mutant clones produced in the first duplication and that the mean fractional mutant clone size would be 0.14 (Green and Krieg 1961). Hence with such a pattern of delayed mutations, $M_p = 0.14 M_s$ and the ratio of the frequency of mutant progeny to the fraction of bases ethylated would be equal to c, even smaller than 0.14.

EMS-induced mutants, produced and detected by methods similar to those employed in this study, occur as delayed mutations (Green and Krieg 1961). Thus we infer that the mechanism accounting for all or most EMS-induced mutations is characterized by a low probability of copy error, and we should regard as most important a mechanism that might have this characteristic. We have indicated that this could be true for (1), based on the ionization of 7-ethylguanine, and that it is probably not true for (3), based on 3-ethyladenine. The implications of (2) are not so clearly predictable, but if gaps were produced by hydrolysis prior to or early in infection it seems unlikely they would produce the observed pattern of delayed mutations. The possibility also exists that a pattern of delayed mutations would be generated if hydrolysis yielded mutants only if it occurred after DNA synthesis had began.

An explanation for the low recovery of EMS-induced revertants in direct selection

The foregoing discussion leads us to an hypothesis to account for the various ratios found between the frequency of revertants observed in direct platings of exposed phage with selective bacteria and the frequency of revertant progeny (Table 2).

When an rII phage infects a bacterium of a selective strain, the chance of any phage being produced is normally quite small; this value is called the transmission coefficient and is about 0.0004 for AP72 in KB. For a revertant frequency of 140 per 10^6 progeny, as is found for our standard exposure of AP72, we infer that the frequency of reversions per survivor is about 1000 per 10^6 ($M_s = M_p/$ 0.14). The frequency of reversions per survivor observed on direct plating with KB bacteria is about 5 per 10^6, so the efficiency of their recovery is about 0.005, considerably larger than the normal transmission coefficient. The corresponding efficiency of recovery for AP275 revertants is about 0.02. These calculations suggest that the production of revertant phage by an infected KB bacterium requires not only the production of a revertant gene but also one or more revertant gene products, and that a revertant gene product is produced in only a small fraction of the bacteria infected by a phage with an ethylated AP72 or AP275 site. If the production of a revertant gene product involves the production of an r^+ messenger RNA molecule by an induced pairing error on the DNA template of the infecting phage, the probability may be directly related to c, the probability per DNA duplication of a pairing error at an ethylated mutant site. (Uracil, like thymine, could pair with 7-ethylguanine.) As we have pointed out, the occurrence of delayed mutations implies that c is much lower than one, and this could be expected for 7-ethylguanine. Hence, a low recovery of induced revertants on direct selection may be a characteristic feature of EMS-induced GC to AT transitions.

If the probability of a copy error from 3-ethyladenine is nearly one per duplication, both during RNA and DNA synthesis, then there may be a characteristically high recovery of EMS-induced AT to GC transition revertants on direct selection. This may account for the high ratio observed for AP156.

Revertants that are not true back mutants

As mentioned with the results, many of the revertants can be distinguished from standard r^+ phage and thus are not true back mutants. We have attempted to record all revertant plaques observed, distinguishing revertant types where possible. We are using reversion analysis primarily in the attempt to characterize representative individual EMS-induced mutational changes, and excluding reversions that are not back mutations could bias the survey. For most purposes it may not be important whether the reversion reverses the original base pair change, is another type of change at the same site, or occurs at another site.

Although the frequencies have not usually been tabulated, partial revertants ("tiny" or "false" revertants) have also been observed in other rII reversion studies (FREESE 1959; FREESE 1962; FREESE, BAUTZ and FREESE 1961; FREESE and FREESE 1961; CHAMPE and BENZER 1962). In some cases they have been shown to be due to suppressor mutations within the rII region (R. P. FEYNMAN, personal communcation; CRICK et al. 1961). A detailed account of suppressor mutations has also been reported for another phage locus (JINKS 1961).

YANOFSKY, HELSINKI, and MALING (1961) found proteins from revertant bacteria that differed from normal tryptophan synthetase by one or two amino acid substitutions but still had partial or even full enzymatic activity. The interpretation was that the original mutant had one base pair substitution and that the partial reverants had either a different base pair substitution at the same site or at another site in the same or a different amino acid coding unit. This interpretation may be applicable to our results, but an alternative is mentioned below with our discussion of proflavin mutants.

The types of base pair changes involved in induced reversion of different rII mutants

GC to AT transitions: The results presented in Table 1 show that base analog-produced mutants are generally revertible by EMS, and that they fall into two fairly distinct groups: those with revertant frequencies close to those of AP72 and AP275 (about 60 to 150 revertants per 10^6 progeny), and those giving about one per 10^6 or less. There are several lines of evidence that indicate that the high EMS-induced revertant frequencies may occur at a GC site: 1. EMS exposure of DNA produces more 7-ethylguanine than other ethylated bases. 2. EMS-induced AP72 reversions, and most EMS-induced r mutations under our conditions, are delayed mutations. This would be expected from 7-ethylguanine but not from 3-ethyladenine. 3. These high revertant frequencies are associated with a low efficiency of recovery on direct selection, as expected from copy errors at the site of ethylated guanines but not from ethylated adenines. 4. These highly EMS-revertible mutants are also highly revertible by hydroxylamine, which reacts with cytosine and 5-hydroxylmethylcytosine more rapidly than with thymine (FREESE, BAUTZ and FREESE 1961; CHAMPE and BENZER 1962).

These lines of evidence would be applicable whether or not the induced revertants were true back mutations at the site originally mutated. The revertants from AP72 and AP275 appear to be true back-mutants, however, and since base analog-induced mutants are thought to be produced by transitions, these highly inducible reversions may be presumed to be GC to AT transitions. The possibility that other types of base pair changes may also be induced at GC sites will be discussed shortly, and we will simply conclude here that EMS-induced GC to AT transitions occur at a high frequency.

AT to GC transitions: Some members of the less revertible class of base analog-produced mutants, like AP156, have a spontaneous reversion frequency so low that it is clear that EMS induces a significant increase in revertants. We tentatively infer that EMS induces AT to GC transitions, at frequencies about two orders of magnitude lower than the highly inducible GC to AT transitions.

The conclusion that both types of transitions are induced by EMS, at widely different rates, is supported by our results for EMS-induced reversion of EMS-produced mutants. As predicted, most EMS-produced mutants are induced to revert at rates of one per 10^6 or less. Occasional mutants would be expected to revert at a much higher rate, if they had been originally produced by an AT to GC transition. BENZER found an EMS-produced mutant (EM126) that mapped at the site of AP72. When we tested this mutant, it was found to give the same high EMS-induced reversion frequency.

Additional evidence that AP156 reverts by mutation of an AT site has been presented by SETLOW (1962). This mutant, and AP61, can be induced to revert by ultraviolet irradiation of infected bacteria, with an action spectrum for induced reversion resembling the absorption spectrum for thymine. Mutants apparently identical to AP156 are regularly produced by AP and by EMS (BENZER 1961), and our data include several at this site (EM9 and AP83) that have the same revertibility. Since the analysis of the revertants indicates that they may be true back-mutants, we infer that these reversions involve transitions.

There are several reasons, therefore, for concluding that EMS-induced AT to GC transitions occur, but at a low frequency.

Proflavin mutants: There is evidence that acridines such as proflavin act as mutagens by adding or deleting single base pairs, and that addition mutants are subject to reversion by closely linked suppressors arising by base pair deletions, and conversely for deletion mutants (BRENNER *et al.* 1961; CRICK *et al.* 1961).

Most proflavin-produced mutants we tested, including P13 which CRICK *et al.* used as the starting point for their analysis (and redesignated FCO), were not particularly subject to EMS-induced reversion. However, P42 and P85 were quite EMS-revertible. It may be that these two mutants are caused by base pair additions and that EMS can induce base pair deletions by the mechanism we have previously outlined.

We have recently found, however, that most of the mutants giving EMS-induced partial revertants (including P85, AP12, AP41, AP50, AP114, BU90 and EM34) are also induced to revert both by 5-bromouracil and 5-aminoacridine (unpublished results). ORGEL and BRENNER (1961) also found a spontaneous mutant that was induced to give apparently standard type revertants by 5-bromouracil and tiny revertants by proflavin. They report that 5-aminoacridine and proflavin have generally the same mutational characteristics. These results are not easily compatible with previous interpretations of mutagen specificity and the action of suppressor mutations. Perhaps there are partly functional genes that differ from the standard base sequence by one base pair substitution and by one base pair addition or deletion.

Transversions: FREESE (1961) concluded that transversions can be induced by

213

ethylating agents and by low pH. The accuracy of this conclusion can be important in distinguishing between the several mechanisms that have been described for the role of 7-ethylguanine in inducing mutations. We shall summarize here the evidence that was offered for this conclusion, together with reasons why we do not find the evidence to be compelling.

1. A small fraction of EES-produced mutants (about 20 percent) and of mutants produced by a heat and acid treatment (about 10 percent) were not inducible to revert by base analogs. This was taken to imply that they could not be transition mutants. This implies the assumption that all transitions should be base analog inducible. However, CHAMPE and BENZER (1962) designate about 30 percent of the BU- and AP-produced mutants that they tested as giving weak, false or no base analog-induced back mutation. This indicates that not all transitions are base analog-inducible and/or that not all base analog-induced mutants are transitions.

2. FREESE (1961) selected five EES-produced and 15 spontaneous mutants that were not base analog revertible and found that about half were induced to revert both by EES and by a pH 4 treatment. Two levels of induced reversion frequencies were observed: about 50 per 10^6 and about four per 10^6 (corrected to treatments that gave 37 percent survival). However, the spontaneous revertant frequency of each of these mutants is correspondingly high. We checked two of these mutants, *r*207 and EES66, by our standard progeny procedure and found that EMS also induced about a 50 percent increase over the spontaneous levels. The increases may be significant, but the analysis and interpretation are difficult. FREESE (1961) also reported that no nitrous acid-induced revertants were detectable for these mutants. However, direct plating of nitrous acid-treated phage is not very efficient in detecting induced revertants of presumed transition mutants (FREESE and FREESE 1961) and low induced rates could have been masked by the high levels of spontaneous revertants.

3. Four of the mutants in the above mentioned group mapped at sites where other distinguishable mutants were also known, and since in two cases one of the other mutants was identified by reversion data as the member of the set that reverted by an AT to GC transition, it was inferred that the EES-revertible mutant could not also be a transition (FREESE 1961). However, the conclusion that two mutants occur at the same site is drawn from their inability to give wild type recombinants, and the high levels of spontaneous revertants characteristic of the mutants in question could mask low recombination frequencies so the site identities of these mutants is only tenative. TESSMAN (1962) reports finding recombination frequencies for some other mutant pairs to be much lower than the level of sensitivity in these cases.

Additional evidence is needed to evaluate the question of EMS-induced transversions. FREESE and coworkers had concluded that ethylation-induced mutations result from the hydrolysis of 7-ethylguanine, the second molecular mechanism we discussed previously. On this basis, transversions might be expected to be at least as frequent as transitions. The first molecular mechanism we discussed would generate only GC to AT transitions, however. The occurrence of a small

214

fraction of EMS mutants that are highly EMS-revertible might be adequately explained by 3-ethyladenine-induced r mutations that revert by GC to AT transitions. More direct tests are needed to evaluate the alternative molecular mechanisms hypothesized for 7-ethylguanine-induced mutation.

SUMMARY

Killing and induced reversion were investigated for bacteriophage T4rII mutants exposed to ethyl methanesulfonate (EMS).

1. The survival curve is approximately a parabola. That is, the frequency of lethal damages is approximately proportional to the square of the duration of exposure of phage to EMS. This could result from a two-step (not a two-hit) mechanism of killing. The frequency of induced revertants from a highly revertible mutant was found to be proportional to the first power of the duration of exposure. It is concluded that ethylated molecules are produced within the phage by a first-order reaction, and that induced mutations result from their presence within the phage DNA. The second step in the mechanism of killing could be the hydrolysis or breakdown of ethylated phage components prior to infection.

2. Many base analog-produced mutants, and some EMS-produced mutants, are induced to revert at high frequencies by EMS. It is concluded that these represent base pair substitutions at a GC site, probably GC to AT transitions.

3. Most base analog and EMS mutants that are not induced to revert at high frequencies by EMS are induced to revert but at much lower frequencies. These are believed to represent AT to GC transitions.

4. Induced revertants from the highly revertible mutants are recovered with low efficiency if the exposed phage are subjected to a selective assay procedure directly, without prior opportunity for nonselective growth. This phenomenon is related to delayed mutation and a low frequency of induced "mispairings" of ethylated guanine with thymine or uracil.

5. The possibility is discussed that EMS may also induce transversions of GC to CG or of GC to TA. The evidence is considered inconclusive.

6. Several proflavin-produced mutants were found to be reverted at fairly high frequencies by EMS. It is possible that these revertants are caused by deletions of single GC pairs.

7. Some of the induced revertants appear to be true back mutants, while others are definitely distinguishable from standard r^+ phage. This further confirms the evidence that a variety of base sequences may specify a functional gene product. Various intracistron suppressor mutation mechanisms are known that may account for these observations.

8. Three possible mechanisms of EMS-induced mutation are described: pairing errors of 7-ethylguanine with thymine, pairing of 3-ethyladenine with cytosine, and replication errors at the site of gaps produced by the hydrolysis of 7-ethyl-guanine.

ACKNOWLEDGMENTS

I am indebted to MISS REA MARIE FULKERSON and MISS CLAIRE J. WITTEN for their able assistance in performing these experiments.

LITERATURE CITED

BAUTZ, EKKEHARD, and ERNST FREESE, 1960 On the mutagenic effect of alkylating agents. Proc. Natl. Acad. Sci. U.S. **46**: 1585–1594.

BENZER, S., 1957 The elementary units of heredity. pp. 70–93. *The Chemical Basis of Heredity.* Edited by W. D. MCELROY and B. GLASS. The Johns Hopkins Press. Baltimore, Maryland.
1961 On the topography of the genetic fine structure. Proc. Natl. Acad. Sci. U.S. **47**: 403–415.

BENZER, S., and S. P. CHAMPE, 1961 Ambivalent *r*II mutants of phage T4. Proc. Natl. Acad. Sci. U.S. **47**: 1025–1038.

BRENNER, S., L. BARNETT, F. H. C. CRICK, and A. ORGEL, 1961 The theory of mutagenesis. J. Mol. Biol. **3**: 121–124.

BROOKES, P., and P. D. LAWLEY, 1960 The methylation of adenosine and adenylic acid. J. Chem. Soc. 539–545.
1961a The alkylation of guanosine and guanylic acid. J. Chem. Soc. 3923–3928.
1961b The reaction of mono- and di-functional alkylating agents with nucleic acids. Biochem. J. **80**: 496–503.
1962 The methylation of cytosine and cytidine. J. Chem. Soc. 1348–1351.

CAVALIERI, L. F., and A. L. STONE, 1955 Studies on the structure of nucleic acids. VIII. Apparent dissociation constants of deoxypentose nucleic acid. J. Am. Chem. Soc. **77**: 6499–6501.

CHAMPE, S. P., and S. BENZER, 1962 Reversal of mutant phenotypes by 5-fluorouracil: An approach to nucleotide sequences in messenger RNA. Proc. Natl. Acad. Sci. U.S. **48**: 532–546.

CRICK, F. H. C., L. BARNETT, S. BRENNER, and R. J. WATTS-TOBIN, 1961 General nature of the genetic code for proteins. Nature **192**: 1227–1232.

FREESE, ELISABETH B., 1961 Transitions and transversions induced by depurinating agents. Proc. Natl. Acad. Sci. U.S. **47**: 540–545.
1962 Further studies on induced reverse mutations of phage T4 *r*II mutants. Biochem. Biophys. Res. Comm. **7**: 18–22.

FREESE, ELISABETH B., and ERNST FREESE, 1961 Induction of reverse mutations and cross reactivation of nitrous acid-treated phage T4. Virology **13**: 19–30.

FREESE, ERNST, 1959 The difference between spontaneous and base-analogue induced mutations of phage T4. Proc. Natl. Acad. Sci. U.S. **45**: 622–633.

FREESE, ERNST, EKKEHARD BAUTZ, and ELISABETH B. FREESE, 1961 The chemical and mutagenic specificity of hydroxylamine. Proc. Natl. Acad. Sci. U.S. **47**: 845–855.

GREEN, D. M., and D. R. KRIEG, 1961 The delayed origin of mutants induced by exposure of extracellular phage T4 to ethyl methanesulfonate. Proc. Natl. Acad. Sci. U.S. **47**: 64–72.

HERSHEY, A. D., and R. ROTMAN, 1949 Genetic recombination between host range and plaque-type mutants of bacteriophage in single bacterial cells. Genetics **34**: 44–71.

JINKS, J. L., 1961 Internal suppressors of the *h*III and *tu*45 mutants of bacteriophage T4. Heredity **16**: 241–254.

KLIGLER, I. J., and E. OLENICK, 1943 Inactivation of phage by aldehydes and aldoses and subsequent reactivation. J. Immunol. **47**: 325–333.

LAWLEY, P. D., 1957 The hydrolysis of methylated deoxyguanylic acid at pH 7 to yield 7-methylguanine. Proc. Chem. Soc. 290–291.

LAWLEY, P. D., and P. BROOKES, 1961   Acidic dissociation of 7:9-dialkylguanines and its possible relation to mutagenic properties of alkylating agents. Nature **192**: 1081–1082.

LOVELESS, A., 1958   Increased rate of plaque-type and host-range mutation following treatment of bacteriophage *in vitro* with ethyl methane sulphonate. Nature **181**: 1212–1213.

   1959   The influence of radiomimetic substances on deoxyribonucleic acid synthesis and function studied in *Escherichia coli*/phage systems. III. Mutation of T2 bacteriophage as a consequence of alkylation *in vitro*: the uniqueness of ethylation. Proc. Royal Soc. London B **150**: 497–508.

ORGEL, A., and S. BRENNER, 1961   Mutagenesis of bacteriophage T4 by acridines. J. Mol. Biol. **3**: 762–768.

PAL, B. C., 1962   Studies on the alkylation of purines and pyrimidines. Biochemistry **1**: 558–562.

PAL, B. C., and C. A. HORTON, 1962   Structure of 3-methyladenine and methylation of 6-dimethylaminopurine (submitted for publication).

PFLEIDERER, W., 1961   Über die Methylierung des 9-Methyl-guanins und die Struktur des Herbipolins. Ann. Chem. **647**: 167–173.

REINER, B., and S. ZAMENHOF, 1957   Studies on the chemically reactive groups of deoxyribonucleic acids. J. Biol. Chem. **228**: 475–486.

SETLOW, J. K., 1962   Evidence for a particular base pair alteration in two T4 bacteriophage mutants. Nature **194**: 664–666.

STRAUSS, B. S., 1961   Specificity of the mutagenic action of the alkylating agents. Nature **191**: 730–731.

TESSMAN, I., 1959   Mutagenesis in phages ΦX174 and T4 and properties of the genetic material. Virology **9**: 375–385.

   1962   The induction of large deletions by nitrous acid. J. Mol. Biol. **5**: 442–445.

YANOFSKY, C., D. R. HELSINKI, and B. D. MALING, 1961   The effects of mutation on the composition and properties of the A protein of *Escherichia coli* tryptophan synthetase. Cold Spring Harbor Symp. Quant. Biol. **26**: 11–24.

Possible Relevance of O–6 Alkylation of Deoxyguanosine to the Mutagenicity and Carcinogenicity of Nitrosamines and Nitrosamides

BIOLOGICAL alkylating agents—"mustards", ethylenimines, epoxides and alkyl alkanesulphonates—have been believed to induce mutations by causing atypical base pairing during DNA replication at sites bearing a guanine residue which has suffered alkylation at the 7(N) position[1,2]. The most abundant reaction product found in acid hydrolysates of treated DNA is 7-alkyl guanine[3,4], although alkylated adenines and cytosine, notably 3-alkyl adenine, are recovered as minor products. Furthermore, genetic studies of alkylation mutants in bacteria and bacteriophages have shown that many point mutational events involve guanine to adenine transitions[5,6].

A major stumbling block to the acceptance of guanine 7(N) alkylation as a universally important pro-mutagenic reaction has been the fact that, whereas both methyl methanesulphonate (MMS) and ethyl methanesulphonate (EMS) react in this way with the DNA of T-even phages[7], only EMS is mutagenic[8]. On the other hand, Hampton and I recently found[9] that both N-methyl-N-nitroso-urea (NMU) and N-ethyl-N-nitroso-urea (NEU) are mutagenic towards T2 bacteriophage.

Reaction of salmon sperm DNA with these two nitrosamides in identical conditions at neutrality, followed by acid hydrolysis and chromatography, revealed that extensive methylation had occurred at the 7(N) position of guanine and, to a lesser extent, at the 3(N) position of adenine moieties; only a trace of ethylated material was, however, detected[9]. These results accorded with those obtained by treating DNA at pH 7 with ethereal solutions of the corresponding diazoalkanes[10]. We therefore concluded that 7(N) alkylation of guanine residues was irrelevant to mutagenesis in this system. Mutation of phage by treatment *in vitro* with reactive substances must result from base modification, and it was therefore supposed that the products leading to mutations were acid-labile, reverting to the parent bases in the conditions habitually used for DNA hydrolysis.

Friedman *et al.*[11,12] obtained deoxy-(O–6)-methylguanosine by reacting deoxyguanosine with diazomethane in methanol/ether; this gave guanine on hydrolysis with perchloric acid. Some ribosides (for example, inosine[13,14]) are known to yield O–6-methylated product with diazomethane. I have now obtained a product from reaction of NMU with deoxyguanosine in phosphate buffer (250 mg deoxyguanosine in 40 ml. 0·4 N phosphate buffer *p*H 7·2, 1·03 g NMU added in 10 ml. dry acetone; 3 h at 37 °C) which has the same chromatographic properties and an ultraviolet spectrum very similar to deoxy-(O–6)-methylguanosine[12] (see Tables 1 and 2). Although the yield of the product was very small, most of the deoxyguanosine being recovered unchanged, its isolation and purification by paper chromatography were facilitated by its brilliant light blue fluorescence in ultraviolet light; the yield from 5 mg deoxyguanosine was easily detected on paper. Preliminary experiments with ³H methyl nitrosourea indicate that it accounts for about 10 per cent of the total methylation products. The radioactivity was progressively lost on heating with perchloric acid.

Reaction of deoxyguanosine with NEU in similar conditions (50 mg deoxyguanosine in 8 ml. 0·4 N phosphate, 234 mg NEU added in 2 ml. dry acetone; 6 h at 37° C) yielded three to four times as much of a similar product. I have also reacted deoxyguanosine with MMS, EMS and dimethyl sulphate each in the conditions used by Lawley[15] for methylation of deoxyguanylic acid. Only EMS gave a fluorescent product with chromatographic behaviour resembling that of the O-alkyl derivatives.

The discovery of O–6 alkylation of guanine residues in DNA treated with NMU, NEU and EMS would contribute to a resolution of the outstanding problem of phage muta-

genesis assuming such reaction leads to atypical base pairing. This is expected because the 1 position which is involved in the normal hydrogen bonding of guanine with cytosine no longer bears a proton after O-alkylation. Dr A. M. Michelson (personal communication) agrees with this view.

The relationship between nucleic acid alkylation and the carcinogenicity of nitrosamines and nitrosamides compared with "classical" alkylating agents is also unsatisfactory[18]. The extent of nucleic acid alkylation, with which a correlation with carcinogenicity has been vainly sought, has been estimated after acid hydrolysis. It is particularly interesting, in this context, that Krüger *et al.*[19] found no ethylation of tissue RNAs following NEU administration. This substance is a powerful carcinogen and of all the reagents discussed here affords the largest yield of O-alkyl-deoxyguanosine. It seems therefore that it may yet be possible to relate carcinogenicity with nucleic acid reaction but only when the extent of O-alkylation can be measured.

It is a pleasure to acknowledge the valuable co-operation of Dr Aubrey Knowles, who determined the ultraviolet spectra of the nitrosamide products. This work has been supported by grants to the Chester Beatty Research Institute (Institute of Cancer Research: Royal Cancer Hospital) from the Medical Research Council and the British Empire Cancer Campaign.

A. LOVELESS

Table 1. CHROMATOGRAPHIC PROPERTIES OF O-ALKYL DERIVATIVES

Substance	1*	Solvent A 3 MM	*Rf* 4	Solvent B 3 MM	Author
Deoxy-(O–6)-methylguanosine			0·79, 0·8		(ref. 12)
NMU product "	0·66	0·72		0·1	Present
" "	0·66	0·72		0·1	"
NEU product	0·73			0·07	"
EMS product	0·74			0·08	"

* No. refers to Whatman paper gradings.
Solvent A: isopropanol : water, 70 : 30[11]; solvent B: saturated aqueous ammonium sulphate : *n*-propanol : 0·1 N phosphate buffer, 79 : 2 : 19[16].

Table 2. SPECTRAL PROPERTIES OF O-ALKYL DERIVATIVES

Substance	in 0·1 HCl λ max λ min	in H₂O λ max λ min	in 0·01 N NaOH λ max λ min	Author
Deoxy-(O–6)-methylguanosine (inflex. 225/235)	284 252	276,244 259,232	275,243 261,231	(ref. 12)
9-(2′,3′,5′-Tri-O-acetyl-β-D-ribofuranosyl)-2-amino-6-methoxy-purine	287		278	(ref. 17)
NMU product	286 252	279,247 262,237	279,245 262	*
NEU product	284 250	280,248 262,238	280,245 262	*

* These spectra were determined by Dr A. Knowles using a model 11 Cary recording spectrophotometer.

[1] Lawley, P. D., and Brookes, P., *Nature*, **192**, 1081 (1961).
[2] Nagata, C., Imamura, A., Saito, H., and Fukui, K., *Gann*, **54**, 109 (1963).
[3] Brookes, P. D., and Lawley, P., *Biochem. J.*, **80**, 496 (1961).
[4] Lawley, P. D., *Prog. Nucleic Acid Res.*, *and Mol. Biol.*, **5**, 89 (1966).
[5] Krieg, D. R., *Genetics*, **48**, 561 (1963).
[6] Osborn, Mary, Person, S., Phillips, S., and Funk, F., *J. Mol. Biol.*, **26**, 437 (1967).
[7] Brookes, P., and Lawley, P. D., *Biochem. J.*, **89**, 138 (1963).
[8] Loveless, A., *Proc. Roy. Soc. B.*, **150**, 497 (1959).
[9] Loveless, A., and Hampton, C. L., *Mutation Res.*, **7**, 1 (1969).
[10] Krieck, E., and Emmelot, P., *Biochim. Biophys. Acta*, **91**, 59 (1964).
[11] Friedman, O. M., Malapatra, G. N., and Stevenson, R., *Biochim. Biophys. Acta*, **68**, 144 (1963).
[12] Friedman, O. M., Malapatra, G. N., Dash, B., and Stevenson, R., *Biochim. Biophys. Acta*, **103**, 286 (1965).
[13] Miles, H. T., *J. Org. Chem.*, **26**, 4761 (1961).
[14] Scheit, K-H., and Holy, A., *Biochim. Biophys. Acta*, **149**, 344 (1967).
[15] Lawley, P. D., *Proc. Chem. Soc.*, 290 (1957).
[16] Markham, R., and Smith, J. D., *Biochem. J.*, **49**, 401 (1951).
[17] Gerster, J. F., Jones, J. W., and Robins, R. K., *J. Org. Chem.*, **28**, 945 (1963).
[18] Swann, P. F., and Magee, P. N., *Biochem. J.*, **110**, 39 (1968).
[19] Krüger, F. W., Ballweg, H., and Maier-Borst, W., *Experientia*, **24**, 592 (1968).

Part III

FRAMESHIFT MUTAGENESIS

Editors' Comments
on Papers 22 Through 26

FRAMESHIFT MUTAGENESIS

Early in his investigations of simple chemical mutagens, Freese postulated that spontaneous mutations, and also those induced by proflavin, consisted of transversions (Freese, 1959, and Paper 13). (Proflavin, by the way, was the first chemical to be shown to be mutagenic in a phage system; DeMars, 1953.) There are no hints, in the literature or elsewhere, that anyone at that time had considered the possibility that mutations might arise by a mechanism other than base pair substitutions. This situation was soon remedied, however, by certain experimental observations which appeared quite inconsistent with the concept that proflavin-induced mutations consisted of transversions. The discrepancies were first pointed out in 1961 by the Cambridge group (Paper 22), who postulated instead that proflavin-induced mutations consisted not of base pair substitutions, but of base pair additions or deletions. This postulate was also of crucial significance for the solution of the genetic code, the major problem which obsessed the Cambridge group at that time. Within the same year this group had worked out the fundamental properties of this new class of mutations (Paper 23), had christened them *frameshift* mutations, and had provided the first proof that the genetic code was composed of nonoverlapping triplets.

(It is interesting to note how transversion mutagenesis got lost in the shuffle. The earlier studies on chemical mutagenesis had explored only transitional mechanisms; the next generation of studies explored only frameshift mutagenesis. Although it is perfectly clear that transversions occur commonly, for instance among mutants of spontaneous or misrepair origin, the chemical mechanisms which generate them remain quite obscure.)

The next obvious question was how frameshift mutations arise, and specifically how they are induced by proflavin. The key to understanding proflavin mutagenesis appeared to lie in Lerman's discovery of the surprising mode of interaction between proflavin and DNA: *intercalation,* or sandwiching between base pairs. The physical evidence supporting the concept of intercalation is most fully presented in Paper 24.

Although Paper 24 appeared in 1963, Lerman's model was known to the Cambridge group in 1961 and contributed significantly to the ideas set forth in Paper 22, as exemplified by the suggestion that an extra base could be inserted during DNA replication at the point of intercalation. This idea turned out not to be very sound, however, for two reasons: (1) proflavin appears able to induce both insertions and deletions of base pairs, and (2) the characteristic intercalation type of interaction between proflavin and DNA does not occur with single-stranded DNA, the presumed intermediate during DNA replication. In an attempt to circumvent these difficulties, Lerman suggested in Paper 24 that proflavin induces errors of recombination, specifically unequal crossing over. Although this idea found support from observations that tended to correlate proflavin mutagenesis with recombination (Magni, 1963; Sesnowitz-Horn and Adelberg, 1968), the concomitant development of ideas about the molecular basis of recombination soon made it clear that the unequal crossing over scheme suffered from similar difficulties, since recombination also presumably involves single-stranded DNA molecules.

This dilemma appears to have been resolved by the Streisinger group (Paper 25), who performed amino acid sequence analyses of bacteriophage T4 lysozymes encoded by cistrons carrying closely linked pairs of mutually suppressing frameshift mutations. These studies achieved three important goals. First, they elegantly confirmed the general hypothesis of the mechanism of mutual suppression by frameshift mutations of opposite sign (as described in Paper 23). Second, they permitted the deduction of just which codons are used in a gene in particular positions, and clearly demonstrated codon degeneracy *in vivo.* Third, and most important to students of mutation, they provided a conceptual basis for the mechanism of frameshift mutagenesis that has thus far survived quite vigorously. The third goal was achieved as a result of an unanticipated observation: close inspection of DNA base sequences deduced from their corresponding amino acid sequences revealed a marked tendency for frameshift mutations to arise in regions of local base sequence redundancy. A model was therefore devised in which bases in the neighborhood of a single-strand interruption became misaligned (owing to the local base sequence redundancy itself), whereupon the misaligned configuration was immortalized by DNA repair. Although the authors also make suggestions about how this process might be promoted by mutagens such as proflavin, the specific role of proflavin in frameshift mutagenesis remains unknown even now.

Extending the analysis of frameshift mutagenesis from phage T4 to bacteria and eucaryotes was somewhat delayed by the apparent inability of the simple acridines, such as proflavin, to induce mutants in purely cellular systems. This failure is now known to be due primarily to permeability problems and can readily be overcome, at least in bacteria (Kohne and Roth, 1974). In the meantime, however, searches were initiated for other agents that might prove to be effective cellular frameshift mutagens. The discovery and characterization by the Ames group of a major group of novel agents of this type is described in Paper 26. The new frameshift mutagens were detected among a large number of potential cancer chemotherapeutic agents synthesized at the Institute for Cancer Research in Philadelphia; all carry the ICR prefix. These compounds appear to be effective mutagens in numerous cellular systems because they contain, in addition to the acridine-like ring system, alkylating side chains that permit the formation of covalent bonds to DNA (mainly to the N-7 atom of guanine).

REFERENCES

DeMars, R. I. 1953. Chemical mutagenesis in bacteriophage T2. *Nature 172:* 964.

Freese, E. 1959. On the molecular explanation of spontaneous and induced mutations. *Brookhaven Symp. Biol.,* *12:* 63.

Kohne, R., and J. R. Roth. 1974. Proflavin mutagenesis of bacteria. *J. Mol. Biol.,* *89:* 17.

Magni, G. E. 1963. The origin of spontaneous mutations during meiosis. *Proc. Natl. Acad. Sci.,* *50:* 975.

Sesnowitz-Horn, S., and E. A. Adelberg. 1968. Proflavin treatment of *Escherichia coli:* generation of frameshift mutations. *Cold Spring Harbor Symp. Quant. Biol.,* *33:* 393.

22

Reprinted from *J. Mol. Biol.,* **3,** 121–124 (1961)

The Theory of Mutagenesis

S. Brenner, Leslie Barnett, F. H. C. Crick, and Alice Orgel

In this preliminary note we wish to express our doubts about the detailed theory of mutagenesis put forward by Freese (1959*b*), and to suggest an alternative.

Freese (1959*b*) has produced evidence that shows that for the r_{II} locus of phage T4 there are two mutually exclusive classes of mutation and we have confirmed and extended his work (Orgel & Brenner, in manuscript). The technique used is to start with a standard wild type and make a series of mutants from it with a particular mutagen. Each mutant is then tested with various mutagens to see which of them will back-mutate it to wild type.

It is found that the mutations fall into two classes. The first, which we shall call the base analogue class, is typically produced by 5-bromodeoxyuridine (BD) and the second, which we shall call the acridine class, is typically produced by proflavin (PF). In general a mutant made with BD can be reverted by BD, and a mutant made with PF can be reverted by PF. A few of the PF mutants do not appear to revert with either mutagen, but the strong result is that no mutant has been found which reverts identically with both classes of mutagens, and that (with a few possible exceptions) mutants produced by one class cannot be reverted by the other.

Freese also showed that 2-aminopurine falls into the base analogue class, and that most (85%) spontaneous mutants at the r_{II} locus were not of the base analogue type. We have confirmed this and shown that they are in fact revertible by acridines. We have also shown that a number of other acridines, and in particular 5-aminoacridine, act like proflavin (Orgel & Brenner, in manuscript).

Freese has produced an ingenious explanation of these results, which should be consulted in the original for fuller details. In brief he postulated that the base analogue class of mutagens act by altering an A—T base-pair on the DNA (A = adenine, T = thymine) into a G—C pair, or *vice versa* (G = guanine, C = cytosine, or, in the T even phages, hydroxymethylcytosine). The fact that BD, which replaces thymine, could act both ways (from A—T to G—C or from G—C to A—T) was accounted for (Freese, 1959*a*) by assuming that in the latter case there was an error in pairing of the BD (such that it accidentally paired with guanine) while *entering* the DNA, and in the former case after it was already in the DNA.

Such alterations only change a purine into another purine, or a pyrimidine into another pyrimidine. Freese (1959*b*) has called these "transitions." He suggested that other conceivable changes, which he called "transversions" (such as, for example, from A—T to C—G) which change a purine into a pyrimidine and *vice versa*, occurred during mutagenesis by proflavin. This would neatly account for the two mutually exclusive classes of mutagens, since it is easy to see that a transition cannot be reversed by a transversion, and *vice versa*.

We have been led to doubt this explanation for the following reasons.

Our suspicions were first aroused by the curious fact that a comparison between the *sites* of mutation for one set of mutants made with BD and another set made with PF (Brenner, Benzer & Barnett, 1958) showed there were no sites in the r_{II} gene, among the samples studied, common to both groups.

Now this result alone need not be incompatible with Freese's theory of mutagenesis, since we have no good explanation for "hot spots" and this confuses quantitative argument. However it led us to the following hypothesis:

that acridines act as mutagens because they cause the insertion or the deletion of a base-pair.

This idea springs rather naturally from the views of Lerman (1960) and Luzzati (in preparation) that acridines are bound to DNA by sliding *between* adjacent base-pairs, thus forcing them 6·8 Å apart, rather than 3·4 Å. If this occasionally happened between the bases on *one* chain of the DNA, but not the other, during replication, it might easily lead to the addition or subtraction of a base.

Such a possible mechanism leads to a prediction. We know practically nothing about coding (Crick, 1959) but on most theories (except overlapping codes which are discredited because of criticism by Brenner (1957)) the deletion or the addition of a base-pair is likely to cause not the substitution of just one amino acid for another, but a much more substantial alteration, such as a break in the polypeptide chain, a considerable alteration of the amino acid sequence, or the production of no protein at all.

Thus one would not be surprised to find on these ideas that mutants produced by acridines were not capable of producing a slightly modified protein, but usually produced either no protein at all or a grossly altered one.

Somewhat to our surprise we find we already have data from two separate genes supporting this hypothesis.

(1) The *o* locus of phage T4 (resistance to osmotic shock) is believed to control a protein of the finished phage, possibly the head protein, because it shows phenotypic mixing (Brenner, unpublished). Using various base analogues we have produced mutants of this gene, though these map at only a small number of sites. We have failed on several occasions to produce any *o* mutants with proflavin. On another occasion two mutants were produced; one never reverted to wild type, while the other corresponded in position and spontaneous reversion rate to a base analogue site. We suspect therefore that these two mutants were not really produced by proflavin, but were the rarer sort of spontaneous mutant (Brenner & Barnett, unpublished).

(2) We have also studied mutation at the *h* locus in T2L, which controls a protein of the finished phage concerned with attachment to the host (Streisinger & Franklin, 1956).

Of the six different spontaneous h^+ mutants tested, all were easily induced to revert to *h* with 5-bromouracil (BU)†. This is especially significant when it is recalled that 85% of the spontaneous r_{II} mutants could not be reverted with base analogues (Freese, 1959*b*).

We have also shown (Brenner & Barnett, unpublished) that it is difficult to produce h^+ mutants from *h* by proflavin, though relatively easy with BU. The production of *r* mutants was used as a control.

It can be seen from Table 1 that if the production of h^+ mutants by BU and proflavin were similar to the production of *r* mutants we would expect to have obtained $\dfrac{57 \times 26}{108} = 13 h^+$ mutants with proflavin, whereas in fact we only found 1, and this may be spontaneous background.

---

† (Added in proof.) Five of these have now been tested and have been shown not to revert with proflavin.

Let us underline the difference between the *r* loci and the *o* and *h* loci. The former appear to produce proteins which are probably *not* part of the finished phage. For both the *o* and the *h* locus, however, the protein concerned forms part of the finished phage, which presumably would not be viable without it, so that a mutant can be picked up only if it forms an *altered* protein. A mutant which deleted the protein could not be studied.

TABLE 1

	r	*h+*
BU	108	57
Proflavin	26	1

It is clear that further work must be done before our generalization—that acridine mutants usually give no protein, rather than a slightly modified one—can be accepted. But if it turns out to be true it would support our hypothesis of the mutagenic action of the acridines, and this may have serious consequences for the naïve theory of mutagenesis, for the following reason.

It has always been a theoretical possibility that the reversions to wild type were not true reversions but were due to the action of "suppressors" (within the gene), possibly very closely linked suppressors. The most telling evidence against this was the existence of the two mutually exclusive classes of mutagens, together with Freese's explanation.

For clearly if the forward mutation could be made at one base-pair and the reverse one at a different base-pair, we should expect, on Freese's hypothesis, exceptions to the rule about the two classes of mutagens. Since these were not found it was concluded that even close suppressors were very rare.

Unfortunately our new hypothesis for the action of acridines destroys this argument. Under this new theory an alteration of a base-pair at one place *could* be reversed by an alteration at a different base-pair, and indeed from what we know (or guess) of the structure of proteins and the dependence of structure on amino acid sequence, we should be surprised if this did not occur.

It is all too easy to conceive, for example, that at a certain point on the polypeptide chain at which there is a glutamic residue in the wild type, and at which the mutation substituted a proline, a further mutation might alter the proline to aspartic acid and that this might appear to restore the wild phenotype, at least as far as could be judged by the rather crude biological tests available. If several base-pairs are needed to code for one amino acid the reverse mutation might occur at a base-pair close to but not identical with the one originally changed.

On our hypothesis this could happen, and yet one would still obtain the two classes of mutagens. The one, typified by base analogues, would produce the substitution of one base for another, and the other, typically produced by acridines, would lead to the addition or subtraction of a base-pair. Consequently the mutants produced by one class could not be easily reversed by the mutagens of the other class.

Thus our new hypothesis reopens in an acute form the question: which back-mutations to wild type are truly to the original wild type, and which only appear to be so? And on the answers to this question depend our interpretation of all experiments on back-mutation.

We suspect that this problem can most easily be approached by work on systems for which the amino acid sequence of the protein can be studied, such as the phage lysozyme of Dreyer, Anfinsen & Streisinger (personal communications) or the phosphatase from *E. coli* of Levinthal, Garen & Rothman (Garen, 1960). Meanwhile we are continuing our genetic studies to fill out and extend the preliminary results reported here.

Medical Research Council Unit
for Molecular Biology
Cavendish Laboratory

Pathology Laboratory
both of Cambridge University
England

S. BRENNER
LESLIE BARNETT
F. H. C. CRICK

ALICE ORGEL

Received 16 December 1960

REFERENCES

Brenner, S. (1957). *Proc. Nat. Acad. Sci., Wash.* **43**, 687.
Brenner, S., Benzer, S. & Barnett, L. (1958). *Nature*, **182**, 983.
Crick, F. H. C. (1959). In *Brookhaven Symposia in Biology*, **12**, 35.
Freese, E. (1959a). *J. Mol. Biol.* **1**, 87.
Freese, E. (1959b). *Proc. Nat. Acad. Sci., Wash.* **45**, 622.
Garen, A. (1960). 10th Symposium *Soc. Gen. Microbiol.*, London, 239.
Lerman, L. (1961). *J. Mol. Biol.* **3**, 18.
Streisinger, G. & Franklin, N. C. (1956). In *Cold Spr. Harb. Sym. Quant. Biol.* **21**, 103.

23

Reprinted from *Nature*, **192**(4809), 1227–1232 (1961)

GENERAL NATURE OF THE GENETIC CODE FOR PROTEINS

By Dr. F. H. C. CRICK, F.R.S., LESLIE BARNETT, Dr. S. BRENNER
and Dr. R. J. WATTS-TOBIN

Medical Research Council Unit for Molecular Biology,
Cavendish Laboratory, Cambridge

THERE is now a mass of indirect evidence which suggests that the amino-acid sequence along the polypeptide chain of a protein is determined by the sequence of the bases along some particular part of the nucleic acid of the genetic material. Since there are twenty common amino-acids found throughout Nature, but only four common bases, it has often been surmised that the sequence of the four bases is in some way a code for the sequence of the amino-acids. In this article we report genetic experiments which, together with the work of others, suggest that the genetic code is of the following general type:

(a) A group of three bases (or, less likely, a multiple of three bases) codes one amino-acid.

(b) The code is not of the overlapping type (see Fig. 1).

(c) The sequence of the bases is read from a fixed starting point. This determines how the long sequences of bases are to be correctly read off as triplets. There are no special 'commas' to show how to select the right triplets. If the starting point is displaced by one base, then the reading into triplets is displaced, and thus becomes incorrect.

(d) The code is probably 'degenerate'; that is, in general, one particular amino-acid can be coded by one of several triplets of bases.

The Reading of the Code

The evidence that the genetic code is not overlapping (see Fig. 1) does not come from our work, but from that of Wittmann[1] and of Tsugita and Fraenkel-Conrat[2] on the mutants of tobacco mosaic virus produced by nitrous acid. In an overlapping triplet code, an alteration to one base will in general change three adjacent amino-acids in the polypeptide chain. Their work on the alterations produced in the protein of the virus show that usually only one amino-acid at a time is changed as a result of treating the ribonucleic acid (RNA) of the virus with nitrous acid. In the rarer cases where two amino-acids are altered (owing presumably to two separate deaminations by the nitrous acid on one piece of RNA), the altered amino-acids are not in adjacent positions in the polypeptide chain.

Brenner[3] had previously shown that, if the code were universal (that is, the same throughout Nature), then all overlapping triplet codes were impossible. Moreover, all the abnormal human hæmoglobins studied in detail[4] show only single amino-acid changes. The newer experimental results essentially rule out all simple codes of the overlapping type.

If the code is not overlapping, then there must be some arrangement to show how to select the correct triplets (or quadruplets, or whatever it may be) along the continuous sequence of bases. One obvious suggestion is that, say, every fourth base is a 'comma'. Another idea is that certain triplets make 'sense', whereas others make 'nonsense', as in the comma-free codes of Crick, Griffith and Orgel[5]. Alternatively, the correct choice may be made by starting at a fixed point and working along the sequence of bases three (or four, or whatever) at a time. It is this possibility which we now favour.

Experimental Results

Our genetic experiments have been carried out on the *B* cistron of the *r*$_{II}$ region of the bacteriophage *T*4, which attacks strains of *Escherichia coli*. This is the system so brilliantly exploited by Benzer[6,7]. The *r*$_{II}$ region consists of two adjacent genes, or 'cistrons', called cistron *A* and cistron *B*. The wild-type phage will grow on both *E. coli B* (here called *B*) and on *E. coli K12* (λ) (here called *K*), but a phage which has lost the function of either gene will not grow on *K*. Such a phage produces an *r* plaque on *B*. Many point mutations of the genes are known which behave in this way. Deletions of part of the region are also found. Other mutations, known as 'leaky', show partial function; that is, they will grow on *K* but their plaque-type on *B* is not truly wild. We report here our work on the mutant *P* 13 (now re-named *FC* 0) in the *B*1 segment of the *B* cistron. This mutant was originally produced by the action of proflavin[8].

We[9] have previously argued that acridines such as proflavin act as mutagens because they add or delete a base or bases. The most striking evidence in favour of this is that mutants produced by acridines are seldom 'leaky'; they are almost always completely lacking in the function of the gene. Since our note was published, experimental data from two sources have been added to our previous evidence: (1) we have examined a set of 126 *r*$_{II}$ mutants made with acridine yellow; of these only 6 are leaky (typically about half the mutants made with base analogues are leaky); (2) Streisinger[10] has found that whereas mutants of the lysozyme of phage *T*4 produced by base-analogues are usually leaky, all lysozyme mutants produced by proflavin are negative, that is, the function is completely lacking.

If an acridine mutant is produced by, say, adding a base, it should revert to 'wild-type' by deleting a base. Our work on revertants of *FC* 0 shows that it usually

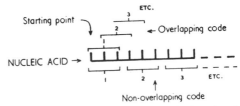

Fig. 1. To show the difference between an overlapping code and a non-overlapping code. The short vertical lines represent the bases of the nucleic acid. The case illustrated is for a triplet code

reverts not by reversing the original mutation but by producing a second mutation at a nearby point on the genetic map. That is, by a 'suppressor' in the same gene. In one case (or possibly two cases) it may have reverted back to true wild, but in at least 18 other cases the 'wild type' produced was really a double mutant with a 'wild' phenotype. Other workers[11] have found a similar phenomenon with r_{II} mutants, and Jinks[12] has made a detailed analysis of suppressors in the h_{III} gene.

The genetic map of these 18 suppressors of FC 0 is shown in Fig. 2, line a. It will be seen that they all fall in the $B1$ segment of the gene, though not all of them are very close to FC 0. They scatter over a region about, say, one-tenth the size of the B cistron. Not all are at different sites. We have found eight sites in all, but most of them fall into or near two close clusters of sites.

In all cases the suppressor was a non-leaky r. That is, it gave an r plaque on B and would not grow on K. This is the phenotype shown by a complete deletion of the gene, and shows that the function is lacking. The only possible exception was one case where the suppressor appeared to back-mutate so fast that we could not study it.

Each suppressor, as we have said, fails to grow on K. Reversion of each can therefore be studied by the same procedure used for FC 0. In a few cases these mutants apparently revert to the original wild-type, but usually they revert by forming a double mutant. Fig. 2, lines b–g, shows the mutants pro-

duced as suppressors of these suppressors. Again all these new suppressors are non-leaky r mutants, and all map within the $B1$ segment for one site in the $B2$ segment.

Once again we have repeated the process on two of the new suppressors, with the same general results, as shown in Fig. 2, lines i and j.

All these mutants, except the original FC 0, occurred spontaneously. We have, however, produced one set (as suppressors of FC 7) using acridine yellow as a mutagen. The spectrum of suppressors we get (see Fig. 2, line h) is crudely similar to the spontaneous spectrum, and all the mutants are non-leaky r's. We have also tested a (small) selection of all our mutants and shown that their reversion-rates are increased by acridine yellow.

Thus in all we have about eighty independent r mutants, all suppressors of FC 0, or suppressors of suppressors, or suppressors of suppressors of suppressors. They all fall within a limited region of the gene and they are all non-leaky r mutants.

The double mutants (which contain a mutation plus its suppressor) which plate on K have a variety of plaque types on B. Some are indistinguishable from wild, some can be distinguished from wild with difficulty, while others are easily distinguishable and produce plaques rather like r.

We have checked in a few cases that the phenomenon is quite distinct from 'complementation', since the two mutants which separately are phenotypically r, and together are wild or pseudo-wild,

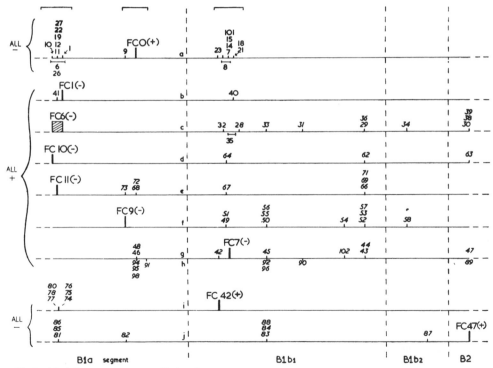

Fig. 2. A tentative map—only very roughly to scale—of the left-hand end of the B cistron, showing the position of the FC family of mutants. The order of sites within the regions covered by brackets (at the top of the figure) is not known. Mutants in italics have only been located approximately. Each line represents the suppressors picked up from one mutant, namely, that marked on the line in bold figures

must be put together in the same piece of genetic material. A simultaneous infection of K by the two mutants in separate viruses will not do.

The Explanation in Outline

Our explanation of all these facts is based on the theory set out at the beginning of this article. Although we have no direct evidence that the B cistron produces a polypeptide chain (probably through an RNA intermediate), in what follows we shall assume this to be so. To fix ideas, we imagine that the string of nucleotide bases is read, triplet by triplet, from a starting point on the left of the B cistron. We now suppose that, for example, the mutant FC 0 was produced by the insertion of an additional base in the wild-type sequence. Then this addition of a base at the FC 0 site will mean that the reading of all the triplets to the right of FC 0 will be shifted along one base, and will therefore be incorrect. Thus the amino-acid sequence of the protein which the B cistron is presumed to produce will be completely altered from that point onwards. This explains why the function of the gene is lacking. To simplify the explanation, we now postulate that a suppressor of FC 0 (for example, FC 1) is formed by deleting a base. Thus when the FC 1 mutation is present by itself, all triplets to the right of FC 1 will be read incorrectly and thus the function will be absent. However, when both mutations are present in the same piece of DNA, as in the pseudo-wild double mutant FC (0 + 1), then although the reading of triplets between FC 0 and FC 1 will be altered, the original reading will be restored to the rest of the gene. This could explain why such double mutants do not always have a true wild phenotype but are often pseudo-wild, since on our theory a small length of their amino-acid sequence is different from that of the wild-type.

For convenience we have designated our original mutant FC 0 by the symbol + (this choice is a pure convention at this stage) which we have so far considered as the addition of a single base. The suppressors of FC 0 have therefore been designated −. The suppressors of these suppressors have in the same way been labelled as +, and the suppressors of these last sets have again been labelled − (see Fig. 2).

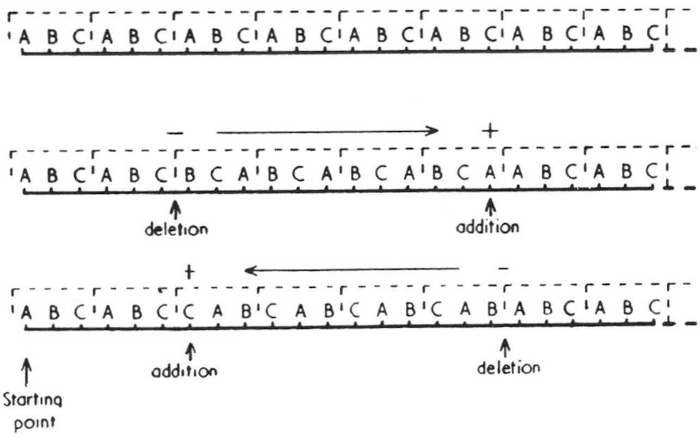

Fig. 3. To show that our convention for arrows is consistent. The letters A, B and C each represent a different base of the nucleic acid. For simplicity a repeating sequence of bases, ABC, is shown. (This would code for a polypeptide for which every amino-acid was the same.) A triplet code is assumed. The dotted lines represent the imaginary 'reading frame' implying that the sequence is read in sets of three starting on the left

Double Mutants

We can now ask: What is the character of any double mutant we like to form by putting together in the same gene any pair of mutants from our set of about eighty? Obviously, in some cases we already know the answer, since some combinations of a + with a − were formed in order to isolate the mutants. But, by definition, no pair consisting of one + with another + has been obtained in this way, and there are many combinations of + with − not so far tested.

Now our theory clearly predicts that all combinations of the type + with + (or − with −) should give an r phenotype and not plate on K. We have put together 14 such pairs of mutants in the cases listed in Table 1 and found this prediction confirmed.

Table 1. DOUBLE MUTANTS HAVING THE r PHENOTYPE

− With −	+ With +	
FC (1 + 21)	FC (0 + 58)	FC (40 + 57)
FC (23 + 21)	FC (0 + 38)	FC (40 + 58)
FC (1 + 23)	FC (0 + 40)	FC (40 + 55)
FC (1 + 9)	FC (0 + 55)	FC (40 + 54)
	FC (0 + 54)	FC (40 + 38)

At first sight one would expect that all combinations of the type (+ with −) would be wild or pseudo-wild, but the situation is a little more intricate than that, and must be considered more closely. This springs from the obvious fact that if the code is made of triplets, any long sequence of bases can be read correctly in one way, but incorrectly (by starting at the wrong point) in two different ways, depending whether the 'reading frame' is shifted one place to the right or one place to the left.

If we symbolize a shift, by one place, of the reading frame in one direction by → and in the opposite direction by ←, then we can establish the convention that our + is always at the head of the arrow, and our − at the tail. This is illustrated in Fig. 3.

We must now ask: Why do our suppressors not extend over the whole of the gene? The simplest postulate to make is that the shift of the reading frame produces some triplets the reading of which is 'unacceptable'; for example, they may be 'nonsense', or stand for 'end the chain', or be unacceptable in some other way due to the complications of protein structure. This means that a suppressor of, say, FC 0 must be within a region such that no 'unacceptable' triplet is produced by the shift in the reading frame between FC 0 and its suppressor. But, clearly, since for any sequence there are two possible misreadings, we might expect that the 'unacceptable' triplets produced by a → shift would occur in different places on the map from those produced by a ← shift.

Examination of the spectra of suppressors (in each case putting in the arrows → or ←) suggests that while the → shift is acceptable anywhere within our region (though not outside it) the shift ←, starting from points near FC 0, is acceptable over only a more limited stretch. This is shown in Fig. 4. Somewhere in the left part of our region, between FC 0 or FC 9 and the FC 1 group, there must be one or more unacceptable triplets when a ← shift is made; similarly for the region to the right of the *FC* 21 cluster. Thus we predict that a combination of a + with a − will be wild or pseudo-wild if it involves a → shift, but that such pairs involving a ← shift will be phenotypically *r* if the arrow crosses one or more of the forbidden places, since then an unacceptable triplet will be produced.

Table 2. DOUBLE MUTANTS OF THE TYPE (+ WITH −)

+	*FC* 41	*FC* 0	*FC* 40	*FC* 42	*FC* 58 *	*FC* 63	*FC* 38
FC 1	*W*	*W*	*W*		*W*		*W*
FC 86		*W*	*W*	*W*	*W*		
FC 9	*r*	*W*	*W*	*W*	*W*		*W*
FC 82	*r*		*W*	*W*	*W*	*W*	
FC 21	*r*	*W*			*W*		*W*
FC 88	*r*	*r*			*W*	*W*	
FC 87	*r*	*r*	*r*	*r*			*W*

W, wild or pseudo-wild phenotype; *W*, wild or pseudo-wild combination used to isolate the suppressor; *r*, *r* phenotype.
* Double mutants formed with *FC* 58 (or with *FC* 34) give sharp plaques on *K*.

We have tested this prediction in the 28 cases shown in Table 2. We expected 19 of these to be wild, or pseudo-wild, and 9 of them to have the *r* phenotype. In all cases our prediction was correct. We regard this as a striking confirmation of our theory. It may be of interest that the theory was constructed before these particular experimental results were obtained.

Rigorous Statement of the Theory

So far we have spoken as if the evidence supported a triplet code, but this was simply for illustration. Exactly the same results would be obtained if the code operated with groups of, say, 5 bases. Moreover, our symbols + and − must not be taken to mean literally the addition or subtraction of a single base.

It is easy to see that our symbolism is more exactly as follows:

$$+ \text{ represents } +m, \text{ modulo } n$$
$$- \text{ represents } -m, \text{ modulo } n$$

where *n* (a positive integer) is the coding ratio (that is, the number of bases which code one amino-acid) and *m* is any integral number of bases, positive or negative.

It can also be seen that our choice of reading direction is arbitrary, and that the same results (to a first approximation) would be obtained in whichever direction the genetic material was read, that is, whether the starting point is on the right or the left of the gene, as conventionally drawn.

Triple Mutants and the Coding Ratio

The somewhat abstract description given above is necessary for generality, but fortunately we have convincing evidence that the coding ratio is in fact 3 or a multiple of 3.

This we have obtained by constructing triple mutants of the form (+ with + with +) or (− with − with −). One must be careful not to make shifts

Table 3. TRIPLE MUTANTS HAVING A WILD OR PSEUDO-WILD PHENOTYPE

FC (0 + 40 + 38)
FC (0 + 40 + 58)
FC (0 + 40 + 57)
FC (0 + 40 + 54)
FC (0 + 40 + 55)
FC (1 + 21 + 23)

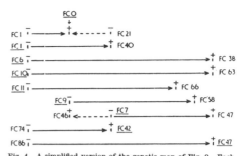

Fig. 4. A simplified version of the genetic map of Fig. 2. Each line corresponds to the suppressor from one mutant, here underlined. The arrows show the range over which suppressors have so far been found, the extreme mutants being named on the map. Arrows to the right are shown solid, arrows to the left dotted

across the 'unacceptable' regions for the ← shifts, but these we can avoid by a proper choice of mutants.

We have so far examined the six cases listed in Table 3 and in all cases the triples are wild or pseudo-wild.

The rather striking nature of this result can be seen by considering one of them, for example, the triple (*FC* 0 with *FC* 40 with *FC* 38). These three mutants are, by themselves, all of like type (+). We can say this not merely from the way in which they were obtained, but because each of them, when combined with our mutant *FC* 9 (−), gives the wild. or pseudo-wild phenotype. However, either singly or together in pairs they have an *r* phenotype, and will not grow on *K*. That is, the function of the gene is absent. Nevertheless, the combination of all three in the same gene partly restores the function and produces a pseudo-wild phage which grows on *K*.

This is exactly what one would expect, in favourable cases, if the coding ratio were 3 or a multiple of 3.

Our ability to find the coding ratio thus depends on the fact that, in at least one of our composite mutants which are 'wild', at least one amino-acid must have been added to or deleted from the polypeptide chain without disturbing the function of the gene-product too greatly.

This is a very fortunate situation. The fact that we can make these changes and can study so large a region probably comes about because this part of the protein is not essential for its function. That this is so has already been suggested by Champe and Benzer[18] in their work on complementation in the *r*_{II} region. By a special test (combined infection on *K*, followed by plating on *B*) it is possible to examine the function of the *A* cistron and the *B* cistron separately. A particular deletion, 1589 (see Fig. 5) covers the right-hand end of the *A* cistron and part of the left-hand end of the *B* cistron. Although 1589 abolishes the *A* function, they showed that it allows the *B* function to be expressed to a considerable extent. The region of the *B* cistron deleted by 1589 is that into which all our *FC* mutants fall.

Joining two Genes Together

We have used this deletion to re-inforce our idea that the sequence is read in groups from a fixed starting point. Normally, an alteration confined to the *A* cistron (be it a deletion, an acridine mutant. or any other mutant) does not prevent the expression of the *B* cistron. Conversely, no alteration within the *B* cistron prevents the function of the *A* cistron. This implies that there may be a region between the two cistrons which separates them and allows their functions to be expressed individually.

We argued that the deletion 1589 will have lost this separating region and that therefore the two (partly damaged) cistrons should have been joined together. Experiments show this to be the case, for now an alteration to the left-hand end of the *A* cistron, if combined with deletion 1589, can prevent the *B* function from appearing. This is shown in Fig. 5. Either the mutant *P*43 or *X*142 (both of which revert strongly with acridines) will prevent the *B* function when the two cistrons are joined, although both of these mutants are in the *A* cistron. This is also true of *X*142 *S*1, a suppressor of *X*142 (Fig. 5, case *b*). However, the double mutant (*X*142 with *X*142 *S*1), of the type (+ with −), which by itself is pseudo-wild, still has the *B* function when combined with 1589 (Fig. 5, case *c*). We have also tested in this way the 10 deletions listed by Benzer[7], which fall wholly to the left of 1589. Of these, three (386, 168 and 221) prevent the *B* function (Fig. 5, case *f*) whereas the other seven show it (Fig. 5, case *e*). We surmise that each of these seven has lost a number of bases which is a multiple of 3. There are theoretical reasons for expecting that deletions may not be random in length, but will more often have lost a number of bases equal to an integral multiple of the coding ratio.

It would not surprise us if it were eventually shown that deletion 1589 produces a protein which consists of part of the protein from the *A* cistron and part of that from the *B* cistron, joined together in the same polypeptide chain, and having to some extent the function of the undamaged *B* protein.

Is the Coding Ratio 3 or 6 ?

It remains to show that the coding ratio is probably 3, rather than a multiple of 3. Previous rather rough estimates[10,14] of the coding ratio (which are admittedly very unreliable) might suggest that the coding ratio is not far from 6. This would imply, on our theory, that the alteration in *FC* 0 was not to one base, but to two bases (or, more correctly, to an even number of bases).

We have some additional evidence which suggests that this is unlikely. First, in our set of 126 mutants produced by acridine yellow (referred to earlier) we have four independent mutants which fall at or

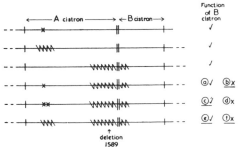

Fig. 5. Summary of the results with deletion 1589. The first two lines show that without 1589 a mutation or a deletion in the *A* cistron does not prevent the *B* cistron from functioning. Deletion 1589 (line 3) also allows the *B* cistron to function. The other cases, in some of which an alteration in the *A* cistron prevents the function of the *B* cistron (when 1589 is also present), are discussed in the text. They have been labelled (*a*), (*b*), etc., for convenience of reference, although cases (*a*) and (*d*) are not discussed in this paper. √ implies function; × implies no function

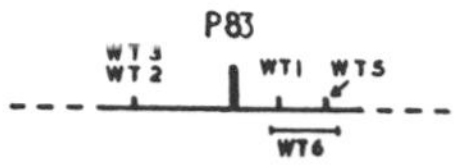

close to the *FC* 9 site. By a suitable choice of partners, we have been able to show that two are + and two are −. Secondly, we have two mutants (*X*146 and *X*225), produced by hydrazine[15], which fall on or near the site *FC* 30. These we have been able to show are both of type −.

Thus unless both acridines and hydrazine usually delete (or add) an even number of bases, this evidence supports a coding ratio of 3. However, as the action of these mutagens is not understood in detail, we cannot be certain that the coding ratio is not 6, although 3 seems more likely.

We have preliminary results which show that other acridine mutants often revert by means of close suppressors, but it is too sketchy to report here. A tentative map of some suppressors of *P* 83, a mutant at the other end of the *B* cistron, in segment *B* 9*a*, is shown in Fig. 6. They occur within a shorter region than the suppressors of *FC* 0, covering a distance of about one-twentieth of the *B* cistron. The double mutant *WT* (2 + 5) has the *r* phenotype, as expected.

Is the Code Degenerate?

If the code is a triplet code, there are 64 (4 × 4 × 4) possible triplets. Our results suggest that it is unlikely that only 20 of these represent the 20 amino-acids and that the remaining 44 are nonsense. If this were the case, the region over which suppressors of the *FC* 0 family occur (perhaps a quarter of the *B* cistron) should be very much smaller than we observe, since a shift of frame should then, by chance, produce a nonsense reading at a much closer distance. This argument depends on the size of the protein which we have assumed the *B* cistron to produce. We do not know this, but the length of the cistron suggests that the protein may contain about 200 amino-acids. Thus the code is probably 'degenerate', that is, in general more than one triplet codes for each amino-acid. It is well known that if this were so, one could also account for the major dilemma of the coding problem, namely, that while the base composition of the DNA can be very different in different micro-organisms, the amino-acid composition of their proteins only changes by a moderate amount[16]. However, exactly how many triplets code amino-acids and how many have other functions we are unable to say.

Future Developments

Our theory leads to one very clear prediction. Suppose one could examine the amino-acid sequence of the 'pseudo-wild' protein produced by one of our double mutants of the (+ with −) type. Conventional theory suggests that since the gene is only altered in two places, only two amino-acids would be changed. Our theory, on the other hand, predicts that a string of amino-acids would be altered, covering the region of the polypeptide chain corresponding to the region on the gene between the two mutants. A good protein on which to test this hypothesis is

Fig. 6. Genetic map of *P* 83 and its suppressors, *WT* 1, etc. The region falls within segment *B* 9*a* near the right-hand end of the *B* cistron. It is not yet known which way round the map is in relation to the other figures

the lysozyme of the phage, at present being studied chemically by Dreyer[17] and genetically by Streisinger[10].

At the recent Biochemical Congress at Moscow, the audience of Symposium I was startled by the announcement of Nirenberg that he and Matthaei[18] had produced polyphenylalanine (that is, a polypeptide all the residues of which are phenylalanine) by adding polyuridylic acid (that is, an RNA the bases of which are all uracil) to a cell-free system which can synthesize protein. This implies that a sequence of uracils codes for phenylalanine, and our work suggests that it is probably a triplet of uracils.

It is possible by various devices, either chemical or enzymatic, to synthesize polyribonucleotides with defined or partly defined sequences. If these, too, will produce specific polypeptides, the coding problem is wide open for experimental attack, and in fact many laboratories, including our own, are already working on the problem. If the coding ratio is indeed 3, as our results suggest, and if the code is the same throughout Nature, then the genetic code may well be solved within a year.

We thank Dr. Alice Orgel for certain mutants and for the use of data from her thesis, Dr. Leslie Orgel for many useful discussions, and Dr. Seymour Benzer for supplying us with certain deletions. We

are particularly grateful to Prof. C. F. A. Pantin for allowing us to use a room in the Zoological Museum, Cambridge, in which the bulk of this work was done.

[1] Wittman, H. G., Symp. 1, Fifth Intern. Cong. Biochem., 1961, for refs. (in the press).

[2] Tsugita, A., and Fraenkel-Conrat, H., *Proc. U.S. Nat. Acad. Sci.*, **46**, 636 (1960); *J. Mol. Biol.* (in the press).

[3] Brenner, S., *Proc. U.S. Nat. Acad. Sci.*, **43**, 687 (1957).

[4] For refs. see Watson, H. C., and Kendrew, J. C., *Nature*, **190**, 670 (1961).

[5] Crick, F. H. C., Griffith, J. S., and Orgel, L. E., *Proc. U.S. Nat. Acad. Sci.*, **43**, 416 (1957).

[6] Benzer, S., *Proc. U.S. Nat. Acad. Sci.*, **45**, 1607 (1959), for refs. to earlier papers.

[7] Benzer, S., *Proc. U.S. Nat. Acad. Sci.*, **47**, 403 (1961); see his Fig. 3.

[8] Brenner, S., Benzer, S., and Barnett, L., *Nature*, **182**, 983 (1958).

[9] Brenner, S., Barnett, L., Crick, F. H. C., and Orgel, A., *J. Mol. Biol.*, **3**, 121 (1961).

[10] Streisinger, G. (personal communication and in the press).

[11] Feynman, R. P.; Benzer, S.; Freese, E. (all personal communications).

[12] Jinks, J. L., *Heredity*, **16**, 153, 241 (1961).

[13] Champe, S., and Benzer, S. (personal communication and in preparation).

[14] Jacob, F., and Wollman, E. L., *Sexuality and the Genetics of Bacteria* (Academic Press, New York, 1961). Levinthal, C. (personal communication).

[15] Orgel, A., and Brenner, S. (in preparation).

[16] Sueoka, N. *Cold Spring Harb. Symp. Quant. Biol.* (in the press).

[17] Dreyer, W. J., Symp. 1, Fifth Intern. Cong. Biochem., 1961 (in the press).

[18] Nirenberg, M. W., and Matthaei, J. H., *Proc. U.S. Nat. Acad. Sci.*, **47**, 1588 (1961).

24

Reprinted from *Proc. Natl. Acad. Sci.*, **49**(1), 94–102 (1963)

THE STRUCTURE OF THE DNA-ACRIDINE COMPLEX*

By L. S. Lerman†

DEPARTMENT OF BIOPHYSICS, UNIVERSITY OF COLORADO

Communicated by Theodore T. Puck, November 29, 1962

It has been proposed that acridines and related compounds bind to DNA by intercalation between normally neighboring base pairs in a plane perpendicular to the helix axis (Lerman, 1961); the space is provided by an extension and local untwisting of the helix. To test this hypothesis further, some other properties of the

DNA-acridine complex have been examined in dilute aqueous solution, including the orientation of the plane of the acridine and the effect of complexing on the orientation of the plane of the purines and pyrimidines, both with respect to the DNA helix axis. The results show that the bound acridine is more nearly perpendicular than tangent to the helix axis, and the perpendicularity of the base pairs to the helix axis is not significantly altered. In a separate study of the effect of complexing on the rates of diazotization of amino acidines, and the effect of complexing on the reactivity of purine and pyrimidine amino groups, it is also shown that the amino groups of amino acridines are relatively inaccessible within the complex, and hydrogen bonding between base pairs remains undisturbed (Lerman, 1963).

Experimental Methods.—Polarized fluorescence was measured at 90° to the incident beam in a conventional instrument constructed in this Department of Biophysics. Monochromatic polarized light was obtained from a tungsten or deuterium arc source by means of a Beckman DU monochromator fitted with a rotatable calcite Glan prism. The sample was examined at 25.0°C in either a 1 cm square fused silica cuvette or a fused silica capillary tube. The intensity of the emission was determined through an appropriate long wave-length pass filter and a calcite Glan-Thompson prism by a chilled, low-noise (EMI 9536) photomultiplier, operated as a photon counter. Flow of the DNA solution was effected by a synchronous motor-driven syringe. All measurements represent roughly 10^5 counts or more (about $100 \times$ background), accumulated in one or more ten-second periods. Flow dichroism was measured in a Cary Model 14 spectrophotometer, using the Glan prism. The solution was driven through a short-path cuvette with planar silica windows, separated by 0.0271 cm, perpendicular to the light beam.

All measurements were carried out with chicken erythrocyte DNA isolated from washed nuclei by phenol extraction, similar to the method of Kirby (1958).

Recognition of the Acridine Plane.—If an array of stationary, randomly oriented fluorescent molecules is illuminated with suitable exciting radiation, the emitted light is polarized; that is, the emitted intensity observed through a polarizing prism will depend on the orientation of the prism. Although polarized emission is also observed when the exciting radiation is unpolarized, the degree of polarization of the emitted light is less. All of the following discussion and experimental results will refer to polarized excitation. Where emission is observed at right angles to the exciting beam, it is appropriate to compare intensities found when the analyzing prism is oriented perpendicular and parallel to this plane. When the exciting beam enters along the X axis, and emission is observed along the Y axis, the degree of polarization is given by the difference between the two intensities measured with the plane of polarization of the analyzing prism respectively parallel to Z and to X, divided by the sum of the same intensities. The probabilities of absorption and of emission with a particular polarization are proportional to the square of the cosine of the angle between the electric vector of the absorbed or emitted light and the direction of charge displacement, the transition moment, corresponding to the electronic transition responsible for absorption or emission. It can be shown that the Z intensity is greater and, consequently, the degree of polarization is positive, when the electronic transition moments for absorption and emission are more or less parallel; when they are more nearly perpendicular, the X intensity will be larger and the degree of polarization will be negative (Pringsheim, 1949). Since we shall be concerned only with singlet $\pi - \pi^*$ transitions, the transition moments for both absorption and emission will lie in the plane of the aromatic rings. It has been found, almost without exception, that the spectral distribution of emission from solutions of fluorescent substances is independent of the wavelength of excitation. Thus, the variation of the degree of polarization of emission with the wavelength of excitation corresponds to changes in direction only of the transition moment for excitation. The actual value of the degree of polarization may be taken to reflect the distribution of excitation probabilities between two or more different transitions that can be excited by the same wavelength (Albrecht, 1960). If the fluorescent molecules are allowed to rotate instead of remaining stationary during the interval between absorption and emission, the observed polarization will either disappear or be diminished depending upon the extent of randomization. In sufficiently concentrated solution, the emission may also be depolarized by migration of the excitation to nearby similar molecules. Polarization spectra for dilute solutions of the acridine deriva-

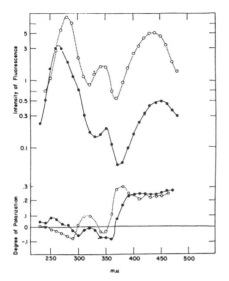

FIG. 1.—Excitation and polarization spectra for quinacrine in sucrose and DNA solution. The upper curves show the relative fluorescence intensities per unit concentration as a function of the wavelength of excitation, uncorrected for the energy distribution in the exciting radiation, for dilute solutions of quinacrine in DNA (solid line) and in 60% sucrose (dotted line). The lower curves show the degree of polarization of the 90° emission at each wavelength. The DNA concentration was about 3.6×10^{-4} M in nucleotides. The quinacrine was 2.65×10^{-6} M in the DNA solution, corresponding to one dye molecule per 68 nucleotide pairs. In the sucrose solution, the quinacrine was 1.51×10^{-7} M.

tive, quinacrine (Atebrin), are shown in Figure 1, together with the corresponding excitation spectra, uncorrected for the energy distribution of the exciting source. Although almost no polarization is observed in simple aqueous solutions, substantial polarization is observed in DNA solution because nearly all of the quinacrine is tightly bound, and DNA, by virtue of its molecular weight and configuration, is characterized by an exceedingly small rotational diffusion constant. (Similarly, the viscosity of a concentrated sucrose solution is sufficient to prevent a large rotational depolarization in the absence of DNA.) It will be noted that the long wave excitation maximum is shifted to somewhat longer wavelengths by attachment to DNA, as is found for the absorption maximum of proflavine (Peacocke and Skerrett, 1956) and other acridines (Morthland et al., 1954). Although the fluorescence yield is generally much less for the bound than the free dye, a relatively increased emission probability can also be seen in the region of ultraviolet absorption by DNA. This presumably corresponds to sensitization of fluorescence and will be treated in a separate communication. From these data the wavelengths of 450 mμ and 300 mμ were selected as providing adequate fluorescence intensities with parallel ($\parallel$) and perpendicular ($\perp$) pairs of transition moments. The directions of these transitions serve to specify the plane of the molecule.

The Effect of Flow.—It will be useful to consider a simplified scheme in which the transition moments for absorption and emission must lie either parallel or perpendicular to the flow axis, and the DNA molecule is assumed to be fully oriented with the helix axis parallel to the flow axis, and randomly rotated around the flow axis. Let us also restrict our attention to just two pairs of prism orientations—both prisms vertical, parallel to flow, and both horizontal, perpendicular to flow.

There are then three ways in which an acridine might be attached to the DNA helix, as indicated by types TL, TS, and P in Table 1 and Figure 2, and a set of four experimental predictions for each. The experimental measurement consists in the comparison of the fluorescence intensities of the flowing and stationary solution, given simply by the relative difference in intensities obtained when the DNA is oriented by flow and when its orientation is random. If a particular transition is that rendered parallel to the relevant prism orientation, the probability of observing transition will increase from the average value for random orientation to the maximum value; if the transition is rendered perpendicular to the prism orientation, it will not be observed at all. It will be seen that where one transition for either absorption or emission is parallel to a prism and the other is perpendicular, the product of the absorption and emission probabilities nevertheless equals zero.

TABLE 1

THE EXPECTED EFFECTS OF FLOW ORIENTATION ON POLARIZED TRANSITION PROBABILITIES
ACCORDING TO THE BINDING ARRANGEMENT

A. Orientation with respect to the helix axis

		TL	TS	P
1.	type of arrangement	TL	TS	P
2.	plane of dye	tangent	tangent	perpendicular
3.	emission transition	perpendicular	parallel	perpendicular
4.	‖ absorption transition	perpendicular	parallel	perpendicular
5.	⊥ absorption transition	parallel	perpendicular	perpendicular

B. Difference in intensities, (flow oriented) − (random); both prisms parallel to flow

6.	absorption only, ‖ transition	−	+	−
7.	absorption only, ⊥ transition	+	−	−
8.	emission only, assuming equal excitation	−	+	−
9.	emission from ‖ absorption	−	+	−
10.	emission from ⊥ absorption	−	−	−

C. Difference in intensities, (flow oriented) − (random); both prisms perpendicular to flow

11.	absorption only, ‖ transition	+	−	+
12.	absorption only, ⊥ transition	−	+	+
13.	emission only, assuming equal excitation	+	−	+
14.	emission from ‖ absorption	+	−	+
15.	emission from ⊥ absorption	−	−	+

The signs, + and −, signify that the flow oriented intensity is either greater or less, respectively, than the intensity for nonoriented molecules. If perfect orientation were achieved (infinitely great shear, etc.), all of the − values would approach zero. Lines 6, 7, 11, and 12 indicate changes that can, in principle, be observed as dichroism due to flow. Lines 9, 10, 14, and 15 indicate the net changes due to the product of the absorption and emission probabilities that would be expected in the fluorescence.

Since the functions are nonlinear, the effect of the decline will usually prevail when orientation is incomplete, and a net decrease in intensity will be observed.

Since the DNA molecule in solution appears to be a statistical coil of limited flexibility, only partial orientation can be expected, and the change in intensities induced by flow will be much less marked than the simplified theory predicts. A qualitative interpretation of the experimental results is possible, however, in terms of the direction of each intensity change induced by flow for each transition at each pair of prism orientations, together with an estimate of the degree of orientation of the DNA. Experimental results for quinacrine, presented in Table 2, should be compared with the expectations presented in lines 9, 10, 14, and 15 of Table 1 for the three hypothetical orientations. Since both intensities decrease during flow when the prisms are parallel to flow and both increase when the prisms are perpendicular, the data are compatible only with arrangement P. The extent of orientation indicated here, about 0.3, is substantially greater than the extent of orientation indicated in the flow dichroism studies discussed below and those of Cavalieri *et al.* (1956), for comparable DNA concentrations. However, both measurements agree when account is taken of the differences in geometry and the form of the functions averaged in the two types of measurement.

TABLE 2

FRACTIONAL CHANGE DUE TO FLOW IN THE FLUORESCENCE OF QUINACRINE BOUND TO DNA

Polarization of emission	X	X	Z	Z
Wavelength of excitation	3000	4500	3000	4500 Å
Exciting radiation Y polarized	+0.24	+0.35	−0.12	−0.11
Exciting radiation Z polarized	−0.10	−0.10	−0.26	−0.29

The DNA (as nucleotides) and total quinacrine concentrations were $3.76 \times 10^{-4} M$ and $5.5 \times 10^{-6} M$, respectively, corresponding to one quinacrine molecule per 34 nucleotide pairs. The shear rate during flow (at the walls of the capillary) was about 1.2×10^4 sec^{-1}. The X and Y polarizations are perpendicular to flow; Z is parallel to flow. The transitions excited at 4,500 Å and 3,000 Å are parallel and perpendicular, respectively, to the transition of emission.

Since the longest wavelength of excitation always corresponds to the same electronic transition as that responsible for emission, it will be possible in general to establish the angle between this transition and the helix axis for any fluorescent molecule that binds to DNA. Perpendicular transitions, however, are not always available at useful intensities, so that complete specification of the orientation of the plane containing the dye is not always possible. For substances having a positively polarized band with a substantially lower degree of polarization than the longest wave-length absorption, the calculation can also be made on the assumption that the lower degree of polarization results from transitions in more than one direction in the plane of the molecule. Acridine yellow, the most powerful mutagen studied by Orgel and Brenner (1961), provides a band at 305 mμ with degree of polarization, 0.10, when bound to DNA, as well as the long wave band in the region of 460 mμ with degree of polarization, 0.35. In flow experiments with acridine yellow like those described above, the fluorescence intensities resulting from excitation of either band again change by the same amount, and in the same directions as found for quinacrine. Since the perpendicular contribution to short wave-length excitation is of course much less, the precision with which the plane can be specified is correspondingly diminished, but the conclusion remains the same; the dye must be nearly perpendicular to the helix axis.

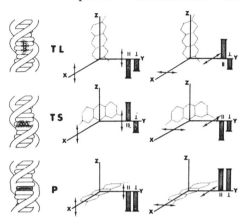

Fig. 2.—Three conjectures for the orientation of acridine bound to DNA. Each arrangement is identified similarly in Table 1. The bar graphs on each set of coordinate axes represent qualitatively the expected change in intensity of fluorescence when the DNA is itself oriented by flow along the Z axis. It is assumed that a beam of monochromatic light along the X axis, polarized as shown in each diagram, is absorbed by a transition in either the long or short axis of the molecule, according to its wavelength, and that the emitting transition is along the short axis. The light emitted in the Y direction is measured through a polarizing prism, directed as shown. The change corresponding to short axis excitation is identified as ∥, and to long axis excitation, as ⊥. (The acridine skeleton in the figure is that of proflavin.)

Orientation of the Plane of the Base Pairs.—The similarity of the configuration of pure, undenatured DNA in solution to the B structure found in fibers at relatively high humidity is usually accepted on the basis of two kinds of evidence: the continuous swelling of fibers at very high humidities, approaching the dissolved state, without any indication of an abrupt structural transition, and the determinations of mass per unit length and helix radius by low-angle X-ray scattering from liquid crystals and concentrated solutions of DNA (Luzzati *et al.*, 1961a). On this assumption the ambiguity of flow dichroism measurements on simple DNA solutions is resolved, and the determinations can be taken to signify only the extent to which the helix is oriented by flow, rather than the inseparable combination of helix orientation and base pair tilt. The question of tilting of the base pairs is again of interest with respect to the DNA-acridine complex, since strong tilting would lower the mass per unit length enough to allow for a net decrease even after the mass of the bound acridines is added (Lerman, 1961). Since the amount of tilting required, roughly 45°, abolishes the dichroism of the helix in stretched DNA fibers (Wilkins *et al.*, 1951), the measurement of the flow dichroism of the DNA-acridine complex can be expected to respond sensitively if this structural change is involved in formation of the complex. If, on the other hand, the intercalation hypothesis is more nearly correct, the DNA dichroism should be about the same in the complex as in the pure state, or even somewhat greater because of the straightening of the helix indicated by the viscosity enhancement at low rates of shear (Lerman, 1961). Flow dichroism measurements on DNA and complexes with quinacrine are presented in Table 3. Similar results have also been obtained with proflavine. Di-

TABLE 3

FLOW DICHROISM OF DNA AND DNA-QUINACRINE

Wavelength	2600	4325	4550 A
DNA only			
Dichroism, Z polarization	−0.19		
Dichroism, Y polarization	+0.086		
DNA-quinacrine			
Dichroism, Z polarization	−0.24	−0.25	−0.24
Dichroism, Y polarization	+0.11	+0.12	+0.11
Absorbance, DNA contribution	0.654	0.000	0.000
Absorbance, DNA plus bound quinacrine	0.989	0.106	0.100
Absorbance, total	1.422	0.240	0.213

The dichroism is tabulated as the fractional change in absorbance of the DNA or the DNA-quinacrine complex when the solution is subjected to flow. The shear rate at the windows was 1.18×10^4 sec^{-1}. The absorbances in the stationary solution show the distribution of absorption between DNA, bound quinacrine, and the unattached dye. The DNA concentration (as nucleotides) when measured alone was $3.2 \times 10^{-3} M$, and in the complex, $3.6 \times 10^{-3} M$. The quinacrine bound to DNA was $0.661 \times 10^{-3} M$, or one quinacrine molecule per 2.7 nucleotide pairs. The Y and Z axes are perpendicular and parallel, respectively, to the direction of flow.

chroism is given as the fractional change in absorbance due to flow, with the polarizing prism either parallel or perpendicular to the flow axis. It should be noted that the geometry is different than that of the flow fluorescence experiments; the flow is between parallel planes (the windows of the cuvette) rather than through a tube of circular cross section. Light will be absorbed in these solutions both by the complex and by unattached quinacrine, which will not contribute to the flow dichroism. Since it was desirable to approach saturation of the binding sites, the concentration of free quinacrine is considerable, and accordingly, it is necessary to deduct its contribution to the absorbance of the solution for correct estimation of the dichroism of the complex. For this purpose, the concentration of bound quinacrine was calculated from the ultraviolet absorption spectra, which differ significantly for the free and bound dye. The dichroism at long wavelengths, due entirely to the dye, will directly indicate the contribution of the dye to the dichroism at 2600 Å, since both transitions are in the plane of the rings. Since the dye accounts for only 0.34 of the absorbance of the complex at 2600 Å, the dichroism due to DNA in the complex must be very nearly the same as that of the dye.

Thus, all strongly tilted models are excluded since DNA is more, rather than less, dichroic in the complex than alone. Also, the close similarity of the dichroism values for the quinacrine and the DNA itself demonstrates that the plane of the purine and pyrimidine pairs must be parallel to the direction of the long wave-length transition of the quinacrine. In view of the preceding considerations, the planes of both must be parallel.

Structure of the Complex.—The fluorescence and dichroism results require that both the base pairs and the acridine lie substantially perpendicular to the molecular axis, but do not necessarily specify the intercalated configuration. The additional restriction that the mass per unit length of the complex is less than that of DNA alone (Luzzati *et al.*, 1961*b*, Lerman, 1961) is in agreement with the properties of the intercalated model; it could be reconciled with arrangements based on external edgewise attachment of the acridine by the *ad hoc* supposition that gaps, filled with solvent, are generated in the stack of base pairs. Roughly one such gap, the thickness of a base pair, would be required for each acridine bound, so that at the saturation region for strong binding of proflavine there would be a gap for every two base pairs. This model is, of course, energetically highly implausible, and the supposition of an almost totally aqueous environment for the acridine is incompatible with the properties of the complex as determined by reaction rate studies (Lerman, 1963). The comparison of the rates of diazotization of the amino groups of proflavine and other acridines when bound to DNA and free in solution shows a severe restriction on the accessibility of the amino groups to attack by the nitrosating reagents. This restriction is not found when the amino acridines are bound to other anionic polymers, even where there is a broad aromatic moiety for face to face contact with acridine. The intercalated structure remains as the only hypothesis fully compatible with the diverse experimental evidence.

A Conjectural Mechanism for Mutagenesis.—A mechanism based on the intercalation of an acridine into a single strand of the DNA molecule, as proposed by Brenner *et al.* (1961) for generating insertion and deletion mutations during DNA synthesis, demands a somewhat different structural basis than the model proposed here, in order to accommodate the pairing of an extended strand containing the acridine with a nonextended strand containing one additional base. It has seemed more useful to consider whether a mechanism might be devised on the basis of the simple intercalated model, and taking into account also some of the other characteristic properties of acridine mutagenesis. Aside from the presumably irrelevant induction of mutations by photodynamic action of the acridines, it appears clear that there is a potent mutagenic effect in the reproduction of the T-even phages, and that there is no significant mutagenic effect on *Escherichia coli* during vegetative growth (as shown by our own observations and other investigators). There are at least two differences between the multiplication of bacteria and the multiplication of phage; the phage DNA contains hydroxymethyl cytosine instead of cytosine, and phage multiplication is accompanied by extensive genetic recombination. It is difficult to attribute any particular significance to the presence of the substituted cytosine that might be relevant to the acridine effect. (We have observed the usual viscosity enhancement in T2 DNA mixed with proflavine.) However, it is easy to attribute a role in mutagenesis to the frequent recombinatorial events that accompany phage multiplication.

Let us suppose that recombination takes place within a short region of the chromosome corresponding to the effective pairing, or switch, region, and that a few acridine molecules are intercalated at random in one or both of the paired chromosomes. Since the length along the chromosome occupied by an acridine is the same as the length required by a base pair (3.4 Å), wherever there is an intercalation in one of the chromosomes without intercalation at exactly the same place in the other, the pairin will be shifted exactly one step out of register at one side of the site of intercalation.   If the chromosomes are paired in perfect homology below the site of intercalation, the chromosome containing the acridine will offer the base pair corresponding to the next earliest neighbor to the other for pairing above the site of intercalation.   After any interval in which the same number of acridines are intercalated into both chromosomes they will return to perfect homology. Within the interval they can be one, two, or more steps out of register in either direction.   If the chromosomes break at the same position, side by side, and reunite with the exchange of partners, it will be seen that the reunion will result in the omission of one or more base pairs in one product and the corresponding duplication of one or more base pairs in the other product of the reunion.   This mechanism thus has the property of generating insertion and deletion mutations, as required in the analysis of the interactions of suppressor mutations by Crick and collaborators (1961), and will be dependent on the frequency of crossing-over.   Mutants will be expected to be recombinant for closely linked outside markers.   The same process, crossing-over between paired chromosomes that are displaced from perfect homology by one, or a few, nucleotide pairs, could also be expected to occur spontaneously with the same results.   It has been suggested (Fresco, personal comm., 1960) that the spontaneous process proceeds through the formation of short loops of one or more nucleotides excluded from the double helix, as suggested by the stoichiometry pairing of synthetic polynucleotides (Fresco and Alberts, 1960).   The occurrence of recombination errors during meiotic division could account for the increased rate of spontaneous mutation found by Magni and Von Borstel (1962) in ascospores as compared with the mutation rate during vegetative growth of yeast.   A formal mechanism for spontaneous mutation by unequal crossing-over has also been advanced independently by Demerec (1962).

Summary.—New physical evidence is presented bearing on the structure of the DNA-acridine complex.   The results are fully compatible with the intercalated model, and incompatible with other plausible models.   A mechanism based on recombination errors is proposed for the production of insertion and deletion mutations by intercalation of acridines.

I should like to acknowledge the assistance of Mr. James Levy in the preparation of DNA, and the help of Mr. James McIntyre with the flow dichroism measurements.

* Contribution No. 202 from the Department of Biophysics, Florence R. Sabin Laboratories University of Colorado Medical Center, Denver, Colorado. This work was supported by a Research Grant (GM 08894) from the National Institutes of Health. Acquisition of the Cary spectrophotometer was made possible by a grant from the National Science Foundation.

† Research Career Development Awardee of the U.S. Public Health Service.

Albrecht, A. C., *J. Mol. Spec.*, **6**, 84 (1961).
Brenner, S., L. Barnett, F. H. C. Crick, and A. Orgel, *J. Mol. Biol.*, **3**, 121 (1961).
Cavalieri, L. F., B. H. Rosenberg, and M. Rosoff, *J. Am. Chem. Soc.*, **78**, 5235 (1956).
Crick, F. H. C., L. Barnett, S. Brenner, and R. J. Watts-Tobin, *Nature*, **192**, 1227 (1961).
Demerec, M., these PROCEEDINGS, **48**, 1696 (1962).
Fresco, J. R., and B. M. Alberts, these PROCEEDINGS, **46**, 311 (1960).
Kirby, K. S., *Biochem. J.*, **70**, 260 (1958).
Lerman, L. S., *J. Mol. Biol.*, **3**, 18 (1961).
Lerman, L. S., (1963) in preparation.
Luzzati, V., A. Nicolaieff, and F. Masson, *J. Mol. Biol.*, **3**, 185 (1961a).
Luzzati, V., F. Masson, and L. S. Lerman, *J. Mol. Biol.*, **3**, 634 (1961b).
Magni, G. E., and R. C. Von Borstel, *Genetics*, **47**, 1097 (1962).

Morthland, F. W., P. P. H. DeBruyn, and N. H. Smith, *Exp. Cell Res.*, **7**, 201 (1954).

Orgel, A., and S. Brenner, *J. Mol. Biol.*, **3**, 762 (1961).

Peacocke, A. R., and J. N. H. Skerrett, *Trans. Far. Soc.*, **52**, 261 (1956).

Pringsheim, P , *Fluorescence and Phosphorescence* (New York: Interscience, 1949)

Wilkins, M. H. F., R. G. Gosling, and W. E. Seeds, *Nature*, **167**, 759 (1951).

25

Reprinted from *Cold Spring Harbor Symp. Quant. Biol.*, **31**, 77–84 (1966)

Frameshift Mutations and the Genetic Code

This paper is dedicated to Professor Theodosius Dobzhansky on the occasion of his 66th birthday.

GEORGE STREISINGER, YOSHIMI OKADA, JOYCE EMRICH, JUDITH NEWTON
AKIRA TSUGITA[1], ERIC TERZAGHI[*,1] AND M. INOUYE[1]

Institute of Molecular Biology, University of Oregon, Eugene, and Institute of Molecular Genetics, University of Osaka, Japan[1].

The genetic message is translated sequentially, three bases at a time, starting from a defined point at the beginning of a cistron (Crick et al., 1961; Terzaghi et al., 1966). The message can thus be separated into groups of three bases (codons) by a "reading frame" which is set in register at the beginning of the message (Crick et al., 1961). Deletions or insertions of a base (or bases) into the genome cause a shift in the reading frame and are thus called *frameshift mutations;* they result in a grossly different translation of the message beyond the site of the mutation. If the deletion of a base is followed by the insertion of a base, the reading frame, and the amino acid sequence of the protein product, might be altered only in the region between the mutations. Such double mutant strains may produce active protein and thus be pseudowild.

We have been examining the lysozyme synthesized by bacteria infected with pseudowild strains of phage T4 carrying certain pairs of frameshift mutations and, in one case, three frameshift mutations, in the lysozyme gene. We find that a short sequence of amino acids is changed in the lysozyme produced by each of these strains.

Utilizing codons proposed on the basis of in vitro studies (Khorana et al., this volume; Nirenberg et al., 1965) we can, in every case, assign a sequence of bases that would code both for the wild-type sequence of amino acids and for the changed sequence of amino acids in the mutant strain. The codons proposed on the basis of in vitro studies are thus compatible with the amino acid replacements in our double-mutant strains. If we assume that these codons *are* in fact utilized in vivo we can specify the *particular* codons that are used in the wild-type and the various mutant strains. Furthermore, we can begin to identify the nature of the events that give rise to frameshift mutations.

THE ANALYSIS OF MUTANT STRAINS

The general procedure we employed was to isolate independent proflavine-induced mutations

in the lysozyme gene. Most of the mutant phage form no plaques when plated on the usual media but do form plaques on special plates supplemented with egg-white lysozyme.

In order to isolate pseudowild double mutant strains, pairwise crosses of mutants were performed. For certain pairs of mutants both wild-type and pseudowild recombinants were observed among the progeny of the cross. The two types of recombinants can be distinguished by the appearance of the halos formed around the plaques upon exposure of the plate to chloroform vapors (Fig. 1). Crosses of the pseudowild double-mutant strain to a wild-type strain yielded the two original mutants, as expected.

In some cases spontaneous pseudowild revertants were isolated from mutant strains. Again, crosses of the pseudowild strain to a wild-type strain yielded

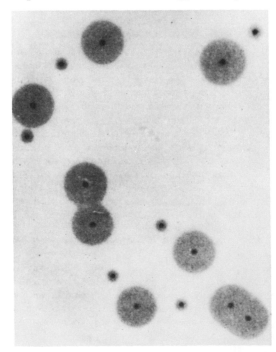

FIGURE 1. Large halos formed by e^+ and small halos formed by $eJ42eJ44$ plaques after exposure to chloroform vapors.

* Present address: Laboratoire de Biophysique, Université de Genève, Geneva, Switzerland.

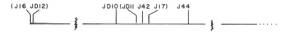

FIGURE 2. A map of the relative positions of *e* mutants described in this study. Mutants with the prefix J were induced by proflavine; mutants with the prefix JD are spontaneous mutants isolated from pseudowild strains. The map is not drawn to scale.

two types of mutants: one identical to the original mutant, the other a new mutant.

A map of the relative positions of the mutations utilized in this study is given in Fig. 2.

The lysozyme from bacteria infected with wild-type phage and with each of the pseudowild strains was purified, oxidized, and digested with trypsin, and the digest was chromatographed on Dowex 50-X2 resin (Terzaghi et al., 1966). In each case the mutant lysozyme was found to differ from the wild-type in one or two peaks out of a total of 18.

The sequence of amino acids in the changed region of the wild-type lysozyme and in each of the mutant lysozymes was determined. Using triplets proposed by Khorana and Nirenberg, it was possible in each case to assign a sequence of bases that could code for the wild-type sequence of amino acids and that, with the proper addition and/or deletion of bases, could code for the changed sequence of amino acids in the mutant strains.

Table 1 shows the sequences of amino acids, and the related base sequences for the lysozymes of the mutant strains analyzed.

The Direction of Translation

The amino acid sequences of the strains *e*+ and *e*J42*e*J44 can be related to one another on the basis of the triplets proposed by Nirenberg and Khorana only by the sequence of bases shown in Table 1.

As is shown in detail elsewhere (Terzaghi et al., 1966) the two amino acid sequences can be related only if the triplets are oriented so that their 5' to 3' polarity parallels the N-terminal to C-terminal polarity of the polypeptide chain. Since the synthesis of a polypeptide chain has been shown to proceed from the N-terminal toward the C-terminal end, our result demonstrates that the translation of a messenger RNA molecule proceeds from the 5' toward the 3' end.

The N-terminal to C-terminal polarity of the lysozyme molecule parallels the genetic direction *e*J17 to *e*J44 (Okada et al., 1966). The order of mutations in the lysozyme gene relative to other genes has been established by genetic crosses (manuscript in preparation). The direction of translation, relative to the genetic map, that has been established for the lysozyme gene is identical to that of the rII gene (Crick et al., 1961; Champe and Benzer, 1962) and opposite to that found for the head-protein gene by Sarabhai et al. (1964).

The Identification of Codons Utilized in Vivo

The results presented in Table 1 suggest that the following codons are in fact utilized in vivo: Ser—*AGU, UCA;* Pro—*CCA;* Leu—*CUU*, UUA; Asn—*AAU;* Val—GUC; His—CAU, CAC; Lys—AAA, AAG; Cys—UGU. The codons that are italicized have been identified in the wild-type strain. It is clear that degeneracy exists and, in addition, that even the code utilized in the DNA of wild-type T4 is degenerate. It is noteworthy that each of the five codons that we have identified in the wild-type phage T4 ends in either A or U. This is not surprising, since the DNA of phage T4 is relatively rich in AT base pairs.

Most of the codons are identified directly by a comparison of a mutant and the wild-type amino acid sequences. The codons for lysine are established by considering together the amino acid sequences of the strains *e*J42*e*J44 and *e*JD10*e*J42*e*J17. *If lysine in the wild-type is coded for by AAG, the mutation *e*J42 must result in the deletion of the terminal G of that codon;* otherwise arginine (AGG) would be observed in the triple mutant strain (Table 1c), rather than lysine. Thus our results establish that, regardless of the terminal base for the codon for lysine in the wild-type strain, the lysine codon in the strain *e*J42*e*J44 (Table 1a) is AAA, and that in the triple mutant strain (Table 1c) is AAG.

The codon for cysteine is established by considering together the amino acid sequences of the strains *e*J17*e*J44 and *e*JD10*e*J42*e*J17. The amino acid sequence of the triple mutant strain would be compatible with the addition of GC, yielding UGC for cysteine; in that case, however, alanine (GCC), rather than valine, would be observed in the strain *e*J17*e*J44. Thus the codon for cysteine in the triple mutant strain can only be UGU.

The codons described here are all based on the analysis of mutant strains in which the amino acid sequences have been firmly established, and do not depend on the tentative sequences described in Table 1(c) and (d).

Two Classes of Frameshift Mutations

It is a corollary of the triplet nature of the code (Crick et al., 1961) that frameshift mutations fall into two classes, members of one class resulting in a shift of the reading frame one base to the right, members of the other class resulting in a shift one base to the left.

The classification of the mutants that we have analyzed is given in Table 2. We define the "+" class as resulting in a frameshift to the left, and have identified members of this class in which one base is added. We define the "−" class as resulting

TABLE 1. CHANGED SEQUENCES OF AMINO ACIDS IN MUTANT STRAINS.

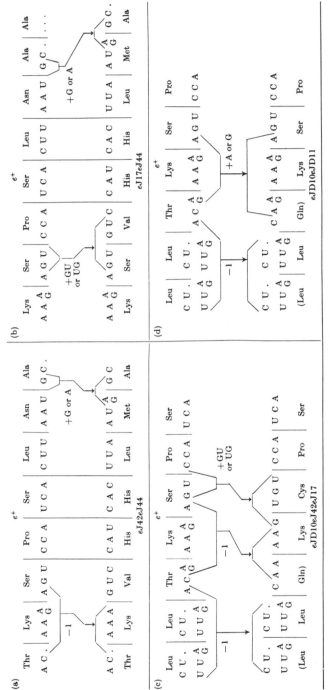

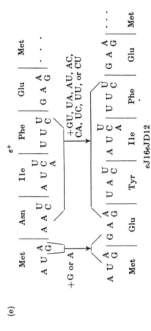

The sequences of amino acids observed, and the base sequences in the mRNA that are compatible with them, are indicated for each mutant strain. All codons that have not been excluded by in vitro studies and that are compatible with our results are given. Thus AUA is indicated for methionine and also iso-leucine, even though the bulk of the evidence at present suggests that AUA codes for isoleucine (Yanofsky et al., 1966). (c) and (d). The amino acid sequences given in parentheses are the ones most compatible with chemical and enzymatic results; but the amino acid sequences, and therefore the base sequences, have not yet been definitively established. Some of the results that establish the sequences indicated have been published: (a) Terzaghi et al., 1966; (b) Okada et al., 1966; the others will be published elsewhere (by Okada et al.).

244

TABLE 2. POSSIBLE NATURE OF MUTATIONAL EVENTS

Mutation	Class	Origin	Possible mutational event*
eJ44	+	Proflavine-induced†	...U A A U G C . G C ... ↓ +G ...U A A U G G C . G C ...
eJ16 or‡ eJD12	+	?	A U G A A U A U ... ↓ +G A U G G A A U A U ...
eJD11	+	Spontaneous	...A C A A A A A G U ... ↓ +A ...A C A A A A A A G U ...
eJD10	−	Spontaneous	cannot be identified
eJ42	−	Proflavine-induced	...A C A A A A A G U ... ↓ −A ...A C A A A A G U C ...
eJ17	−	Proflavine-induced	...A C A A A A A G U C ... +GU or UG ...A C A A A A A G U G U C ...
eJ16 or‡ eJD12	−	?	A U G A A U A U $\begin{smallmatrix}U\\A\end{smallmatrix}$ U U U G A A ... +AU, UA or +UU A U G A A U A U A U $\begin{smallmatrix}U\\A\end{smallmatrix}$ U U U G A A ...

* In order to facilitate comparison with Table 1 the mutational event is shown as it affects the sequence of bases in the mRNA molecule.

In describing the possible nature of the mutational event, the base sequences presented in Table 1 have been modified according to the following assumptions: (1) The last base in each codon of the wild-type DNA (except for the methionine codon) is A or U, as has been observed for all the codons that we have identified in the wild-type DNA. (2) The codon for methionine is AUG, as is suggested by the bulk of the evidence at this time. (3) The base or base doublet that is added is identical to one already present in the wild-type DNA. This last assumption influences *only* the sequences of bases proposed for the mutant strain. The sequences of bases presented for the wild-type strain have been modified from the results presented in Table 1 by the application of assumptions (1) and (2) presented above.

† Growth in the presence of proflavine increased the frequency of mutants about 60-fold in the sample from which these phage were selected.

‡ In this case it has not yet been possible to isolate the spontaneous mutation from the double mutant strain, or to identify which mutation occurred spontaneously and which was induced by proflavine. This sequence codes for the N-terminal end of the lysozyme molecule (Inouye and Tsugita, manuscript in prep.). The sequence of bases to the left of AUG cannot therefore be assigned.

in a frameshift to the right and have identified some members of this class in which one base is deleted and others in which two bases are added.

We find that double mutant strains carrying a "+" and a "—" mutation produce active proteins which differ from that of the wild-type strain by a short sequence of amino acids, while the only double-mutant strain that we have isolated that carries two mutations of the same class (eJ42eJ17) produces no measurable lysozyme activity. A triple mutant strain carrying three mutations of the same class (eJ42eJ17eJD10) was observed to produce an active protein which differs from that of the wild-type strain by a short sequence of amino acids. These results are thus in complete accord with the general nature of the genetic code as proposed by Crick and his co-workers (1961).

THE GENERAL MECHANISM OF FRAMESHIFT MUTATION

A plausible mechanism of frameshift mutation would involve the insertion of a base, or a base doublet, identical to an adjacent one already present in the wild-type DNA. The insertion would be most

FIGURE 3. Possible mode of origin of some frameshift mutations. In the case of each mutation, a DNA molecule is shown with a gap in one of the two chains and a mispaired set of bases near the end of this chain. The critical area for each mutation is outlined by a dashed square. The appearance of the molecule after synthesis (and mutation) is shown in the second line, for each case.

In each case, the mutation could have occurred in the strand complementary to the one shown here by the same mechanism that is illustrated.

* One of the two mutations in this double mutant strain has not yet been isolated. As shown in Table 1 the mutation in this case could be either the insertion of TA (diagrammed here) or the insertion of AT, TT, TC, CT, CA, AC, or GT. The mechanism in the case of AT or TT would be identical to the one shown here. The proposed model would not be compatible with the other insertions.

** Note that in this case the two terminal bases adjacent to the gap would have to be assumed not to be hydrogen bonded.

likely to occur in a region of repeating bases or base doublets through the pairing of a set of bases in one chain of the DNA molecule with the wrong, but complementary, set in the other chain. Frameshift mutations could then occur through the following sequence of events: (1) The presence of a gap in one of the two chains of a DNA molecule at or near a region of a repeating set of bases. (2) The mispairing of bases at the repeating sequence. (3) New synthesis that fills in the gap and results in the addition or deletion of a base or bases.

Because of the degeneracy of the code, many of the mutations and base sequences that we have studied cannot be identified unambiguously (Table 1), making a definitive test of these ideas difficult. Likely base sequences can be assigned, however, by assuming that the last base in each codon is A or U (as is the case for each of the five codons that *have* been identified in the wild-type DNA). It is noteworthy that, with these sequence assignments, four out of the five mutations identified occur in a region of a repeating sequence of a base or base doublet, (the sequence to the left of the sixth mutation not being identified) (Table 2). Furthermore, in each case we have studied, our results are compatible with the addition of a base, or a base doublet, that is identical to an adjacent one already present in the wild-type DNA (Table 2). The general model of frameshift mutagenesis we have described would thus be adequate to account for the mutations we have studied. The possible mode of origin of some of these mutations is diagrammed in Fig. 3.

THE MECHANISM OF FRAMESHIFT MUTATION IN PHAGE T4

In order to discuss the origin of frameshift mutations in phage T4 we need to review some features concerning the life cycle of this phage.

The DNA in mature phage T4 is a single linear molecule (Rubenstein, Thomas, and Hershey, 1961; Davison et al., 1961) with a region of terminal redundancy, the ends of a population of molecules being randomly distributed over the genome (Streisinger et al., 1964, Séchaud et al., 1965).

There is now a great deal of evidence that recombination in phage T4 and other bacteriophages occurs through the formation of "internal heterozygotes" (Séchaud et al., 1965). An intermediate in recombination has been identified as a "joint molecule" in which parts of parental molecules are joined by hydrogen bonds (but not covalent bonds) in a heterozygous region. There is further evidence that such a joint molecule is transformed into a hybrid molecule in which the fragments originating from different parents are covalently bonded (Anraku and Tomizawa, 1965). The transformation

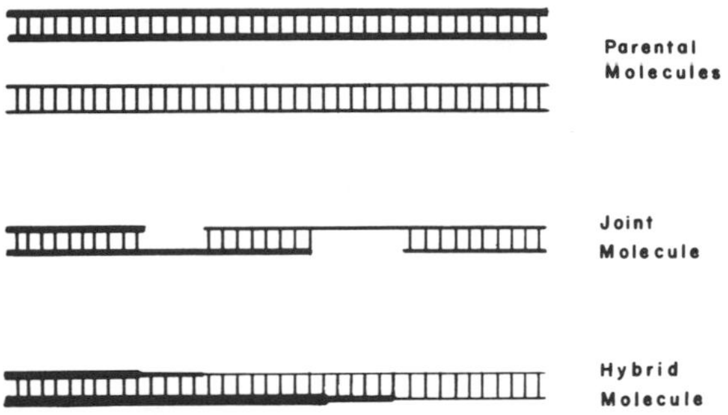

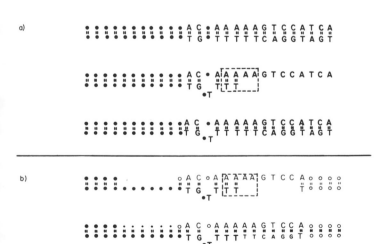

Parental
Molecules

Joint
Molecule

Hybrid
Molecule

FIGURE 4. Currently accepted scheme of recombination in phage T4. The heavy as compared to light lines represent the contributions from the two parental molecules, intermediate lines represent newly synthesized material, vertical lines represent hydrogen bonding.

from a joint to a hybrid molecule presumably occurs through new synthesis in which single-strand gaps are filled and eventually all nucleotides are covalently bonded (Meselson, 1964). The whole process is diagrammed in Fig. 4.

Frameshift mutations in phage T4 could occur either at the ends of DNA molecules, or else internally in heterozygous regions, by the general mechanism described in the previous section. At the end of a molecule an exonuclease is assumed to digest one of the two chains, the digested material being resynthesized later; the mutation in this case occurs during resynthesis and is a consequence of mispairing, as illustrated in Fig. 5a. In a heterozygous region the mutation is also a consequence of mispairing and occurs during the synthesis that transforms a joint molecule into a hybrid molecule, as illustrated in Fig. 5b.

Neither of these alternatives can be excluded at this time. Nevertheless the following results, some of them apparently contradictory, can best be accounted for if it is assumed that most frameshift mutations in phage T4 occur at ends of molecules.

Most newly arisen frameshift mutations are heterozygous (Drake, 1964), and the majority of these are also heterozygous for neighboring markers. The neighboring markers may be as far away as 10 recombination units, or 1% to 2% of the total genetic map length of phage T4 (Strigini, 1965). The length of the region of terminal redundancy is about 1% to 2% of the length of the genetic map of phage T4, whereas the regions of internal heterozygosity are considerably shorter (Berger, 1965). If it is assumed that each mutation is included within a heterozygous region that may also include markers situated at a distance of 1% to 2% of the map, these regions, because of their length, must represent regions of terminal redundancy. Newly arisen frameshift mutations found in mature particles of phage T4 thus probably occur within the region of terminal redundancy.

Frameshift mutation in phage T4 has been observed to occur both in the presence of fluorodeoxyuridine, where only limited DNA synthesis occurs (Drake, 1964), and in the presence of high concentrations of acridines that greatly decrease the frequency of recombinants (Lerman, pers. commun.). The occurrence of mutations near ends would not be expected to depend on net DNA synthesis or recombination. In contrast, the

FIGURE 5. (a) Origin of a frameshift mutation at the end of a molecule. Line 1 shows the normal end of a molecule, line 2 shows an end in which one chain has been digested by an exonuclease followed by mispairing, and line 3 shows the appearance of the molecule after resynthesis of the digested chain.

(b) Origin of a frameshift mutation in a heterozygous region. The lengths of the various regions of overlap are meant to be indicated schematically only. The contribution of the two parental DNA molecules to the heterozygote are distinguished by light vs. heavy print and newly synthesized material is indicated by smaller print. The heterozygote is shown as a joint molecule with a set of mispaired bases in the first line and, after synthesis, as a hybrid molecule in which a mutation has occurred, in the second line.

The possible origin of the mutation eJD11 (addition of a T) is illustrated in each case. The critical area for the mutation is outlined by a dashed square.

inhibition of recombination would be expected to decrease the frequency of frameshift mutations if these arise during the formation of internal heterozygotes.

It should be emphasized that ends of DNA molecules are expected to undergo recombination avidly in the vegetative pool of DNA. Thus, after a sojourn in the vegetative pool, a molecule carrying a frameshift mutation would be expected to be recombinant for neighboring markers, even though the mutation occurred independently of the subsequent recombinational event. This prediction needs to be tested.

Frameshift Mutation in Yeast

Magni (1963; Magni et al., 1964) has found that frameshift mutations in yeast occur during meiosis but not during mitosis, and that they are associated with genetic recombination. Magni (1963) as well as Lerman (1963) have suggested that these mutations occur through unequal crossing over.

Although the steps that occur in recombination in a higher organism such as yeast are not yet understood in detail, it has been suggested that breaks occur, and that rejoining in some respects resembles the formation of heterozygotes in phage T4 (Whitehouse, 1963). Either of the mechanisms illustrated in Fig. 5 could give rise to frameshift mutations in yeast and be compatible with Magni's observations. The mechanism illustrated in Fig. 5a could well be described as unequal crossing over.

We assume that the mispairing we describe can occur concomitantly with the synthesis of short stretches of a single strand of a DNA molecule. The fact that spontaneous frameshift mutations are rare in bacteria, or in other organisms during mitosis, suggests that this type of mispairing does not usually occur during normal replication.

The Nature of a Mutational Hot Spot

On the basis of the model we have presented it may be expected that the frequency of frameshift mutations at a given site would depend on the sequence of bases at that site. If these mutations occur through mispairing in regions of repeated bases (or base doublets) we would expect that the frequency of mutation would be higher in longer stretches of identical bases, since a longer stretch would increase the chances of mispairing.

The spontaneous mutant eJD11 was observed to revert with a frequency that is at least 100-fold higher than the frequency of the forward mutation at that site. The reversion occurs at the same site as the original mutation, as judged by genetic experiments. As can be seen (Table 1d), this mutation is compatible with the addition of an A to a region in the DNA that normally has a stretch of (most probably) five A residues (Table 2). Thus a spontaneous mutational "hot spot" is in this case probably due to a sequence of six identical bases.

The Role of Acridines in Mutagenesis

Acridines have been shown to be intercalated into the DNA molecule, increasing the distance between previously adjacent base pairs from 3.4 A to 6.8 A (Lerman 1963).

The models that have been presented for acridine mutagenesis have in general invoked the fact that each intercalated acridine stretches the molecule by the distance that is occupied by a single base pair. Lerman (1963) has suggested that two double-stranded DNA molecules are paired along a part of their length by some mechanism that assures the juxtaposition of identical base sequences of the two molecules, and that an exchange takes place distal to the locus of insertion of an acridine. This process would result in frameshift mutations in the recombinant molecules. An earlier proposal by Brenner et al. (1961) involved the insertion of an acridine into one, but not the other chain of a DNA molecule. In this case a base would be added (or deleted) if the insertion occurred during replication, or else during the synthesis that occurred in a heterozygous region during recombination.

While each of these models is compatible with our results, and cannot be rejected, neither is entirely satisfactory: Additional details would have to be provided to make Lerman's model compatible with the known role of heterozygotes in recombination and with the observation that acridine-induced mutations occur as heterozygotes (Drake, 1964), while the binding of acridines to a single (random coil) DNA chain, as required by Brenner's model, is now known to be very weak (Lerman, 1963).

We propose that acridines may be mutagenic not because they stretch the DNA molecule but rather because they stabilize it. It is known that intercalated acridines substantially increase the thermal denaturation temperature of DNA (Lerman, 1964). If acridines were intercalated between base pairs in the short regions of mispairing illustrated in Figs. 4 and 5, they would be expected to increase substantially the half life of the hydrogen bonding of the mispaired regions, and would thus increase the probability of synthesis occurring before the regions melt out.

Proflavine and similar acridines are highly mutagenic in phage T4, but have not been found to be mutagenic in several strains of bacteria. Acridines or acridine-like substances with half-mustard or polyamine side chains, in contrast, are highly mutagenic in bacteria and other organisms (Ames and Whitfield, this volume). We suggest that the mechanism of mutagenesis in bacteria may be very similar to the one proposed here for phage T4, except that the mispairing and new synthesis would occur at the site of a mutagen-induced break.

ACKNOWLEDGMENTS

We are grateful to F. H. C. Crick and F. W. Stahl for numerous discussions.

This work was supported by a grant (to G. S.) from the National Science Foundation, and by grants (to A. T.) from the Jane Coffin Memorial Fund for Medical Research, and the National Institutes of Health.

E. T. was supported by a U.S. Public Health Service Training Grant while at Eugene, and by a U.S. Public Health Service Post-Doctoral Fellowship while at Osaka.

REFERENCES

ANRAKU, N., and J. TOMIZAWA. 1965. Molecular mechanisms of recombination in bacteriophage. V. Two kinds of joining of parental DNA molecules. J. Mol. Biol. *12:* 805–815.

BERGER, H. 1965. Genetic analysis of T4D phage heterozygotes produced in the presence of 5-fluorodeoxyuridine. Genetics *52:* 729–746.

BRENNER, S., L. BARNETT, F. H. C. CRICK, and A. ORGEL. 1961. The theory of mutagenesis. J. Mol. Biol. *3:* 121–124.

CHAMPE, S. P., and S. BENZER. 1962. An active cistron fragment. J. Mol. Biol. *4:* 288–292.

DAVISON, P. F., D. FREIFELDER, R. HEDE, and C. LEVINTHAL. 1961. The structural unity of the DNA of T2 bacteriophage. Proc. Natl. Acad. Sci. *47:* 1123–1129.

DRAKE, J. W. 1964. Studies of the induction of mutations in bacteriophage T4 by ultraviolet irradiation and by proflavine. J. Cell. and Comp. Physiol. *64:* (Suppl. 1) 19–32.

CRICK, F. H. C., L. BARNETT, S. BRENNER, and R. J. WATTS-TOBIN. 1961. General nature of the genetic code for proteins. Nature *192:* 1227–1232.

LERMAN, L. S. 1963. The structure of the DNA-acridine complex. Proc. Natl. Acad. Sci. *49:* 94–102.

LERMAN, L. S. 1964. Acridine mutagens and DNA structure. J. Cell. and Comp. Physiol. *64:* (Suppl. 1) 1–18.

MAGNI, G. E. 1963. The origin of spontaneous mutations during meiosis. Proc. Natl. Acad. Sci. *50:* 975–980.

MAGNI, G. E., R. C. VON BORSTEL, and S. SORA. 1964. Mutagenic action during meiosis and antimutagenic action during mitosis by 5-aminoacridine in yeast. Mutation Research *1:* 227–230.

MESELSON, M. 1964. On the mechanism of genetic recombination between DNA molecules. J. Mol. Biol. *9:* 734–745.

NIRENBERG, M., P. LEDER, M. BERNFIELD, R. BRIMACOMBE, J. TRUPIN, F. ROTTMAN, and C. O'NEAL. 1965. RNA codewords and protein synthesis. VII. On the general nature of the RNA code. Proc. Natl. Acad. Sci. *53:* 1161–1168.

OKADA, Y., E. TERZAGHI, G. STREISINGER, J. EMRICH, M. INOUYE, and A. TSUGITA. 1966. A frame shift mutation involving the addition of two bases in the lysozyme gene of phage T4. Proc. Natl. Acad. Sci. *56:* in press.

RUBENSTEIN, I., C. A. THOMAS, JR., and A. D. HERSHEY. 1961. The molecular weights of T2 bacteriophage DNA and its first and second breakage products. Proc. Natl. Acad. Sci. *47:* 1113–1122.

SARABHAI, A. S., A. O. W. STRETTON, and S. BRENNER. 1964. Co-linearity of the gene with the polypeptide chain. Nature *201:* 13–17.

SÉCHAUD, J., G. STREISINGER, J. EMRICH, J. NEWTON, H. LANFORD, H. REINHOLD, and M. M. STAHL. 1965, Chromosome structure in phage T4. II. Terminal redundancy and heterozygosis. Proc. Natl. Acad. Sci. *54:* 1333–1339.

STREISINGER, G., R. S. EDGAR, and G. HARRAR-DENHARDT. 1964. Chromosome structure in phage T4, I. Circularity of the linkage map. Proc. Natl. Acad. Sci. *51:* 775–779.

STRIGINI, P. 1965. On the mechanism of spontaneous reversion and genetic recombination in bacteriophage T4. Genetics *52:* 759–776.

TERZAGHI, E., Y. OKADA, G. STREISINGER, J. EMRICH, M. INOUYE, and A. TSUGITA. 1966. Change of a sequence of amino acids in phage T4 lysozyme by acridine-induced mutations. Proc. Natl. Acad. Sci. *56:* 500–507.

WHITEHOUSE, H. L. K. 1963. A theory of crossing over by means of hybrid deoxyribonucleic acid. Nature *199:* 1034–1040.

YANOFSKY, C., E. C. COX, and V. HORN. 1966. The unusual mutagenic specificity of an *E. coli* mutator gene. Proc. Natl. Acad. Sci. *55:* 274–281.

26

Reprinted from *Cold Spring Harbor Symp. Quant. Biol.*, **31**, 221–225 (1966)

Frameshift Mutagenesis in Salmonella

BRUCE N. AMES AND HARVEY J. WHITFIELD, JR.

Laboratory of Molecular Biology, National Institute of Arthritis and Metabolic Diseases,
National Institutes of Health, Bethesda, Maryland

In bacteriophage, acridine mutagens such as proflavine add or delete nucleotides from the DNA. If the addition or deletion of nucleotides is not a multiple of 3 bases, the translation of the genetic message which occurs in unpunctuated triplets is shifted out of frame (Crick, Barnett, Brenner, and Watts-Tobin, 1961). These mutations are called frameshift mutations. In contrast to the results with bacteriophage systems, attempts to mutate bacteria with proflavine or similar acridines have met with little success (Orgel, 1965).

This paper describes the ICR mutagens, a group of new acridine-like compounds which are powerful mutagens in bacteria, and presents evidence suggesting that these compounds add and/or delete nucleotides from DNA.

These ICR compounds (Fig. 1) were synthesized as anti-tumor agents at the Institute for Cancer Research (hence ICR) by Dr. H. J. Creech and his associates. The first of these compounds shown to be a mutagen, ICR 170 (previously called ICR 100) (Fig. 1), is a strong mutagen for Drosophila (Carlson and Oster, 1962; Snyder and Oster, 1964; Southin, 1966), and also for Neurospora (Brockman and Goben, 1965). In the latter case, the authors concluded that, because of the noncomplementing nature of the mutants made with ICR 170, the mutations were either of the frameshift type or were base-pair substitutions giving rise to nonsense mutations. Further work on the compound in Neurospora was done by Dr. H. V. Malling (pers. commun.), who concluded that both the acridine ring and the alkylating chain were necessary for activity.

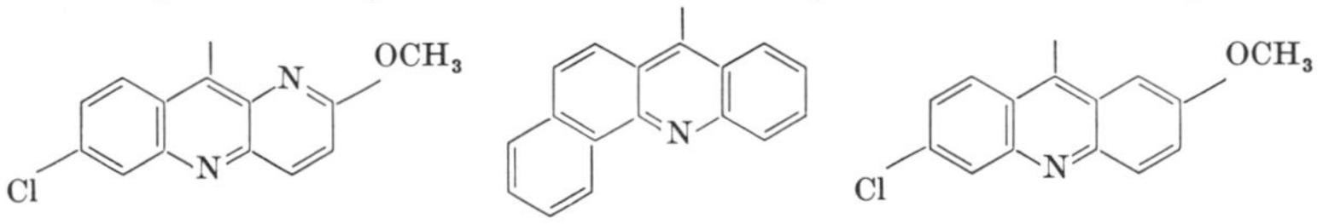

Side Chain	Aza-quinacrine ring	Benzacridine ring	Quinacrine ring
—NHCH$_2$CH$_2$CH$_2$NHCH$_2$CH$_2$Cl	ICR 372 300	ICR 370 375	ICR 191 400
—NHCH$_2$CH$_2$NHCH$_2$CH$_2$Cl	ICR 364 24	ICR 312 56	ICR 171 19
—NHCH$_2$CH$_2$CH$_2$N(CH$_2$CH$_3$)(CH$_2$CH$_2$Cl)	ICR 340 151	ICR 292 63	ICR 170 8
—NHCH$_2$CH$_2$CH$_2$NHCH$_2$CH$_2$OH	ICR 372-OH 37	ICR 370-OH +	ICR 191-OH 0
—NHCH$_2$CH$_2$NHCH$_2$CH$_2$OH	ICR 364-OH 15		
—NHCH$_2$CH$_2$NHCH$_2$CH$_2$NH$_2$	ICR 364-NH$_2$ +		
—NHCHCH$_3$CH$_2$CH$_2$CH$_2$N(CH$_2$CH$_3$)—CH$_2$CH$_3$			Quinacrine +

FIGURE 1. The number of revertants produced by 5 μg of different ICR mutagens on frameshift mutant *C207*. Several experiments have been averaged and the spontaneous blank has been subtracted in each case. The technique is described in the legend to Fig. 2. In the case of ICR 372, 370, and 191, 1 μg has been used and the results have been multiplied by five. Those compounds that were negative at 5 μg were tested at higher concentrations by using about 0.5 mg of crystals directly on the plate: a + indicates a positive response.

OPTIMUM CHEMICAL STRUCTURE FOR FRAMESHIFT MUTAGENESIS

A class of presumed frameshift mutants not mutable by standard mutagens (Whitfield, Jr., Martin, and Ames, 1966) was found to be mutable by ICR 170. (The properties of these mutants will be discussed in the next section.) We have examined over 60 compounds related to ICR 170 (Peck, O'Connell, and Creech, 1966; Preston, Peck, Breuninger, Miller, and Creech, 1964; Peck, Breuninger, Miller, and Creech, 1964) in a study of the optimum chemical structure for inducing reversions of these frameshift mutants. A more detailed study of structure versus mutagenicity will be published separately in collaboration with these chemists at the Institute for Cancer Research.

ICR 170 (Fig. 1) is a monofunctional mustard, closely related to quinacrine (atabrine), one of the standard antimalarial drugs. Figure 1 shows that

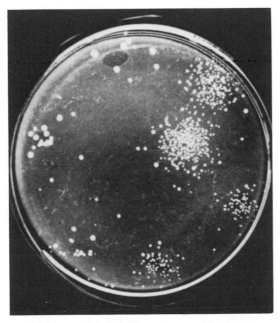

FIGURE 2. Revertant colonies on a minimal plate with a lawn of frameshift mutant *C207* (0.1 ml of a full grown culture) to which 5 μg of each of the indicated ICR mutagens had been added (as 5 μl of a 1 mg/ml aqueous solution). The plates were incubated 2 days at 37°. ICR 170 is at 12 o'clock and ICR 191, ICR 312, ICR 292, ICR 217, and ICR 171 follow (clockwise). The cluster of colonies slightly off center is from 5 μg of ICR 191 (from a solution that had been stored in the freezer for a month). All compounds were hydrochlorides and dissolved readily in sterile distilled water, and the solutions were checked to show they were essentially sterile. All operations were done in dim light, as the acridines are known to be very light-sensitive.

A trace of histidine (0.2 μmole) is added to the plate so that the background lawn can grow slightly and so that any inhibition of the compounds can be seen. A control plate with no mutagen gave about 15 revertant colonies.

the aza-quinacrine, benzacridine, and quinacrine rings are all quite effective, and that the best of the side chains is the one in ICR 372, 370, and 191. Of the compounds available so far, ICR 372, ICR 370, and ICR 191 are about equally effective. Of the nonalkylating compounds, ICR 364-OH and ICR 372-OH are the most effective of those tested. None of the compounds are very inhibitory to the bacteria and all give more revertants if used at higher concentrations.

The action of a number of these compounds as mutagens is shown in Fig. 2.

EVIDENCE THAT THE ICR COMPOUNDS ADD OR DELETE NUCLEOTIDES

1. The Presumed Frameshift Class of Polar *C* Gene Mutants

A group of 65 randomly selected mutants in the *C* gene of the histidine operon in *Salmonella typhimurium* have been separated into nonsense (*amber* or *ochre*), missense, and frameshift classes (Whitfield, Jr. et al., 1966). The major criteria for this classification were mutagenesis, suppression data from *amber* and *ochre* suppressors, and phenotypic curing. Among the *C* mutants that did not fit into the *amber* or *ochre* classes (as defined by suppressors), some were reverted by NG (N-methyl-N′-nitro-N-nitrosoguanidine) and some were not. All of the mutants reverted by NG were nonpolar and were classified as missense mutants. All of the mutants *not reverted* by NG (16 mutants) were polar (Martin, Silbert, Smith, and Whitfield, Jr., 1966) and were classified (Whitfield, Jr. et al., 1966) as frameshift mutants on the basis of the following evidence:

(a) *Response to mutagens.* The presumed frameshift mutants were not induced to revert by alkylating or base-substituting mutagens. None of the frameshift mutants were induced to revert by NG, diethylsulfate, 2-aminopurine, β-propiolactone, or nitrogen mustard. In contrast, NG induced reversion of all mutants of the *amber*, *ochre*, and missense classes. Typical missense and frameshift revertant plates using NG as a mutagen are shown in Fig. 3.

As would be expected for point addition or deletion mutants, the spontaneous reversion frequencies of the frameshift mutants are very low.

ICR 191 (Whitfield, Jr. et al., 1966) and also ICR 364-OH (a nonalkylating aza-quinacrine) reverted 11 out of 14 (79%) of the frameshift mutants. Each mutant is reverted to a different degree by a particular ICR compound, but in general the effectiveness of the different ICR mutagens parallels the results shown in Fig. 1.

Two additional mutants not revertible by NG

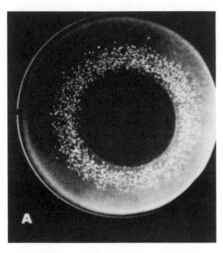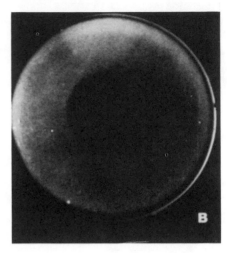

FIGURE 3. Action of NG on a missense mutant *C464* (Fig. 3a) or frameshift mutant *C122* (Fig. 3b). The method is the same as in Fig. 2 except that a small crystal of NG was added to the center of a plate. The zone of inhibition by the NG can be seen.

or by the ICR compounds do not revert spontaneously. These mutants could contain small deletions of a sufficient number of bases to make reversion unlikely.

(b) *Origin.* Small and large deletions are known to be frequent among spontaneous mutants (Crick et al., 1961), X-ray-induced mutants (Freese, 1963) and nitrous acid-induced mutants (Tessman, 1962). The *C*-gene frameshift mutants were obtained in all three of these ways. In contrast, none of the frameshift mutants were produced by the base-substituting agents, 2-aminopurine or 5-bromouracil, although most of the *amber* and *ochre* mutants and some missense mutants were obtained by using these mutagens.

(c) *Phenotypic curing.* As expected, none of the frameshift mutants were phenotypically cured by kanamycin, neomycin, or streptomycin (Gorini and Kataja, 1965). All of the nonsense mutants except one, and about one-third of the missense mutants, were phenotypically cured by at least one of these agents.

(d) *Isolation of suppressors.* As expected, none of 106 spontaneous revertants examined from the frameshift mutants were found to contain external suppressors. This is in marked contrast to the results obtained for revertants of the nonsense and missense mutants in which external suppressors were very frequent.

(e) *Suppression by suppressors from the nonsense or missense class.* None of the frameshift mutants was suppressed by any *amber*, *ochre*, or missense suppressor.

(f) *Polarity.* All of the frameshift mutants are polar. Presumably, frameshift mutants are polar because they give rise to *amber* and *ochre* codons (which cause polarity) in the new frame (Martin, Whitfield, Jr., Berkowitz, and Voll, this volume). Missense mutants are not polar (Whitfield, Jr. et al., 1966).

(g) *Lack of altered proteins.* As would be expected of frameshift mutants, none of these mutants is leaky and none of those examined produces cross-reacting protein (CRM).

2. PRESUMED FRAMESHIFT MUTANTS REVERTIBLE BY ICR

Dr G. R. Fink has developed a selection technique for polar histidine-requiring mutants, starting with an operator constitutive mutant (unpubl. results). He has isolated 54 presumed frameshift mutants of spontaneous origin that are not revertible by NG and have some spontaneous rate of reversion. ICR 191 (and also in every case ICR 364-OH) reverted 49 (or 91%) of these mutants.

Dr. J. Loper (pers. commun.) has analyzed the *D* gene of the histidine operon by the techniques of Whitfield, Jr. et al., (1966) and has obtained similar results. Almost all NG nonrevertible mutants are revertible by ICR 191.

3. PROPERTIES OF REVERTANTS OF FRAMESHIFT MUTANTS

Crick et al., (1961) have shown that mutants may arise by the deletion (or addition) of one or two bases from the DNA. As the genetic code is an unpunctuated triplet code, such point deletions result in a shift of the reading frame of the mRNA, and thus give rise to the synthesis of a polypeptide chain of radically altered amino acid composition.

By a second mutational event within the same cistron, function can be restored to a frameshift mutant. This second event may be a base addition (or deletion) to a pre-existing base deletion (or addition) mutation near, *but not necessarily at*, the site of the first mutation. Such a + — type contains an "internal suppressor" which restores the proper reading frame, but will have a short segment of protein with an altered amino acid sequence (between the point addition and deletion) (Streisinger, 1965; Streisinger et al., this volume).

Most of the revertants of *C*-gene frameshift mutants, both those that appear spontaneously and those induced by the ICR compounds, have poor growth rates and are not true revertants, and presumably are comparable to these + — types of Crick et al., (1961). For example, Martin, Silbert, Smith, and Whitfield, Jr. (1966) have examined 34 spontaneous revertants of one of these frameshift-type mutants (*C202*). Of these mutants, 24 had a *C*-enzyme that was less active than the wild type, as judged by a decreased growth rate or by physiological derepression on minimal medium.

In all of these presumed + — types in the *C*-gene, along with the restoration of frame due to the reversion, the polarity was abolished, as would be expected if frameshift mutants are polar because they give rise to nonsense codons beyond the point of mutation (Whitfield, Jr. et al., 1966).

Because revertants of frameshifts are mostly + — types, the percentage of the frameshifts revertible by ICR depends on the stringency of the protein of the particular gene involved, and also on how much the enzyme can be derepressed. In different genes of the histidine operon the percentage of the frameshift types revertible by the ICR mutagens varies considerably. It is about 80% (Whitfield, Jr. et al., 1966) in the *C*-gene and 90% in the *D*-gene (Dr. J. Loper, pers. commun.). None of the 17 NG⁻ CRM⁻ mutants in the *A*-gene were revertible by ICR 91 (Drs. M. Margolies and R. G. Goldberger, pers. commun.), though many of these mutants could have extended deletions. The *A* protein may be more stringent than the *C* or *D* proteins.

4. More Direct Evidence that the ICR Mutagens Cause Additions and Deletions

Studies to show that the nonpolar revertants of the presumed frameshift types in the *C*-gene are true + — types, by separating out the two polar histidine-requiring mutants, are in progress (R. G. Martin, unpubl.). These results give more direct evidence that this class of mutants is indeed a frameshift class. The indirect evidence, however, by itself seems reasonably convincing. The case is also strengthened by the following more direct evidence. Dr. Ronald Sederoff (pers. commun.)

has tested the ICR 170 against some known addition and deletion frameshift mutations in the *r*II bacteriophage system (Crick et al., 1961) and found them to be potent mutagens in reverting these strains. Berger, Brammar, and Yanofsky (in prep.) have found that the ICR 191 reverts a mutant of *E. coli* known to have a frameshift mutation, as determined from the protein chemistry of spontaneous revertants.

SPECIFICITY OF THE ICR MUTAGENS

We suspect that even though the ICR 191 has an alkylating side chain it is not acting by causing base pair substitutions, but solely by causing additions and/or deletions. Most of our testing, however, has been done at low concentrations. The ICR compounds such as 364-OH and 372-OH, which are not alkylating agents, presumably would specifically cause additions and/or deletions at all concentrations. The following evidence supports this view.

(1) Out of 22 *amber* mutants in the *hisC* gene which are revertible by a variety of agents causing base pair substitutions (e.g. NG, diethylsulfate, 2-aminopurine, etc.) none is revertible by ICR 191, even at the high concentrations around some crystals of the material (Whitfield, Jr. et al. 1966). A number of the other compounds tested, of less potency than ICR 191 (e.g. the chloroquine half mustard, ICR 180), did mutate, when at high concentration, a number of the *amber* mutants in addition to mutating the same set of frameshift mutants.

(2) Whitfield, Jr. et al. (1966) found that 5 *ochre* and 19 missense mutants which also are revertible by a variety of alkylating agents were not revertible by ICR 191.

It would also be of interest to study the reversion by ICR and NG of ICR-caused mutants. Such a study is in progress by N. S. Oeschger.

Mutation of nonsense suppressors by ICR. Whitfield, Jr. et al. (1966) did report, however, that 3 mutants (1 *ochre* and 2 missense mutants) did revert with ICR 191. Dr. Fink has also isolated several *amber* mutants (in an Oᶜ-*amber* double mutant) that are reverted by ICR 191 and ICR 364-OH. We believe that one explanation is that the ICR is causing an addition or deletion in a transfer RNA which can act as a suppressor for the nonsense or missense codon. In support of this, it has been possible to show that the ICR compounds are mutating this *amber* mutant (*hisOᶜ1242 hisC2132*) at a suppressor locus. If the ICR compounds cause only additions and deletions, then mutagenesis of known *amber* and *ochre* mutations with ICR should be useful in studies on *amber* and *ochre* suppressors. Perhaps it is possible to add a fourth base to a 3 base anticodon in a transfer RNA, for example, so as to cause some ambiguity and consequently suppression.

Magni, von Borstel, and Steinberg (1966) have reported that a super-suppressor in yeast (the yeast *amber*-type suppressors) can be caused by additions and deletions, and they suggested that this may be in a gene for a transfer RNA.

RECOMBINATION AND FRAMESHIFT MUTAGENESIS

Several studies suggest that the acridines cause mutations only during recombination in phage (Streisinger et al., this volume) and in yeast (Magni, 1964; Magni and Puglisi, this volume). Recent studies in bacteria indicate that proflavine is mutagenic if the bacteria are treated under conditions of recombination, i.e. when a transduction has to occur (Dr. H. Boyer, pers. commun.).

It is not clear whether the ICR compounds are also working during some type of recombination, for example, during operation of a repair system. It is conceivable that the greater effectiveness of the alkylating ICR compounds is because of a stimulation of some repair system, though it should be pointed out that the nonalkylating ICR compounds are also quite active. A number of experiments to show mutagenic activity of proflavine in Salmonella frameshift mutants by combining it with various alkylating agents have been negative.

ANTIMALARIALS AND MUTAGENESIS IN THE HUMAN POPULATION

Quinacrine (atabrine) itself is a weak mutagen in reverting frameshift mutant *C207* (Fig. 1). This raises the possibility that the standard antimalarials chloroquine, quinine, and quinacrine, which are known to bind to DNA strongly, are causing frameshift mutations in the human population.

ACKNOWLEDGMENTS

We would like to thank Dr. H. J. Creech, of the Institute for Cancer Research (Philadelphia), for his generous gifts of compounds and Dr. Laura Stewart for calling our attention to these ICR compounds. We are indebted to Drs. P. E. Hart-man, R. G. Martin, G. R. Fink, and D. B. Berkowitz for helpful discussions.

REFERENCES

BROCKMAN, H. E. and W. GOBEN. 1965. Mutagenicity of a monofunctional alkylating agent derivative of acridine in Neurospora. Science *147:* 750–751.

CARLSON, E. A. and I. I. OSTER. 1962. Comparative mutagenesis of the dumpy locus in *Drosophila melanogaster*. Genetics *47:* 561–576.

CRICK, F. H. C., L. BARNETT, S. BRENNER, and R. J. WATTS-TOBIN. 1961. General nature of the genetic code for proteins. Nature *192:* 1227–1232.

FREESE, E. 1963. Molecular mechanism of mutations, p. 207–269. *In* Molecular Genetics. J. H. Taylor, ed., Academic Press, New York.

GORINI, L. and E. KATAJA. 1965. Suppression activated by streptomycin and related antibiotics in drug sensitive strains. Biochem. Biophys. Res. Commun. *18:* 656–663.

MAGNI, G. E. 1964. Origin and nature of spontaneous mutations in meiotic organisms. J. Cell. Comp. Physiol. *64:* Sup. 1, 165–172.

MAGNI, G. E., R. C. von BORSTEL, and C. M. STEINBERG. 1966. Supersuppressors as addition-deletion mutations. J. Mol. Biol. *16:* 568–570.

MARTIN, R. G., D. F. SILBERT, D. W. E. SMITH, and H. J. WHITFIELD, JR. 1966. Polarity in the histidine operon. J. Mol. Biol., in press.

ORGEL, L. E. 1965. The chemical basis of mutation, p. 290–346. *In* F. F. Nord, [ed.] Advances in enzymology. Interscience Publishers, New York.

PECK, R. M., E. R. BREUNINGER, A. J. MILLER, and H. J. CREECH. 1964. Acridine and quinoline analogs of nitrogen mustard with amide side chains. J. Medicinal Chem., *7:* 480–482.

PECK, R. M., A. P. O'CONNELL, and H. J. CREECH. 1966. Heterocyclic derivatives of 2-chloroethyl sulfide with antitumor activity. J. Medicinal Chem., *9:* 217–221.

PRESTON, R. K., R. M. PECK, E. R. BREUNINGER, A. J. MILLER, and H. J. CREECH. 1964. Further investigation of heterocyclic alkylating agents. J. Medicinal Chem., *7:* 471–480.

SOUTHIN, J. L. 1966. An analysis of eight classes of somatic and gonadal mutation at the dumpy locus in *Drosophila melanogaster*. Mutation Res., *3:* 54–65.

STREISINGER, G. 1965. Gene-protein relationships, p. 1–7. *In* National Cancer Institute Monograph 18. J. I. Valencia, R. F. Grill, and R. M. Valencia, ed., Nat. Cancer Inst., Bethesda.

SNYDER, L. A. and I. I. OSTER. 1964. A comparison of genetic changes induced by a monofunctional and a polyfunctional alkylating agent in *Drosophila melanogaster*. Mutation Res. *1:* 437–445.

TESSMAN, I. 1962. The induction of large deletions by nitrous acid. J. Mol. Biol. *5:* 442–445.

WHITFIELD, H. J., JR., R. G. MARTIN, and B. N. AMES. 1966. Classification of aminotransferase (*C* gene) mutants in the histidine operon. J. Mol. Biol., in press.

Part IV

MISREPAIR
MUTAGENESIS

Editors' Comments
on Papers 27 Through 29

MISREPAIR MUTAGENESIS

The preceding papers provide much of the flavor of the early studies on specifically induced mispairing during DNA synthesis. A number of other important mutagens, however, had failed to yield up the secrets of their mutagenic mechanisms; highly visible among these was ultraviolet (UV) irradiation, which had long been known to be a potent mutagen for a wide range of organisms, encompassing *Drosophila*, bacteria, and viruses. The fact that UV is far less energetic than X-rays had led some workers as early as the 1930s to hope that it would produce more localized damages, specifically gene alterations instead of gene deficiencies. However, even the discovery in the late 1950s of the most remarkable UV-induced photoproduct of DNA, the pyrimidine dimer (Beukers and Berends, 1960), failed to reveal the mutagenic mechanism of UV irradiation. The crucial observation came instead from an observation by Evelyn Witkin, a long-time student of ultraviolet mutagenesis, that certain repair-deficient mutants of *Escherichia coli* were simultaneously *more* sensitive to UV-induced inactivation and *less* sensitive to UV-induced mutagenesis. Her first observations were reported in a review article (Witkin, 1967), but we have chosen two papers from 1969 for this collection (partly because the 1967 paper is so long). Papers 27 and 28 show specifically that, although *excision* repair systems appear to be at least as accurate as is normal DNA replication itself, certain *postreplication* repair systems appear to be very inaccurate.

Misrepair mutagenesis was quickly shown to be stimulated not only by irradiation, but also by numerous chemical agents, as first demonstrated by Kondo (Paper 29). (Kondo's results are described in much greater detail in another long paper; see Kondo et al., 1970. However, the essential conclusions appear in Paper 29.) Because these error-prone repair systems operate in genetic recombination as well as in the repair of numerous types of DNA damage, early descriptions of misrepair tended to emphasize the possibility that these repair systems possessed elements in common with recombination itself. This supposition was supported by studies, quoted in Paper 29, which indicated that primary DNA lesions often cause the appearance of gaps in progeny DNA strands, and that these gaps are eventually filled by DNA strand-transfer processes. At present, however, it is not at all clear whether strand-transfer processes are in fact error prone, since misrepair mutagenesis also appears to occur among the single-stranded DNA viruses (which lack the required donor strand). Instead, an alternative gap-filling process probably operates, one that synthesizes DNA past the parental strand damage when the strand transfer process fails to occur.

Two important aspects of misrepair mutagenesis are not revealed by Papers 27 through 29. First, the mutagenic specificity of misrepair mutagenesis is very wide, so that base pair substitutions, frameshifts, and even deletions are induced (Drake, 1963; Green and Drake, 1974; J. W. Drake, unpublished results). Second, misrepair mutagenesis is a very generally occurring mechanism, and is observed in most (but not all) systems where it has been sought (Green and Drake, 1974), including at least some eucaryotes (Prakash, 1975).

REFERENCES

Beukers, R., and W. Berends. 1960. Isolation and identification of the irradiation product of thymine. *Biochim. Biophys. Acta, 41:* 550.

Drake, J. W. 1963. Properties of ultraviolet-induced *r*II mutants of bacteriophage T4. *J. Mol. Biol., 6:* 268.

Green, R. R., and J. W. Drake. 1974. Misrepair mutagenesis in bacteriophage T4. *Genetics, 78:* 81.

Kondo, S., H. Ichikawa, K. Iwo, and T. Kato. 1970. Base-change mutagenesis and prophage induction in strains of *Escherichia coli* with different DNA repair capacities. *Genetics, 66:* 187.

Prakash, L. 1975. Lack of chemically induced mutation in repair-deficient mutants of yeast. *Genetics, 78:* 1101.

Witkin, E. M. 1967. Mutation-proof and mutation-prone modes of survival in derivatives of *Escherichia coli* B differing in sensitivity to ultraviolet light. *Brookhaven Symp. Biol., 20:* 17.

27

Reprinted from *Proc. XII Intern. Congr. Genetics,* **3,** 225–245 (1969)

THE ROLE OF DNA REPAIR AND RECOMBINATION
IN MUTAGENESIS

EVELYN M. WITKIN

State University of New York, Downstate Medical Center
Brooklyn, New York, U.S.A.

During the last five years, the study of DNA repair has taken new directions. Not only do repair enzymes monitor the DNA for damaged bases, cross-links and single strand breaks, a great many of which can be successfully repaired, but some of them appear to participate also in genetic recombination, and perhaps even in normal replication of DNA (for recent reviews, see Haynes, Baker and Jones 1967, Hanawalt 1967, Setlow 1968). This report considers the role of DNA repair in still another major biological process, mutagenesis, especially as it affects the induction of mutations in bacteria by ultraviolet light.

Exposure of *Escherichia coli* to ultraviolet light (UV) results in the production in its DNA of pyrimidine dimers (Wacker, Dellweg and Jacherts 1962), photoproducts in which neighboring pyrimidines in the same strand of DNA are covalently linked via carbon to carbon bonds. Dimers of thymine form most readily, but cytosine-thymine and cytosine-cytosine dimers are also produced (Wacker 1963, Smith 1963, Setlow, Carrier and Bollum 1965a,b). All strains of *E. coli,* whatever their UV sensitivity, are about equal in their susceptibility to the production of pyrimidine dimers in their DNA, approximately five dimers being formed for each dose increment of one erg per mm^2 (Swenson and Setlow 1966). Strains of *E. coli* may differ by a factor of more than a thousand in their sensitivity to UV, but only because they differ either in the ability to *repair* pyrimidine dimers, or in the ability to *tolerate* unrepaired dimers, or both. A mutation that eliminates the ability to repair pyrimidine dimers can increase UV sensitivity about twentyfold. Other mutations, affecting in one way or another the ability to tolerate unrepaired pyrimidine dimers (*i.e.,* to divide and form colonies despite their presence in the DNA) can further increase sensitivity from two to one hundred-fold, the most sensitive strains of all being unable to form colonies of even one or two dimers are formed in the DNA, since they can neither repair nor tolerate even so few (Howard-Flanders and Boyce 1966). The most resistant strains can survive doses that produce several thousand pyrimidine dimers in their DNA, repairing most of them, and tolerating as many as one hundred that have not been repaired (Witkin 1967).

Two kinds of repair mechanisms have been identified in *E. coli* whereby pyrimidine dimers are eliminated from the DNA: photoreactivation and excision repair. In photoreactivation (Kelner 1949), which takes place only in the presence of visible light, pyrimidine dimers are split, or monomerized, *in situ,* (see Figure 1) by a "photoreactivating enzyme" that is activated by light of wave lengths around 4,000Å (Wulff and Rupert 1962). Excision repair, which takes place in the dark, is a more complex process, probably involving the sequential action of several enzymes, and is not yet understood in detail. According to one model, the "cut-and-patch" model (see Figure 2),

PHOTOREACTIVATION

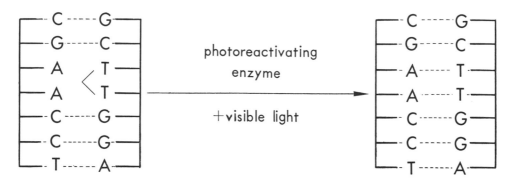

Figure 1. Photoreactivation, or splitting of pyrimidine dimers produced by UV in DNA by photoreactivating enzyme in the presence of visible light.

EXCISION REPAIR

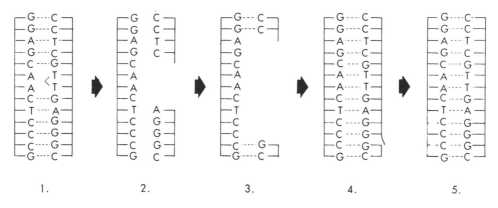

Figure 2. Excision repair of pyrimidine dimers produced by UV in the DNA of *E. coli* ("cut-and-patch" model). 1) Thymine dimer in DNA; 2) section of DNA strand bearing dimer excised; 3) gap enlarged by exonuclease degradation; 4) gap filled by repair synthesis; 5) single-strand break sealed by DNA ligase.

a short, single-stranded segment of DNA, bearing the pyrimidine dimer and a small number of additional bases, is first excised from the DNA, presumably by the action of specific endonucleases. The gap thus produced in the DNA may be enlarged somewhat by degradative enzymes, and is then "patched" by repair synthesis, in which DNA polymerase (or a similar enzyme) utilizes the intact information on the opposite strand as a template for accurate restoration of the original sequence. The final step is presumably the sealing of the sugar-phosphate backbone by a DNA ligase (Gellert 1967, Olivera and Lehman 1967, Gefter, Becker and Hurwitz 1967, Weiss and Richardson 1967). Other models of excision repair differ in some respects, but there is good evidence for both the removal of oligonucleotides bearing the pyrimidine dimers (Setlow and Carrier 1964, Boyce and Howard-Flanders 1964b), and for repair

259

synthesis (Pettijohn and Hanawalt 1964).

Photoreactivation is specific for pyrimidine dimers, no other known kind of DNA damage being repairable by photoreactivating enzyme and visible light (Setlow 1966). Excision repair is less specific, since there is evidence that DNA damage caused by nitrogen mustard (Bridges and Munson 1966) and mitocycin C (Boyce and Howard-Flanders 1964a), both DNA cross-linking agents, and by 4-nitroquinoline 1-oxide (Kondo and Kato 1966b), a powerful carcinogen, are repaired by mechanisms sharing at least some steps with the one that removes pyrimidine dimers from the DNA. It is possible that excision enzymes recognize distortions in the helical structure of DNA, rather than specific kinds of lesions producing such distortions (Hanawalt and Haynes 1965).

The specificity of enzymatic photoreactivation as a repair mechanism that seems to act only on pyrimidine dimers has been of great value in assessing the extent to which pyrimidine dimers contribute to the lethal and mutagenic effects of ultraviolet light. The demonstration that a UV effect is reversed by visible light does not, in itself, establish that pyrimidine dimers are responsible for the effect, since there are "indirect" mechanisms of photoreversal that have nothing to do with the splitting of pyrimidine dimers (Witkin, Sicurella and Bennett 1963, Jagger and Stafford 1965). A rigorous test for the involvement of pyrimidine dimers in UV killing or mutation induction is to compare the photoreversibility of the effect in a strain which possesses the photoreactivating enzyme (a Phr$^+$ strain) with that in a Phr$^-$ mutant strain lacking this enzyme (Harm and Hillebrandt 1962). Any effect of UV that is photoreversible in the Phr$^+$ strain, but not in the Phr$^-$ strain, must depend upon the more or less persistent presence in the DNA of pyrimidine dimers. By this criterion, essentially all of the killing in *E. coli* B/r caused by UV doses up to about 200 ergs per mm^2, and most of the killing caused by doses up to about 600 ergs per mm^2 must be due to pyrimidine dimers (Witkin, Sicurella and Bennett 1963). By the same criterion, at least 90% of the UV-induced mutations to streptomycin-resistance obtained, in the dark, at doses up to about 1,000 ergs per mm^2, and at least 90% of the UV-induced suppressor mutations obtained at doses below 100 ergs per mm^2 in some strains of *E. coli* do not occur unless pyrimidine dimers remain in the DNA for some time after irradiation (Witkin, Sicurella and Bennett 1963, Witkin 1966a, Kondo and Kato 1966a). Substantial fractions of various other kinds of UV-induced mutations have also been shown to depend upon pyrimidine dimers (Kaplan 1963), and there is now little doubt that the pyrimidine dimer is a major cause of both lethal and mutagenic effects of UV in *E. coli*, especially at low doses. Pyrimidine dimers are not excluded as the possible causes of UV effects that are *not* photoreversible by the photoreactivating enzyme. Some pyrimidine dimers may not be split because of their location (*i.e.*, in temporarily single-stranded regions of the DNA, or in regions that are for some other reason protected against the action of repair enzymes), or because the supply of photoreactivating enzyme is limiting. The possibility remains open, therefore, that virtually all of the lethal and mutagenic effects of low doses of UV in *E. coli* may be due to pyridimine dimers, the enzymatically photoreversible fraction indicating only a minimum estimate.

THE ACCURACY OF PHOTOREACTIVATION

Since the splitting of a pyrimidine dimer should restore the pre-irradiation structure of the DNA, enzymatic photoreacitvation should be an accurate repair mechanism that does not, in itself, introduce errors into the DNA. This expectation is supported by the great reduction in mutation frequencies observed after photoreactivation, *e.g.*, to about 10% of the dark yield in the case of induced streptomycin-resistance. However, since the nonphotoreactivable fraction could include mutations caused by the repair mechanism itself, the data permit us only to state that if the splitting of a pyrimidine dimer can cause a mutation, an unsplit dimer remaining in the DNA is at least ten times more likely to cause one.

THE ACCURACY OF EXCISION REPAIR

In the absence of visible light, *E. coli* can repair pyrimidine dimers, as far as we know, only by excision repair. UV-induced mutations caused by unsplit pyrimidine dimers must therefore come about either as errors introduced into the DNA in the course of excision repair, or as errors due to the continued presence of unrepaired pyridimine dimers remaining in the DNA, after excision repair (if any) is completed. If most UV-induced mutations are due to the inaccuracy of excision repair, then strains incapable of performing excision repair should show greatly decreased mutation frequencies after UV. On the other hand, if excision repair is accurate, and UV-induced mutations are caused by unrepaired dimers, strains lacking excision repair should show much higher yields of induced mutations, when compared at the same dose with strains capable of removing most of the dimers by excision. These expectations are diagrammed in Figure 3. Comparisons of otherwise isogenic strains, differ-

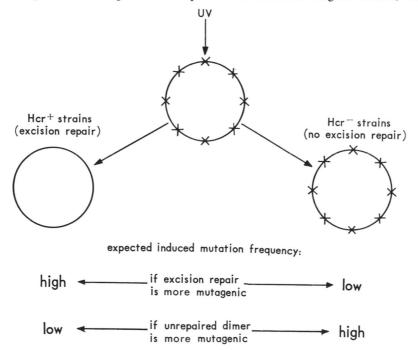

Figure 3. Expected yields of UV-induced mutations in Hcr⁺ and Hcr⁻ strains if excision repair is more mutagenic than unrepaired pyrimidine dimer and *vice versa*. Pyrimidine dimers represented as crosses.

ing only in their ability to excise pyrimidine dimers from their DNA, and from the DNA of irradiated bacteriophages with which they are infected, have yielded an unequivocal answer. As illustrated in Figure 4, UV-sensitive, excision-defective (Hcr⁻ strains produce large yields of UV-induced mutations at doses far below those neces-

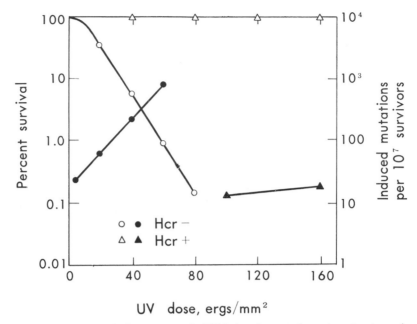

Figure 4. Survival and frequency of UV-induced mutations to streptomycine-resistance in an excision-defective Hcr⁻ strain and its Hcr⁺ derivative at low doses of UV. No induced mutations were detected in the Hcr⁻ strain at doses below 90 ergs per mm² (see Witkin, 1966a).

sary to produce a detectable number of mutations in Hcr⁻ strains, which excise at least 99.9% of the pyrimidine dimers produced at low UV doses (Hill 1965, Witkin 1966a, Kondo and Kato 1966a, Witkin 1967). Thus, far fewer mutations are caused by mistakes in the excision repair of pyrimidine dimers than by unrepaired UV photo-products which remain in the DNA after excision repair (if any) is completed.

THE UV-STABILITY OF EXR– STRAINS

Unrepaired pyrimidine dimers are mutagenic in most strains of *E. coli*, but not in any strain having a mutation at the *exr* (Bs1) or *exr* (Bs2) locus (Witkin 1967). Exr⁻ strains do not differ from otherwise isogenic strains in their ability to produce or repair pyrimidine dimers, but they are two to three times more sensitive to UV, and are entirely UV-stable, yielding no detectable UV-induced mutations to auxotrophy, to streptomycin-resistance or to prototrophy. Figure 5 shows survival curves of a typical pair of Exr⁺ and Exr⁻ strains, the Exr⁺ strain having been derived in a single mutational step from the Exr⁻ strain. The Exr⁺ derivative exhibits normal UV-mutability, producing mutatoins to streptomycin-resistance abundantly, while the number of stereptomycin-resistant mutants obtained in the Exr⁻ strain does not exceed the number due to spontaneous mutation. No strain carrying the *exr⁻* allele derived from strains Bs1 or Bs2 yields UV-induced mutations of any kind thus far screened, at any dose. The lethal effect of a pyrimidine dimer is approximately doubled by an *exr⁻* mutation, but its mutagenic effect appears to be entirely eliminated.

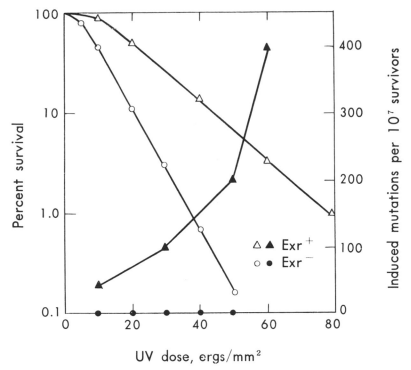

Figure 5. Survival and frequency of UV-induced mutations to streptomycin-
resistance in an Exr⁻ strain and its Exr⁺ derivative at low doses of UV. Both strains
are excision-defective (Hcr⁻),   and were derived initially from strain Bsl (see
Witkin, 1967).

PYRIMIDINE DIMERS IN DNA REPLICATION

Before considering how a pyrimidine dimer persisting in the DNA might cause a
mutation, and why the product of the *exr⁺* allele is necessary if it is to do so, one
must first ask how an unrepaired pyrimidine dimer behaves in DNA replication. The
suggestion that some strains of *E. coli* can replicate their DNA and form colonies
despite the presence of a fairly large number of unrepaired pyrimidine dimers in the
DNA (Witkin 1966b) has been substantiated by the studies of Rupp and Howard-
Flanders (1968), who have examined the sedimentation properties after replication
of DNA in excision-defective strains of *E. coli* K12 containing about 100 unrepaired
pyrimidine dimers per DNA molecule. Their findings indicate that the daughter
strands are discontinuous, *i.e.*, that they exist in a number of relatively short pieces.
The number and size of the pieces suggests that there is a gap opposite each pyrimidine
dimer in the parental template strand, a conclusion that is firmly supported by additional evidence obtained by an entirely different method (Howard-Flanders, Wilkins
and Rupp 1968). If the daughter strands are examined after about an hour of incubation, instead of immediately after replication, they aιe found to behave like continuous strands of full length. The disappearance of the discontinuities or gaps is
interpreted by Rupp and Howard-Flanders as the operation of a new kind of dark
repair process, which they call "post-replication repair" (see Figure 6). Post-replication repair does not act on the pyrimidine dimers themselves, which may remain in
the DNA of excision-defective strains for several cell divisions (Bridges and Munson

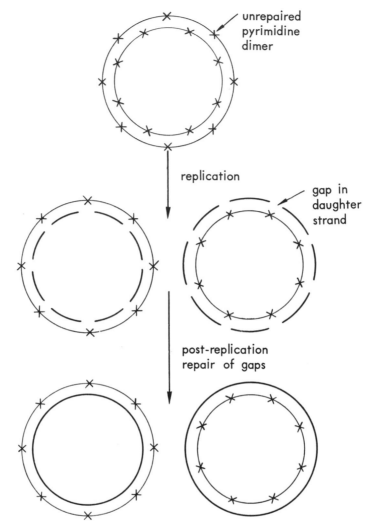

Figure 6. Diagramatic representation of post-replication repair of gaps opposite pyrimidine dimers in *E. coli*. Light lines are parental strands of DNA, heavy lines are daughter strands. Pyrimidine dimers are represented as crosses. No particular mechanism of repair is implied here. (See Rupp and Howard-Flanders, 1968).

1968); it repairs the gaps or discontinuities induced by unrepaired pyrimidine dimers in replicating DNA. Thus, pyrimidine dimers in the DNA do not block replication permanently, but they do appear to prevent it from proceeding continuously past the dimer, so that daughter molecules are produced in which one strand contains pyrimidine dimers and the other strand contains gaps, which may be regarded as secondary UV lesions. The size of the gaps is not known, but there must be at least one base missing, and there may be a great many, for if the discontinuities were simply single-strand breaks, there is no reason why they should not be rapidly closed by DNA ligase, after repair synthesis, if necessary, to replace any nucleotides removed by exonuclease degradation. Further replication of a DNA molecule containing discontinuities is probably not possible (Cairns and Davern 1966), and it is almost certainly not possible if the discontinuity is opposite a pyrimidine dimer (Howard-

Flanders, Wilkins and Rupp 1968). Survival, therefore, requires that all gaps in at least one of the daughter strands produced by the first post-irradiation replication be repaired before colony-forming ability is irreversibly lost, *i.e.,* within the "critical time" of about five hours after UV (Witkin 1967). Current ideas about normal DNA replication suggest that newly synthesized DNA may be formed initially in short fragments, which are rapidly connected by repair synthesis (Okazaki *et al.* 1968). If an unrepaired pyrimidine dimer is unable to serve as a template for Watson-Crick base pairing (whether in replication or in repair synthesis), a pyrimidine dimer in the template strand may serve as an end-point either in the synthesis of a fragment, or in the repair synthesis that would otherwise connect two fragments. In either case, the result would be a daughter strand having a persistent gap opposite each dimer.

UV-INDUCED MUTATIONS AS ERRORS IN THE POST-REPLICATION REPAIR OF GAPS

The evidence that pyrimidine dimers in replicating DNA induce the formation of gaps in the daughter strands, which are then slowly repaired, is obviously relevant to the mechanism of UV mutagenesis. Knowing that mutations are caused by unrepaired pyrimidine dimers, the simplest assumption is that such dimers cause replication errors via the insertion of the wrong bases (or the wrong number of bases) into the daughter strand, opposite the dimer, as the dimer passes through the replication point (Setlow 1964). This cannot be true, however, if a pyrimidine dimer stops replication, which must then be reinitiated at some point beyond the dimer, as is strongly suggested by the evidence of Rupp and Howard-Flanders. If gaps, with one or more bases missing, are produced opposite each pyrimidine dimer, it is much more likely that UV-induced mutations are *errors in the post-replication repair of the gaps.*

Thus, the pursuit of understanding of UV mutagenesis leads us directly to ask *how* gaps opposite pyrimidine dimers are repaired. If UV-induced mutations are due to errors in the post-replication repair of gaps, there must be two distinct kinds of post-replication repair in *E. coli,* one of which (Exr$^+$ repair) is relatively efficient (since Exr$^+$ strains are about twice as resistant to UV as Exr$^-$ strains), but very inaccurate (since Exr$^+$ strains are UV-mutable). The second kind of post-replication repair (Exr$^-$ repair) must be capable of repairing only half as many daughter-strand gaps within the critical time, resulting in reduced UV-survival, but it must do so accurately, without introducing any more errors into the DNA than occur in normal replication.

Rupp and Howard-Flanders (1968) have suggested a mechanism for the repair of gaps opposite pyrimidine dimers, postulating the occurence of a series of recombination-like events after replication. In each of these events, the daughter strand containing a gap at a given level pairs with the other daughter strand, which is its complement, and which may contain gaps elsewhere, but never at the same level. Such pairing would permit repair synthesis to restore the correct sequence of the region within each gap, utilizing as a template the corresponding intact region of the other daughter strand. Occurence of a series of such recombinational events at the level of each gap could ultimately reconstitute an intact DNA molecule, capable of further replication. There is now compelling evidence supporting the hypothesis that post-replication repair of gaps opposite pyrimidine dimers does involve recombination. Mutations in *E. coli* K12 which reduce recombination ability invariably also reduce UV resistance, without affecting the ability to repair pyrimidine dimers by excision (Clark and Margulies 1965, Howard-Flanders and Boyce 1966). This indicates that the ability to tolerate unrepaired pyrimidine dimers depends, at least to some extent, on the same gene products that are utilized in recombination. More recently, it has been established that there are three distinct loci in *E. coli*, *rec*A, *rec*B and *rec*C, a mutation at any one of which will drastically reduce or eliminate recombination ability (Willets, Clark and Law 1968). The extent to which UV-sensitivity is increased, in Rec⁻ mutants having a defect at any of the three *rec* loci, or in double mutants with defects at two *rec* loci, is proportional to the extent to which recombination ability is reduced (Clark, personal communication). Mutations at the *rec*A locus, for example, completely eliminate recombination ability, and at the same time increase UV-sensitivity to such a degree that a single unrepaired pyrimidine dimer may be a lethal event. A reasonable conclusion is that the post-replication repair of even a single gap opposite a pyrimidine dimer cannot be effected unless the strain has some recombination ability.

It has also been demonstrated (Howard-Flanders, Wilkins and Rupp 1968) that post-replication repair does not occur, even in a Rec⁺ strain fully able to carry on recombination, unless both daughter molecules of DNA, produced by the first post-UV replication, are present. This was by shown transferring an irradiated *Flac⁺* episome into a Rec⁺ recipient, an effective way to separate the two daughter molecules, since the transferred DNA in bacterial conjugation is a single replica (*i.e.*, one

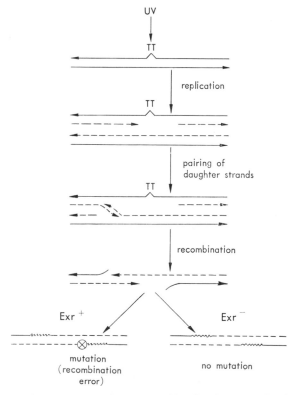

Figure 7. One possible mechanism of recombinational post-replication repair of gaps opposite pyrimidine dimers in Exr⁺ and Exr⁻ strains. Solid lines are parental strands of DNA, dotted lines are daughter strands. Polarity of strands is indicated by arrows. Wavy lines are areas filled by repair synthesis following pairing of daughter strands and breakage and degradation of parental strands. Exact location of recombination error in Exr⁺ strains is not necessarily as shown here. An alternative mechanism might involve sister strand exchanges in which the parental strands remain intact.

daughter molecule) of the original donor DNA (Gross and Caro 1966). The donor DNA, in this case, was UV-irradiated episomal DNA and therefore the transferred element was presumably a daughter molecule consisting of one preformed strand containing pyrimidine dimers, and one daughter strand containing gaps opposite the dimers. Although . the Rec$^+$ recipient strain (which was also Exr$^+$) contained all the enzymatic equipment necessary to perform both recombination and post-replication repair of gaps, the data indicate that the episomal DNA is *not* repaired, which leads to the conclusion that post-replication repair requires not only recombination ability, but also the physical presence of both daughter molecules of DNA. It is very difficult to avoid the conclusion that post-replication repair of gaps opposite pyrimidine dimers depends upon genetic recombination.

If post-replication repair of gaps involves recombination, UV-induced mutations, which are errors in post-replication repair, must ultimately be errors associated with the recombinational process. This leads to the proposition that recombination, itself, is mutagenic in Exr$^+$ strains, including wild type strains, but not in Exr$^-$ strains, as diagrammed in Figure 7. Mutations at the *exr* locus, in this case, would presumably alter the recombination mechanism in some way, reducing its efficiency to about half (and thus doubling UV-sensitivity), and at the same time eliminating an error-producing feature that is responsible for the UV-mutability of Exr$^+$ strains. It now seems probable that UV induces mutations by the following steps: 1) pyrimidine dimers are produced in the DNA; 2) dimers not repaired by excision repair or photoreactivation induce the formation of gaps, one gap opposite each dimer, in the daughter strands produced in the first post-irradiation replication; 3) in Exr$^+$ strains having normal recombination ability, the gaps are repaired by a recombinational mechanism having a high probability of introducing errors into the DNA; in Exr$^-$ strains, the same steps occur, but the recombinatoin mechanism is error-free because of the mutation at the *exr* locus, and no mutations are produced.

It has also been proposed (Witkin 1967, Bridges, Dennis and Munson 1967) that the relatively efficient but highly mutagenic mechanism of post-replication repair, utilized by Exr$^+$ strains to repair gaps opposite pyrimidine dimers could involve the random insertion of bases into the gaps, by a process of enzymatic "end-addition" which does not use a template. In this case, *exr$^-$* mutations would eliminate the ability to close gaps by random insertion of nucleotides, and would leave Exr$^-$ strains able to repair gaps only by recombination, which, according to this hypothesis, is invariably accurate. Closure of gaps by a mechanism independent of recombination, however, could not explain the striking correlation between recombination and UV-resistance, in both Exr$^+$ and Exr$^-$ strains, a correlation which is further strengthened by the finding that Exr$^-$ strains, which are two to three times more sensitive to UV than Exr$^+$ strains, are also two to three times less efficient as recipients in recombination (Witkin, unpublished observation).

UV-INDUCED MUTATIONS IN THE TRYPTOPHANE
SYNTHETASE A GENE

In considering how recombination might generate mutations, it would be helpful to know what kinds of mutations are actually induced by UV (*i.e.*, frameshifts or base substitutions, transitions or transversions, etc.). In *E. coli*, the detailed study of mutagenesis in a particular gene, the trytophane synthetase A gene, accompanied by studies of the amino acid replacements caused by particular mutations in the tryptophane synthetase A protein, has been carried on extensively by Yanofsky and his coworkers. Fortunately, this work includes some mutations induced by UV, and these data provide the best available material for analysis of the kinds of changes produced by UV. Analysis of 28 missense mutations induced by UV in this gene (Yanofsky, Ito and Horn 1966, Yanofsky, personal communication) shows that all but one are probably single-base substitutions, with both transitions and transversions represented. On the whole it is estimated that about 80% of all UV-induced mutations in this gene are single-base substitutions, the remainder being frameshifts (Yanofsky, personal communication). The only analysed frameshift (Brammar, Berger and Yanofsky 1967) proved to be a single-base deletion, and it is quite possible (but not certain) that all UV-induced frameshifts are of this type, since none are revertible by UV. Data on UV-induced reversion to prototrophy in missense mutants (Yanofsky, personal communication) show that, at least for one site in the tryptophane synthetase A gene, any of the three other bases can replace the one responsible for the auxotrophy as the result of a UV-induced mutation (*i.e.*, G, C or U can replace A in the codon position corresponding to this site), with about equal probability. UV thus appears to cause mainly unrestricted single-base substitutions, plus a substantial minority of frameshift mutations, possibly all of the one-base deletion type. The tryptophane synthetase A data are compatible with the idea that alteration of the pairing specificity of a single base is responsible for most UV-induced mutations in *E. coli*.

THE MUTAGENIC STEP IN RECOMBINATION

According to current models of recombination, it is not obvious why it should be mutagenic. In some cases, two preformed fragments of DNA are joined (Meselson 1964). It is probable that base pairing between overlapping single-stranded ends is an early step in the process, and that gaps on either side are filled by repair synthesis as the final step (see for example Howard-Flanders and Boyce 1966). The repair synthesis which acts in recobination should, in principle, be as accurate as the repair synthesis step in excision repair, since in both cases an intact and correct template is available. Since excision repair is not responsible for UV-induced mutations, it is unlikely that repair synthesis, or any other step that is common to both recombination and excision repair, is the mutagenic step in the recombination mechanism of Exr$^+$ strains.

The idea that recombination is mutagenic has been proposed (Magni and von Borstel 1962, Magni 1963, 1964), to account for the higher spontaneous mutation rates observed in meiosis than in mitosis in yeast. Magni has provided clear evidence that the increased mutation frequencies in meiosis are associated with genetic exchange. He considers the mechanism of "exchange-associated" mutation to be unequal crossing over, and his prediction that mutations of the frameshift type should be generated exclusively by this process now seems to have been borne out (Magni, these Proceedings). If UV-induced mutations are recombination errors, the mutagenic effect of recombination in *E. coli*, however, is not due to inequality of exchange, since the vast majority of UV-induced mutations are single-base substitutions.

Another phenomenon suggesting the mutagenicity of recombination in bacteria is "selfing" (Demerec 1962) in which some auxotrophic strains can be "transduced" to prototrophy by phages grown on the same auxotroph. Demerec first interpreted this as an example of unequal exchange, but later (Demerec 1963) ruled out this interpretation by showing that "selfing" occurs even if the donor carries a deletion covering a large part of the marker gene. He concluded that the mutability of some genes is increased by the proximity of a paired transducing fragment, but a more likely explanation now is that the reversions are induced by a mutagenic step in the process of recombination itself.

Exr$^+$ and Exr$^-$ strains have been found to differ in the amount of spontaneous DNA breakdown, which is significantly more extensive in Exr strains (Hurwitz, personal communication). DNA breakdown after exposure to UV is also much more extensive in Exr$^-$ strains, assuming that the *exr* locus in *E. coli* B and the *lex* locus in *E. coli* K12 are identical, which appears likely (Howard-Flanders and Boyce 1966). Assays of crude extract of Exr$^+$ and Exr$^-$ strains have revealed no significant differences in the activity of Exonuclease III (Richardson, personal communication) or of Exonuclease I (Lehman, personal communication). It therefore seems possible that the heightened degradative activity in Exr$^-$ strains is due to a defect in a regulator substance involved in preventing or limiting DNA breakdown, as has been proposed for some Rec$^-$ and Lex$^-$ strains (Howard-Flanders and Boyce 1966). In this case, the mutagenic step in the Exr$^-$ recombination mechanism may be one in which a single-strand end is stabilized against degradation, perhaps in a way that prepares it specifically for recombination. This may involve an enzymatic modification of the base which alters its pairing specificity at the same time that it is rendered refractory to degradation, thus increasing the chance of a subsequent replication error. Exr$^-$ strains may stabilize single-strand ends against degradation somewhat less effectively, but without altering the pairing properties of the modified base. Whatever the mutagenic step in recombination may prove to be, we can say now that it generates single-base substitutions primarily; its mutagenicity can be eliminated by a mutation that reduces recombination ability only slightly, and the elimination of its mutagenicity is associated with an increase in DNA breakdown.

THE SPECIFICITY OF RECOMBINATION-INDUCED MUTATIONS

Demerec (1964) reported that only certain auxotrophs were selfers. This suggests a degree of specificity in the mutagenic activity of recombination which may, however, have a simple explanation. If recombination causes mainly single-base substitutions, plus some single-base deletions, as the data in *E. coli* indicate, the chance of detecting recombination-induced reversion to prototrophy should depend upon the allele involved, and on the properties of the gene product involved. A protein like the tryptophane synthetase A protein, for example, once inactivated by a missense mutation, can be restored to activity not only by a true reversion to the original wild type base sequence (an extremely rare event), but also by a variety of primary-site mutations to a different form of missense (Yanofsky, Ito and Horn 1966), by second-site missense mutations resulting in a second amino-acid substitution (Helinski and Yanofsky 1963), or even by frameshift mutations (Berger, Brammar and Yanofsky 1968). Since the activity of this protein is so readily restored by changes in many different positions of the kind caused by UV (*i.e.*, by recombination errors), one would predict that missense mutants with defects in the tryptophane synthetase A gene and others coding gene products which share this property should be relatively susceptible to selfing, as well as to other manifestations of recombination-induced mutability, including UV-mutability. On the other hand, alleles coding a product that cannot be restored to activity except by true reversion to wild type, or by changes other than those produced by recombination errors, should be refractory both to selfing and to UV-mutagenesis. This should include mutants owing their auxotrophy to frameshifts that are not correctible by single-base deletions, if these are indeed the only kinds of frameshifts induced by UV (*i.e.*, by recombination). It is also to be expected that *exr*$^-$ mutations should eliminate selfing and other manifestations of recombination-induced mutation, as well as UV-mutability. Most of these predictions are readily testable.

RECOMBINATION-INDUCED MUTATION AND
NEGATIVE INTERFERENCE

The possibility that a recombinational event may itself generate an induced mutation is relevant also to the phenomenon of negative interference (Chase and Doermann 1958, Pritchard 1960). Negative interference is the apparent clustering of recombinational events in very small regions, corresponding approximately to about 1,200 nucleotide pairs (Amati and Meselson 1965), or the approximate length of a single cistron. It seems possible that recombination-induced reversions, scored as recombinations, could distort estimates of negative interference. This, of course, would apply only if the locus involved is initially of a type susceptible to recombination-induced reversion, if the strain involved is capable of producing such mutations, and if the probability of induced mutation per recombinational event is very high. This could be tested by comparing negative interference in the same region in Exr$^+$ and Exr$^-$ strains. The same considerations apply to fine-structure mapping.

THE ROLE OF REPAIR IN MUTAGENESIS INITIATED
BY MUTAGENS OTHER THAN UV

Information about the role of repair in mutagenesis as initiated by mutagens other than UV is still too fragmentary to permit many definite conclusions, but some tentative statements can be made. As in the case of UV, the most useful approach is to compare the lethal and mutagenic effects of the agent in question in selected pairs of strains, the members of each pair differing in a single, well-defined, genetically controlled aspect of repair capability. The interpretation of the complex pattern of responses now emerging in such analyses can give valuable clues to the mechanism of mutagenesis.

4-NITROQUINOLINE-1-OXIDE

Thus far, 4-nitroquinoline-1-oxide (NQO), a carcinogen, is the only mutagen showing a pattern similar to UV (Kondo 1968). Excision-defective strains are much more sensitive to NQO than their Hcr$^+$ counterparts, indicating that the predominant type of lethal damage produced by this agent is exciseable; excision-defective strains also yield much higher frequencies of NQO-induced mutations than Hcr$^+$ strains treated with the same doses, indicating that these mutations are caused primarily by unrepaired NQO damage, and that errors in the excision repair process, as in the case of UV, are not responsible for their induction. Exr$^-$ strains have not yet been tested for sensitivity to NQO, but the increased sensitivity to NQO of strains with reduced recombination ability (Kondo 1968) suggests that NQO-induced mutations, like UV-induced mutations, may prove to be errors in a recombinational repair mechanism.

MITOMYCIN C

Excision-defective strains are sensitive to mytomycin C (MMC) compared to their Hcr$^+$ counterparts (Boyce and Howard-Flanders 1964a), suggesting that MMC damage is also repaired by excision, or by a mechanism utilizing excision enzymes. However, here the resemblance to UV and NQO stops. An excision-defective strain, far from producing more mutations than Hcr$^+$ strains, as was found for UV and NQO, was found to produce no detectable MMC-induced mutations at all (Kondo 1968). This clearly implies that the repair of MMC damage by excision, or by a mechanism sharing at least one enzymatic step with excision repair, is itself responsible for MMC-induced mutations. Since MMC is known to cause cross-linking of the two strands

270

of DNA (Iyer and Szybalski 1963, 1964), it is possible that this damage is invariably lethal if not at least partially repaired, and thus cannot cause mutations in Hcr⁻ strains, in which the only survivors may be initially undamaged cells. There are two possible interpretations of the origin of the MMC-induced mutations in Hcr⁺ strains. The action of one or more enzymes also required for the excision of pyrimidine dimers may free one side of the cross-link, permitting survival, but leaving a modified base with altered pairing specificity in the other strand. Alternatively, MMC damage may be repaired completely by the excision repair mechanism, and the absence of any unrepaired mutagenic lesions compatible with survival may permit the detection of low-level mutagenic effects of the excision repair mechanism itself, masked in the case of UV and NQO by the much higher mutagenicity of residual unrepaired damage. Quantitative studies, as well as additional efforts to determine whether MMC damage is actually removed from the DNA by excision, are necessary before these questions can be resolved.

X-RAYS AND OTHER IONIZING RADIATIONS

Excision-defective strains are no more sensitive to X-rays than their Hcr⁺ counterparts (Bridges and Munson 1966, Howard-Flanders and Boyce 1966), from which it may be concluded that the predominant type of potentially lethal X-ray damage is not repairable by excision, or at least not by a mechanism involving all steps in the excision repair mechanism. Loss of excision ability does not significantly alter the yield of X-ray-induced mutations (Bridges and Munson 1966, Kondo 1968), indicating that the predominant type of potentially mutagenic X-ray damage is also nonexciseable. On the other hand, sensitivity to X-rays is significantly increased by Rec⁻ mutations (Howard-Flanders and Boyce 1966, van de Putte, Zwenk and Rörsch 1966), and by mutations at the *exr* locus (Mattern, Zwenk and Rörsch 1966), suggesting that some potentially lethal primary or secondary X-ray damage may be repairable by recombination, or by a mechanism utilizing recombination enzymes. There are preliminary indications that most X-ray-induced mutations may be caused by errors in recombinational repair, since no X-ray-induced mutations are produced in one Rec⁻ strain (Kondo 1968), and the yield of gamma-ray induced mutations in an Exr⁻ strain is only 5% of that observed in its Exr⁺ derivative (Bridges, Law and Munson, 1968).

X-rays are known to cause single-strand breaks in DNA (Peacocke and Preston 1960) which are subject to repair (McGrath and Williams 1966). In principle, such breaks should be repairable simply by ligase sealing, or by a combination of repair synthesis and ligase sealing, if the breaks are enlarged into gaps by exonuclease activity, and there is no reason why such repair should require recombination. However, some of the single-strand breaks may be produced, and may be enlarged into gaps, in regions of the DNA that are not repairable by simple repair synthesis (for instance, in temporarily single-stranded regions). Breaks occurring in such regions, if they are also located in parts of the DNA which have replicated before X-irradiation, may be subject to repair by recombination, and may thus be the only breaks capable of giving rise to X-ray-induced (*i.e.*, recombination-induced) mutations. Such recombinational repair would not differ essentially from that occurring after UV treatment, except that it would occur *before* rather than *after* the first post-irradiation DNA replication. This would be consistent with the observation of Kada and Marcovitch (1963) that X-ray induced mutations in *E. coli,* unlike UV-induced mutations, can be transferred by conjugation immediately after irradiating the donor. Since the transferred DNA is now known to be a *replica* of the DNA present in the donor at the time of irradiation (Gross and Caro 1966), a current reinterpretation of the Kada and Marcovitch result would be that X-ray-induced mutations are already present, in a complete and expressible form, in the immediate product of the first post-irradiation DNA replication, while UV-induced mutations are not. Whether the mutagenic component of X-ray damage is a special fraction of the single-strand breaks, or some other class of lesions not yet identified, the current limited evidence indicates that it is repairable (or that it induces a gap which is repairable) by recombination, thus generating recombination-induced mutations, at some time before the first post-irradiation DNA replication.

NITROSOGUANIDINE

Loss of ability to excise pyrimidine dimers does not significantly increase sensitivity to nitrosoguanidine (NTG) (Witkin 1967), indicating that the predominant type of potentially lethal damage produced by NTG is nonexciseable, or at least that its repair does not require all the same gene products needed for excision of pyrimidine dimer. This contradicts reports that NTG damage and UV damage are repaired by the same mechanism (Cerdá-Olmedo and Hanawalt 1967). This conclusion was based on comparisons of strains Bs1 and B/r, which differ not only in excision ability, but also in two other mutational differences affecting radiosensitivity (Mattern, Zwenk and Rörsch 1966). The greater sensitivity of strain Bs1 to nitrosoguanidine than that of B/r has been shown to be due entirely to its *fil*$^+$ and *exr*$^-$ mutations, and not at all to its *hcr*$^-$ mutation (Witkin 1967). Excision-defective strains also produce NTG-induced mutations with about the same frequency as Hcr$^+$ strains (Kondo 1968), showing, in this regard, the same pattern as X-rays: neither the predominant type of potentially lethal damage nor the predominant type of potentially mutagenic damage is exciseable. Exr$^-$ mutations and at least some Rec$^-$ mutations increase NTG-sensitivity (Witkin 1967, Kondo 1968), indicating that some of the lethal NTG damage may be repairable by recombination, or by a mechanism utilizing some enzymes in common with recombination. Unlike X-ray induced mutations, however, NTG-induced mutations are produced abundantly in Exr$^-$ strains (Witkin 1967) and in at least some Rec$^-$ strains (Kondo 1968), from which one might conclude that most NTG-induced mutations are probably not errors in recombinational repair.

Alkylating Agents

Bifunctional alkylating agents, such as the nitrogen and sulfur mustards, are known to cause inter-strand crosslinks in DNA. Strains lacking excision repair are more sensitive to nitrogen mustard than are their Hcr$^+$ parent strains (Bridges and Munson 1966), which, together with biochemical evidence suggesting that the cross-linked residues are actually removed from the DNA in strains capable of excision but not in repair-deficient strains (Lawley and Brooks 1965), indicates that damage produced by nitrogen and sulfur mustards is exciseable. Nothing is yet known of the mutagenic response to these compounds in Hcr$^-$ or in Exr$^-$ strains, and there is therefore no basis for discussing the possible mutagenic pathway initiated by them.

Another alkylating agent, methyl methane sulfonate (MMS) produces single-strand breaks in DNA (Strauss 1968). Sensitivity to MMS is not increased by a mutation eliminating excision repair (Bridges and Munson 1966), nor is sensitivity to a similar agent, ethyl methane sulfonate (EMS) (Kondo 1968). Kondo has also shown that EMS-induced mutagenesis is not affected by mutations that eliminate excision ability. These and other indications that UV and MMS damage are repaired by distinct mechanisms (Strauss, Reiter and Searashi 1966) suggest that MMS and EMS damage are not exciseable. Information about the behaviour of Rec$^-$ and Exr$^-$ strains toward EMS and MMS is still too meager to permit speculation as to whether these mutagens produce damage that is repairable by a recombinational mechanism. As indicated above in the discussion of X-ray mutagenesis, it is possible that some single-strand breaks can generate gaps that are repairable only by a recombinational mechanism. In that case, MMS and EMS-induced mutations may also prove to be errors in recombinational repair.

Thymineless Mutagenesis

Mutations induced by thymine starvation (Coughlin and Adelberg 1956) might easily be imagined to come about in the same way as UV-induced mutations, as recombinational errors produced in the course of post-replication repair of gaps. DNA in bacteria incubated without thymine is found to exist in very short fragments (Mennigmann and Szybalski 1963), as would be expected if replication occurred with a gap opposite each adenine residue. Such DNA would differ from the daughter strands produced after UV only in that the parental strand is normal, presumably, and that the number of gaps is much greater, since adenine residues in the parent strand are undoubtedly much more frequent than pyrimidine dimers produced even at extremely high doses. The nature of the gaps produced, however, is the same in that in neither case can they be filled by repair synthesis. In the case of UV, this is because there is a noncoding lesion opposite the gap; in the case of thymine starvation, it is because the correct partner for one base is not available. Since recombinational repair probably operates to repair any gap which cannot be filled by repair synthesis, assuming that there is an intact complementary strand nearby, one would expect gaps produced by thymine starvation to be repaired by recombination, and to generate recombination-induced mutations. Preliminary evidence suggesting that this may be true has recently been obtained. Treated with 5-fluorodeoxyuridine to induce a phenotypic requirement for thymine, two different Exr⁻ strains failed to produce mutations, while their Exr⁺ derivatives treated in the same way did (Witkin unpublished observation, Bridges, Law and Munson 1968). More rigorous experiments with strains having a genetically determined thymine requirement are needed before a firm conclusion can be drawn.

SPONTANEOUS MUTATIONS

Since Exr strains do not differ greatly from their Exr⁺ parent strains in spontaneous mutability, at least at some loci (Witkin 1967), it is not likely that recombination errors are the most important source of spontaneous mutations in *E. coli*, although more complete and systematic investigation is required before this can be established. It has been repeatedly found (Witkin unpublished observation) that Exr⁺ revertants, obtained from Exr⁻ mutant strains by selecting for UV-resistance accompanied by the reacquisition of UV-mutability, frequently show higher spontaneous mutation rates for many different kinds of mutations than their Exr⁻ parent strains. Sometimes, this high spontaneous mutability is high enough to be described as "mutator" activity, although this characteristic is unstable, and in such strains there is strong selection pressure for a variant with lower spontaneous mutability. This suggests that recombination may produce high frequencies of spontaneous mutations in some Exr⁺ strains, if not in the wild type.

Thus, the role of repair mechanisms is complex and far from well understood. Mutations may be prevented by some repair mechanisms (photoreactivation, excision repair of pyrimidine dimers), but they may also be caused by others (excision of MMC damage, post-replication recombinational repair of gaps opposite pyrimidine dimers in Exr⁺ strains). Some mutations (such as most of those induced by NTG) are not affected by any known repair mechanism. The systematic study of lethal and mutagenic effects induced by all categories of mutagens in all categories of repair-deficient mutant strains must be completed before these problems can be further clarified. Interpretation of the patterns that emerge, in the light of increasing knowledge of the genetic and biochemical basis of each repair mechanism, and of the nature of the primary and secondary lesions produced by mutagenic agents, should ultimately permit a reasonably complete description of the interlacing pathways of mutagenesis and repair.

SUMMARY AND CONCLUSIONS

The repair of pyrimidine dimers by photoreactivation or by excision repair prevents many mutations by eliminating the potentially mutagenic dimers from the DNA. Photoreactivation is at least ten times less likely to cause a mutation than is an unrepaired dimer, and excision repair is at least one hundred times less like to do so.

The post-replication repair of daughter-strand gaps in the DNA opposite unrepaired pyrimidine dimers is effected by recombination, which is mutagenic in strains of *E. coli* having the wild type allele at the *exr*(Bs1) and *exr*(Bs2) locus, but is nonmutagenic in strains having mutational defects at this locus. Therefore, UV-induced mutations are actually recombination-induced mutations, produced as a consequence of inaccurate recombinational repair of secondary UV damage (gaps opposite pyrimidine dimers) in Exr$^+$ strains.

In *E. coli*, most UV-induced mutations (*i.e.*, recombination-induced mutations) are single-base substitutions, although a small fraction are frameshifts, possibly all of the single-base deletion type. Recombination, therefore, does not cause mutations in *E. coli* primarily by unequal exchange, but by the alteration of the pairing specificity of a single base.

Preliminary evidence suggests that mutations induced by X-rays and by thymine starvation may also be largely due to errors in recombinational repair of gaps which cannot be repaired by simple repair synthesis. In the case of X-rays, recombinational repair probably occurs *before* the first post-irradiation replication. In some strains, recombination errors may also play an important part in spontaneous mutability.

ACKNOWLEDGMENT

This report was prepared under grant 5-R01-A1-01240 from the National Institute of Allergy and Infectious Diseases.

LITERATURE CITED

AMATI, P. and M. MESELSON 1965 Localized negative interference in bacteriophage λ. *Genetics* **51**: 369–379.

BERGER, H., W. J. BRAMMAR and C. YANOFSKY 1968 Analysis of amino acid replacements resulting from frameshift and missense mutations in the tryptophane synthetase A gene in *E. coli*. *J. Mol. Biol.* **34**: 219–238.

BOYCE, R. P. and P. HOWARD-FLANDERS 1964a Genetic control of DNA breakdown and repair in *E. coli* K12 treated with mitomycin C or ultraviolent light. *Z. Vererbungsl.* **95**: 345–350.

BOYCE, R. P. and P. HOWARD-FLANDERS 1964b The release of UV-induced thymine dimers from DNA in *E. coli* K12. *Proc. Nat. Acad. Sci. U.S.* **51**: 293–300.

BRAMMAR, W. J., H. BERGER and C. YANOFSKY 1967 Altered amino acid sequences produced by reversion of frameshift mutants of tryptophane synthetase A gene of *E. coli. Proc. Nat. Acad. Sci. U.S.* **58**: 1499–1506.

BRIDGES, B. A., R. E. DENNIS and R. J. MUNSON 1967 Differential induction and repair of ultraviolet damage leading to true reversions and external suppressor mutations of an ochre codon in *Escherichia coli B/r WP2. Genetics* **57**: 897–908.

BRIDGES, B. A. and R. J. MUNSON 1966 Excision-repair of DNA damage in an auxotrophic strain of *E. coli. Biochem. Biophys. Res. Commun.* **22**: 268–273.

BRIDGES, B. A. and R. J. MUNSON 1968 Mutagenesis in *Escherichia coli*: evidence for the mechanism of base change mutation by ultraviolet radiation in a strain deficient in excision-repair. *Proc. Roy. Soc. B* **171**: 213–226.

CAIRNS, J. and C. I. DAVERN 1966 Effect of ^{32}P decay upon DNA synthesis by a radiation-sensitive strain of *Escherichia coli. J. Mol. Biol.* **17**: 418–427.

CERDÁ-OLMEDO, E. and P. C. HANAWALT 1967 Repair of DNA damaged by N-methyl-N′-nitro-N-nitrosoguanidine in *Escherichia coli. Mutation Res.* **4**: 369–371.

CHASE, M. and A. H. DOERMANN 1958 High negative interference over short segments of the genetic structure of bacteriophase T4. *Genetics* **43**: 332–353.

CLARK, A. J. and A. D. MARGULIES 1965 Isolation and characterization of recombination deficient mutants of *E. coli* K12. *Proc. Nat. Acad. Sci. U.S.* **53**: 451–459.

COUGHLIN, C. A., and E. A. ADELBERG 1956 Bacterial mutation induced by thymine starvation. *Nature* **178**: 531–532.

DEMEREC, M. 1962 "Selfers" attributed to unequal crossovers in *Salmonella. Proc. Nat. Acad. Sci. U.S.* **48**: 1696–1704.

DEMEREC, M. 1963 Selfer mutants of *Salmonella typhimurium. Genetics* **48**: 1519–1531.

GEFTER, M. L., A. BECKER and J. HURWITZ 1967 The enzymatic repair of DNA, I. Formation of circular λDNA. *Proc. Nat. Acad. Sci. U.S.* **58**: 240–247.

GELLERT, M. 1967 Formation of covalent circles of lambda DNA by *E. coli* extracts. *Proc. Nat. Acad. Sci. U.S.* **57**: 148–155.

GROSS, J. D. and L. C. CARO 1966 DNA transfer in bacterial conjugation. *J. Mol. Biol.* **16**: 269–284.

HANAWALT, P. C. 1968 Cellular recovery from photochemical damage. *Photophysiology,* **4**: pp. 203–251. (A. C. Giese, ed.) Academic Press, Inc., New York.

HANAWALT, P. C. and R. H. HAYNES 1965 Repair replication of DNA in bacteria: irrelevance of chemical nature of base defect. *Biochem. Biophys. Res. Commun.* **19**: 462–467.

HARM, W. and B. HILLEBRANDT 1962 A non-photoreactivable mutant of *E. coli* B. *Photochem. Photobiol.* **1**: 271–272.

HAYNES, R. H., R. M. BAKER and G. E. JONES 1968 Genetic implications of DNA repair. in *Energetics and Mechanisms in Radiation Biology* (G. Phillips, ed.) Academic Press, London and New York, pp. 425–465.

HELINSKI, D. R. and C. YANOFSKY 1963 A genetic and biochemical analysis of second site reversion. *J. Biol. Chem.* **238**: 1043–1048.

HILL, R. F. 1965 Ultraviolet-induced lethality and reversion to prototrophy in *Escherichia coli* strains with normal and reduced dark repair ability. *Photochem. Photobiol.* **4**: 563–568.

HOWARD-FLANDERS, P. and R. P. BOYCE 1966 DNA repair and genetic recombination: studies on mutants of *Escherichia coli* defective in these processes. *Radiation Research Supplement* **6**: 156–184.

HOWARD-FLANDERS, P., B. M. WILKINS and W. D. RUPP 1968 Genetic recombination induced by ultraviolet light. In *Molecular Genetics* (H. G. Wittmann and H.

Schuster) Springer-Verlag, Berlin and Heidelberg, pp. 161–173.

IYER, V. N. and W. SZYBALSKI 1963 A molecular mechanism of mitomycin action: linking of complementary DNA strands. *Proc. Nat. Acad. Sci. U.S.* **47**: 950–955.

IYER, V. N. and W. SZYBALSKI 1964 Mitomycins and porfiromycin: chemical mechanism of activation and cross-linking of DNA. *Science* **145**: 55–58.

JAGGER, J. and R. S. STAFFORD 1965 Evidence for two mechanism of photoreactivation in *Escherichia coli* B. *Biophys. J.* **5**: 75–88.

KADA, T. and H. MARCOVICH 1963 The initial site of the mutagenic action of X- and UV-rays in *Escherichia coli*. *Ann. Inst. Pasteur* **105**: 989–1006.

KAPLAN, R. W. 1963 Photoreversion von vier Gruppen UV-dyzierter Mutationen zur Giftresistenz im Nicht photoreacktivierbaren *E. coli*. *Photochem. Photobiol.* **2**: 461-470.

KATO, T. and S. KONDO 1967 Two types of X-ray-sensitive mutants of *Escherichia coli* B: Their phenotypic characters compared with UV-sensitive mutants. *Mutation Res.* **4**: 253–263.

KELNER, A. 1949 Photoreactivation of ultraviolet-irradiated *Escherichia coli* with special reference to the dose-reduction principle and to ultraviolet-induced mutation. *J. Bacteriol.* **58**: 511–522.

KONDO, S. and T. KATO 1966a Action spectra for photoreactivation of killing and mutation to prototrophy in UV-sensitive strains of *Escherichia coli* possessing and lacking photoreactivating enzyme. *Photochem. Photobiol.* **5**: 827–837.

KONDO, S. and T. KATO 1966b Photoreactivation of mutation and killing in *Escherichia coli*. *Adv. Biol. Med. Physics* **12**: 283–298.

KONDO, S. 1968 Mutagenicity versus radiosensitivity in *Escherichia coli*. Proc. XIIth Int. Congress of Genetics **2**: 126–127.

LAWLEY, P. D. and P. BROOKES 1965 Molecular mechanism of the cytotoxic action of difunctional alkylating agents and resistance to this action. *Nature* **206**: 480–483.

MAGNI, G. E. 1963 The origin of spontaneous mutations during meiosis. *Proc. Nat. Acad. Sci. U.S.* **50**: 975–980.

MAGNI, G. E. 1964 Origin and nature of spontaneous mutations in meiotic organisms. *J. Cell. Comp. Physiol.* **64** Supplement **1**: 165–172.

MAGNI, G. E. and R. C. VON BORSTEL 1962 Different rates of spontaneous mutation during mitosis and meiosis in yeast. *Genetics* **47**: 1047–1108.

MATTERN, I. E., H. ZWENK and A. RÖRSH 1966 The genetic constitution of the radiation-sensitive mutant *Escherichia coli* Bs-₁. *Mutation Res.* **3**: 374–380.

McGRATH, R. A. and R. W. WILLIAMS 1966 Reconstruction *in vivo* of irradiated *Escherichia coli* deoxyribonucleic acid; the rejoining of broken pieces. *Nature* **212**: 534–535.

MENNIGMANN, H. and W. SZYBALSKI 1962 Molecular mechanism of thymine-less death. *Biochem. Biophys. Res. Comm.* **9**: 398–404.

MESELSON, M. 1964 On the mechanism of genetic recombination between DNA molecules. *J. Mol. Biol.* **9**: 734–745.

OKAZAKI R., T. OKAZAKI, K. SAKABE, K. SUGIMOTO and A. SUGINO 1968 Mechanism of DNA chain growth, I. Possible discontinuity and unusual secondary structure of newly synthesized chains. *Proc. Nat. Acad. Sci. U.S.* **59**: 598–605.

OLIVERA, B. and I. R. LEHMAN 1967 Linkage of polynucleotides through phosphodiester bonds by an enzyme from *Escherichia coli*. *Proc. Nat. Acad. Sci. U.S.* **57**: 1426–1433.

PEACOCKE, A. R. and B. N. PRESTON 1960 The action of γ-rays on sodium deoxyribonucleate in solution II. Degradation. *Proc. Roy. Soc.* **153B**: 90–110.

PETTIJOHN, D. and P. C. HANAWALT 1964 Evidence for repair replication of ultraviolet damaged DNA in bacteria. *J. Mol. Biol.* **9**: 395–410.

PRITCHARD, R. H. 1960 Localised negative interference and its bearing on models of gene recombination. *Genet. Res., Camb.* **1**: 1–24.

VAN DE PUTTE, P., H. ZWENK and A. RÖRSCH 1966 Properties of four mutants of *Escherichia coli* defective in genetic recombination. *Mutation Res.* **3**: 381–392.

RUPP, W. D. and P. HOWARD-FLANDERS 1968 Discontinuities in the DNA synthesized in an excision-defective strain of *Escherichia coli* following ultraviolet irradiation. *J. Mol. Biol.* **31**: 291–304.

SETLOW, J. 1966 Photoreactivation. *Radiation Research* Supplement **6**: 141–155.

SETLOW, R. B. 1964 Physical changes and mutagenesis. *J. Cell. Comp. Physiol.* **64**: supplement **1**: 51–68.

SETLOW, R. B. 1968 The photochemistry, photobiology and repair of polynucleotides. *Progr. Nucleic Acid Res. and Mol. Biol.* **8**: 257–295.

SETLOW, R. B. and W. L. CARRIER 1964 The disappearance of thymine dimers from DNA: an error correcting mechanism. *Proc. Nat. Acad. Sci. U.S.* **51**: 226–231.

SETLOW, R. B., W. L. CARRIER and F. J. BOLLUM 1965a Pyrimidine dimers in UV-irradiated poly dI: dC. *Proc. Nat. Acad. Sci. U.S.* **53**: 1111–1118.

SETLOW, R. B., W. L. CARRIER and F. J. BOLLUM 1965b Dimers of cytosine and thymine residues in UV-irradiated polydeoxyribonucleotides. *Abstr. Biophys. Soc.*, p. 111.

SMITH, K. C. 1963 Photochemical reactions of thymine, uracil, uridine, cytosine and bromouracil in frozen solution and in dried films. *Photochem. Photobiol.* **2**: 503–517.

STRAUSS, B., H. REITER and T. SEARASHI 1966 Recovery from ultraviolet- and alkylating-agent-induced damage in *Bacillus subtilis*. *Radiation Research*. Supplement **6**: 201–211.

STRAUSS, B. 1968 DNA repair mechanisms and their relation to mutation and recombination. *Current Topics Microbiol. Immunol.* **44**: 1–85.

SWENSON, P. A. and R. B. SETLOW 1966 Effects of ultraviolet radiation on macromolecular synthesis in *Escherichia coli*. *J. Mol. Biol.* **15**: 201–219.

WACKER, A. 1963 Molecular mechanism of radiation effects. *Progr. Nucleic Acid Res.* **1**: 369–399.

WACKER, A., H. DELLWEG and D. JACHERTS 1962 Thymine dimerization and survival of bacteria. *J. Mol. Biol.* **4**: 410–412.

WEISS, B. and C. C. RICHARDSON 1967 Enzymatic breakage and joining of deoxyribonucleic acid, I. Repair of single-strand breaks in DNA by an enzyme system from *Escherichia coli* infected with T4 bacteriophage. *Proc. Nat. Acad. Sci. U.S.* **57**: 1021–1028.

WILLETS, N. S., A. J. CLARK and B. LOW 1969 Genetic location of certain mutations conferring recombination deficiency in *Escherichia coli*. *J. Bact.* **97**: 244–249.

WITKIN, E. M. 1966a Radiation-induced mutations and their repair. *Science* **152**: 1345–1353.

WITKIN, E. M. 1966b Mutation and the repair of radiation damage in bacteria. *Radiation Research* Supplement **6**: 30–53.

WITKIN, E. M. 1967 Mutation-proof and mutation-prone modes of survival in derivatives of *Escherichia coli* B differing in sensitivity to ultraviolet light. *Brookhaven Symp. Biol.* **20**: 17–55.

WITKIN, E. M., N. A. SICURELLA and G. M. BENNETT 1963 Photoreversibility of induced mutations in a nonphotoreactivable strain of *Escherichia coli*. *Proc. Nat. Acad. Sci. U.S.* **50**: 1055–1056.

277

Wulff, D. L. and C. S. Rupert 1962 Disappearance of thymine photodimer in ultraviolet irradiated DNA upon treatment with a photo-reactivating enzyme from baker's yeast. *Biochem. Biophys. Res. Commun.* **7**: 237–240.

Yanofsky, C., J. Ito and V. Horn 1966 Amino acid replacements and the genetic code. *Cold Spring Harbor Symp. Quant. Biol.* **31**: 151–162.

28

Reprinted from *Mutation Res.*, **8**, 9–14 (1969)

THE MUTABILITY TOWARD ULTRAVIOLET LIGHT OF RECOMBINATION-DEFICIENT STRAINS OF *ESCHERICHIA COLI*

EVELYN M. WITKIN

State University of New York, Downstate Medical Center, Brooklyn, N.Y. (U.S.A.)

(Received December 24th, 1968)

SUMMARY

Strains having *rec*A and *rec*C mutations were tested for UV mutability. A *rec*A strain is UV-stable at doses that are demonstrably mutagenic in its Rec+ derivative. Strains of *rec*C genotype are not UV-stable, but produce UV-induced mutations at frequencies significantly lower than those produced at the same doses in Rec+ strains. The results support the hypothesis that most UV-induced mutations originate as errors in recombinational repair of single-strand gaps in DNA.

Mutations at the *exr* locus of *Escherichia coli* B eliminate all detectable mutability toward ultraviolet light[15]. In addition to being UV-stable, *exr*— strains are about twice as sensitive to UV[15] and only about half as efficient as recipients in genetic recombination (WITKIN, unpublished observation) as their otherwise isogenic Exr+ counterparts. It has been postulated[16] that most UV-induced mutations in *E. coli* originate as errors in recombinational repair of daughter-strand gaps, produced opposite unexcised pyrimidine dimers in the first post-irradiation DNA replication[12], and that the *exr*+ gene product, while necessary for maximal efficiency of recombinational repair, is also responsible for its inaccuracy. BRIDGES *et al.*[1] and KONDO[9] have also suggested that inaccurate recombinational repair, among other possibilities, could account for UV-induced mutations.

The *lex* locus[7] of *E. coli* K12 may be identical with the *exr* locus of *E. coli* B, since it has the same approximate map location, and produces the same cluster of phenotypic effects, including UV stability (WITKIN, unpublished observation). Other UV-sensitive strains of *E. coli* K12 which are also recombination-deficient have been described[2,3,6,7], and their genetic defects mapped at three distinct loci (*rec*A, *rec*B and *rec*C)[4,10]. The UV-sensitivity of these strains has been ascribed to their reduced ability to effect post-replication recombinational repair of gaps opposite unexcised pyrimidine dimers[5,8]. If UV-induced mutations are errors in recombinational repair, *rec*A strains, which lack recombination ability entirely, should be UV-stable; *rec*B and *rec*C strains, in which some recombination ability is retained, should show reduced UV mutability. On the other hand, if the *exr*+ or *lex*+ gene product introduces errors into the DNA through an inaccurate gap-repair process which is independent of recombination (*e.g.*, random insertion of bases into gaps[1,15]), *rec*— strains should exhibit normal UV mutability. This report describes preliminary results, using strains carrying the *rec*A1 and the *rec*C22 alleles, in a study of the UV mutability of recombination-deficient strains.

MATERIAL AND METHODS

Table I lists the strains used, their relevant characteristics and their sources.

Mutations from streptomycin-sensitivity to streptomycin-resistance (Str-s to Str-r) were studied using methods and media previously described[17], with modifica-

TABLE I

DESCRIPTION OF STRAINS USED[a]

Strain	Rec genotype	Other relevant markers	UV-sensitivity (1/e dose in erg/mm²)	Derived from strain	Obtained from
KL1699	*rec*A1	*str*-s	4	K12	W. Maas
KL1699-1	+	*str*-s	450	1699	—
DM37	*rec*C22	*str*-s *mal*—	100	B/r[b]	D. Mount
B/r	+	*str*-s *mal*—	450	B	—
DM34	*rec*C22	—	40	K12	D. Mount
DM34-5	*rec*C22	*pur*—	40	DM34	—
DM34-5-1	+	*pur*—	450	DM34-5	—

[a] All strains listed are *exr+ hcr+* (B) or *lex+ uvr+* (K12).
[b] *rec*C22 transduced into B/r from JC5474 *via* P₁ phage.

tions involving only the number of cells plated and the time of adding streptomycin. In all experiments, streptomycin was added after incubation for a time previously determined to permit a 10-fold increase of the plated inoculum. This time varied from 2–6 h.

Mutations from purine requirement to non-requirement (Pur— to Pur+) were scored using methods and media previously described for mutations from auxotrophy to prototrophy[13]. The same methods were used for the study of mutations from inability to ability to ferment maltose (Mal— to Mal+), except that cultures were initially grown in minimal "E" medium containing 0.4% glucose, and the selective plating medium used to score Mal+ mutants was minimal "E" agar containing 0.4% maltose and 0.04% glucose. The *mal* marker in strains B/r and DM37 is derived from the wild-type strain B, which is naturally *mal*—.

Methods of growing cultures, preparing them for irradiation and exposing them to ultraviolet light were all as previously described[13].

RESULTS

All results are presented in Table II.

Tests with *rec*A1 strain KL1699 were limited to UV doses no higher than 20 erg/mm² because of its extreme UV-sensitivity. At this low dose, mutations from Str-s to Str-r were detected in the Rec+ derivative of KL1699, strain KL1699-1, but none were found among the survivors in the *rec*A1 strain, although the irradiated suspensions were concentrated as much as 100 times to increase the size of the population screened. If this *rec*A strain is capable of producing UV-induced mutations to

TABLE II

MUTATIONS INDUCED BY ULTRAVIOLET LIGHT IN Rec— AND Rec+ STRAINS OF *E. coli*

Strain	Rec genotype	Mutation scored	UV dose (erg/mm²)	Survival %	Total number of survivors screened	Total number of induced mutations	Number of induced mutations per 10^8 survivors
KL1699	recA1	Str-r	20	0.9	$1.2 \cdot 10^{10}$	0	<0.008
KL1699-1	+	Str-r	20	100	$1.4 \cdot 10^{9}$	56	4
DM37	recC22	Str-r	225	12.5	$1 \cdot 10^{9}$	140	14
B/r	+	Str-r	225	100	$8.2 \cdot 10^{8}$	1742	212
DM37	recC22	Mal+	225	11.6	$4.4 \cdot 10^{7}$	640	1450
B/r	+	Mal+	225	100	$6.8 \cdot 10^{7}$	3680	5410
DM34-5	recC22	Pur+	80	12.6	$4.2 \cdot 10^{7}$	312	742
DM34-5-1	+	Pur+	80	100	$1 \cdot 10^{8}$	1650	1650

streptomycin-resistance, it must do so with a frequency at least 500 times lower than that found in its Rec+ revertant exposed to the same UV dose, and below the practical limits of detectability.

Strains carrying the *recC22* allele are not as UV-sensitive as *recA* strains, and were tested at doses ranging from 80 to 225 erg/mm². Mutations from Str-s to Str-r and from Mal— to Mal+ were induced in strain DM37, but with frequencies respectively 1/15 and 1/4 as high as those obtained at the same dose in its Rec+ parent strain B/r. Mutations from Pur— to Pur+ were also induced in a *recC22* strain, DM34-5, with a frequency slightly less than half that observed in a Rec+ revertant, DM34-5-1, obtained from it by selection for a UV-resistant variant. Strains carrying the *recC22* allele are clearly UV-mutable, but the frequency of all three classes of induced mutations, expressed as the number per 10^8 survivors, is markedly lower than that produced at the same dose in Rec+ strains.

DISCUSSION

UV-mutability is either eliminated or significantly reduced in strains carrying a *recA1* or a *recC22* mutation. Preliminary results with *recB* strains indicate that UV-mutability is also greatly reduced by *recB* mutations. If UV-induced mutations were caused primarily by an error-prone gap-filling mechanism operating independently of recombination, *rec* mutations would not be expected to affect the UV-mutability of Exr+ strains. The requirement for all three functional *rec* gene products, as well as for the product of the *exr* gene, for normal UV-mutability, greatly strengthens the likelihood that the induction of a mutation by UV actually requires a recombinational event.

If UV-induced mutations are indeed errors in the recombinational repair of single-strand gaps in DNA, *recA* strains should be UV-stable, but not for the same reason that *exr*— strains are UV-stable. *Exr*— strains are UV-stable because their recombinational repair appears to have been converted from an error-prone to an error-free mechanism, with relatively little reduction in efficiency[15]. *RecA* strains should be UV-stable, *a priori*, because they are unable to effect any recombinational repair, whether accurate or inaccurate. A process that does not happen cannot make mistakes.

The extreme sensitivity to UV of those *recA* strains which also lack excision ability because of a *uvr* mutation indicates that such strains can tolerate no more than 1–2 unexcised pyrimidine dimers[5]. Since loss of excision ability does not change the tolerance of a strain for unexcised dimers, but only its ability to repair them[15], recA Uvr+ strains, like strain KL1699, probably have the same low tolerance for unexcised pyrimidine dimers (1–2), although their mean lethal dose is about 20 times higher (4 erg/mm²). At this dose, about 24 pyrimidine dimers are produced per genome, but it is likely that only 1–2 remain in the DNA after excision repair is completed. This implies that any recA strain, whether or not it has excision ability, can survive UV irradiation only if at least one strand of its DNA is free of pyrimidine dimers by the time of DNA replication. Only then will at least one intact daughter molecule, requiring no recombinational repair, be produced. The predicted UV-stability of recA strains, therefore, arises from the expectation that only initially undamaged strands of DNA, or strands which have been completely repaired by the excision–resynthesis repair system (which is known to be highly accurate[14]), can contribute genetic information to the descendents of the original survivor. Since such DNA strands are essentially like those of untreated cells, they would not be expected to yield UV-induced mutations. Although the greatly reduced UV-mutability of strain KL1699 is compatible with the prediction that recA strains should be UV-stable, the doses that can be used with such sensitive strains are only weakly mutagenic even in Rec+ strains. The data permit us to conclude, therefore, only that UV-induced mutations to streptomycin resistance, in the recA strain KL1699, are produced (if at all) with a frequency at least 500 times lower than that obtained in its Rec+ derivative. This result supports the hypothesis[16] that UV-induced mutations in *E. coli* originate as errors in the recombinational repair of gaps. The recA gene product is required for *any* recombinational repair; the exr gene product is responsible for the inaccuracy of this repair.

Strains having a mutation in the recC locus, unlike recA strains, are able to effect a considerable amount of recombination, and their sensitivity to UV is usually no more than about 10 times greater than that of Rec+ strains[4,5]. (However, the effect of a given recB or recC marker on UV-sensitivity and recombination ability varies considerably with the genetic background. The same recC allele (C22), for instance, which reduces UV-resistance and recombination ability more than 10-fold in strain DM34, causes only a 4-fold reduction when it is transduced into strain B/r, as in strain DM37. Even within the K12 family of strains, the variable expression of some recB and recC alleles is very marked, indicating that recombination ability is probably influenced by many factors not yet identified). Since recC strains do perform some recombinational repair, although less efficiently than Rec+ strains, their UV-mutability indicates that the recombinational repair that does occur is inaccurate, as it is in Rec+ strains which are also Exr+(Lex+). Thus, while the recC22 mutation reduces the efficiency of recombinational repair, it does not eliminate its error-prone feature.

Although recC strains are mutable toward UV, it is difficult to make a quantitative statement about their UV-mutability. There is no doubt that the frequency of UV-induced mutations, expressed as the number per 10^8 survivors, is from 2 to 15 times lower in recC strains than in Rec+ strains. However, in comparing strains which differ in their tolerance of unexcised pyrimidine dimers, this kind of comparison can be very misleading. The important value to compare (at least in the case of mutations known to be caused by unexcised pyrimidine dimers, such as the enzymatically photoreversible mutations induced in Hcr— strains[14]) is the number of mutations *per unexcised pyrimidine dimer*.   Two strains may have the same survival level at the same dose of UV, and yet may differ greatly in the average number of unexcised pyrimidine dimers per survivor. For example, a Rec+ Uvr— strain (owing its UV-sensitivity to inability to excise pyrimidine dimers) and a Rec— Uvr+ strain (owing its UV-sensitivity to reduced recombinational repair) may have the same mean lethal dose of about 20 erg/mm². The survivors at this dose in the first strain, however, will contain about 100 unexcised pyrimidine dimers per genome; those in the second strain, only about five. The first strain cannot excise dimers, but can tolerate many; the second strain can excise most of the dimers, but can tolerate few. The result may be similar survival but the first strain would yield about 20 times more induced mutations per survivor, *even if the probability of mutation per unexcised dimer is exactly the same in the two strains*. Because of such considerations, the reduced mutation frequencies per survivor observed in the *recC* strains, as compared to Rec+ strains, do not necessarily mean that the accuracy of recombinational repair is greater in the recombination-deficient strains. It can be concluded only that *recC* mutations, unlike *exr* mutations, do not eliminate UV-mutability, and that they result in reduced yields of UV-induced mutations per survivor. A meaningful statement about their quantitative effects on the probability of error per unexcised pyrimidine dimer (*i.e.*, per act of recombinational repair) must await experiments, now in progress in this laboratory, designed to provide this information.

The mutagenicity of recombinational repair in Exr+ strains is probably not due to unequal exchange, since most UV-induced mutations in *E. coli* are single base substitutions[16]. It has been suggested[16] that the *exr* gene codes a modifying enzyme, active in the alteration of an end base so that the modified base is: (*1*) refractory to further exonuclease breakdown; (*2*) specifically prepared to participate in a recombinational event, and (*3*) altered in its pairing specificity, so that the probability of a subsequent replication error is greatly increased. Utilization of a less efficient alternative modifying mechanism by Exr— strains, which does not alter the pairing specificity of the modified base, could account for their UV-stability, as well as for their reduced recombination ability and increased levels of DNA breakdown.

Although data on spontaneous mutability are not included in this report, spontaneous rates of mutation from Mal— to Mal+ and from Pur— to Pur+ were found to be greatly reduced in *recC22* strains as compared to their Rec+ counterparts. Mutations at the *exr* locus do not reduce spontaneous mutability[15]. KONDO[9] has reported decreased spontaneous mutability in one Rec— strain and increased spontaneous mutability in another, and has proposed that minor modifications in the recombination machinery may greatly affect spontaneous mutability. PUGLISI[11] has suggested that spontaneous mutability is reduced by agents which inhibit DNA degradation. If so, the low spontaneous mutability of *recC* strains may be related to their "cautious" (*i.e.*, lower than normal) breakdown of DNA[5].

NOTE ADDED IN PROOF

After this article was submitted for publication, a paper appeared (MIURA, A., AND J.TOMIZAWA, Studies on radiation-sensitive mutants of *E. coli* III, Participation of the Rec system in induction of mutation by ultraviolet irradiation, *Mol. Gen. Genetics*,

103 (1968) 1–10) in which it is shown that Lac− mutations in bacteria (as well as clear mutations in λ phage) are induced by UV only in strains having the *rec*A+ genotype.

ACKNOWLEDGEMENTS

This research was conducted under grant No. 5-R01-A1-01240 from the National Institute of Allergy and Infectious Diseases, with the efficient assistance of Mr. E. FARQUHARSON.

REFERENCES

1 BRIDGES, B. A., R. E. DENNIS AND R. J. MUNSON, Differential induction and repair of ultra-violet damage leading to true reversions and external suppressor mutations of an ochre codon in *Escherichia coli* B/r WP2, *Genetics*, 57 (1967) 897–908.
2 CLARK, A. J., The beginning of a genetic analysis of recombination proficiency, *J. Cellular Physiol.*, 70 (1967) 165–180.
3 CLARK, A. J., AND A. D. MARGULIES, Isolation and characterization of recombination deficient mutants of *E. coli* K12, *Proc. Natl. Acad. Sci. (U.S.)*, 53 (1965) 451–459.
4 EMMERSON, P. T., Recombination deficient mutants of *Escherichia coli* K12 that map between *thy*A and *arg*A, *Genetics*, 60 (1968) 19–30.
5 HOWARD-FLANDERS, P., DNA repair, *Ann. Rev. Biochem.*, 37 (1968) 175–200.
6 HOWARD-FLANDERS, P., AND L. THERIOT, Mutants of *Escherichia coli* K12 defective in DNA repair and in genetic recombination, *Genetics*, 53 (1966) 1137–1150.
7 HOWARD-FLANDERS, P., AND R. P. BOYCE, DNA repair and genetic recombination: Studies on mutants of *Escherichia coli* defective in these processes, *Radiation Res., Suppl.*, 6 (1966) 156–184.
8 HOWARD-FLANDERS, P., W. D. RUPP AND B. M. WILKINS, The replication of DNA containing photoproducts and UV-induced genetic recombination, in W. F. PEACOCK (Ed.), *Replication and Recombination in Genetic Material*, Australian Academy of Sciences, Canberra, Australia, 1968, pp. 142–151.
9 KONDO, S., Mutagenicity *versus* radiosensitivity in *Escherichia coli*, *12th Intern. Congr. Genetics*, II (1968) 126–127 [Abstract].
10 LOW, B., N. S. WILLETS AND A. J. CLARK, in preparation.
11 PUGLISI, P. P., Antimutagenic activity of actinomycin D and basic fuchsin in *Saccharomyces cerevisiae, Mol. Gen. Genetics*, in the press.
12 RUPP, W. D., AND P. HOWARD-FLANDERS, Discontinuities in the DNA synthesized in an excision-defective strain of *Escherichia coli* following ultraviolet irradiation, *J. Mol. Biol.*, 31 (1968) 291–304.
13 WITKIN, E. M., The effect of acriflavine on photoreversal of lethal and mutagenic damage produced in bacteria by ultraviolet light, *Proc. Natl. Acad. Sci. (U.S.)*, 50 (1963) 1055–1058.
14 WITKIN, E. M., Radiation-induced mutations and their repair, *Science*, 152 (1966) 1345–1353.
15 WITKIN, E. M., Mutation-proof and mutation-prone modes of survival in derivatives of *Escherichia coli* B differing in sensitivity to ultraviolet light, *Brookhaven Symp. Biol.*, 20 (1967) 17–55.
16 WITKIN, E. M., The role of repair and recombination in mutagenesis, *Proc. 12th Intern. Congr. Genetics*, III (1968) in the press.
17 WITKIN, E. M., AND E. C. THEIL, The effect of posttreatment with chloramphenicol on various ultraviolet-induced mutations in *Escherichia coli*, *Proc. Natl. Acad. Sci. (U.S.)*, 50 (1960) 226–231.

29

Reprinted from *Proc. XII Intern. Congr. Genetics*, **2**, 126–127 (Aug. 1968)

MUTAGENICITY VERSUS RADIOSENSITIVITY IN ESCHERICHIA COLI

Sohei Kondo

Dept. of Fundamental Radiology, Faculty of Med., Osaka Univ., Osaka, Japan

Many lines of evidence are accumulating to provide information on the kinds of final alterations of DNA corresponding to mutations. However, the processes leading to the final fixation of mutagenic alterations are not yet elucidated. The author wishes to discuss the possible processes of mutagenesis and to present suggestive evidence for the proposed models which was obtained from the mutation work with E. coli done in his laboratory.

Fundamental Hypothesis and Models of Mutation Induction: We assume that 1) errors exist in biochemical reactions and that biological systems possess 2) repair (or error-correcting), 3) recombination (or variability producing) and 4) replication mechanisms. This hypothesis implies that mutation can be caused by

A) induction of errors in DNA structure,

B) action of recombination mechanism,

C) repair error, D) recombination error, and/or E) replication error.

After the occurrence of one or combination of these events, it may take one or more rounds of DNA replication to reach the final step of mutation fixation.

Methods and Materials: The method adopted to check the proposed models consists of observation and comparison of differences in mutagenic sensitivity to spontaneous, physical and chemical agents between a wild-type strain of E. coli and its radiosensitive derivatives defective in repair or recombination ability. Strains used were radioresistant strain H/r30(arg⁻phr⁻) of E. coli B and its revertant H/r30-R(arg⁻phr⁺) (1) and radiosensitive derivatives H$_s$30-R(arg⁻phr⁺hcr⁻), RII(arg⁻phr⁺hcr⁻), RI5(arg⁻phr⁺Rec⁻)) and NG 30 (arg⁻phr⁻Rec⁻) (2,3).

The mutation scored in all these strains was the phenotypic reversion of the same auxotrophic marker (arg⁻); it was at least partly a suppressor mutation. Mutants and survivals were scored by plating on semi-enriched agar medium. Acriflavine was added to the medium when necessary to reduce the repair of premutational and lethal damage induced by UV, 4NQO or MMC. Mutagens used were UV(ultraviolet radiation), X(X rays), 4NQO(4-nitroquinoline l-oxide), MMC(mitomycin C), EMS(ethyl methanesulfonate), MMS (methyl methanesulfonate) and NTG(N-methyl-N'-nitro-N-nitrosoguanidine). Overnight cultures were washed once, resuspended and then starved for l hr at 37°C in phosphate buffer. The buffer suspensions were exposed to UV or X rays. Or they were kept in dark at 30°C or room temperature for l to 3 hrs after addition of one of the above chemical mutagens and then the mutagen was removed by washing twice in buffer. Sensitivities to prophage induction by some of the mutagens were examined using derivatives of the above-mentioned strains lysogenized by phage Ø80, H/r30-R(80), RII(80), RI5(80) and NG30(80). Details of the biological procedures have been previously reported (2, 3, 4).

Results and Discussions: Experimental results are summarized in Table 1. The degree of sensitivity is given numerically as reciprocal ratio of dose D$_{37}$(37% survival) or D$_m$ (dose required for giving mutation frequency of l0⁻⁷); e.g., figures 0 and 4 mean, respectively, negligible inducibility and 4 times high sensitivity compared with figure l.

A) Mutation by induction of errors in DNA structure : UV- and 4NQO-produced lesions in DNA have killing as well as mutagenic effect because parallel increase in killing and mutation sensitivity occurred in H$_s$30-R, which lacks the excision-repair ability for UV and 4NQO damages, when compared with those in repair-proficient strain H/r30.

Table I. Comparison of differences in sensitivity to killing (K), mutation (M) and prophage induction (Ø) between different strains.

Mutagen / Strain	SPONT**			UV			4NQO		MMC		X		NTG		EMS		
	K	M	Ø	K	M	Ø	K	M	K	M	K	M	K	M	K	M	Ø
H/r30(Hcr⁺Rec⁺)	±	1$^{a)}$	+	1	1	1	1	1	1	1	1	1	1	1	1	1	0
H$_s$30 (Hcr⁻Rec⁺)	±	1.2$^{a)}$	+*	25	33	10*	26	30	10	0	1.2	1.2	1.4	1	1	1.4	0*
RI5 (Hcr$^{(+)}$Rec$^{(-)}$)	+	3$^{a)}$	++	12	1	10	10	1	1.5	1	3.3	1	10	1	10	0.7	
NG30(Hcr⁺Rec⁻)	+	0.3$^{a)}$	±	100	0	0	14	0	4	0	2.5	0	10	0.4	7	0.7	1
D$_{37}$/D$_m$ for H/r30 (dose units)	a) relative frequency			580/8 (erg/mm²)			130/8 (ɣ·hr/ml)		0.3/0.1 (ɣ·hr/ml)		5/1.6 (kR)		50/1.5 (ɣ·hr/ml)		3×10⁴/10³ (ɣ·hr/ml)		

** spontaneously very low (±), low (+) and high (++).   * Data with RII(80).

B) Mutation due to the action of recombination mechanism : Spontaneous mutation rate was low in the Rec⁻(recombination deficient) strain NG30 but was normal in the repair-deficient strains H$_s$30-R and RII. This suggests that spontaneous mutation is not due to spontaneous DNA damage of the type reparable by the excision-repair system but is presumably due, at least partly, to the action of recombination mechanism.

C) Mutation by repair errors : Repair errors are responsible, at least partly, for MMC-induced mutation in H/r30 because it was greatly suppressed by posttreatment with acriflavine and no MMC mutation was detected in H$_s$30-R and three other Hcr⁻ strains lacking the repair ability for MMC-induced damage in DNA (cross-linking).

D) Mutation by recombination errors : Recombination errors due to a minor defect in the recombination machinery may be responsible for the high spontaneous mutation rate in RI5. This strain has a slightly reduced genetic recombination frequency (Rec$^{(-)}$) when compared with normal Rec⁺ strains, and a reduced and normal host-cell-reactivation abilities (Hcr$^{(+)}$) for X-rayed and UV-irradiated T1 phage, respectively.

E) Mutation by replication errors : Replication errors by mutagen-induced damage in the replication machinery may be responsible for NTG- and EMS-induced mutation in NG30. This strain was extremely sensitive to killing by   all the mutagens tested but was not susceptible to mutation induction by UV, 4NQO, MMC and X rays which are well-known to react with DNA. Therefore, the ability of NTG and EMS to induce mutation in NG30 indicates that these agents primarily react NOT with DNA but with replication machinery, e.g., polymerase or other essential components, to the effect that the damage inudced in the machinery leads to production of daughter DNA full of errors even from parental intact DNA. This model was supported by the findings that NTG and EMS showed low and negligible efficiencies, respectively, for killing and prophage induction in the Rec⁺ strains used while they were able to induce prophage in Rec⁻ strain NG30(80) whose prophage was not inducible by UV irradiation (cf. Table I).

References
1. Witkin, E. M. Mutation Res. 1, 22 (1964).
2. Kondo, S. and Kato, T. Photochem. Photobiol. 5, 827 (1966).
3. Kato, T. and Kondo, S. Mutation Res. 4, 253 (1967).
4. Takebe, H., Ichikawa, H., Iwo, K. and Kondo, S. Virology 33, 638 (1967).

Part V

THE GENETIC DETERMINATION AND EVOLUTION OF MUTATION RATES

Editors' Comments
on Papers 30 Through 33

HOT SPOTS AND COLD SPOTS

Numerous observations accumulating from about 1920 on had established that some genes were intrinsically much more mutable than others, and even that mutation rates at specific loci could vary considerably among organisms with in the same species. It was therefore clear that gene mutability was itself genetically controlled. These observations were strikingly reproduced at the intragenic as well as intergenic level by Benzer's extensive fine-scale analyses of the mutability of the *rII* genes of bacteriophage T4. Benzer's studies, of which Paper 30 is characteristic, were primarily aimed at elucidating the topography of the gene itself, and depended upon the isolation and mapping of literally thousands of mutants. The two T4*rII* genes are now known to contain about 3,600 base pairs (O'Farrell et al., 1973), but only a small fraction of these potential sites were ever detected among all the mutants mapped by Benzer, whether of spontaneous or of induced origin. Furthermore, the numbers of spontaneous mutants per detected site varied enormously; some sites were represented by only a single occurrence and others by hundreds of independent recurrences. Such mutational "hot spots," in fact, completely dominate the T4*rII* fine-scale map; fully 50 percent of all spontaneous mutants arise in only two such regions.

The data of Paper 30, then, raised two fundamental questions about the genetic determination of mutation rates at specific base pair sites. The first was why so few sites were ever detected by mutation: although it is thermodynamically nonsensical to suppose that replication of any base pair is ever completely error free, it is certainly conceivable that some base pairs mutate perhaps even thousands of times less frequently than the average base pair. The second question was why some base pairs clearly do mutate at least hundreds of times more frequently than does the average base pair.

There are several reasons to anticipate the relative immutability of at least many sites within a gene. First, the extensive degeneracy of the codon catalogue ensures that many base pair substitutions arising in those portions of a gene which encode polypeptide sequences go undetected because they do not produce amino acid substitutions. This is particularly true in the case of transitions and almost always true of transitions at the third positions of codons. Second, many amino acid substitutions may be quite innocuous (see, for instance, Perutz and Lehmann, 1968). This possibility is particularly likely in the case of the T4*rII* locus, since the ratio of missense to nonsense mutants is far less than 20, and since nonsense mutants are well suppressed by very inefficient suppressors. How, then, might one determine experimentally whether the missing missense mutants were really arising? The answer, described in Paper 31, was to make the gene sufficiently sick so that even small increments of additional mutational damage would result in readily observable phenotypic changes.

Mutational hot-spotting and large variations in site-specific mutability generally are now known to occur in all genes that have been sufficiently closely examined. They are commonly attributed to chemically mysterious phenomena called *nearest-neighbor* or *neighboring-base* effects. The main argument for the existence of neighboring-base effects is simply that there are many more different site-specific mutation rates (ranging over many orders of magnitude) than there are different types of base pairs (four), so that the mutation rate of a given base pair must be influenced by its neighbors. Even if such influences extended only from the first two neighbors on either side of a particular base pair, for example, the possible number of different mutation rates would be increased from 4 to 1,024. The possibility always existed, however, that these supposedly *neighboring* base effects were in fact determined by base pair sequences at a considerable distance from the affected site. Formal proof of truly neighboring base effects therefore required that neighboring base pair changes be shown to bring about site-specific mutation rate changes. The first such proof is described in Paper 32.

Thus far we have considered site-specific mutability only in terms of base pair substitution pathways. Most of the mutations depicted on the fine-scale map of Paper 30, however, actually consist of frameshift mutations, including the two giant hot spots. Since frameshift mutants are rarely leaky and are detected with a very high efficiency, large differences in site-specific frameshift mutability clearly reflect direct differences in local base sequence composition. An elegant insight into the genetic determination of site-specific frameshift mutability is provided in Paper 33 from the Streisinger group. According to the theory of frameshift mutagenesis laid out in Paper 25, the frameshift mutation rate should be strongly influenced by the density of opportunities for misalignment in the neighborhood of a single-strand DNA interruption. Paper 25 shows that an extremely highly frameshift mutable region of the T4 lysozyme gene does in fact contain a high density of local base sequence redundancy. (The two giant *rII* hot spots presumably also contain similar repeating sequences.) Furthermore, a mutation that increased the amount of local base sequence redundancy in the lysozyme gene hot spot greatly increased the frameshift mutation rate in that region.

REFERENCES

O'Farrell, P. Z., L. M. Gold, and W. M. Huang. 1973. The identification of prereplicative bacteriophage T4 proteins. *J. Biol. Chem., 248:* 5499.

Perutz, M. F., and H. Lehmann. 1968. Molecular pathology of human haemoglobin. *Nature, 219:* 902.

30

Reprinted from *Proc. Natl. Acad. Sci.*, **47**(3), 403–415 (1961)

ON THE TOPOGRAPHY OF THE GENETIC FINE STRUCTURE

By Seymour Benzer

DEPARTMENT OF BIOLOGICAL SCIENCES, PURDUE UNIVERSITY

Read before the Academy, April 27, 1960*

In an earlier paper,[1] a detailed examination was made of the structure of a small portion of the genetic map of phage T4, the *r*II region. This region, which controls the ability of the phage to grow in *Escherichia coli* strain K, consists of two adjacent cistrons, or functional units. Various *r*II mutants, unable to grow in strain K, have mutations affecting various parts of either or both of these cistrons. The topology of the region; i.e., the manner in which its parts are interconnected, was intensively tested and it was found that the active structure can be described as a string of subelements, a mutation constituting an alteration of a point or segment of the linear array.

This paper is a sequel in which inquiry is made into the topography of the structure, i.e., local differences in the properties of its parts. Specifically, are all the subelements equally mutable? If so, mutations should occur at random throughout the structure and the topography would be trivial. On the other hand, sites or regions of unusually high or low mutability would be interesting topographic features.

The preceding investigation of topology was done by choosing mutants showing no detectable tendency to revert. This avoided any possible confusion between recombination and reverse mutation, so that a qualitative (yes-or-no) test for recombination was possible. The class of non-reverting mutants automatically included those marked by relatively large alterations, which will be referred to as "deletions." Such a mutant is defined for the present purposes as one which inter-

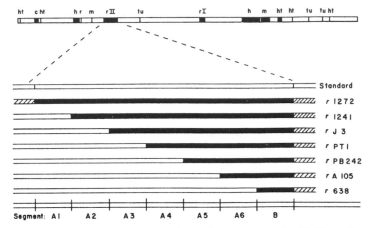

FIG. 1a.—At the top, the *r*II region is shown compared with the entire genetic map of the phage. This map is a composite[15] of markers mapped in T4 and the related phage T2. Seven segments of the *r*II region are defined by a set of "deletions" beginning at different points and extending to the right-hand end (and possibly beyond, as indicated by shading).

sects (fails to give recombination with) two or more mutants that do recombine with each other. Deletions provided overlaps of the sort needed to test the topology and to divide the map into segments.

The present investigation of topography, however, is concerned with differentiation of the various points in the structure. For this purpose mutants which do revert are of the greater interest, since they are most likely to contain small alter-

FIG. 1b.—Mapping a mutation by use of the reference deletions. If mutant x has a mutation in segment 1, it is overlapped by *r*1272, but not by *r*1241. Therefore, standard-type recombinants (as indicated by the dotted line) can only arise when x is crossed with *r*1241.

ations. As a rule (there are exceptions) an *r*II mutant that reverts behaves as if its alteration were localized to a point. That is to say, mutants that intersect with the same mutant also intersect with each other. In a cross, recombination can be scored only if it is clearly detectable above the spontaneous reversion noise of the mutants involved. Therefore, the precision with which a mutation can be mapped is limited by its reversion rate. The detailed analysis of topography can best be done with mutants having low, non-zero reversion rates.

Some thousands of such *r*II mutants, both spontaneous and induced, have been analyzed and the resultant topographic map is presented here.

Assignment of Mutations to Segments.—To test thousands of mutants against one another for recombination in all possible pairs would require millions of crosses. This task may be greatly reduced by making use of deletions. Each mutant is first tested against a few key deletions. The recombination test gives a negative result if a deletion overlaps the mutation in question and a positive result if it does not overlap. These results quickly locate a mutation within a particular segment of the map. It is then necessary to test against each other only the group of mutants having mutations within each segment, so that the number of tests needed is much smaller. In addition, if the order of the segments is known, the entire set of point mutations becomes ordered to a high degree, making use of only qualitative tests.

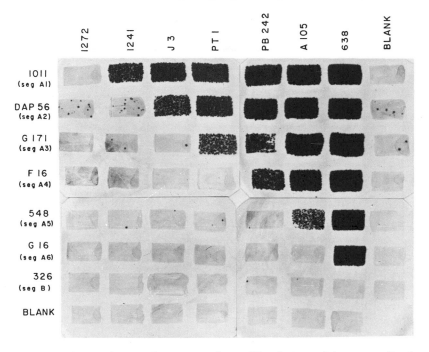

FIG. 2.—Crosses for mapping *r*II mutations. The photograph is a composite of four plates. Each row shows a given mutant tested against the reference deletions of Figure 1*a*. Plaques appearing in the blanks are due to revertants present in the mutant stock. The results show each of these mutations to be located in a different segment.

Procedure for crosses—The broth medium is 1% Difco bacto-tryptone plus 0.5% NaCl. For plating, broth is solidified with 1.2% agar for the bottom layer and 0.7% for the top layer. Stocks are grown in broth using *E. coli* BB which does not discriminate between *r*II mutants and the standard type. To cross two mutants, one drop of each at a titer of about 10^9 phage particles/ml is placed in a tube and cells of *E. coli* B are added (roughly 0.5 ml of a 1-hour broth culture containing about 2×10^8 cells/ml). The *r*II mutants are all able to grow on strain B and have an opportunity to undergo genetic recombination. After allowing a few minutes for adsorption, a droplet of the mixture is spotted (using a sterile paper strip) on a plate previously seeded with *E. coli* K. If the mutants recombine to produce standard type progeny, plaques appear on K. A negative result signifies that the proportion of recombinants is less than about 10^{-3}% of the progeny.

Within any one segment, however, the order of the various sites remains undetermined. This order can still be determined, if desired, by quantitative measurements of recombination frequencies.

In order to facilitate this project many more deletions have been mapped than were described in the previous paper. These suffice to carve up the structure into 47 distinct segments. By virtue of the proper overlaps, the order of almost all of these segments is established.   Observe first the seven large mutations in Figure 1*a*. These are of a kind which begin at a particular point and extend all the way to one end.   Thus, they serve to divide the structure into the seven major segments shown.

Consider a small mutation located in the segment A1, as indicated in Figure 1*b*. It is overlapped by *r*1272 and therefore when crossed with it cannot give rise to standard type recombinants.   It will, however, give a positive result with *r*1241 or any of the others, since, with them, recombinants can form as indicated by the

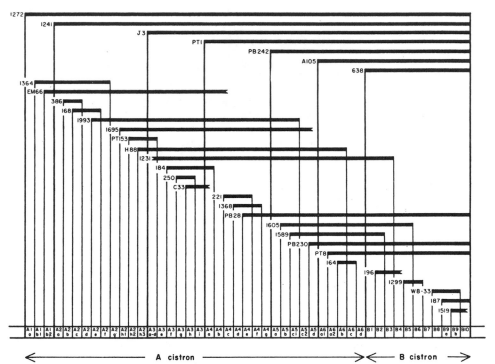

←——————————— A cistron ————————————→ ←— B cistron —→

FIG. 3.—Deletions used to divide the main segments of Figure 1 into 47 smaller segments. (Some ends have not been used to define a segment, and are drawn fluted.)   The A and B cistrons, which are defined by an independent functional test, coincide with the indicated portions of the recombination map.   Most of the mutants are of spontaneous origin.   Possible exceptions are EM66, which was found in a stock treated with ethyl methane sulfonate, and the PT and PB mutants, which were obtained from stocks treated with heat at low pH.   The PT mutants were contributed by Dr. E. Freese.

dotted line.   A point mutation located in the second segment will give zero with mutants *r*1272 and *r*1241 but not with the rest, and so on.   Thus, if any point mutant is tested against the set of seven reference mutants in order, the segment in which its mutation belongs is established simply by counting the number of zeros. Figure 2 shows photographs of the test plates for seven mutants, each having its mutation located in a different segment.

Only these seven patterns, with an uninterrupted row of zeros beginning from the left, have ever been observed for thousands of mutants tested against these seven deletions.   The complete exclusion of the other 121 possible patterns confirms the linear order of the segments.

Now a given segment can be further subdivided by means of other mutations having suitable starting or ending points. Figure 3 shows the set used in this study and the designation of each segment. Each mutant is first tested against the seven which have been chosen to define main segments. Once the main segment is known,

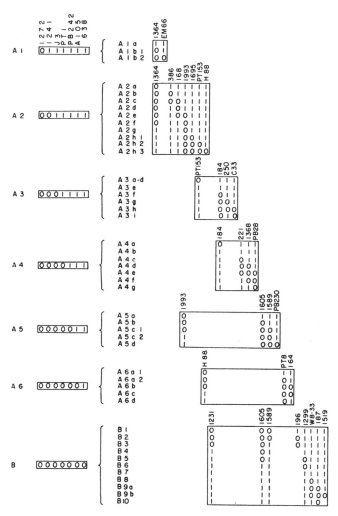

FIG. 4.—The test pattern which identifies the location of a point mutation in each of the segments of Figure 3. The test is done in two stages. An unknown mutant is first crossed with the "big seven" of Figure 1 in order. *Zero* signifies no detectable recombination and *one* signifies some, and the number of zeros defines the major segment. Once this is known, the mutant is crossed with the pertinent selected group of deletions to determine the small segment to which it belongs.

the mutant is tested against the appropriate secondary set. Figure 4 shows the pattern which identifies the location of a point mutation within each of the small segments. Thus, in two steps, a point mutation is mapped into one of the 47 segments.

The order of the first 42 segments, Ala through B6, is uniquely defined.   Unfortunately, there remains a gap between r1299 and rW8-33.   Therefore the order of segments B8 through B10, although fixed among themselves, could possibly be the reverse of that shown.[17]   Also if there exists space to the right of segment B10, a mutation in that segment might map as if it were in segment B7, so that the latter segment must be tentatively regarded as a composite.

In the previous topology paper, the possibility that the structure contains branches was not eliminated.   As pointed out by Delbrück, the existence of a branch would not lead to any contradiction with a linear topology if loss of a segment containing the branch point automatically led to loss of the entire branch.

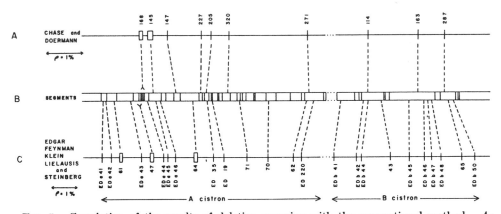

FIG. 5.—Correlation of the results of deletion mapping with the conventional method. A: The map constructed by Chase and Doermann[2] for ten rII mutants of phage T4B, using quantitative measurements of recombination frequency.   The interval between adjacent mutations is drawn proportional to the frequency of recombination in a cross between the two.   C: The map constructed in similar fashion by Edgar et al. (personal communication) with rII mutants of the very closely related phage T4D.   The procedure used by Edgar et al. gives higher recombination frequencies.   Therefore, the scales of the two maps are adjusted in the figure to produce a good over-all fit.   Some of the mutations cover several sites and are drawn as having a corresponding length.   A gap is left between the two cistrons because crosses between mutations in different cistrons give abnormally high frequencies due to the role of heterozygotes[16].

All of these mutations have also been mapped by the deletion method, and dotted lines indicate their locations in the various segments (B).   The length of each segment is drawn in proportion to the number of distinct sites that have been found within it.

To show that a given segment is *not* a branch, it is required to find a mutation which penetrates it partially.   From the mutations shown in Figure 3, it can be concluded that no branch exists that contains more than one of the 47 segments.

Comparison of Deletion Mapping by Recombination Frequencies.—The conventional method of genetic mapping makes use of recombination frequency as a measure of the distance between two mutations and requires careful quantitative measurements of the percentage of recombinant type progeny in each cross.   By the method of overlapping deletions the order of mutations can be determined entirely by qualitative yes-or-no spot tests.   Maps obtained independently by the two methods are compared in Figure 5.   The upper part of the figure (A) shows the order obtained by Chase and Doermann[2] for a set of ten mutants, the distance between adjacent mutations being drawn proportional to the percentage of standard-type recombinants occurring among the progeny of a cross between the two.   The central part of the figure (B) shows the rII region divided into the segments of

Figure 3, with the size of each segment drawn in proportion to the number of distinct sites which have been discovered within it (see below). As indicated by the dotted lines, there is perfect correlation in the order. In the lower part of the figure, a similar comparison is made for a set of *r*II mutations in the closely related phage strain T4D, which have been mapped, using recombination frequencies, by Edgar, Feynman, Klein, Lielausis, and Steinberg. Again the order agrees perfectly with that obtained by the use of deletions.

Topography for Spontaneous Mutations.—We now proceed to map reverting mutants of T4B which have arisen independently and spontaneously. The procedure is exactly as in Figure 4: first localizing into main segments, then into smaller segments. Finally mutants of the same .small segment are tested against each other. Any which show recombination are said to define different sites. If two or more reverting mutants are found to show no detectable recombination with each other, they are considered to be repeats and one of them is chosen to represent the site in further tests. A set of distinct sites is thus obtained, each with its own group of repeats.

This procedure is based on the assumption that revertibility implies a point mutation. While this is a good working rule for *r*II mutants, a few exceptions have been found which appear to revert (i.e., give rise to some progeny which can produce detectable plaques on stain K) yet fail to give recombination with two or more mutants that do recombine with each other. If a mutant chosen to represent a "site" happens to be of this kind, mutations it overlaps will appear to be at the same site. Therefore, a group of "repeats" remains subject to splitting into different groups when they are tested against each other. This has not yet been done for all of the sites described here. It is, of course, in the nature of the recombination test that it is meaningful to say that two mutations are at different sites, while the converse conclusion is always tentative.

Figure 6 shows the map obtained for spontaneous mutants, with each occurrence of a mutation at a site indicated by a square. Within each segment the sites are drawn in arbitrary order. Other known sites are also indicated even though no occurrences were observed among this set of spontaneous mutants.

That the distribution is non-random leaps to the eye. More than 500 mutations have been observed at the most prominent "hotspot," while, at the other extreme, there are many sites at which only a single occurrence, or none, has so far been found.

To decide whether a given number of recurrences is significantly greater than random, the data may be compared with the expectation from a Poisson distribution. Figure 7 shows a distribution calculated to fit the least hot of the observed spontaneous sites, i.e., those at which one or two mutations have occurred, on the assumption that these sites belong to a uniform class of sites of low mutability. Comparing the observations with this curve, it would seem that if a site has four occurrences, there is a two-thirds probability that it is truly hotter than the class of sites of low mutability. Those having five or more are almost certainly hot. It can be concluded that at least sixty sites belong in a more mutable class than the coolest spots. Whether the hot sites can be divided into smaller homogeneous groups, assuming a Poisson distribution within each class, is difficult to say. Each of the two hottest sites is obviously unique.

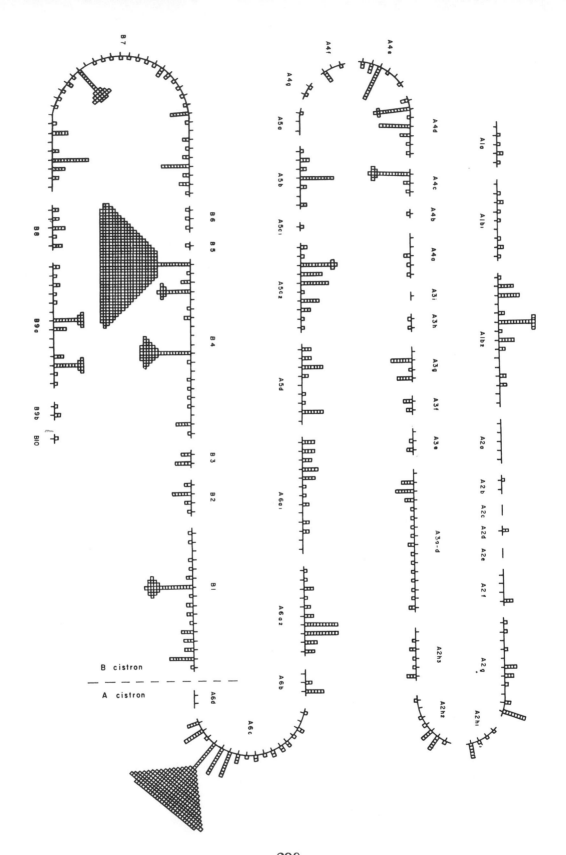

From the distribution it can be predicted that there must exist at least 129 spontaneous sites not observed in this set of mutants.   This is a minimum estimate since it is calculated on the assumption that the 2-occurrence sites are no more mutable than the 1-occurrence sites.   If this is not correct, the predicted number of

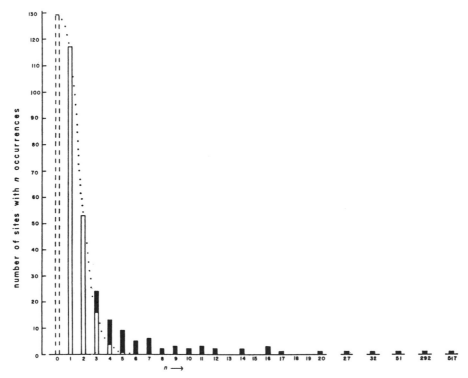

Fig. 7.—Distribution of occurrences of spontaneous mutations at various sites.   The dotted line indicates a Poisson distribution fitted to the numbers of sites having one and two occurrences.   This predicts a minimum estimate for the number of sites of comparable mutability that have zero occurrences due to chance (dashed column at n = 0).   Solid bars indicate the minimum numbers of sites which have mutation rates significantly higher than the one- and two-occurrence class.

0-occurrence sites will be larger.   Also, of course, there could exist a vast class of sites of much lower mutability.   With 251 spontaneous sites identified and at least 129 more to be found, the degree of saturation of the map achieved with this set of 1,612 spontaneous mutants can be no greater than 66 per cent.

←

Fig. 6.—Topographic map of the *r*II region for spontaneous mutations.   Each square represents one occurrence observed at the indicated site.   Sites with no occurrences indicated are known to exist from induced mutations and from a few other selected spontaneous ones.   The order of the segments is known for A1a through B7, but is only tentative for B7 through B10.[17] The arrangement of sites within each segment is arbitrary.

Each mutant arose independently in a plaque of either standard-type T4B or, in somewhat less than half of the cases, revertants of various *r*II mutants.   All revertants (except F) gave results very similar to T4B.   The pattern for *r*II mutants isolated from revertant F differs noticeably only in a reduced rate at the hotspot 117 (the site of its original *r*II mutation) and therefore does not significantly alter the topography.   All the data for mutants isolated from standard type and from revertants are pooled in this figure.

Topography for Induced Mutations.—By the use of specific mutagens, new topographic features are revealed. This has been shown for *r*II mutants induced during reproduction of the phage inside the bacterial host cell with 5-bromouracil (Benzer and Freese[3]), proflavine (Brenner, Barnett, and Benzer[4]), and 2-aminopurine (Freese[5]). Other effective mutagens are 2,6-diaminopurine (Freese[5]) and 5-bromodeoxycytidine (Gregory, personal communication). Mutations may also be induced *in vitro*, i.e., in extracellular phase particles, by ethyl methane sulfonate (Loveless[6]) and nitrous acid (Vielmetter and Wieder;[7] Freese;[8] Tessman[9]).

*r*II mutants induced by all of these mutagens have now been mapped with respect to each other and spontaneous ones, and the results are given in Figure 8 (facing page 416) which shows the locations of over 2,400 induced and spontaneous mutations. Only *r*II mutants that have low reversion rates and are not too "leaky" on K have been included.

Each "spectrum" differs obviously from the spontaneous one. While the specificities of the various mutagens overlap in many respects, each differs significantly from the others at specific points. In making the comparison it must be borne in mind that the total number of mutants mapped is not the same for each mutagen and also that each induced set inevitably includes some proportion of spontaneous mutants. (An upper limit to this background can be set from the number of occurrences at the hottest spontaneous sites.) Also, none of the spectra are "saturated." Therefore, even if two mutagens act similarly upon a given site, it is possible, due to chance, that a few occurrences would be observed in one spectrum and not the other. Within these limitations, the map shows the comparative response at each site to each mutagen as well as the locations of various kinds of hotspots in various segments of the *r*II region.

The study of the induced mutations has added 53 new sites to the 251 identified by the spontaneous set alone, bringing the total to 304. (Four sites more are shown in Figure 8, but they come from a selected group of mutants outside this study.) Thus, a closer approach toward saturation of all the possible sites must have been made. By lumping together all the data, both spontaneous and induced, one can again make an estimate of the number of sites which must be detectable if one were to continue mapping mutants in the same proportion for the same mutagens. The result is that there must exist still a minimum of 120 sites not yet discovered. This appears discouragingly similar to the estimate based on spontaneous mutations alone. However, it need not be surprising if the use of mutagens brings into view some sites which have extremely low spontaneous mutability. With 308 sites identified and at least 120 yet to be found, the maximum degree of saturation of the map is 72 per cent.

Discussion.—One topographic feature, non-random mutability at the various sites, is obvious. Another question is whether mutable sites are distributed at random, or whether there exist portions of the map that are unusually crowded with or devoid of sites. The mapping technique used here defines only the order of sites from one segment to another (but not within a given segment). The distance between sites remains unspecified. However, all mutations in a segment more distal to a given point must be farther away than those in a more proximal segment. If the number of sites in a segment is used as a measure of its length, as in Figure 5, it can be seen that there is no major discrepancy between these distances and those defined in terms of another measure of distance, recombination frequency. On a gross scale, therefore, there is no evidence for any large portion of the rII region that is unusually crowded or roomy with respect to sites. This does not necessarily mean that some other measure of distance would not reveal such regions, since it is at least conceivable that mutable sites coincide with points highly susceptible to recombination. The distribution of sites on a finer scale, within a small segment, remains to be investigated.

The number of points at which mutations can wreck the activity of a cistron is very large. This would be expected if a cistron dictates the formation of a poly peptide chain and "nonsense" mutations[10] are possible which interrupt the completion of the chain. Such mutations would be effective at any point of the structure, whereas ones which lead to "missense," i.e., the substitution of one amino acid for another, might be effective at relatively special points or regions which are crucial in affecting the active site or folding.

It would be of interest to compare the number of genetic sites to the material embodiment of the rII region in terms of nucleotides. Unfortunately, the size of the latter is not well known. Estimates based upon its length, in units of recombination frequency compared to the length of the entire genetic structure, are uncertain. A more direct attempt has been made using equilibrium sedimentation in a cesium chloride gradient and looking for a change in density of mutants known by genetic evidence to have portions of the rII region deleted (Nomura, Champe, and Benzer, unpublished). This technique has been successful in characterizing defective mutants of phage λ (Weigle, Meselson, and Paigen[11]) and is sufficiently sensitive to detect a decrease of 1 per cent in the amount of DNA per phage particle, but has so far failed with rII mutants. Although other explanations are possible, this result may suggest that the physical structure corresponding to the rII region represents less than 1 per cent of the total DNA of the phage particle, or less than 2,000 nucleotide pairs. If this is so, the number of possible sites would be of the order of at least one-fifth of the number of nucleotide pairs.

The data show that, if each site is characterized by its spontaneous mutability and response to various mutagens, the sites are of many different kinds. Some response patterns are represented only once in the entire structure. According to the Watson-Crick model[12] for DNA, the structure consists of only two types of elements, adenine-thymine (AT) pairs and guanine-hydroxymethylcytosine (GC) pairs. This does not mean, however, that there can only be two kinds of mutable sites, even if a site corresponds to a single base pair. Considering only base pair substitutions, a given AT pair can undergo three kinds of change: AT can be replaced by GC, CG, or TA. Certain of these changes may lead to a mutant phenotype, but some may not. The frequency of observable mutations at a particular AT pair will be determined by the sum of the probabilities for each type of change, each multiplied by a coefficient (either one or zero) according to whether that specific alteration at that particular pair does or does not represent a mutant type. Thus, if the probability that a base pair will be substituted is independent of its neighbors, the various AT sites may have seven different mutation rates. Similarly, there are seven rates possible for the various GC sites, so that it would be possible to account for fourteen classes by this mechanism. Some of these may have (total) spontaneous mutation rates that are similar. If a mutagen induces only certain substitutions, it will facilitate further discriminations between sites but there should still be no more than fourteen classes.

If one allows for interactions between neighbors, the number of possible classes increases enormously. Such interactions are to be expected. As an example, consider the fact that AT pairs are held together much less strongly than are GC pairs.[13] If several AT pairs occur in succession, this segment of the DNA chain will be relatively loose, making it easier to consummate an illicit base pairing during replication. Thus, guanine and adenine, which make a very satisfactory pair of hydrogen bonds but require a larger than normal separation between the backbones, could be more readily accommodated. This would lead, in the next replication, to a replica in which one of the AT pairs has been substituted by a CG pair, with the orientation of purine and pyrimidine reversed. Thus, a region rich in AT pairs will tend to be more subject to substitution. If the same (standard-type) phenotype can be achieved by alternative sequences, the ones containing long stretches of AT pairs would tend to be lost because of their high mutability. In other words, cistrons ought to have evolved in such a way as to eliminate hotspots. The spontaneous hotspots that are observed would be remnants of an incomplete ironing-out process. In fact, a map of the *rII* region of the related phage T6 (Benzer, unpublished) also shows hotspots at locations corresponding to *r*131 and *r*117. However, while the first of these has a mutability similar to that in T4, the second is lower by a factor of four.

This point is emphasized by the data on reverse mutations. It is not uncommon for an *rII* mutant to have a reverse mutation rate that is greater than the total forward rate observed for the composite of at least 400 sites. That some of these high-rate reverse mutations represent true reversion (and not "suppressor" mutations) has been established in several cases by the most stringent criteria, including the demonstration that the revertant has exactly the same forward mutation rate at the same site as did the original standard type (Benzer, unpublished). It would therefore appear that certain kinds of highly mutable configurations are systematically excluded from the standard form of the *rII* genetic structure, and a mutation may recreate one of these banned sequences.

In the attempt to translate the genetic map into a nucleotide sequence, the detection of the various sites by forward mutation is necessarily the first step. By studies on the specificity of induction of reverse mutations,[14] one site at a time can be analyzed in the hope of identifying the specific bases involved.

Summary.—A small portion of the genetic map of phage T4, the two cistrons of the *rII* region, has been dissected by overlapping "deletions" into 47 segments. If any branch exists, it cannot be larger than one of these segments. The overlapping deletions are used to map point mutations and the map order established by this method is consistent with the order established by the conventional method that makes use of recombination frequencies. Further dissection has led to the identification of 308 distinct sites of widely varied spontaneous and induced mutability. The distributions throughout the region for spontaneous mutations and those induced by various chemical mutagens are compared. Data are included for nitrous acid and ethyl methane sulfonate acting *in vitro*, and 2-aminopurine, 2,6-diaminopurine, 5-bromouracil, 5-bromodeoxycytidine, and proflavine acting *in vivo*. The characteristic hotspots reveal a striking topography.

It is a pleasure to thank Mrs. Karen Sue Supple, Mrs. Joan Reynolds, and Mrs. Lynne Bryant for their indefatigable assistance in mapping mutants and bookkeeping.   I am indebted to Dr. Robert S. Edgar and his associates for permission to make use of their unpublished data in Figure 5 and to Dr. Ernst Freese for several deletions as well as mutants induced with 2-aminopurine and 5-bromodeoxyuridine.   This research was supported by grants from the National Science Foundation and the National Institutes of Health.

* Given by invitation of the Committee on Arrangements for the Annual Meeting as part of a Symposium on Genetic Determination of Protein Structure, Robley C. Williams, Chairman.

[1] Benzer, S., these PROCEEDINGS, **45**, 1607 (1959).
[2] Chase, M., and A. H. Doermann, *Genetics*, **43**, 332 (1958).
[3] Benzer, S., and E. Freese, these PROCEEDINGS, **44**, 112 (1958).
[4] Brenner, S., L. Barnett, and S. Benzer, *Nature*, **182**, 983 (1958).
[5] Freese, E., *J. Molec. Biol.*, **1**, 87 (1959).
[6] Loveless, A., *Nature*, **181**, 1212 (1958).
[7] Vielmetter, W., and C. M. Wieder, *Z. Naturforsch*, **14b**, 312 (1959).
[8] Freese, E., *Brookhaven Symposia in Biol.*, **12**, 63 (1959).
[9] Tessman, I., *Virology*, **9**, 375 (1959).
[10] Crick, F. H. C., J. S. Griffith, and L. E. Orgel, these PROCEEDINGS, **43**, 416 (1957).
[11] Weigle, J., M. Meselson, and K. Paigen, *J. Molec. Biol.*, **1**, 379 (1959).
[12] Watson, J. D., and F. Crick, *Cold Spring Harbor Symposia Quant. Biol.*, **18**, 123 (1953).
[13] Doty, P., J. Marmur, and N. Sueoka, *Brookhaven Symposia in Biol.*, **12**, 1 (1959).
[14] Freese, E., these PROCEEDINGS, **45**, 622 (1959).
[15] Brenner, S., in *Advances in Virus Research* (New York: Academic Press, 1959), pp. 137–158.
[16] Edgar, R. S., *Genetics*, **43**, 235 (1958).
[17] The terms topology and topography are used here in the following senses (Webster's New Collegiate Dictionary, 1959)—*topology:* the doctrine of those properties of a figure unaffected by any deformation without tearing or joining; *topography:* the art or practice of graphic and exact delineation in minute detail, usually on maps or charts, of the physical features of any place or region.
[18] *Note added in proof.*   Recent data have established that the orientation shown for segments B8 through B10 is the correct one.

31

Reprinted from *Genetics,* **65,** 379–390 (July 1970)

CRYPTIC MUTANTS OF BACTERIOPHAGE T4

ROBERT E. KOCH AND JOHN W. DRAKE

Department of Microbiology, University of Illinois, Urbana, Illinois 61801

Received March 9, 1970

THE apparent mutation rate depends upon a number of contingencies: the primary error rate, the probability of repair, error avoidance by the apparatus of DNA replication, and the probability of detection. The probability of detection of those base pair substitutions which produce an amino acid substitution appears to be highly variable, and presumably reflects the sensitivity of polypeptide functions to small variations in primary structure.

Fine-scale mapping has often revealed a highly nonrandom distribution of mutations among sites. In the T4*rII* system, for instance, 809 of the first 1609 independently isolated mutants were observed to map into only two sites (BENZER 1961). Genetic data indicate that the number of *rII* sites capable of mutation by base pair substitution is about 1700 (EDGAR *et al.* 1962; STAHL, EDGAR and STEINBERG 1964), and this estimate is supported by quasi-chemical measurements (GOLDBERG 1966). When frameshift mutations are excluded from consideration, however, fewer than 3% of these potential sites have been identified (FREESE 1959; BENZER 1961; ORGEL and BRENNER 1961; see Chapter 5 of DRAKE 1970). What then is the nature of the apparently immutable sites?

Two possible answers deserve serious consideration: many sites are extremely weakly mutable, or else many amino acid substitutions are undetected in standard screening systems because they fail to exert significantly deleterious effects upon protein function. The first possibility is supported by the observation that T4*rII* amber and ochre mutations, which are likely to be detected with uniformly high efficiencies, nevertheless appear at very different frequencies at different sites during the course of hydroxylamine mutagenesis (BRENNER, STRETTON and KAPLAN 1965). The second possibility is supported by the observation that a disproportionately large fraction of T4*rII* base pair substitution mutants contain amber codons, compared to the fraction expected from random base pair substitutions (BENZER and CHAMPE 1961; CHAMPE and BENZER 1962). The second possibility is also supported by the observation that approximately one-quarter of the beginning of the *rIIB* cistron is dispensable (DRAKE 1963). Furthermore, the *rII* function is overproduced by the (possibly trivial) criteria of the standard screening system, since many *rII* ochre mutants are well suppressed even though the efficiency of chain propagation is well below 10% (BRENNER and BECKWITH 1965).

We have approached the problem of the missing sites by designing procedures to increase the efficiency of detection of leaky mutants. It was already clear that temperature-sensitive (*ts*) mutations of the *rII* region are readily obtained

(DRAKE and McGUIRE 1967a,b; DRAKE, unpublished results), and nineteen such mutations are sufficiently inactive at 42°C to be mapped into sites. In the *rII* system, the wild-type plaque is relatively small with a fuzzy edge, whereas the mutant plaque is large with a sharp edge. The *ts* mutants of this system frequently produce semi-*r* plaques at low temperatures. We supposed that these partially defective mutants might be rendered fully defective by secondary amino acid substitutions which would otherwise be innocuous. Nitrous acid was adopted as a mutagen, first because it induces both transitions (see Chapter 13 of DRAKE 1970), and second because the mutations which it induces show relatively strong clustering into sites (BENZER 1961; KRIEG 1963), so that new sites should be easily recognized. It will be convenient to introduce two terms to describe mutations at these new sites. A leaky mutation which is used to enhance the recovery of normally undetectable mutants will be called a *sensitizing mutation*. (Our sensitizing mutations contain the designation *SM*. This is fortuitous, the symbol merely indicating the spontaneous origin of the mutation during the storage of a T4 stock.) The mutations which are exposed by a sensitizing mutation, and which are otherwise poorly detectable, will be called *cryptic mutations*.

MATERIALS AND METHODS

All experiments were carried out using *Escherichia coli* and bacteriophage T4B. Media, procedures for growing stocks, selection of mutants, scoring of phenotypes, and genetic crosses have all been described previously (BENZER 1961; EDGAR *et al.* 1962; DRAKE and McGUIRE 1967a,b). Plaque morphology was scored on B cells, and ability to grow in a lambda lysogen was scored using OP or KB cells. Stocks were grown on BB cells, which do not discriminate between *rII* mutants and the wild type.

Nitrous acid mutagenesis was performed by exposing stocks to 0.08 or 0.09 M $NaNO_2$ in 0.2 M sodium acetate buffer at p.H 4.0 at room temperature. The reaction was terminated by a 10-fold dilution into 1.0 M Tris buffer at pH 8.0. Mutagenesis of T4*rSM104* was performed by exposing the stock to approximately 20 lethal hits using 0.08 M $NaNO_2$. The treated particles were immediately adsorbed to BB cells at an average of eight particles per cell for a single cycle of growth in liquid culture. The resulting average burst size of two indicated the occurrence of strong multiplicity reactivation. The progeny particles were plated on B cells, and after 18 hr at 30°C an *r* mutant frequency of 0.7% was observed, whereupon mutants were collected. Under these conditions the selection of more than one member of a clone will occur very rarely. The spontaneous background in the untreated stock was 0.025%, or 3.6% of the induced mutant frequency. Control experiments at lower levels of inactivation, and without a cycle of growth in BB cells, indicated that about 2% to 3% of *r* mutants should have been observed. Multiplicity reactivation therefore appears to select against the induced mutants, a result also observed after ultraviolet mutagenesis (DRAKE, unpublished observation). Mutagenesis of T4*rSM122* was performed by exposing the stock to approximately 6.5 lethal hits using 0.09 M $NaNO_2$. In this case the particles were plated directly on B cells. The *r* mutant frequency was 1.0%, compared to a spontaneous background of 0.05% (5% of the induced level).

Complementation tests were performed by adsorbing an average of three particles of each parental type to OP cells in L broth, and completing lysis at 40 min with chloroform. The test was performed at 37°C unless a *ts* mutant was involved, in which case it was performed at 41°C.

RESULTS

T4 wild type produces plaques on the order of 1.5 mm diameter on B cells and on OP cells. Its *rII* mutants, if fully defective, produce plaques on the order of

3 mm diameter on B cells, and these plaques exhibit much sharper edges than do r^+ plaques because of the absence of lysis inhibition. No plaques are produced on OP cells. A continuous range of intermediate types of r mutants can also be obtained. These typically produce intermediate plaque morphologies on B cells, and may produce small plaques on OP cells. The wild-type plaque morphology on B cells is mutationally altered much more readily than is the ability to grow on lambda lysogens, and many mutants which produce r plaques on B cells also produce wild-type plaques on OP cells. The two sensitizing mutants which we have studied produce plaques on B cells at 32°C which are larger and less turbid than are r^+ plaques, but which are clearly distinct from the r plaques produced by fully defective mutants.

A collection of mutants exhibiting much more of the mutant character than did the parental type was collected after nitrous acid mutagenesis of two strains carrying sensitizing mutations. A few derivatives of $rSM122$ were collected which were not completely defective, but which were obviously more defective than the parental type. These mutants produced small plaques on OP cells at 32°C, but were unable to produce visible plaques at 37°C. (Both $rSM104$ and $rSM122$ produce nearly normal plaques on OP cells at 37°C.) Of 390 r mutants obtained from $rSM104$, 168 or 43% were classified as rII. Of 342 r mutants obtained from $rSM122$, 154 or 45% were classified as rII. These frequencies are typical of r mutants produced by base pair substitutions.

The two sets of mutants were mapped into segments and then into sites using recombination spot tests and the mapping deletions of BENZER (1961). (If a newly induced mutation were cryptic, then deletion mapping would often produce ambiguous results. We therefore constructed deletion-plus-sensitizer double mutants for use in many of the mapping experiments; see Figure 2.) The resulting mutational spectra, together with BENZER's (1961) spectrum of mutations obtained by nitrous acid mutagenesis of the wild type, appear in Figure 1. The horizontal line segments refer to map segments defined by deletions, and are well ordered. However, segments containing no mutants are not included in these maps. The heavy line segments identify the regions containing the sensitizing mutations, A6c for $rSM104$ and A2g for $rSM122$. The A cistron is on the left, and the B cistron is on the right. When sites on our two spectra are aligned vertically, they are identical. The sites $r131$, $rN24$ and $r117$ from the BENZER map are also known to coincide with the corresponding sites on our maps, but the rest of the alignment between our mutants and the BENZER set is conjectural. Sites are not ordered within a segment. Each independently arising mutant is represented by a box, the filled boxes representing the cryptic mutants to be described below. Some cryptic mutants could not be unequivocally assigned to a particular segment, but could be assigned to a group of segments. These sites appear on line segments floating above the main axis of the map.

Few deletions appeared in either set of rII mutants. The set obtained from $rSM104$ contained three unequivocal multisite mutations, while the set obtained from $rSM122$ contained only one. A number of mutants produced stocks containing fewer than 10^{-8} revertants, but their mutations all behaved like point

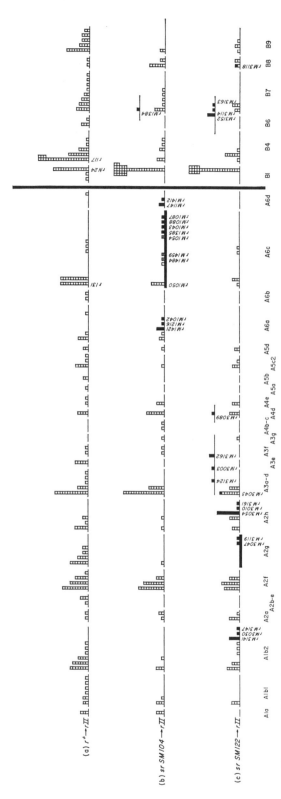

FIGURE 1.—Mutational spectra produced by nitrous acid mutagenesis of T4B (top, from BENZER 1961) and two of its semi-r derivatives. The A cistron is on the left of the vertical bar, and the B cistron is on the right. Solid squares represent cryptic mutants. The sensitizing mutations fall in the segments denoted by heavy horizontal lines. Segments defined by the standard mapping deletions are identified at the bottom of the Figure. "The three dotted boxes in segment A6c of the middle spectrum represent putative but unconfirmed cryptic mutants."

lesions in mapping experiments, and could contain at most only very short deletions. The overall large deletion frequency was therefore 1.25%. Since about 10% of most collections of spontaneous *rII* mutants contain deletions, correcting our observed frequency for the spontaneous background indicates that only about 0.8% of the induced mutants contain deletions. This frequency is much lower than the 8% observed by TESSMAN (1962) in similar experiments.

The two spectra obtained by us are very similar except for cryptic mutants, and are also similar to the 260-mutant spectrum obtained by BENZER. Certain hot spots (sites containing many independently arising mutations) which appeared in the BENZER spectrum in segments A2g, A3e and A6c, however, seem not to have appeared in our spectra, and quantitative differences have appeared among the three largest hot spots. The hot spots *r131* and *r117*, which account for 50% of all spontaneous *rII* point mutations and contain frameshift mutations (DRAKE and McGUIRE 1967a,b), are relatively more strongly represented in the 1961 spectrum, which also clearly contained a much higher frequency of mutants from the spontaneous background than do either of the present collections. The *rN24* site, which is strongly induced by all mutagens which generate transitions, is much more weakly represented in the 1961 spectrum than in the new spectra. It should be noted that the conditions of mutagenesis used by BENZER (1.8 M NaNO₂, pH approximately 6.5) are very different from those employed by us.

The *rSM104* spectrum contains a much higher density of sites in the A6 region than does either of the other two spectra. These sites, and many sites from the *rSM122* spectrum within segments A1, A2 and A3, frequently contained cryptic mutations. Mutants were classified as cryptic if they produced recombinants able to grow on OP cells in crosses with a mapping deletion, but not in crosses with the same mapping deletion coupled to the sensitizing mutation (Figure 2). When the appropriate deletion mutant was not available for this test, the mutant was crossed against deletions which approached the mutant pair (sensitizer plus putative cryptic) from one or both sides, and 10 to 18 isolates able to grow on OP cells were isolated (Figure 2). Each isolate was examined for its plaque morphology on B cells, and was also crossed with the appropriate sensitizing mutation in order to attempt the reconstruction of the fully defective double mutant. The results of these tests are summarized in Table 1 as well as in Figure 1. Three mutants from the *rSM104* spectrum could not be unequivocally analyzed in any of these tests. These mutants are likely to contain cryptic mutations because they map in a region relatively devoid of base-pair substitution sites, but they must be considered "unconfirmed" at the present time.

A considerable number of crosses was performed during the characterization of the cryptic mutations within the A6 segment of the *rSM104* spectrum. The data are summarized in Figure 3. Analysis of the map distances by means of the data of STAHL, EDGAR and STEINBERG (1964) and of GOLDBERG (1966) suggests that the interval surrounding *rSM104* in which cryptic mutations are exposed corresponds to 130 to 170 amino acids. The cryptic mutations are approximately evenly distributed throughout this interval.

Limited tests were performed of the specificity of interaction between cryptic

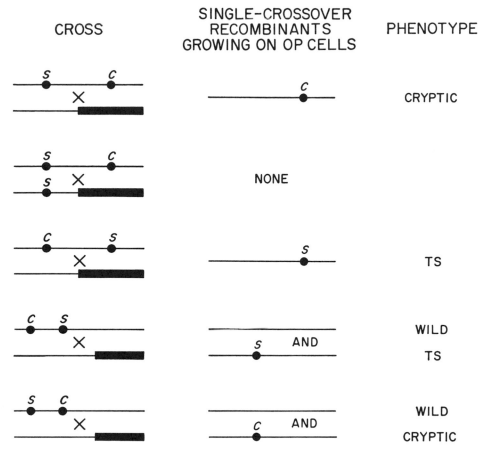

FIGURE 2.—Tests to identify and isolate cryptic mutants. In cases where the position of the putative cryptic site was not known relative to the deletion terminus, multiple tests were performed. The "cryptic" phenotype consists either of a semi-r plaque morphology (with or without temperature sensitivity), or of the ability to reform a fully mutant derivative when recombined with a sensitizing mutation.

and sensitizing mutations. The cryptic mutation *rM1366* (separated from *rSM104*, and mapping at the *rM1421* site) was crossed against *rSM122*, and the cryptic mutation *rM3054* (separated from *rSM122*) was crossed against *rSM104*. Approximately 3500 progeny from each cross were screened for the fully defective *r* plaque morphology on B cells at 30°C. No *r* isolates were observed which contained the appropriate double mutant configuration. The interaction between sensitizing mutation and cryptic mutation therefore appears to be allele-specific.

We have determined the plaque morphologies of nine isolated cryptic mutants from each spectrum. These are arranged downwards in Table 2 in approximate order of increasing defectiveness. Their behavior depends upon temperature. On B cells, several mutants appear wild type at 32°C but produce *r* plaques at 43°C. On OP cells, nearly all mutants grow at 43°C, but many produce tiny

TABLE 1

Identification of cryptic mutants

Method of identification	Sensitizer	
	rSM104	rSM122
Crosses with mapping deletion ± sensitizer*	M1042	M3003
	M1216	M3030
	M1384	M3043
	M1421	M3089
		M3114
		M3118
		M3124
		M3141
		M3147
		M3152
		M3162
		M3163
Splitting and reconstruction of double mutant	M1043	M3010
and/or analysis of phenotype	M1087	M3047
	M1088	M3054
	M1147	M3119
	M1385	M3161
	M1412	
Unconfirmed (see text)	M1054	
	M1459	
	M1484	

* *M3030, M3141* and *M3147* were identified by their ability to recombine with the deletion *r1241* (which covers *rSM122* but not the cryptic site) and with the deletion *r1364* (which covers the cryptic site but not *rSM122*), but not with the deletion *rEM66* (which covers both sites). This test is the diagnostic equivalent of the "deletion ± sensitizer" test.

plaques. There is a slight tendency for the cryptic modifiers of *rSM122* to be less defective than the cryptic modifiers of *rSM104*. This difference probably results from the deliberate selection of obviously leaky double mutants from the *rSM122* stock, but not from the *rSM104* stock.

Despite the fact that the sensitizing mutations are located in the *A* cistron, both spectra contain cryptic mutants which map in the *B* cistron (Figure 1). This result, together with the observation of intracistronic complementation between *rIIA* mutants (CHAMPE and BENZER, described in FINCHMAM 1966), suggested the existence of complex quaternary interactions among the *A*-cistron and *B*-cistron polypeptides. We sought further evidence for such interactions by means of two tests, intracistronic complementation among *B* mutants and mutual suppression between *A* and *B* mutants. Eighteen slightly leaky or temperature-sensitive *rIIB* mutants were tested for complementation in a total of 39 pairwise combinations. These mutants were scattered throughout the *B* cistron. No evidence for complementation was observed, the largest factor of increase over either single-parent control being less than two-fold. CHAMPE and BENZER, on the other hand, observed increases of more than 400-fold with pairs of *A*-cistron

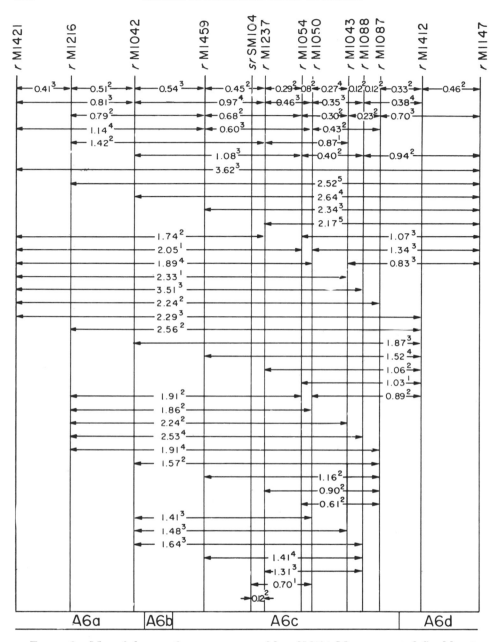

FIGURE 3.—Map of the cryptic mutants exposed by *rSM104*. Map segments defined by standard deletions are indicated at the bottom. Distances represent twice the mean percentage of wild-type recombinants, and superscripts represent the number of independent crosses performed. The mutants *rM1050* and *rM1237* are not cryptic. The unconfirmed cryptic mutant *rM1484* and the cryptic mutant *rM1385* were not mapped in these experiments. The cryptic mutant *rM1384* falls in the *B* cistron.

TABLE 2

Plaque morphology of cryptic mutants in the absence of the sensitizing mutation

Mutant	30°C	Appearance on B cells 32°C	37°C	43°C	Appearance on OP cells 30°–37°C	43°C
rM3124	w	w	sr	r	w	w
rM3162	w	w	sr	r	w	w
rM1385	w	w	fr	r	w	w
rM3054	w	w	sr	r	w	ti
rM3113	w	w	sr	r	w	ti
rM3161	w	sr	r	r	w	w
rM1421	sr	sr	fr	r	w	w
rM3152	sr	sr	sr	r	w	ti
rM1147	sr	sr	r	r	w	w
rM1216	sr	sr	r	r	w	w
rM1087	fr	fr	r	r	w	w
rM3010	sr	fr	r	r	w	ti
rM3089	sr	fr	r	r	w	ti
rM1042	sr	r	r	r	w	ti
rM1043	fr	r	r	r	w	ti
rM3114	sr	fr	r	r	w	0
rM1412	fr	r	r	r	w	ti
rM1088	fr	r	r	r	w	0
rSM104	sr	sr	r	r	w	ti
rSM122	sr	sr	r	r	w	ti

w = wild type, maximum lysis inhibition.
sr = semi-*r*, some lysis inhibition, partially lysed halo-like edge.
fr = fuzzy-*r*, less lysis inhibition, but fuzzy edge.
r = fully mutant, no lysis inhibition, sharp edge.
ti = tiny plaque.
0 = no visible plaque.

mutants. In the second test, the reversion properties of 18 *A*-cistron mutants were examined. All of these mutants were missense, being reverted by base analogues and not suppressed by any of a set of suppressors of UAA, UAG and UGA codons. An average of four independent revertants of each mutant was selected, and each revertant was backcrossed to the wild type to identify suppressors mapping in the *B* cistron. None was found.

DISCUSSION

Although cryptic mutants are easily discovered, they usually would not be harvested during routine screening tests, first because many are insufficiently *r*-like at 32° or 37°C to attract the eye, and second because all grow well on lambda lysogens, at least at 37°C. Even when separated from the sensitizing mutation, however, all of the cryptic mutations studied are deleterious under at least some conditions. It is of course still possible that other sensitizing mutations might expose cryptic mutations which were indistinguishable from the wild type, or that some of the cryptic mutations which we did not separate from the sensitizing mutation would turn out to be similarly invisible.

About 6% to 8% of the isolates obtained from the *rSM104* background were cryptic, and 11% from the *rSM122* background were also cryptic. (The higher frequency from *rSM122* probably resulted from the deliberate selection of some mutants which were more *r*-like than the sensitizing mutant, but still were not fully mutant.) Our limited tests failed to indicate any synergism between a sensitizing mutation and a cryptic mutation isolated by means of a different sensitizing mutation. If this is generally the case, then a set of several dozen sensitizing mutations might well expose hundreds of different cryptic mutants. This procedure might be made even more efficacious by using mutagens which induce both transitions and transversions, and which exhibit less hot spotting than does nitrous acid. It is therefore reasonable to suppose that a large fraction of the unidentified T4*rII* base-pair substitution sites are not intrinsically immutable. This conclusion is supported by the observation that hot spotting is at least as intense among cryptic mutants as it is among ordinary mutants (Figure 1). A large fraction of amino acid substitutions in the *rII* polypeptides apparently fail to reduce gene function below what is required to pass through the standard screening tests. With the possible exception of the purple-adenine *ad3* mutants of Neurospora (DE SERRES, personal communication), this plasticity of primary structure seems to be rather widely observed. The underlying protein structural chemistry has been elegantly probed in studies of mutant vertebrate hemoglobins (PERUTZ *et al.* 1968; PERUTZ and LEHMANN 1968).

If cryptic base pair substitutions do in fact occur at typical rates at most sites, then most screening systems will tend to underestimate forward mutation rates. A reasonable estimate of the fraction of readily detected spontaneous T4*rII* mutations which contain base pair substitutions is 14% (FREESE 1959; DRAKE 1970). Since the fraction of detected base-pair substitution sites is below 3%, the true *rII* mutation rate is likely to be several-fold higher than previously estimated. It is also likely, however, that the *rII* region is somewhat less sensitive to base pair substitutions than is the average gene, since *rII* ochre mutants are more easily suppressed than are ochre mutants in many other cistrons. A reasonable corrected mutation rate for T4 would therefore be about 10^{-7} per base pair duplication, or about 10^{-4} per cistron duplication, or about 2% per chromosome duplication (DRAKE 1969).

We observed fewer than 1% deletions among induced mutants, whereas TESSMAN (1962) reported that 8% of *rII* mutants induced by nitrous acid contained large deletions. This large difference is not readily explicable on the basis of different experimental conditions. The TESSMAN mutants and the *rSM122* mutants were collected under similar or identical conditions of temperature (20°C *vs.* room temperature), pH (3.7 *vs.* 4.0), buffer, nitrite (0.2 M KNO_2 *vs.* 0.09 M $NaNO_2$), and survival. Some critical difference seems to have existed between the two experiments, however, for which we have no explanation. The differences between the BENZER (1961) and the present spectra, on the other hand, are very minor, and are largely explicable on the basis of a much larger contamination by the spontaneous background in the earlier collection.

One unexpected outcome of these experiments was the observation of mutational synergism between lesions in the epistatic *A* and *B* cistrons. It therefore seems likely that the **A** and **B** polypeptides, encoded by the corresponding cistrons, interact rather directly. Since intracistronic complementation is demonstrable for *A* mutants but not for *B* mutants, and since this kind of complementation is also probably the result of subunit interactions in an oligomeric protein (FINCHAM 1966), it is likely that the minimum *rII* gene product consists of A_2B. Neither the intercistronic nor the intracistronic mutational synergism is likely to become explicable in chemical terms until the three-dimensional structures of **A** and **B** are known. Other examples exist, however, of intracistronic interactions between moderately distant sites, one of the most impressive having been observed in the *E. coli* tryptophan synthetase A protein (YANOFSKY, HORN and THORPE 1964).

Many of these data were obtained by Miss LYNNE BARTENSTEIN, whose patient persistence has been crucial to the entire project. Some of the early mapping experiments were performed by Mr. ROBERT PASSOVOY. The work was supported by grant E59 from the American Cancer Society, grant GB6998 from the National Science Foundation, and grant AI-04886 from the National Institutes of Health, USPHS. R.E.K. was supported as a Predoctoral Trainee by USPHS Training Grant GM-510.

SUMMARY

About a tenth of the T4*rII* mutants induced by nitrous acid within a leaky *rII* background contain a second very leaky (cryptic) *rII* mutation which specifically interacts with the original (sensitizing) mutation to produce gene inactivation. The cryptic mutations usually map in the vicinity of the sensitizing mutation. However, they occasionally map in a different cistron altogether. Hot spotting is as prevalent among cryptic mutations as it is among ordinary base-pair substitution mutations. There is no need to suppose that as yet undetected base-pair substitution sites are particularly immutable. Instead, mutations at these sites probably escape detection because of the plasticity of composition of the corresponding polypeptide. In contrast to an earlier report, we detected hardly any deletions after nitrous acid mutagenesis. Although intracistronic complementation has been observed among *A* cistron mutants, we could detect none among *B* cistron mutants.

LITERATURE CITED

BENZER, S., 1961 On the topography of the genetic fine structure. Proc. Natl. Acad. Sci. U.S. **47**: 403–415.

BENZER, S. and S. P. CHAMPE, 1961 Ambivalent *r*II mutants of phage T4. Proc. Natl. Acad. Sci. U.S. **47**: 1025–1038.

BRENNER, S. and J. R. BECKWITH, 1965 *Ochre* mutants, a new class of suppressible nonsense mutants. J. Mol. Biol. **13**: 629–637.

BRENNER, S., A. O. W. STRETTON and S. KAPLAN, 1965 Genetic code: the 'nonsense' triplets for chain termination and their suppression. Nature **206**: 994–998.

CHAMPE, S. P. and S. BENZER, 1962 Reversal of mutant phenotypes by 5-fluorouracil: an approach to nucleotide sequences in messenger-RNA. Proc. Natl. Acad. Sci. U.S. **48**: 532–546.

DRAKE, J. W., 1963 Mutational activation of a cistron fragment. Genetics **48**: 767–773. ——, 1969 Comparative rates of spontaneous mutation. Nature **221**: 1132. ——, 1970 *The Molecular Basis of Mutation*. Holden-Day, San Francisco.

DRAKE, J. W. and J. McGUIRE, 1967a Characteristics of mutations appearing spontaneously in extracellular particles of bacteriophage T4. Genetics **55**: 387–398. ——, 1967b Properties of r mutants of bacteriophage T4 photodynamically induced in the presence of thiopyronin and psoralen. J. Virology **1**: 260–267.

EDGAR, R. S., R. P. FEYNMAN, S. KLEIN, I. LIELAUSIS and C. M. STEINBERG, 1962 Mapping experiments with r mutants of bacteriophage T4D. Genetics **47**: 179–186.

FINCHAM, J. R. S., 1966 *Genetic Complementation*. W. A. Benjamin, New York.

FREESE, E., 1959 On the molecular explanation of spontaneous and induced mutations. Brookhaven Symp. Biol. **12**: 63–73.

GOLDBERG, E. B., 1966 The amount of DNA between genetic markers in phage T4. Proc. Natl. Acad. Sci. U.S. **56**: 1457–1463.

KRIEG, D. R., 1963 Specificity of chemical mutagenesis. Prog. Nucleic Acid Res. **2**: 125–168.

ORGEL, A. and S. BRENNER, 1961 Mutagenesis of bacteriophage T4 by acridines. J. Mol. Biol. **3**: 762–768.

PERUTZ, M. F. and H. LEHMANN, 1968 Molecular pathology of human hemoglobin. Nature **219**: 902–909.

PERUTZ, M. F., H. MUIRHEAD, J. M. COX and L. C. G. GOAMAN, 1968 Three-dimensional Fourier synthesis of horse oxyhaemoglobin at 2.8 Å resolution: the atomic model. Nature **219**: 131–139.

STAHL, F. W., R. S. EDGAR and J. STEINBERG, 1964 The linkage map of bacteriophage T4. Genetics **50**: 539–552.

TESSMAN, I., 1962 The induction of large deletions by nitrous acid. J. Mol. Biol. **5**: 442–445.

YANOFSKY, C., V. HORN and D. THORPE, 1964 Protein structure relationships revealed by mutational analysis. Science **146**: 1593–1594.

32

Reprinted from *Proc. Natl. Acad. Sci.*, **68**(4), 773–776 (1971)

The Influence of Neighboring Base Pairs upon Base-Pair Substitution Mutation Rates

ROBERT E. KOCH

Department of Microbiology, University of Illinois, Urbana, Ill. 61801

Communicated by Alexander Hollaender, January 28, 1971

ABSTRACT   The 2-aminopurine-induced transition, $A \cdot T \rightarrow G \cdot C$, was studied at particular sites in bacteriophage T4 as a function of the nearby base-pair composition of the DNA. Changing a base pair changed the transition rate at the adjacent base pair up to 23-fold, and at the next base pair by a lesser amount. Destabilization was achieved by replacing an $A \cdot T$ base pair by a $G \cdot C$ base pair.

The base-pair substitution mutation rate at a particular genetic site is often thought to depend not only upon the base pair at that site, but also upon the molecular environment determined by nearby base pairs. This idea seems to have originated as an explanation of the "hot spots" observed in the pioneering studies of mutability at the *r*II locus of bacteriophage T4 (1, 2). The main difficulty in interpreting the spectra of forward- and reverse-mutation rates arises from indefinite knowledge of which degenerate codon occurs at a given DNA locus, and about what amino acids are acceptable at particular positions in the polypeptide. Determinations of the amino acid compositions of many revertants of amber mutations located in the T4 head-protein gene, however, revealed differing rates of specific base-pair substitution at different positions within the cistron (3). Furthermore, T4*r*II ochre mutants were induced by hydroxylamine (CAA → UAA) at rates that varied by as much as 20-fold at different sites (4). The ultraviolet-induced reversion of ochre mutants within the yeast *iso*-1-cytochrome *c* gene differs at different positions: although glutamine is acceptable at two particular positions, for instance, it was induced in 11 of 11 revertants at one, and in 0 of 11 at another (5).

While strongly suggestive of neighboring-base effects, results of this type are also compatible with an alternative interpretation, namely, that mutation rates vary within a cistron because of extrinsic factors such as direction of gene replication, direction of transcription, proximity to control elements, and so on. In fact, only a small number of externally determined gradients of mutability would be necessary to explain the variations in specific base-pair substitution mutation rates thus far observed. A convincing demonstration that nearby base pairs do in fact influence base-pair substitution mutation rates therefore requires that a mutation rate be experimentally altered by a nearby base-pair substitution.

RATIONALE

The amber (UAG) and ochre (UAA) chain-terminating codons are mutationally interconvertible by a transition in the third position (Fig. 1). Mutation rates at the first two positions of such homologous codons can be measured, providing that the

phenotype corresponding to each possible genotype can be recognized with confidence. These measurements are possible using the T4*r*II system.

First, the possible types of base-pair substitutions have been restricted by using the mutagen 2-aminopurine, which produces exclusively transitions. It is already clear that 2-aminopurine readily produces transitions, and the following evidence shows that it does not produce transversions at an appreciable rate (6). Revertants of mutants of both the tryptophan synthetase A protein of *Escherichia coli* and the head protein of T4, when induced by 2-aminopurine and characterized by amino acid analysis, have been shown to arise by transversions no more often than expected from contamination by the spontaneous background. Furthermore, many T4*r*II mutants are reverted neither by proflavin (which induces frameshift mutations) nor by base analogues (including 2-aminopurine), but do revert spontaneously, and behave like point mutations in mapping experiments. Such mutants could exist only if 2-aminopurine cannot revert transversions (or, less likely, if many transitions are not reverted by base analogues, and/or many frameshift mutants are not reverted by proflavin). Finally, unpublished experiments (L. Ripley, personal communication) in this laboratory have shown that 2-aminopurine-induced revertants of numerous T4*r*II mutants exhibit only one phenotype, whereas spontaneous revertants of the same mutants usually exhibit several phenotypes; the induced revertants, therefore, are exclusively transitions.

Second, each possible revertant genotype has been identified, as follows. (*a*) Transition revertants from the first and

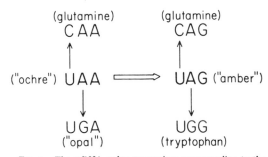

FIG. 1.   The mRNA codon conversions corresponding to the transitions measured in this study. The ochre, amber, and opal codons produce chain termination and the mutant phenotype, unless suppressed. Glutamine and tryptophan are fairly acceptable amino acids at the position represented by *r*UV200.

second positions of the ochre codon produce glutamine (UAA → CAA) and "opal" (UAA → UGA) codons, respectively. The opal codon is recognized by the ability of the phage to grow on an rII-restrictive host carrying an opal-specific suppressor. Providing that glutamine is acceptable, the glutamine revertant is recognized by its ability to grow on an rII-restrictive host. (b) Transition revertants from the first and second positions of the amber codon produce glutamine (UAG → CAG) and tryptophan (UAG → UGG) codons, respectively. These two revertants can often be distinguished phenotypically, as follows. Many T4rII mutants are sufficiently leaky to grow on the restrictive host, while producing a mutant or semi-mutant plaque morphology on certain permissive hosts. A codon was adopted for study at which 2-aminopurine produces two types of revertants of an amber mutant, one of which is virtually wild type while the other is semi-mutant. Since the glutamine revertant from the homologous ochre mutant is wild type, the wild-type revertant from the amber mutant must also be the glutamine revertant, and the semi-mutant amber revertant must therefore be the tryptophan revertant.

MATERIALS AND METHODS

E. coli and bacteriophage T4B were used throughout. Most strains and methods have been described (2–4, 7–9); K38 is an rII-restrictive strain that is unusually restrictive for opal mutants, and was kindly provided by Dr. J. Speyer.

The rII mutant whose reversion was studied is rUV200, which maps in the A6a2 segment. This mutant contains an ochre codon by the following criteria (L. Ripley, personal communication): it is suppressed exclusively by CA165, an rII-restrictive strain carrying an ochre suppressor; it is convertible by 2-aminopurine mutagenesis to an rII mutant that is suppressed by QA1, an rII-restrictive strain carrying an amber suppressor; it is also convertible to an rII mutant that is suppressed by CAJ64, an rII-restrictive strain carrying an opal suppressor; and finally, the amber and opal convertants, when crossed, produce semi-mutant recombinants at the expected low frequency of 10^{-6}, whereas the original ochre mutant fails to recombine ($<10^{-8}$) with the amber convertant.

Mutagenized stocks of the amber convertant of rUV200 produce two distinct plaque types on K38 cells at 43°C (the tryptophan and the glutamine revertants), whereas mutagenized stocks of rUV200, itself, produce only one of these plaque types (the glutamine revertant). On the opal suppressor, CAJ64, mutagenized stocks of rUV200 produce two distinct plaque types that are easily distinguished as the glutamine revertant and the opal convertant. All of these plaque types were further confirmed by plating on B cells. The glutamine revertants are virtually wild type. The tryptophan revertant is decidedly r-like on B cells at 37°C, but is clearly distinct from the r phenotype at 32°C. The opal convertant produces the completely mutant r phenotype. Plaque morphologies were routinely determined by adsorbing the particles before plating, and control plates were prepared containing predetermined mixtures of the genotypes to be scored.

Mutagenesis was conducted in M9 medium containing 1 mg/ml of 2-aminopurine. Log-phase BB cells were infected with an average of 10 phage particles and were shaken at 37°C. At 10 min, an additional 10 particles were added per cell. At 180 min, lysis was completed with chloroform. Total particles were assayed on BB cells, and revertants or convertants of the rII allele were assayed on the appropriate rII-restrictive cell strain. Acridine-resistant mutants were assayed by plating with B cells on plates containing 0.05 μg/ml of neutral acridine. The rI mutant frequency was measured on BB cells.

Crosses were performed by infecting log-phase BB cells with five particles of each parent with shaking at 37°C. An average of five additional particles were added at 6 min. The complexes were diluted at 10 min, and chloroform was added at 40 min or at 60 min to complete lysis. This procedure increases recombination frequencies somewhat compared to standard crosses.

RESULTS

Table 1 lists the frequencies of each of the transitions produced by 2-aminopurine mutagenesis. Although all tubes were run in parallel, a further check that the UAA and the UAG stocks were equally mutagenized was obtained by measuring the mutation frequency at two other loci, ac and rI, and accepting data only when both of these loci responded equally in the two stocks. Burst sizes were also very similar in parallel tubes. The spontaneous background revertant or convertant fre-

TABLE 1.   *Transition frequencies at the first two codon positions as influenced by the base composition at the third position*

| | Expt. | At the first position: | | | At the second position: | | |
		UAG ↓ CAG	UAA ↓ CAA	Ratio	UAG ↓ UGG	UAA ↓ UGA	Ratio
Immediate plating	1	1.3	1.3	1.0	14	1.5	9.2
	2	2.7	3.3	0.81	53	5.3	9.9
Corrected values	1	1.4	1.5	0.94	8.7	1.6	5.4
	2	3.0	4.1	0.74	32	5.6	5.8
After passaging	1	3.2	1.3	2.6	32	5.3	6.1
	2	5.0	2.3	2.2	51	2.7	19
Corrected values	1	3.7	0.89	4.2	38	5.1	7.4
	2	5.8	1.7	3.5	59	2.6	23

Revertant (or convertant) frequencies are per 10^6 mutagenized particles; the spontaneous background is negligible. "Ratio" indicates (UAG → CAG)/(UAA → CAA), etc., or the factor of change in the mutation rate at a given site as a function of the nearby base-pair substitution. "Corrected values" are described in the text.

quency was always about 1% of that induced by 2-aminopurine.

Transition frequencies were measured under two conditions: upon direct plating, and after a single additional cycle of growth in BB cells in the absence of 2-aminopurine. The additional round of replication was performed to segregate mutational heterozygotes, which may exhibit various plating efficiencies on selective strains (10). These results are also listed in Table 1.

The plating efficiency of each revertant or convertant type, relative to its plating efficiency on B cells at 37°C, was determined under selective conditions. These measurements were performed in parallel with the measurements of the transition frequencies. In order to closely approximate conditions on the selection plates, the plating efficiency controls were performed in the presence of a suitable concentration (10^7–10^8 per plate) of a nonreverting *r*II deletion mutant. These corrections are included in all of the data of Table 1, and were always small in comparison with the magnitude of the changes associated with the base composition of the third position.

Reconstruction selection controls were also performed, in which mixtures of the mutant plus 10^{-4} of the appropriate revertant or convertant were mutagenized and cycled through BB cells under experimental conditions. Correction factors were obtained as the average of three such control experiments, and used to produce the "corrected values" of Table 1. These reconstruction controls are not strictly comparable to the experimental situation, however, since in the experimental situation, most revertants arise rather late in the growth cycle, and are exposed to the selective conditions for less time than in the control cultures.

The recent report (11) of a UAG suppressor arising within the T4 genome suggested a possible influence of extracistronic suppression on the interpretation of these results. Five independent isolates of each of the CAG, CAA, and UGG revertants were therefore backcrossed against the wild type, and the progeny were screened for the original *r*UV200 ochre mutant or its amber convertant. No suppressors were detected within a map distance corresponding to about 10 base pairs from the original mutational lesion.

DISCUSSION

The very marked increase in the transition rate produced by changing an adjacent base pair from $A \cdot T$ to $G \cdot C$ demonstrates unequivocally, and for the first time, the role of local base-pair composition in determining base-pair substitution mutation rates. In addition, the effect is seen to be markedly stronger on the adjacent than on the next-to-adjacent base.

It is very difficult to choose among the many possible mechanisms that might produce the observed effects. Among the unknowns are the direction of chain growth during DNA replication, the assignment of these mutations to DNA replication or to DNA repair, and the mechanism of 2-aminopurine transition mutagenesis. 2-Aminopurine behaves like guanine in some enzymatically catalyzed reactions, and like adenine in others (12). The extent of the phenotypic lag after formation of the mutational heterozygote is also unclear (6). For reasons such as these, it will suffice simply to indicate some of the factors that might be important in the interpretation of neighboring base-pair effects.

First, the effect might be mediated either by direct base-to-base interactions associated with stacking, or else by indi-

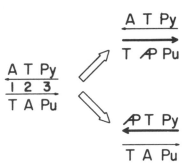

FIG. 2. The DNA base pairs corresponding to homologous ochre and amber codons. The top strand is the transcribed strand. Arrows indicate the $5' \rightarrow 3'$ polynucleotide polarity. "Py" indicates either T or C, "Pu" indicates either A or G, and "AP" indicates 2-aminopurine. Newly synthesized strands are indicated by the heavier lines.

rect effects whereby the perturbation is transmitted through one of the many enzymes that operate upon DNA. If the effect is mediated through base-stacking interactions, then it may be relevant that the cytosine residue (really 5-hydroxymethylcytosine in the case of T4) of the genetically less stable configuration may stack less strongly than the homologous thymine residue (13), thus producing a relative relaxation of local structural rigidity. On the other hand, it would not be surprising to find the effect mediated through an enzyme of DNA metabolism, since, for instance, the T4 DNA polymerase is deeply involved in the avoidance of mutation (14).

Secondly, even though the type of effect observed here would probably be expected *a priori* to diminish rapidly with distance, the present data do not unequivocally define such a polarity. This conclusion arises from the fact that there are two rare events that must both occur to bring about the observed transitions (6): the insertion of a 2-aminopurine residue, and its subsequent mispairing with cytosine. These events necessarily occur on different DNA strands. In Fig. 2, the base-pair positions are labeled 1, 2, and 3, the "ATPy" strand being the transcribed strand. Assuming that all DNA synthesis proceeds in the $5' \rightarrow 3'$ direction, and that the neighboring base effects observed here propagate only in the direction of DNA synthesis and away from position 3, consider first the replication of the nontranscribed "TAPu" strand. Here, position 3 will be filled prior to the insertion of 2-aminopurine into position 1, and position 3 could affect the (rather small) probability that 2-aminopurine is actually inserted, but it could not affect the probability that the inserted base analogue acts mutagenically at some later replication. In the case of the replication of the transcribed "ATPy" strand, however, 2-aminopurine incorporation must already have occurred at position 2 before position 3 is filled. In this case, position 3 would have its effect directly on the mutation probability during some subsequent replication (not shown), since the base analogue at position 2 mispairs with C only after position 3 has been filled. Thus, the diminished effect at position 1 compared to position 2 could simply result from a mixed mechanism, rather than from a single polar mechanism.

It is a pleasure to thank Miss Lynne Bartenstein for her expert technical assistance. I also thank Mrs. Lynn Ripley, who provided the pertinent phage stocks and for helpful discussions concerning the conception of this work. I am especially grateful to Dr. John W. Drake, who was constantly available for advice and who assisted in the preparation of this manuscript. This study was supported by grants E59 from the American Cancer Society, GB15139 from the National Science Foundation, and AIO4886 from the National Institutes of Health.

1. Benzer, S., and E. Freese, *Proc. Nat. Acad. Sci. USA*, **44**, 112 (1958).
2. Benzer, S., *Proc. Nat. Acad. Sci. USA*, **47**, 403 (1961).
3. Stretton, A. O. W., S. Kaplan, and S. Brenner, *Cold Spring Harbor Symp. Quant. Biol.*, **31**, 173 (1966).
4. Brenner, S., A. O. W. Stretton, and S. Kaplan, *Nature*, **206**, 994 (1965).
5. Sherman, F., J. W. Stewart, M. Cravens, F. L. X. Thomas, and N. Shipman, *Genetics*, **61**, s55 (1969).
6. Drake, J. W., *The Molecular Basis of Mutation* (Holden-Day, Inc., San Francisco, Calif., 1970).
7. Brenner, S., and J. R. Beckwith, *J. Mol. Biol.*, **13**, 629 (1965).
8. Sambrook, J. F., D. P. Fan, and S. Brenner, *Nature*, **214**, 452 (1967).
9. Drake, J. W., *J. Mol. Biol.*, **o**, 268 (1963).
10. Lindstrom, D. M., and J. W. Drake, *Proc. Nat. Acad. Sci. USA*, **65**, 617 (1970).
11. McClain, W. H., *FEBS Lett.*, **6**, 99 (1970).
12. Rogan, E. G., and M. J. Bessman, *J. Bacteriol.*, **103**, 622 (1970).
13. Shugar, D., and W. Szer, *J. Mol. Biol.*, **17**, 174 (1966); Browne, D. T., J. Eisinger, and N. J. Leonard, *J. Amer. Chem. Soc.*, **90**, 7302 (1968).
14. Drake, J. W., E. F. Allen, S. A. Forsberg, R.-M. Preparata, and E. O. Greening, *Nature*, **221**, 1128 (1969).

33

Reprinted from *Nature*, **236**(5346), 338–341 (1972)

Molecular Basis of a Mutational Hot Spot in the Lysozyme Gene of Bacteriophage T4

YOSHIMI OKADA

Institute for Plant Virus Research, Aoba-cho, Chiba, Japan

GEORGE STREISINGER, JOYCE (EMRICH) OWEN & JUDITH NEWTON

Institute of Molecular Biology, University of Oregon, Eugene, Oregon 97403

AKIRA TSUGITA & MASAYORI INOUYE*

Laboratory of Molecular Genetics, University of Osaka, Osaka, Japan

Investigation of reversion rates and sequencing of amino-acids has shown that one mutational hot spot in bacteriophage T4 is probably a sequence of six identical bases.

It has been known for a long time that mutation frequencies differ greatly at different sites within a gene[1], but the reasons for these differences are not well understood. Here we de-

* Present address: Department of Biochemistry, State University of New York at Stony Brook, Stony Brook, New York.

scribe a frame shift mutation[2] in the lysozyme gene[3] of bacteriophage T4 which reverts with an unusually high frequency, and describe the likely base sequence at the mutant site. We believe that the high reversion frequency of this mutant is due to the presence of six identical base pairs. Previously[3] we proposed a model for the origin of frame shift mutations. In general we believe that such mutations are likely to occur in regions with repeating bases or base doublets. The results presented here are compatible with that model and in fact, together with other results, motivated that model.

Reversion Frequency of *e*JD11

The mutation *e*JD11 arose originally in the mutant strain *e*JD10[4] giving rise to the pseudowild double mutant strain *e*JD10*e*JD11. The mutant *e*JD11 was isolated from the progeny of a cross of *e*JD10*e*JD11 to *e*⁺ (wild type). Crosses of *e*JD11 to *e*JD10 gave rise to pseudowild double mutant recombinants as well as to *e*⁺ phage.

Table 1 Frequency of Revertants

Mutant strain	Frequency of e^+-like revertants* after one cycle of selective growth† of mutant
*e*JD11	1.6×10^{-4}, 1.2×10^{-4} ‡
*e*J42	6×10^{-7}
*e*J17	$< 10^{-8}$

* To determine the titre of e^+-like phage, an aliquot of the lysate obtained after one cycle of selective growth † was added to *Escherichia coli* B grown to about 2×10^8 ml.$^{-1}$ in broth ('Difco' bacto-tryptone, 10 g; NaCl, 5 g; H₂O, 1 l.) and the mixture was incubated for about 8 min at 37° C. After this period of time for adsorption, five volumes of melted top agar ('Difco'-agar, 7 g; broth, 1 l.) at 45° C were added and 3 ml. aliquots of the mixture were plated on bottom agar ('Difco'-agar, 11 g; broth, 1 l.) in glass Petri plates. After 20 min at room temperature the plates were incubated at 43° C for 4 h and then at 37° C for 2 h. The plates were inverted, 1 ml. of CHCl₃ was poured into the lid of each plate and the plates were examined after 6–12 h at room temperature. Control plates were prepared in a similar fashion with e^+ phage. More than 90 % of the plaques on the control plates exhibited haloes that were 7–8 mm in diameter. Plaques with haloes greater in diameter than 7 mm on the experimental plates were classified as e^+-like.

† *E. coli* B were grown to a concentration of 10^8 ml.$^{-1}$ in broth supplemented by 0.1 volume of buffer (Na₂HPO₄, 3 g; KH₂PO₄, 1.5 g; NaCl, 4 g; K₂SO₄, 5 g; H₂O, 1 l.; 1 M MgSO₄, 1 ml.; 1 % gelatin, 10 ml.; 0.25 M CaCl₂, 0.4 ml.) and were centrifuged in the cold, resuspended in buffered broth and adjusted to a concentration of 4×10^8 ml.$^{-1}$. The bacteria were then brought to 37° C and aerated and were infected with a multiplicity of about five phage per bacterium. Two and a half minutes after infection antiphage serum ($k = 15$) was added and the culture was diluted 1 : 4 in broth. At 22 min after infection, 0.1 volume of chloroform was added and the bacteria (in the aqueous phase) were chilled and sedimented in the cold by centrifugation. The bacteria were washed twice with broth and were resuspended in broth. Chloroform was again added followed by 10 μg ml.$^{-1}$ of egg-white lysozyme. The bacterial debris was sedimented by centrifugation and the lysate was filtered through 'Celite' (Johns-Manville and Co.), centrifuged in the cold at about 20,000g for 70 min and the phage in the pellet were resuspended by incubation overnight in the cold with buffer. The titre of total phage was determined by adsorbing to *E. coli* B as described above for e^+ phage and plating on citrate bottom agar ('Difco'-noble agar, 11 g; broth, 1 l.; supplemented after autoclaving with: Tris(hydroxymethyl) aminomethane (1.0 M, pH 8.0), 50 ml.; sodium citrate 2H₂O (25 % solution), 10.0 ml.) with citrate top agar (same as citrate bottom agar except with 7.0 gm 'Difco'-agar) supplemented with egg white lysozyme (500 μg egg white lysozyme per 2.5 ml. top agar added just before plating). Plaques were counted after overnight incubation at 37° C.

‡ Values derived from independently prepared cultures.

On plating stocks of *e*JD11 under restrictive conditions (bottom agar plates (Table 1) incubated at 43°), generally more than 10^{-3} wild type plaques were observed. Wild type phage are known to have an advantage over mutants during the growth that occurs in the course of stock preparation, and thus the high frequency of revertants in these stocks is not an accurate index of the reversion frequency.

To obtain a more accurate measure of the frequency of revertants in a single cycle of growth, stocks were subjected to a selective procedure that efficiently eliminates pre-existing revertants: a culture of bacteria infected with mutant phage is exposed to chloroform, a procedure that causes the lysis of bacteria in which a sufficient concentration of lysozyme is present. Anti-phage serum is also added to the culture. Thus bacteria infected with wild type phage are lysed, liberating phage that become inactivated by antiserum; bacteria that contain predominantly mutant phage are not lysed. The un-lysed bacteria are sedimented by centrifugation, resuspended, and then lysed by the addition of egg-white lysozyme. The frequency of revertants for *e*JD11 treated in this way is given in Table 1. Also included in this table, for the sake of com-

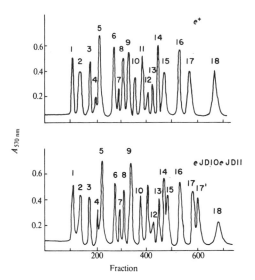

Fig. 1 Elution patterns of tryptic digests of e^+ and *e*JD10*e*JD11 lysozyme. Tryptic digests of lysozyme oxidized by performic acid were applied to 'Dowex 50' columns (0.9 cm × 150 cm) and eluted by a gradient of pyridine acetate buffer of increasing pH and pyridine concentration by means of an eight-chamber Varigrad apparatus; fractions of 4.0 ml. were collected and portions from alternate tubes were hydrolysed with alkali and treated with ninhydrin and the absorbance at 570 nm was measured as described previously[7].

Table 2 Crosses to Determine whether Reversion is at the Site of the *e*JD11 Mutation

Cross and control infections*	Frequency of phage that produce plaques that appear mutant†	Maximum No. of base pairs possible between known sites of mutation
Revertant No. 1 × e^+	4.3×10^{-6}	
Revertant No. 2 × e^+	2.0×10^{-5}	
Revertant No. 3 × e^+	1.5×10^{-5}	
Revertant No. 1	1.5×10^{-5}	
Revertant No. 2	1.3×10^{-5}	
Revertant No. 3	1.5×10^{-5}	
e^+	4.0×10^{-6}	
*e*JD16*e*JD12 ‡ × e^+	3.5×10^{-3}, 2.0×10^{-3} §	14
*e*JD10*e*JD11 ‡ × e^+	3.0×10^{-3}	12
*e*J42*e*J44 ‡ × e^+	8.0×10^{-3}	20

* *E. coli* B phr⁻ were grown in broth at 37° C to a concentration of 10^8 ml.$^{-1}$, chilled, sedimented by centrifugation, resuspended in cold broth and adjusted to a concentration of 2×10^9 ml.$^{-1}$. 0.1 ml. aliquots of the concentrated bacteria in 16 mm tubes were infected at 37° C with a multiplicity of five of each of the parental phages per bacterium; 9 min later the infected bacteria were diluted 1 : 10^2 in cold buffer and were irradiated in the cold with a dose of ultraviolet light equivalent to twenty-five phage-lethal hits. The irradiated bacteria were diluted in broth, and after 90 min at 37° C were lysed by the addition of 0.1 volume of chloroform.

Control infections were performed in the same way except that a multiplicity of ten phage particles per bacterium was used.

† In order to measure low frequencies of mutant phage, lysates of crosses and control infections were subjected to selective growth (for the elimination of e^+ or pseudowild phage) exactly as described in Table 1 except that the infection was with about one phage particle per 10^3 bacteria. The titre of mutants in the lysate was calculated from a comparison of the titre of mutants recovered after selective growth with that observed after selective growth of known mixtures of e^+ and mutant phage.

‡ These strains are pseudowild.

§ Values derived from independently prepared cultures.

parison, are frequencies of revertants in stocks of eJ42 and eJ17 similarly treated; these two strains had the highest and lowest reversion frequencies respectively of eleven lysozyme frame shift mutants for which base sequences have been proposed. Thus the frequency of reverse mutation at the eJD11 site is more than one hundred times greater than that at any other site yet examined.

The revertants of eJD11 that were observed seemed identical to e^+ phage in every respect. They gave rise to plaques under restrictive conditions, and the haloes that were produced around the plaques after exposure to chloroform vapours[3] were identical to those of e^+ at temperatures ranging from 30° to 43° C. To determine whether the reversion occurred at the site of the original (eJD11) mutation, three revertants of independent origin were crossed to e^+ phage, ultraviolet irradiation being used to enhance the recombination frequency. If the revertants had been pseudowild, containing the original mutation

Table 3 Frequency of Revertants in Stocks of eJD10eJD11

Strain	Frequency of phage that produce plaques that appear mutant†
eJD10eJD11 *	1.7×10^{-4}, 1.1×10^{-4}, 1.6×10^{-4} ‡
eJ42eJ44 *	1.1×10^{-5}, 2.5×10^{-6}
e †	6.5×10^{-6}, 4.2×10^{-6}

* These strains are pseudowild.
† These frequencies were measured as described in Table 2.
‡ Values derived from independently prepared cultures.

Table 4 Mapping of Revertants

Parent strain	No. of mutants tested	No. of mutants in each map region						No. of mutants giving no recombination with eJD10
		1	2	3	4	5	6	
eJD10eJD11	16							14
eJ42eJ44	11	0	0	1	7	3	0	
e^+	17	0	1	0	9	7	0	

Mutant phage were picked from the parent pseudowild and e^+ strains described in Table 3. In each case mutants were isolated from two independent cultures. The mutants from eJ42eJ44 and e^+ were mapped by spot-test crosses to a set of overlapping deletions which define six map regions[5]. The mutants from eJD10eJD11 were crossed by use of spot-tests to eJD10 (which is in map region 2). Spot-test crosses were performed as follows: 1×10^7 phage particles of each deletion mutant strain and eJD10 were plated on *E. coli* B and top agar on separate plates containing bottom agar. After 20 min, a small drop (containing about 0.005 ml. of phage at a concentration of 1×10^8 ml.$^{-1}$) of each of the mutants to be tested was deposited on each plate. After the spots had dried, the plates were incubated at 43° C. Control plates were also prepared, with spotted phage but no test strains on the plates. After overnight incubation, the spots on the control plates were compared with those on the test plates. In most cases the control spots contained no plaques or very few plaques. Recombination was inferred when test spots were clear or contained many more plaques than the control spot.

described in Table 1, a high frequency of small plaques with small haloes was observed; such plaques contain e^+ phage and are interpreted as being due to revertants arising during the course of phage growth on the plate. The frequency of reversion (from eJD11 to eJD11$^+$) was therefore measured in

e^+

Thr-Ile-Gly-Ile-Gly-His-Leu-Leu-Thr-Lys   Ser-Pro-Ser-Leu-Asn-Ala-Ala-Lys
Peak 18   Peak 10

Thr(Ile,Gly,Ile,Gly)His-Leu-Leu-Gln-Lys   Ser-Pro-Ser-Leu-Asn-Ala-Ala-Lys
Peak 17′   Peak 10

eJD10eJD11

Fig. 2 Amino-acid sequence differences between *e* and eJD10eJD11 lysozymes.

eJD11 and a second mutation at another site, mutant phage would have been recovered among the progeny of the cross. As Table 2 shows, no mutant recombinants were observed. For comparison, mutant frequencies from control crosses of known pseudowild double mutants to e^+ phage are included in Table 2.

Even though the selective procedure described above eliminates pre-existing revertants it seemed possible that a high reversion frequency was only apparent; wild type revertants might, for instance, occur on the Petri plate after plating and produce apparently normal plaques. In fact, in the experiment

the strain eJD10eJD11: in this case the double mutant produces plaques under restrictive conditions (it is pseudowild) whereas phage which have reverted at the eJD11 site contain only the single mutation eJD10, and make plaques only on media supplemented with egg-white lysozyme. As shown in Table 3 the frequency of mutants (due to reversion of eJD11, as shown below) observed under these conditions is high, and similar to the reversion frequency observed more directly, described in Table 1.

In order to demonstrate that the mutants arising in the strain eJD10eJD11 were in fact the revertant single mutant

Table 5 Sequence of Amino-acids of Peptide 17′ from eJD10eJD11 Lysozyme

Methods	Amino-acid composition of recovered material							Amino-acid sequence deduced
	Thr	Ile	Gly	His	Leu	Glu	Lys	
Acid hydrolysis	1.2	1.7	2.0	0.7	1.8	1.1	1.0	
Edman degradation, Step 1	0.4	1.7	2.0	0.6	1.9	1.3	N.D.	Thr(Ile,Gly,Ile,Gly,His,Leu,Leu,Glx,Lys)
Carboxypeptidase B *							0.8	Thr(Ile,Gly,Ile,Gly,His,Leu,Leu,Glx)Lys
Carboxypeptidase A + B *			0.2		1.8	1.1†	1.1	Thr(Ile,Gly,Ile,Gly)His(Leu,Leu,Gln)Lys
Chymotrypsin + carboxypeptidase A *			0.2		1.7	0.1	0.1	Thr(Ile,Gly,Ile,Gly)His-Leu-Leu-Gln-Lys ‡

N.D., not determined.
* Molar recovery of liberated amino-acids.
† Glutamine.
‡ The residues within parentheses are ordered to correspond with the known sequence of the peak 18 peptide of e^+ lysozyme[8].

eJD10 rather than triple mutants involving new mutations at other sites, a number of the mutants were crossed to eJD10: in crosses of the putative triple mutant strain some pseudowild eJD10eJD11 recombinants would be recovered while crosses of a revertant (eJD10) strain would yield no recombinants. Table 4 illustrates the results of these and some further control experiments: most of the mutants are in fact eJD10 and hence revertants at the eJD11 site. This result is particularly noteworthy because it demonstrates directly that the frequency of reversion at the eJD11 site is considerably higher than the total forward mutation frequency summed over all other sites within the lysozyme gene.

Amino-acid Sequences

Lysozyme produced in bacteria infected with eJD10eJD11 phage was purified as described previously[6], and peptides obtained on tryptic digestion of this lysozyme, as well as e^+ lysozyme, were separated by chromatography on 'Dowex-50'[7]. The only differences observed in the two chromatographic patterns (Fig. 1) are the presence of peak 17′ and the smaller size of peak 18 in the tryptic digest of eJD10eJD11.

The amino-acid sequence of the peptide isolated from peak 17′ is Thr-(Ile,Gly,Ile,Gly)-His-Leu-Leu-Gln-Lys (Table 5). Thus the amino-acid sequence of the tryptic peptide isolated from peak 17′ appears identical to that of a tryptic peptide isolated from the triple mutant pseudowild strain eJD10eJ17eJ42 and described previously[4].

The sequences of the wild type lysozyme peptides have been established previously and the peak 18 (=T5–II[8], a product of cleavage by chymotrypsin activity contained in the trypsin preparation) and peak 10 (=T6[8]) peptides have been shown to occupy adjacent positions on the polypeptide chain[9]. Thus the difference in amino-acid sequence between the e^+ and eJD10-eJD11 strains is as seen in Fig. 2.

Possible base sequence differences between e^+ and eJD10eJD11.

The amino-acid sequences described in Fig. 2 are compatible with the base sequences of Fig. 3, using codon assignments proposed by Nirenberg[10] and Khorana[11].

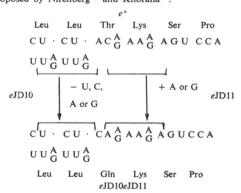

Fig. 3 Base sequences compatible with amino-acid sequences described in Fig. 2.

Earlier analyses of the triple mutant strain eJD10eJ17eJ42 were compatible only with the assignment of the single base deletions indicated here for the mutation eJD10[4]. Thus the mutation eJD11 must consist of the addition of either A or G as shown above. The Ser and Pro codons were identified unambiguously earlier[7].

Molecular Basis of Hot Spot

Of fifteen codons cited by Okada et al.[12] which have been identified in e^+ lysozyme, fourteen ended in A or U and only one in C (disregarding the codons for Trp and Met). Therefore the most likely codons for Thr and Lys in the above base sequence of the e^+ strain are ACA and AAA respectively. A likely interpretation for the mutation e^+ to eJD11, and for the reverse mutation, is thus as follows:

The high frequency of the reverse mutation (from eJD11 to e^+) would then be due to a sequence of six consecutive identical base pairs in the DNA.

We have not yet examined the amino-acid sequence of a revertant of eJD11 to confirm that it is identical to the e^+ strain. As described above, however, the revertants appear identical to e^+ as judged by sensitive physiological and genetic tests.

One of us has subsequently found five different sites in the lysozyme gene at which mutations with high reversion frequencies occur. These five sites are all best interpreted as sequences of five identical base pairs[5].

As discussed above, stretches of five or more GC base pairs are very unlikely to exist in the DNA of bacteriophage T4 and mutation frequencies in such stretches cannot be measured directly at present. One of us has shown, however, that mutations which revert with very high frequencies are not found in stretches that most likely consist of alternating or random sequences of five or more adjacent A and T residues[5]. Thus we believe that the high mutation frequency described above is due to the presence of a sequence of identical bases, rather than being due to a region rich in AT residues.

We thank D. Thomas and T. Warner for help in the preparation of lysozyme samples and Dr J. Crasemann for permitting us to report her measurements of the reversion frequencies of eJ17, eJ42 and eJD11. This study was supported by grants from the Jane Coffin Memorial Fund for Medical Research to Y. O. and A. T., by a grant from the US National Science Foundation to G. S. and by a grant from the US National Institutes of Health to A. T.

Received November 5, 1971.

1 Benzer, S., *Proc. US Nat. Acad. Sci.*, **47**, 403 (1961).
2 Crick, F. H. C., Barnett, L., Brenner, S., and Watts-Tobin, R. J., *Nature*, **192**, 1227 (1961).
3 Streisinger, G., Okada, Y., Emrich, J., Newton, J., Tsugita, A., Terzaghi, E., and Inouye, M., *Cold Spring Harbor Symp. Quant. Biol.*, **31**, 77 (1966).
4 Okada, Y., Streisinger, G., Emrich, J., Tsugita, A., and Inouye, M., *J. Mol. Biol.*, **40**, 299 (1969).
5 Owen (Emrich), J., thesis, Univ. Oregon (1971).
6 Inouye, M., Tsugita, A., Terzaghi, E., and Streisinger, G., *J. Biol. Chem.*, **243**, 391 (1968).
7 Terzaghi, E., Okada, Y., Streisinger, G., Emrich, J., Inouye, M., and Tsugita, A., *Proc. US Nat. Acad. Sci.*, **56**, 500 (1966).
8 Inouye, M., Okada, Y., and Tsugita, A., *J. Biol. Chem.*, **245**, 3439 (1970).
9 Inouye, M., Imada, M., Akaboshi, E., and Tsugita, A., *J. Biol. Chem.*, **245**, 3455 (1970).
10 Nirenberg, M., Leder, P., Bernfield, M., Brimacombe, R., Trupin, J., Rottman, F., and O'Neal, C., *Proc. US Nat. Acad. Sci.*, **53**, 1161 (1965).
11 Khorana, H. G., Buchi, H., Ghosh, H., Gupta, N., Jacob, T. M., Kossel, H., Morgan, R., Narang, S. A., Ohtsuka, E., and Wells, R. D., *Cold Spring Harbor Symp. Quant. Biol.*, **31**, 39 (1966).
12 Okada, Y., Amagase, S., and Tsugita, A., *J. Mol. Biol.*, **54**, 219 (1970).

Printed in Great Britain by Flarepath Printers Ltd., St. Albans, Herts.

Editors' Comments
on Papers 34 Through 38

THE EVOLUTION OF FIDELITY

Site-specific mutability is far from the only source of variation in mutation rates, as is well documented by this final group of papers. The genetic control of average mutation rates is in fact well illustrated by two broad phenomena: (1) large differences in average mutation rates per base pair in diverse oraganisms, and (2) mutations in specific organisms, which drastically alter average mutation rates. Mutations of this latter type are called *mutator* or *antimutator* mutations, and have been observed in numerous organisms, probably including man.

Mutator mutations were very poorly understood until 1965, when Speyer (Paper 34) reported that a temperature-sensitive mutation of the gene coding the bacteriophage T4 DNA polymerase produced a huge increase in the average mutation rate. It had long been obvious to chemists that the strengths of the hydrogen bonds within base pairs were grossly insufficient to ensure the observed high fidelity of DNA synthesis. Base hydrogen bonding to water, for instance, is energetically nearly equivalent to base hydrogen bonding to a complementary base. Furthermore, the frequencies of the mispairing tautomers of the normal bases are probably far higher than observed mutation rates. It was therefore reasonable to anticipate that the enzymes involved in DNA synthesis, and DNA polymerase in particular, were likely to actively reject incorrect base pairings. Speyer's observation gave powerful impetus to this belief.

Shortly thereafter, however, a much less predictable mutant T4 DNA polymerase was discovered. As described in Paper 35, this mutant, also originally detected because of its temperature sensitivity, exhibited antimutator properties. In addition to confirming the crucial role of DNA polymerase in maintaining fidelity, therefore, this observation indicated that phage T4 has adopted an optimal mutation rate (see also Paper 38) which is greater than that which could be achieved by a simple modification of the apparatus of DNA synthesis. (The temperature sensitivity of these strains can be repaired by mutation without abolishing their antimutator activities.) Furthermore, later studies from this group showed that not only spontaneous but also at least some kinds of chemically induced mutations are strongly influenced in their rates by the T4 DNA polymerase (Drake and Greening, 1970). Other genes besides that coding for the T4 DNA polymerase are also involved in maintaining the fidelity of T4 DNA replication and repair (Drake, 1973).

Mutator mutations have also been analyzed in bacteria, and Paper 36 from the Yanofsky group demonstrates that the powerful *E. coli* mutator mutation *mutT* promotes a specific base pair substitution pathway. This pathway happens to be a transversion $(A:T \to C:G)$, and this is one of the very few cases in which a transversional mechanism is specifically enhanced; unfortunately, the mechanism of action of *mutT* and, in particular, the normal function of the mutated gene remain unknown even now. Paper 36 also demonstrates a method of deducing mutagenic specificities that depended originally upon amino acid sequence analyses of mutated proteins, but which thereafter depends only upon the relatively simple characterization of revertant phenotypes. This system has since been used for several analyses of mutational specificity in *E. coli*.

Numerous temperature-sensitive mutants of the T4 DNA polymerase have been characterized with respect to their effects upon mutation rates; they are classified as mutators, antimutators, or neutrals, and, in some cases, as to the specific mutational pathways affected (Drake et al., 1969). What is really desired, however, is an understanding of the enzymological basis for these specificities. The first major advance in this direction is described in Paper 37 from Bessman's group. It had been known previously that the T4 DNA polymerase and several other procaryotic DNA polymerases as well contain, in addition to their polymerizing activity, a 3'-exonuclease activity which can directly reverse the polymerization step. This nuclease activity acts preferentially on incorrectly inserted bases and has been assigned a probable copy-editing or proofreading function. Paper 37 demonstrates that the fidelity of DNA replication results at least in part from a balance between the polymerase and the nuclease activities, and further shows that the mutator or antimutator properties of mutant DNA polymerases can be described, at least to a first approximation, as the result of perturbations of this balance.

The final paper (Paper 38) of our collection is, like Paper 5, a simple survey of species-specific mutation rates. It demonstrates a remarkable constancy in spontaneous mutation rates: the total mutation rate *per genome* remains approximately constant throughout the microbial world over a range of genome sizes varying by about 1,000-fold. Although the selective forces that maintain this particular mutation rate are very poorly understood, they are sufficiently powerful to have directed the evolution of a 1,000-fold decrease in mutation rate *per base pair* as one proceeds from phage lambda through *Neurospora.* By the time that genomes have evolved to the size of the *Drosophila* genome, however, this evolutionary trend appears to have been halted, a result that is of considerable potential significance for man himslef.

REFERENCES

Drake, J. W. 1973. The genetic control of spontaneous and induced mutation rates in bacteriophage T4. *Genetics Suppl., 73:* 45.

——, and E. O. Greening. 1970. Suppression of chemical mutagenesis in bacteriophage T4 by genetically modified DNA polymerases. *Proc. Natl. Acad. Sci., 66:* 823.

——, E. F. Allen, S. A. Forsberg, R.-M. Preparata, and E. O. Greening. 1969. Genetic control of mutation rates in bacteriophage T4. *Nature, 221:* 1128.

34

Reprinted from *Biochem. Biophys. Res. Commun.*, **21**(1), 6-8 (1965)

MUTAGENIC DNA POLYMERASE

Joseph F. Speyer
Cold Spring Harbor Laboratory
of Quantitative Biology L.I., N.Y.

Received August 23, 1965

An experiment will be described which shows that DNA polymerase helps select the base in DNA replication.

After elucidating the structure of DNA (1953a), Watson and Crick proposed a template mechanism for the replication of DNA (1953b). This hypothesis invokes the same hydrogen bonds which hold the finished DNA strands together as the agents for the selection of precursor nucleotides to complement those of the parental DNA strand. The enzyme for this polymerization process was subsequently discovered, Kornberg (1960), and was not previously known to play a selective role. Nor was it implicated in mutagenesis.

In another selective polymerization directed by a nucleic acid, protein synthesis, the ribosomes-which may be considered as enzymes-affect the specificity of the process, Gorini and Kataja (1964). This suggested that the enzymes involved in nucleic acid synthesis might also affect the specificity. Since enzymes may change their substrate specificity when their gene is mutated, Hotchkiss and Evans (1960), a mutant DNA polymerase might therefore cause errors in DNA synthesis. These errors can be detected as an increase in the frequency of mutations.

When a mutant DNA polymerase was discovered a direct test became possible. This discovery was made by Epstein and Edgar and their colleagues (1963) who mapped the genes of coliphage T4. They have found about 70 genes, of which several are involved in DNA replication. One of these, gene 43, (Edgar *et al*, 1964) has been identified as a structural gene of DNA polymerase by de Waard *et al* (1965). Dr. Edgar generously gave us thirteen temperature sensitive mutants of this gene. We have studied the effect of two ts alleles of this gene on mutagenesis.

Materials and Methods

The rII system of coliphage T4 was described by Benzer (1955). The reversion frequency of rII mutants is readily determined since rII will not grow or plate on E. coli K 12 (λ) − (K112-12(λh). This characteristic was used in the identification of double mutants containing rHB118 and either of two ts DNA polymerase mutations, L-56 and L-141. The double mutants rHB118/L-56 and rHB118/L-141 were constructed by genetic crosses. The rII and the ts mutation of these double mutants was characterized by genetic recombination. Crosses of the double mutant with hetero alleles of cistrons rIIA and 43 gave wild type. No recombination was observed in crosses with the alleles purported to be in the double mutants.

Results

A highly variable plaque morthology was the first indication of the high mutagenicity of one of these DNA polymerase mutants. The mutant L-56 when grown and plated on E. coli B produces mostly mottled plaques and a scattering, 3% each, of rI and rII plaques. Lysozyme mutations were also seen. However, another gene 43 mutant, L-141, gave homogenous appearing plaques.

To quantitate mutagenesis, the back mutation to r^+ of rHB118/L-56 and rHB118/L-141 was studied and compared to rHB118 with the normal DNA polymerase gene. The results, Table I, show that as with the forward mutations L-56 is highly mutagenic: the reversion frequency increased 2000 fold. The temperature at which the lysate was prepared had an effect. Higher temperature, $37^\circ C$ vs. $27^\circ C$, decreased the yield of plaque forming units and increased the reversion frequency.

In contrast to this rHB118/L-141 had an almost normal reversion frequency. Thus one ts allele is highly mutagenic, another is not. This indicates that mutagenesis is not due to impaired DNA synthesis or temperature sensitivity as such.

A cross of rHB118/L-56 with wild type T4 allowed the reisolation of rHB118 as judged by recombination, amber character and the now normal reversion frequency. Independent isolates of rHB118/L-56 and rHB118/L-141 gave similar results to those above. In order to check that the plaques of rHB118/L-56 on E. coli K-12(λ) represented true reversion we analysed thirty plaques. These were replated on E. coli K-12(λ) and their plaque morphology was studied on E. coli S/6. All were r^+.

Table I

Mutant	Grown at	Plaques per ml at $30^\circ C$ on		Reversion frequency
		K 12 (λ)	B	$\frac{K}{B}$ x 10^7
rHB 118	$37^\circ C$	95	1.72×10^{10}	0.06
rHB 118/L-56	$27^\circ C$	1.82×10^5	2.19×10^{10}	83
rHB 118/L-56	$37^\circ C$	1.75×10^5	0.83×10^{10}	210
rHB 118/L-141	$27^\circ C$	140	0.4×10^{10}	0.35
rHB 118/isolated from rHB 118/1-56 x T4	$37^\circ C$	100	1.01×10^{10}	0.10

Preliminary results show that L-56 increases the reversion frequency of other base analogue revertible rII mutatious but has no effect on the reversion of deletions or of a frame shift rII mutation. Other gene 43 ts mutants also are mutagenic. The rII mutants produced by gene 43 ts mutants are being studied.

Discussion

The results show that one mutant of gene 43 (DNA polymerase) is mutagenic, another is not. Thus a DNA polymerase mutant has the properties of a mutator gene. This implies that the polymerase is involved in the selection of the base in DNA replication

In the template hypothesis this may mean that a defective enzyme will cause polymerization of imperfect base pairs more often than a normal enzyme with stricter requirements.

However, the results may be an indication that the replicating enzyme is involved more directly in the selection of the base. It is possible that the information of the parental DNA strand is transmitted sequentially by the enzyme to an allosteric site where selection of the nucleotide bases and synthesis of the daughter strand occurs. Such an enzymic mechanism may permit selection by criteria other than the relatively weak hydrogen bonds postulated in the template hypothesis and account for the high accuracy of DNA replication. The effect of mutagens may be on the polymerase to induce errors in transmission and thus replication. This would be partly similar to the effect of streptomycin which acts on the ribosomes to produce errors in protein synthesis.

From this unproven model several predictious may be made. One is that different DNA polymerase mutatious may cause different types of mutation.

Acknowledgments

I thank Dr. R.S. Edgar of California Institute of Technology for the mutants of gene 43 and good advice. I thank Dr. S.P. Champe, of Purdue University, for rII mutants, and from this laboratory, Drs. C.I. Davern and P. Lengyel for encouragement, J.D. Karam for help with some experiments

This research was supported by research grant GM 12371-01 from the National Institutes of Health, United States Public Health Service.

Benzer, S., Proc. Natl. Acad. Sci. U.S., 41 344 (1955).

Benzer, S., and Champe, S.P., Proc. Natl. Acad. Sci. U.S., 48 1114 (1962).

deWaard, A., Paul, A.V., and Lehman, I.R., Proc. Natl. Acad. Sci. U.S. in press 1965.

Edgar, R.S., Denhardt, G.H. and Epstein, R.H., Genetics, 49, 635 (1954).

Edgar, R.S., and Lielausis, A., Genetics, 49,649 (1954).

Epstein, R.H., Bolle, A., Steinberg,C.M., Kellenberger, E., Boy de la Tour, E., Chevally, R., Edgar, R.S., Denhard, G.H. and Lielausis, A., Cold Spring Harbor Symposia on Quantitative Biology, 28, 375 (1963).

Gorini, L., and Kataja, E., Proc. Natl. Acad. Sci. U.S., 51, 487 (1964).

Hotchkiss, R.C., and Evans, A.H., Federation Proc., 19 912 (1960).

Kornberg, A., Science, 131ᵤ 1503 (1960).

Watson, J.D., and Crick, F.H.C., Nature, 171, 694 (1953a).

Watson, J.D, and Crick, F.H.C., Nature, 171, 964 (1953b).

Antimutagenic DNA Polymerases of Bacteriophage T4

JOHN W. DRAKE AND ELIZABETH F. ALLEN

Department of Microbiology, University of Illinois, Urbana, Illinois

Mutations whose phenotype consists of an increased mutation rate, either throughout the genome or within some more localized region, have been discovered in a wide variety of organisms, including both procaryotes and eucaryotes. Only very recently, however, has any appreciable understanding been gained concerning mechanisms of destabilization in any of these systems. The mutations produced in *E. coli* by the Treffers mutator *mutT* have been shown to result from a highly specific transversional pathway, AT → CG (Yanofsky et al., 1966; Cox and Yanofsky, 1967). A mutator in *S. typhimurium* which maps near the *purA* locus induces miscellaneous base pair substitutions, and may produce a mutagenic base analog (Kirchner, 1960; Kirchner and Rudden, 1966).

Speyer (1965) observed that temperature-sensitive (*ts*) lesions in gene 43 of bacteriophage T4 may produce powerful mutator effects. Studies of two such mutants, *ts*L56 and *ts*L88, suggested that they produce both transitions and transversions (Speyer et al., 1966). These mutants were of considerable general interest because of the identification of the product of gene 43 as a component in the apparatus of DNA synthesis, namely a polymerase responsible for chain propagation in the 5′ → 3′ direction in the presence of single-stranded template (de Waard et al., 1965; Warner and Barnes, 1966; Goulian et al., 1968). Mutator activity resulting from mutationally modified DNA polymerases has sometimes been taken as evidence for a direct role for the enzyme in base selection (Speyer et al., 1966). This suggestion has been challenged by Freese and Freese (1967) on the grounds that mutator activity does not interact synergistically with chemical mutagenesis, suggesting the absence of base binding sites on the enzyme: were they present, the sites on the mutant enzyme would presumably exhibit a decreased ability to reject unusual bases compared to the wild-type enzyme.

It is clear from Speyer's results that DNA polymerases can play significant roles in determining the characteristic error frequencies in DNA replication. It is therefore probable that polymerases from different species produce characteristically different error frequencies. When mutation rates are measured in systems where most mutational lesions can be detected, bacteriophages T4 and λ suffer about 2×10^{-5} mutations per cistron per DNA replication (Luria, 1951; Drake, 1969; M. W. Walker and W. F. Dove, pers. commun.) whereas *E. coli* and *S. typhimurium* experience mutations about 100 times less frequently (Lieb, 1951; Weinberg and Boyer, 1965; Drake, 1969). Because of the large difference in the sizes of viral and bacterial genomes, the net result in all four cases is the appearance of about 0.2% of newly mutant individuals in each generation. Since a considerable fraction of new mutations in any organism may persist for many generations before being eliminated by selection, the equilibrium frequency of mutants is undoubtedly much larger than 0.2%. This result suggests that mutation rates are usually about as large as can be tolerated. Conversely, DNA polymerases can be expected to be more accurate, the larger the genome, even though other factors may also be acting to reduce error frequencies in genetically more sophisticated organisms. This conclusion is supported indirectly by the observation that mutation rates in bacteriophage λ are 20 to 100 times lower when the virus is replicating as prophage than when it is replicating lytically (Walker and Dove, pers. commun.).

These considerations raise the interesting possibility that antimutagenic DNA polymerases may arise by mutation. They would be expected to occur most readily among the viral DNA polymerases. Two such antimutators will be described here.

THE *ts* MUTANTS OF GENE 43

We have studied the Caltech collection of *ts* mutants of gene 43, kindly donated by Professor R. S. Edgar. A contemporary map of these mutants appears in Fig. 1. The accuracy of this map is limited by two factors: it was constructed exclusively from two-factor crosses, and it probably contains sites which produce marker effects (Bernstein, 1967), that is, are themselves recombinogenic or antirecombinogenic. According to our mapping data, and to the best mapping function of Stahl et al. (1964), gene 43 would encode a polypeptide of mol wt approximately 340,000,

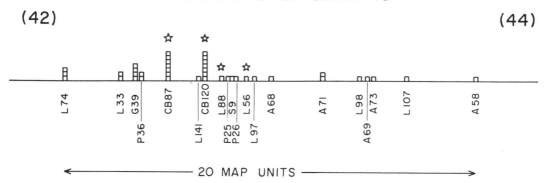

FIGURE 1. Map of the *ts* mutants of gene 43, oriented with respect to genes 42 and 44. The L74 site also contains L91 and L106; the L33 site, N24; the G39 site, G40, G42 and G43; the P36 site, P39; the CB87 site, CB78, CB79, CB82, CB86, CB88 and L42; the CB120 site, CB90, CB121, CB122, CB123, CB124, and CB126; and the A71 site, G37.

assuming that we have detected markers near both ends of the cistron. This value should be contrasted with the mol wt of 112,000 determined by Goulian et al. (1968) by equilibrium centrifugation of purified T4 DNA polymerase. Recombination events within the T4 chromosome are known to be nonrandomly distributed on a per nucleotide basis (Berger, 1965; Mosig, 1966; Rottländer et al., 1967), and gene 43 might be unusually susceptible to recombination. The map, however, reveals no large open regions which would correspond to recombinational hot spots. An alternative possibility is that the polypeptide is cleaved following synthesis, as occurs in insulin biosynthesis (Steiner and Oyer, 1967).

Forty gene 43 *ts* mutations map into 20 sites. Four sites are starred (Fig. 1). Two of these, *ts*L56 and *ts*L88, are the strong mutator alleles described by Speyer. The other two, *ts*CB87 and *ts*CB120, will be discussed in detail below.

A preliminary determination was made of mutation rates in strains containing each of these *ts* lesions, by sampling young plaques grown at 30° for their frequencies of *r* mutants. The lowest value from two or more measurements was adopted as a measure of the mutation rate. Approximately 30% of the sites contain mutations which produce increased frequencies of *r* mutants compared to *ts*+. Another 20% of the sites contain mutations which were later shown to increase reversion rates of specific *r*II mutants. By extrapolation from measurements already performed, we estimate that over 80% of the *ts* sites in Fig. 1 contain mutator alleles. In some cases, however, different members of a site appear to differ in their mutator activities.

A few *ts* mutants produced unusually low frequencies of *r* mutants. Since plaque morphologies of *ts* mutants are often abnormal even at low tem-

peratures, further studies of putative antimutator alleles were conducted by measuring their effects upon the reversion frequencies of *r*II mutants.

ANTIMUTATOR SPECIFICITIES

Both *ts*CB87 and *ts*CB120 exhibit powerful antimutator activities in reversion tests. The pathways affected can be estimated by comparing the strength of an antimutator effect with the responsiveness of the tester mutant to chemically induced reversion. The data of Table 1 reveal that significant antimutator effects upon spontaneous reversion rates are limited to mutants which are induced to revert by base analogs and not by the cytosine-specific mutagen, hydroxylamine. (Mutator activities are suggested by some of the data of Table 1, but these have not yet been carefully re-examined. A weak antimutator effect may also occur with the frameshift mutant *r*UV6.) These results suggest that only the transitional pathway AT → GC is inhibited by *ts*CB87 and *ts*CB120. However, if each member of a group of tester mutants usually reverts spontaneously by a different pathway from that suggested by its chemically induced reversion responses, then the specificity of the antimutators could be broader than it presently appears.

The antimutator effect is not limited to *r*II mutations, since amber mutants carrying lesions in genes 23 and 44 can also be affected. The preferred spontaneous reversion pathways from the amber codon have been determined for the gene 23 mutants (Stretton et al., 1966). Two mutants which frequently revert to tyrosine (by transversions from the GC base pair) are insensitive to antimutator action. Two mutants which revert to amino acids other than tyrosine *may* follow the

TABLE 1. ANTIMUTATOR SPECIFICITIES

Tester Mutant	Gene	Reversion Response			Revertants per 10^8 Phages		
		PF	BA	HA	ts^+	tsCB87	tsCB120
rUV4	rIIA	−	+	−	4.2	0.7	0.5
rUV109	rIIA	−	+	−	69	2.6	1.9
rUV183	rIIA	−	+	−	106	4.1	6.5
rUV199	rIIB	−	+	−	80	1.0	0.3
rUV248	rIIA	−	+	−	5.4	1.0	0.9
rSM7	rIIA	−	+	+	3.9	2.4	4.1
rUV13	rIIA	−	+	+	17	20	26
rUV363	rIIB	−	+	+	8	22	17
rUV6	rIIB	+	−	−	34	16	13
rUV58	rIIA	+	−	−	8.0	10	3.0
rUV353	rIIB	+	−	−	94	320	420
rSM18	rIIB	−	−	−	105	50	65
rSM49	rIIA	−	−	−	68,000	87,000	143,000
rSM58	rIIA	−	−	−	2,700	2,300	4,250
rSM94	rIIB	−	−	−	4	11	11
amA489	23		+		600	30	21
amB278	23		+		14	31	59
amE161	23		+		78	33	—
amH36	23		+		9	61	32
amN55	42		+		43	55	80
amN82	44		+		120	220	3

Tester mutants: rUV are described in Drake (1963), rSM in Drake and McGuire (1967), and *am* in gene 23 in Stretton et al. (1966); *am* in genes 42 and 44 were provided by R. S. Edgar.

Reversion responses to chemical mutagens: PF = proflavin, BA = 2-aminopurine and/or 5-bromouracil, HA = hydroxylamine.

Revertants: Several stocks (usually 4) were grown to approximately the same titer from small inocula, r^+ or am^+ revertant frequencies were determined by selective platings [K(λ)/BB and BB/CR63, respectively] at 30°, and the average of the 3 lowest values was taken (unless two obvious 'jackpot' lysates appeared). This measurement can overestimate the revertant frequency by about 2-fold, but rarely underestimates it. Typical data (for rUV4, top line): ts^+ = 2.2, 3.5, 7.1, 10.0; tsCB87 = 0.37, 0.73, 0.90, 1.00; tsCB120 = 0.32, 0.33, 0.84, 0.90.

Amber mutants: One-third of amA489 and amE161 spontaneous revertants contain tyrosine (G → pyrimidine), and two-thirds of their revertants *could* be AT → GC transitions. Virtually all amB278 and amH36 spontaneous revertants contain tyrosine, excluding the AT → GC transition.

AT → GC pathway, and are sensitive to antimutator action.

The interaction of base analog mutagenesis and antimutator activity is described in Table 2. Base analog mutagenesis is strongly inhibited, in confirmation of the conclusion that these antimutators suppress the AT → GC transition. Preliminary

TABLE 2. ANTIMUTATOR INHIBITION OF BASE ANALOG MUTAGENESIS

Tester Mutant	Base Analog	Revertants per 10^8 Phages		
		ts^+	tsCB87	tsCB120
rUV183	none	82	3.9	5.8
	5BU	660	7.2	52
	2AP	5,900	23	476
rUV199	none	49	2.7	0.3
	5BU	600	8.0	4.2
	2AP	2,000	260	43

Measurements were made following a single cycle of growth in M9-casamino acids medium containing 5-bromodeoxyuridine (5BU) at 0.5 mg/ml of 2-aminopurine (2AP) at 1 mg/ml.

evidence also suggests that GC → AT transitions induced by base analogs may also be inhibited by these antimutators.

IDENTIFICATION OF THE ts MUTATIONS AS THE ANTIMUTATORS

The tsCB87 and tsCB120 mutations were originally induced in T4D by 5-bromouracil (Edgar, pers. commun.), and may carry both spontaneous and induced mutational differences from T4B, the strain in which most of the tester mutations reside. A number of tests have therefore been performed to establish the unitary basis of temperature sensitivity and antimutator activity in the ts mutants. Backcrosses of the ts mutants to T4B (four times, with 5-fold excesses of T4B) did not alter their antimutator activities qualitatively or quantitatively. We have also frequently recovered the rII tester mutations from r-ts double mutants by backcrosses to T4B, whereupon the rII mutants always regain their characteristic 'normal' reversion rates.

TABLE 3. ANTIMUTATOR ACTIVITIES OF *ts* ALLELES AND VARIOUS *ts*+ REVERTANTS

Tester Mutant	*ts* Allele	Revertants per 10^8 Phages	Relative Burst Size at 42.5°C
*r*UV183	wild type	134	
	CB87	4	
	CB87-R1	41	
	CB120	7	
	CB120-R1	45	
*r*UV199	wild type	136	
	CB87	1	
	CB87-R1	33	
*r*UV109	wild type	48	1.0
	CB120	1.3	0.03
	CB120-R1	1.6	1.2
	-R2	3.4	1.2
	-R3	9.4	1.1
	-R4	42	0.9
	-R5	54	1.3
	-R6	55	1.1
	-R7	74	1.1

In a number of instances, *r-ts* double mutants constructed by recombination were converted to *r-ts*+ by spontaneous mutation. (Although the *ts* mutants appeared after 5-bromouracil mutagenesis, we have not been able to revert them with base analogs.) The burst sizes of the revertants at 42.5° suggest that they have fully recovered from their temperature-sensitive condition, and backcrosses of the revertants to wild type have thus far failed to reveal the existence of temperature-sensitivity suppressors located more than a few map units away from the *ts* sites themselves. However, the *r*II mutations in the *ts*+ revertants now exhibit a wide spectrum of reversion rates (Table 3). Since the accuracies assignable to these reversion rates are factors of approximately two, the number of different *ts*CB120+ revertants revealed by different residual antimutator activities is approximately four.

These experiments clearly identify the *ts* mutations themselves as the sources of antimutator activity. In addition, they demonstrate that the antimutator phenotype need not be associated with temperature sensitivity. To the extent that burst size is a measure of fitness, therefore, it is possible to construct a genetically improved DNA polymerase. However, it should be emphasized that other significant parameters, such as latent period, have yet to be examined in the *ts*+ revertants.

DO ANTIMUTATORS AFFECT MUTATION RATES DIRECTLY?

A decreased mutant frequency can result either from a decreased mutation rate, or from a decreased efficiency of detecting mutants. The later

situation would arise if certain *r*+ revertants which exhibit an *r*+ phenotype in a *ts*+ background were to exhibit an *r* phenotype in a *ts* background. This possibility can be tested by a reconstruction experiment. Ten independent *r*+ revertants were collected from *r*UV183-*ts*+, and each was crossed against both *ts*CB87 and *ts*CB120. The *r*II cistrons and gene 43 are sufficiently separated so that the *ts* and *r*+ sites should be combined in about 20% of the progeny; these *r*II mutants do not revert by extracistronic suppressors. However, the only progeny from the cross which exhibited the *r* phenotype (either plaque morphology or inability to grow in K(λ) cells) consisted of rare new spontaneous *r* mutations. Revertants of *r*UV183 therefore do not become defective in the presence of either *ts* allele.

Another possibility exists for which we have not yet devised a control experiment: the conversion of a mutational event to a lethal event in the presence of the *ts* allele. This mechanism of antimutator action appears implausible.

OTHER PROPERTIES OF THE ANTIMUTATOR MUTANTS

The one-step growth curve parameters of T4D and its *ts* mutants are given in Table 4. The *ts* mutants are incompletely active at 32°, and also produce somewhat smaller plaques than does the wild type. They remain partially active at 42° (also see Table 3). Both the eclipse periods and the latent periods of the mutants are considerably extended compared to the wild type, even at 32°. As a result, we have sometimes found them to be difficult experimental material.

Neither *ts* mutation at 32° affects recombination frequencies in the *r*II region. Crosses of *r*UV183 × *r*UV199 produced 3.7%, 3.6%, and 3.4% *r*+ recombinants in *ts*+, *ts*CB87, and *ts*CB120 backgrounds, respectively.

Dominance relationships have been investigated in single growth cycles, at high temperature for the

TABLE 4. GROWTH PARAMETERS

Temp.	Gene 43 Allele	Eclipse Period	Latent Period	Relative Burst Size
32°C	*ts*+	17	39	1.0
	*ts*CB87	29	42	0.60
	*ts*CB120	31	45	0.38
42°C	*ts*+	15	23	1.0
	*ts*CB87	>20	36	0.05
	*ts*CB120	20	28	0.09

All measurements were made in a T4D*r*+ background. Entries for the eclipse and latent periods are minimal values, in minutes. The burst size of the wild type at 42°C is 73% of that at 32°C.

TABLE 5. RECESSIVENESS OF ANTIMUTATOR ACTIVITY
AND TEMPERATURE SENSITIVITY

Infecting Genotypes	Revertants per 10^8 Phages		Relative Burst Size at 42.5°C
	Control	2-Aminopurine	
ts^+	58	3,800	1.00
tsCB87	1.2	226	0.01
tsCB87 + ts^+	20	2,900	0.96
tsCB120	0.3	80	0.19
tsCB120 + ts^+	42	4,000	0.92

An average of 10 phages infected each cell (5 + 5 in mixed infections). The tester mutant rUV199 was used for revertant measurements; complexes were incubated at 32°C in M9CA ± 2-aminopurine at 1 mg/ml. T4Dr^+ and L broth were used for burst size measurements.

ts characteristic and in the presence of 2-aminopurine for the antimutator characteristic. As indicated in Table 5, both antimutator activity and temperature sensitivity are recessive. The recessive nature of antimutator activity is expected if most successful DNA replication ensuing in mixed infections is conducted by wild type enzyme, even at low temperatures.

The antimutator effect is independent of the composition of the medium, and of temperature over the range where measurements can be performed (Table 6).

TABLE 6. INVARIANCE OF ANTIMUTATOR ACTIVITY

Tester Mutant	Conditions	Revertants per 10^8 Phages		
		ts^+	tsCB87	tsCBl20
rUV183	25°	49	3.2	5.0
	32°	63	3.4	10.7
	37°	101	1.5	10.6
	43°	36	—	—
rUV199	M9	32	0.7	0.4
	M9CA	36	1.0	0.5
	LB	55	0.4	0.6

Conditions: rUV183 stocks were grown in L broth; rUV199 stocks were grown at 32°C. M9 = salts-glucose medium; M9CA = M9 + casamino acids; LB = nutrient broth + yeast extract + glucose.

THE SIGNIFICANCE OF ANTIMUTATOR MUTATIONS

The inverse correlation between genome size and mutation rate in various species suggests that selection pressure for polymerase accuracy may be relatively relaxed in viruses. It should be noted that there is no significant increase in the rate of DNA synthesis per replication fork in T4-infected cells (Werner, 1968) which might correlate with decreased polymerase accuracy. Furthermore, a large fraction of spontaneous mutations in T4 consist of frameshift mutations, whose production may not be very closely correlated with polymerase accuracy (Streisinger et al., 1966; Drake,

1969; but see Table 1). Increased polymerase accuracy alone might therefore not appreciably assist bacteriophage T4.

A comparison of the properties of the strong mutator and antimutator alleles of gene 43 may help to throw into relief certain problems of DNA synthesis. In particular, it is unknown whether DNA polymerase merely attaches a base to the end of the growing chain after the base has been selected by stereochemical factors such as hydrogen bonding; whether the polymerase intercedes completely between the incoming base and the template strand, being fully responsible for base selection; or whether it acts in an intermediate capacity. Both mutator and antimutator activities are compatible with either of the extreme possibilities, however, since the probability of polymerizing an inserted base could depend quite strongly upon the exact configurations both of the base itself and of the sugar-phosphate moiety, which could in turn be subtly modified if an abnormal base were inserted. For these reasons, neither mutator nor antimutator activities by themselves directly answer the question of how accuracy is achieved in DNA synthesis. The limited chemical evidence now available for the E. coli DNA polymerase indicates the presence of but a single binding site for all four nucleoside triphosphates (Englund et al., this volume). This result, however, also fails to answer the central question, since the measured binding site may be primarily concerned with the sugar-phosphate moiety, with additional sites capable of discriminating among the bases yet to be detected.

Despite these theoretical limitations, it is presently possible to interpret the characteristics of the strong mutator and antimutator alleles in terms of a model in which DNA polymerase exercises a direct role in base selection. Such an enzyme would have to possess binding sites for the bases, and would also have to maintain lines of communication between pairs of binding sites (for instance, by orientation of the molecule within the replication fork, by allosteric rearrangements, or by assumption of a specific oligomeric configuration). Two extreme types of mutational modifications of polymerase accuracy could then occur: either the binding sites themselves, or their lines of communication, could be altered. It is tempting to surmise that binding site modifications would change the frequency of transitions, but not of transversions, since the insertion of purines into pyrimidine sites is difficult to imagine. Binding site modifications could also interact with chemical mutagenesis, since the probability of accepting an altered base could be changed. On the other hand, modifications of the lines of communication between binding sites might alter both transition and transversion frequencies, but be unable to interact with chemical mutagenesis. By these criteria, the antimutators *ts*CB87 and *ts*CB120 would modify base binding sites, whereas the mutators *ts*L56 and *ts*L88 would modify lines of communication between binding sites.

ACKNOWLEDGMENTS

It is a pleasure to acknowledge the expert assistance of Mrs. Sue Forsberg, who conducted virtually all of the experiments with the antimutator mutants. This work has been supported by grant E59 from the American Cancer Society, grant AI04886 from the National Institutes of Health, United States Public Health Service, and grant GB6998 from the National Science Foundation. E. F. A. was supported by an N.S.F. Predoctoral Fellowship.

REFERENCES

BERGER, H. 1965. Genetic analysis of T4D phage heterozygotes produced in the presence of 5-fluorodeoxyuridine. Genetics 52: 729.

BERNSTEIN, H. 1967. The effect on recombination of mutational defects in the DNA-polymerase and deoxycytidylate hydroxymethylase of phage T4D. Genetics 56: 755.

COX, E. C., and C. YANOFSKY. 1967. Altered base ratios in the DNA of an *Escherichia coli* mutator strain. Proc. Nat. Acad. Sci. 58: 1895.

deWAARD, A., A. V. PAUL, and I. R. LEHMAN. 1965. The structural gene for deoxyribonucleic acid polymerase in bacteriophages T4 and T5. Proc. Nat. Acad. Sci. 54: 1241.

DRAKE, J. W. 1963. Properties of ultraviolet-induced *r*II mutants of bacteriophage T4. J. Mol. Biol. 6: 268.

——. 1969. An introduction to the molecular basis of mutation. Holden-Day, San Francisco (in press).

DRAKE, J. W., and J. McGUIRE. 1967. Characteristics of mutations appearing spontaneously in extracellular particles of bacteriophage T4. Genetics 55: 387.

FREESE, E. B., and E. FREESE. 1967. On the specificity of DNA polymerase. Proc. Nat. Acad. Sci. 57: 650.

GOULIAN, M., Z. J. LUCAS, and A. KORNBERG. 1968. Enzymatic synthesis of deoxyribonucleic acid. XXV. Purification and properties of deoxyribonucleic acid polymerase induced by infection with phage T4+. J. Biol. Chem. 243: 627.

KIRCHNER, C. E. J. 1960. The effects of the mutator gene on molecular changes and mutation in *Salmonella typhimurium*. J. Mol. Biol. 2: 331.

KIRCHNER, C. E. J., and M. J. RUDDEN. 1966. Location of a mutator gene in *Salmonella typhimurium* by cotransduction. J. Bacteriol. 92: 1453.

LIEB, M. 1951. Forward and reverse mutation in a histidine-requiring strain of *Escherichia coli*. Genetics 36: 460.

LURIA, S. E. 1951. The frequency distribution of spontaneous bacteriophage mutants as evidence for the exponential rate of phage reproduction. Cold Spring Harbor Symp. Quant. Biol. 16: 463.

MOSIG, G. 1966. Distance separating genetic markers in T4 DNA. Proc. Nat. Acad. Sci. 56: 1177.

ROTTLÄNDER, E., K. O. HERMANN, and R. HERTEL. 1967. Increased heterozygote frequency in certain regions of the T4-chromosome. Mol. Gen. Genet. 99: 34.

SPEYER, J. F. 1965. Mutagenic DNA polymerase. Biochem. Biophys. Res. Commun. 21: 6.

SPEYER, J. F., J. D. KARAM, and A. B. LENNY. 1966. On the role of DNA polymerase in base selection. Cold Spring Harbor Symp. Quant. Biol. 31: 693.

STAHL, F. W., R. S. EDGAR, and J. STEINBERG. 1964. The linkage map of bacteriophage T4. Genetics 50: 539.

STEINER, D. F., and P. E. OYER. 1967. The biosynthesis of insulin and a probable precursor of insulin by a human islet cell adenoma. Proc. Nat. Acad. Sci. 57: 473.

STREISINGER, G., Y. OKADA, J. EMRICH, J. NEWTON, A. TSUGITA, E. TERZAGHI, and M. INOUYE. 1966. Frameshift mutations and the genetic code. Cold Spring Harbor Symp. Quant. Biol. 31: 77.

STRETTON, A. O. W., S. KAPLAN, and S. BRENNER. 1966. Nonsense codons. Cold Spring Harbor Symp. Quant. Biol. 31: 173.

WARNER, H. R., and J. E. BARNES. 1966. Deoxyribonucleic acid synthesis in *Escherichia coli* infected with some deoxyribonucleic acid polymerase-less mutants of bacteriophage T4. Virology 28: 100.

WEINBERG, R., and H. W. BOYER. 1965. Base analogue induced arabinose-negative mutants of *Escherichia coli*. Genetics 51: 545.

WERNER, R. 1968. Distribution of growing points in DNA of bacteriophage T4. J. Mol. Biol. 33: 679.

YANOFSKY, C., E. C. COX, and V. HORN. 1966. The unusual mutagenic specificity of an *E. coli* mutator gene. Proc. Nat. Acad. Sci. 55: 274.

36

Reprinted from *Proc. Natl. Acad. Sci.*, 55(2), 274–281 (1966)

THE UNUSUAL MUTAGENIC SPECIFICITY OF AN E. COLI MUTATOR GENE*

By Charles Yanofsky, Edward C. Cox,† and Virginia Horn

DEPARTMENT OF BIOLOGICAL SCIENCES, STANFORD UNIVERSITY

Communicated by Victor C. Twitty, December 13, 1965

Genes which lead to an appreciable increase in the rate of mutation of other genes have been recognized in several organisms.[1-6] On the basis of our current understanding of mutagenesis, the following mechanisms of mutator gene action could be envisaged: a polymerase alteration which results in errors during the replication of normal nucleotide sequences; production of mutagenic base analogues which lead to replication or incorporation errors; modification of one or more of the different bases in DNA, thereby leading to replication errors. Whether all of these mechanisms are operative *in vivo* is not known. Recently, however, Speyer has shown[7] that an alteration in the gene controlling T4 DNA polymerase results in a mutator effect on other genes.

The present investigation was initiated as the first stage in an attempt to determine the mechanism of mutagenesis of the *Escherichia coli* mutator gene described by Treffers *et al.*[3] The experiments that are reported were designed to examine the mutagenic specificity of this mutator gene. As will be seen, the mutator gene is highly specific, unidirectional in its action, and it causes an unexpected base-pair change.

The system selected for study of mutator-gene specificity is the tryptophan synthetase A gene-A protein system of *E. coli*. Extensive mutational studies with specific A protein mutants and a variety of mutagenic agents have led to a reasonably complete survey of the amino acids that can replace one another as a result of single mutational events.[8-12] Since multiple amino acid changes are observed at several positions in the A protein, the mutator gene can be tested to determine whether it favors any one of the alternative amino acid changes. With this information, and knowledge of the RNA codons that probably correspond to the relevant amino acids, it is possible to designate the base-pair change that is favored by the mutagenic agent. This approach was employed in an examination of the mutagenic specificity of the mutator gene.

Materials and Methods.—The tryptophan synthetase A protein mutants studied were strains A23, A46, A58, A78, and A223. Each of these mutants produces an inactive A protein which has a single amino acid difference from the wild-type A protein. A *tryp* deletion mutant (lacking the tryptophan synthetase A gene) was isolated in the mutator stock and this deletion was replaced by each altered A gene by transduction with phage Plkc. Reversion experiments were performed with each stock carrying a mutant A gene in the mutator background. To ensure that many independent reverse mutational events were being examined, several independent cultures were sampled in each experiment, and a small revertant-free inoculum was employed. In addition, a short growth period was used so that differential growth rates and differential survival of revertant types would not be complicating factors. If both fast- and slow-growing revertant colonies were observed, representatives of both types were picked for further testing. After purification by streaking, the revertants were characterized by previously described procedures.[13] Revertants selected for protein primary structure studies were transduced out of the mutator background into a T3 *tryp ABC*[deletion] stock and T3-"A reversion" stocks were selected. Each T3-"A reversion" stock carried the reverted A gene from the corresponding mutator stock. The T3 marker was introduced into these stocks to permit derepression of the *tryp* enzymes and the formation of high levels of A protein. Mutant T3 lacks anthranilate synthetase.

Characterization of revertants: Revertants were classified[13] as full revertants (FR) or partial revertants (PR). Full revertants are strains which have approximately the same growth rate as the parental wild-type strain on minimal medium, do not accumulate indoleglycerol, and are as sensitive to DL-5-methyl tryptophan inhibition as the wild-type strain. Partial revertants form a less-active A protein and are distinguishable from the wild type and FR's on the basis of the characteristics mentioned above. In some cases genetic tests were performed to distinguish between "true" reversions and suppressor mutations.

Examination of the A proteins of revertants: The A protein of each revertant examined was purified as described elsewhere.[14] The purified proteins were digested with the appropriate proteolytic enzyme(s) to release the peptide presumed to have an amino acid change. Tryptic digestion was used with the A proteins of revertants of mutants A58 and A78, and peptide TP6 was isolated by chromatography on Dowex 1-X2 columns as previously described.[10] Chymotryptic digests were employed for the isolation of peptide CP2 from the A proteins of revertants of mutants A23 and A46. The peptide was separated and eluted from two-dimensional peptide patterns as described elsewhere.[11] Digestion with trypsin and chymotrypsin was employed for the release of peptide TP4 of the A proteins of revertants of mutant A223.[12] This peptide was also isolated from peptide patterns. All eluted peptides were hydrolyzed in 5.7 N HCl for 48 hr, and the amino acid composition was determined with a Beckman/Spinco amino acid analyzer.

Results.—Mutator-induced reversion of mutant A223: The results of two typical reversion experiments with the A223 A gene in a mutator background are summarized in Table 1. It can be seen that the reversion frequency is very high;

TABLE 1

MUTATOR-INDUCED REVERSION OF MUTANT A223

	Tube	Approximate number of viable cells per plate	Number of tryp+ colonies per plate	Colonies picked and examined	Characteristics	Amino acid change in peptide TP4
Expt. 1	1	1040	1	1	All FR	(1)* Ileu → Ser
	2	1520	3	3	"	
	3	10³	0	0	—	
	4	"	1	1	All FR	"
	5	"	8	8	"	(1) Ileu → Ser
	6	"	2	2	"	
	7	"	5	5	"	(1) Ileu → Ser
	8	"	0	0	"	
	9	"	0	0	"	
	10	"	2	2	"	(1) Ileu → Ser
Expt. 2	1	2000	2,2	4	All FR	
	2	2790	2,2	4	"	
	3	3510	3,1	4	"	
	4	1970	2,1	3	"	
	5	1950	3,4	4	"	
	6	2360	3,2	3	"	

In all tables, M* = presence of the active mutator gene.

Five ml L-broth tubes were inoculated with 50–100 cells each of strain A223 M* and grown at 37° with shaking for 4–5 hr. The cells were centrifuged, washed with saline, and resuspended in saline. Aliquots were plated on minimal and minimal + indole agar plates. The inoculum for each experiment contained < 1 tryp+ cell/10³ viable cells.

* The number in parentheses refers to the number of revertant proteins examined for amino acid changes.

approximately one revertant colony was observed per 10^3 cells plated. The spontaneous reversion frequency of mutant A223 (in a nonmutator background) is about $1/10^8$ cells plated. In view of the high frequencies observed and the fact that microcolonies probably could develop from each plated cell in the absence of a tryptophan supplement, many of the reversion events probably occurred after plating. For this reason it is felt that the values reported represent maximum values and are not reliable for estimating mutation rates. It can also be seen in Table 1 that only FR colonies were recovered in mutator experiments although mutant A223 is known to give PR types spontaneously as well.[12] Four revertants selected at random were all found to form functional A proteins with Ser at the position in peptide TP4 occupied by Ileu in mutant A223. The Ileu → Thr change was not observed, although it is a common spontaneous change and, in fact, is the only change favored by 2-aminopurine mutagenesis.[12]

Mutator-induced reversion of mutant A78: The results of a typical reversion experiment with mutant A78 in the mutator background are shown in Table 2.

TABLE 2

MUTATOR-INDUCED REVERSION OF MUTANT A78

Tube	Approximate number of viable cells per plate	Number of tryp$^+$ colonies per plate	Colonies picked and examined	Characteristics	Amino acid change in peptide TP6
1	1060	3	2	2FR	
2	1260	4	2	2FR	(1) Cys → Gly
3	10^3	2	2	1 FR, 1 su*	
4	"	1	1	su	
5	"	4	2	1 FR, 1 su	
6	"	8	2	2FR	(1) Cys → Gly
7	"	4	2	1 FR, 1 su	
8	"	11	2	1 FR, 1 su	

Eight tubes of L-broth, each containing 5 ml, were inoculated with approximately 50 cells of A78 M* each. The tubes were shaken at 37° for 4.5 hr and the cells collected by centrifugation. The cells were suspended in saline and aliquots were plated on minimal and minimal + indole agar plates. The inoculum contained approximately 1 tryp$^+$ cell/10^4 viable cells.
  * su = suppressed mutant.

Both FR and su types are recovered, and at high frequency. The same types are observed in spontaneous mutation experiments but the frequency is no greater than 0.5 per 10^8 plated cells when the A78 gene is in a mutator-free background. When the mutator gene is present, the frequency of both types of changes is obviously increased considerably. Two of the FR strains were selected for protein primary structure studies. The Cys introduced in peptide TP6 by the A78 mutation was replaced by Gly, the amino acid occupying this position in the wild-type protein. Neither the Cys → Gly change, nor mutation to the FR type, was observed in mutagenesis experiments with 2-aminopurine.[12]

Mutator-induced reversion of mutant A58: The results of two reversion experiments with a mutator stock containing the A58 A gene are summarized in Table 3. FR types were not observed although, in the two experiments reported, over 1000 similar tryptophan-independent colonies were recovered. Primary structure studies were performed with the A proteins isolated from two revertants selected at random, and the analyses showed that the Asp residue introduced in peptide TP6 as a result of the A58 mutation was replaced by Ala. The change Asp → Ala occurs spontaneously, but only the Asp → Gly change is favored by 2-aminopurine treatment.[12] Since Gly is the amino acid at the relevant position in TP6 of the

TABLE 3

MUTATOR-INDUCED REVERSION OF MUTANT A58

Expt.	Number of tubes	Approximate number of viable cells plated from each tube	Number of tryp+ colonies per plate	Characteristics	Amino acid change in peptide TP6
1*	10	10^4	30–90	All PR or su‡	—
2†	9	Not counted	Not counted	All PR or su‡	(2) Asp → Ala

* Cells of A58 M* (17 viable cells/tube) were inoculated into 5 ml L-broth tubes. The cultures were incubated with shaking at 37° for 5 hr. The cells were collected by centrifugation, suspended in saline, and aliquots plated on minimal and minimal + indole agar plates. The inoculum contained approximately 1 tryp+ cell for every 10^3 tryp⁻ cells.
† Procedure essentially the same as in experiment 1.
‡ All of the colonies, without exception, were slower growing than wild-type colonies on minimal agar. Twenty were tested for the accumulation of indoleglycerol and all accumulated this compound, indicating that they were either PR or su strains. Genetic tests were not performed to determine the relative proportion of PR and su types, but the two colonies selected for primary structure studies both proved to be PR types. Four prototrophs selected from other experiments also arose by reversion rather than suppression.

wild-type protein, the absence of FR types in the mutator reversion experiments with mutant A58 is particularly important and strongly suggests that the change Asp → Gly cannot be brought about by the action of the mutator gene.

Mutator-induced reversion of mutants A23 and A46: The A23 and A46 mutations are in the same codon since the same Gly residue in peptide CP2 is replaced by Arg (A23) or Glu (A46). Both mutants spontaneously give FR and PR types at low frequency, approximately $1/10^8$. Experiments with these mutant alleles in the mutator background are summarized in Table 4. It can be seen that the reversion frequency of both mutant alleles is increased considerably and only FR types are obtained. Six mutator-induced revertants of A23 and 8 of A46 were examined in primary structure studies. Only the changes Arg → Ser and Glu → Ala were observed (Table 4). These changes were at the same position in peptide CP2 that was affected by the A23 and A46 mutations. A23 also reverts spontaneously to a second FR type in which Arg is replaced by Gly, but this change was not observed in the mutator experiments. The change Arg → Gly but not Arg → Ser is favored by 2-aminopurine treatment.[8, 9] Mutant A46 also spontaneously reverts to a second FR type, in which the amino acid change is Glu → Gly. This change was not observed in the mutator experiments, although this change is favored by 2-aminopurine treatment.[8, 9]

Examination of mutator-induced reversion of mutator-induced auxotrophs: The stability of mutants appearing in a mutator background was examined by determining the reversion frequencies of a group of mutator-induced auxotrophs isolated by penicillin selection. The frequency of appearance of mutants in mutator cultures was extremely high compared with nonmutator cultures, ensuring that

TABLE 4

MUTATOR-INDUCED REVERSION OF MUTANTS A23 AND A46

Expt.	Number of tubes	Approximate number of viable cells plated from each tube	Number of tryp+ colonies per plate	Colonies picked and examined from each plate	Characteristics	Amino acid change in peptide CP2
1 (A23 M*)	10	2.5×10^4	10–100	4	All 40 FR	(4) Arg → Ser
2 (A23 M*)	10	10^4	10–50	5	All 50 FR	(2) Arg → Ser
1 (A46 M*)	10	4×10^4	10–200	4	All 40 FR	(5) Glu → Ala
2 (A46 M*)	10	10^4	10–70	5	All 50 FR	(3) Glu → Ala

Cells (50–100) of each parent strain (A23 M* and A46 M*) were inoculated into tubes containing 5 ml of L-broth, and the cultures were grown with shaking at 37° for 4–6 hr. The cells were collected by centrifugation, resuspended in saline, and aliquots were plated on minimal and minimal + indole agar plates. The inocula contained less than 1 tryp+ cell/10^3 viable cells.

TABLE 5

Reversion Frequencies of Mutator-Induced Mutants

Mutant	Reversion frequency*	Mutant	Reversion frequency*	Mutant	Reversion frequency*
1	1×10^{-5}	9	7×10^{-5}	17	1×10^{-8}
2	2×10^{-8}	10	1×10^{-5}	18	2×10^{-7}
3	3×10^{-8}	11	5×10^{-5}	19†	$<2 \times 10^{-9}$
4†	$<2 \times 10^{-9}$	12†	$<2 \times 10^{-9}$	20	1×10^{-7}
5	1×10^{-7}	13	2×10^{-6}	21†	$<2 \times 10^{-9}$
6	3×10^{-7}	14	4×10^{-8}	22	8×10^{-8}
7	2×10^{-7}	15	9×10^{-9}	23	$<1 \times 10^{-8}$
8	8×10^{-5}	16	2×10^{-7}		

Mutants are grouped according to their nutritional requirements. Mutant 1 responds to a mixture of purines and pyrimidines; mutants 2–5 to a B vitamin mixture; mutants 6 and 7 to a mixture of phenylalanine, tyrosine, and tryptophan; mutants 8–11 to a mixture of arginine, methionine, and proline; mutants 12–16 to isoleucine plus valine; mutant 17 to a mixture of cysteine, lysine, and histidine; mutants 18–22 to acid-hydrolyzed casein plus tryptophan; and mutant 23 to tryptophan. The specific nutritional requirements were not determined.

* Ratio of the number of colonies found on minimal plates to the total number of cells plated as determined by plate counts on nutrient agar.

† Not determined whether a single or double mutant.

the mutants examined were mutator-induced. Five cultures, started from different single colonies, were treated with penicillin. Aliquots of each culture were plated on a complete medium, and mutants were detected and classified by replication to differently supplemented media. Twenty-three mutants were selected, representing different nutritional classes from each culture. Throughout the isolation and classification procedure, care was exercised not to discard mutants which appeared

⌐——— **2nd nucleotide** ———⌐

1st	U	C	A	G	3rd
	Phe	Ser	Tyr	Cys	U
	Phe	Ser	Tyr	Cys	C
U		Ser			A
	Leu	Ser		Try	G
		Pro	His	Arg	U
	Leu	Pro	His	Arg	C
C		Pro	Gln	Arg	A
	Leu	Pro	Gln	Arg	G
	Ileu	Thr	Asn	Ser	U
	Ileu	Thr	Asn	Ser	C
A		Thr	Lys	Arg	A
	Met	Thr	Lys		G
	Val	Ala	Asp	Gly	U
	Val	Ala	Asp	Gly	C
G	Val	Ala	Glu	Gly	A
	Val	Ala	Glu	Gly	G

(a)

Fig. 1.—(a) Ordered RNA codons described by Nirenberg et al.[15] and Khorana et al.[16] This method of presentation was suggested by Dr. F. H. C. Crick.

to be unstable.   Each mutant was grown overnight in L-broth.   Suitable aliquots
were then spread on minimal and on completely supplemented plates, and the
frequency of reversion to prototrophy scored.   The results obtained are presented
in Table 5.   It is evident that of the 23 isolates studied, only 4 revert with a frequency that can be considered to reflect a pronounced effect of the mutator gene.
We have also examined the stability of the mutator gene itself, on a more limited
scale, by testing colonies derived from single streptomycin-sensitive cells for the
presence of appreciable numbers of resistant cells.   Of approximately 4,600 colonies
examined by this technique none had lost the mutator gene.

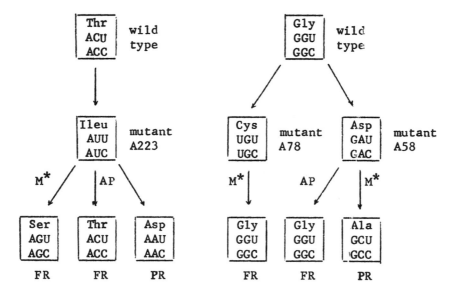

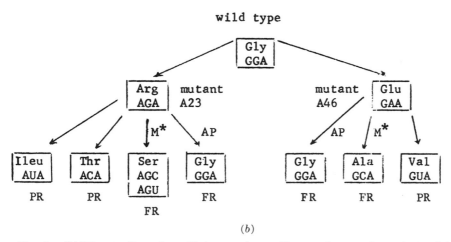

(b)

Fig. 1.—(b) Primary-site amino acid changes observed in mutation experiments[8‑12] and the probable corresponding RNA codon changes.   *FR*, full revertants; *PR*, partial revertants; *AP*, only
this change of those possible was found following 2-aminopurine mutagenesis;[8, 9, 12] *M**, only this
change was observed in stocks with the mutator gene.

Discussion.—The *E. coli* mutator gene studied appreciably increased the reversion frequencies of all five of the tryptophan synthetase A protein mutants examined. In each case, mutation to a particular revertant type was favored, although it was known from spontaneous mutation experiments that each mutant gives rise to several distinguishable revertant types. Of particular importance is the fact that the reversion events favored by the mutator gene were not the same ones observed following 2-aminopurine mutagenesis. The amino acid replacements resulting from mutator action, 2-aminopurine treatment, and spontaneous reversion are summarized in Figure 1*b*.

The RNA codons for all 20 amino acids, based on the studies of Nirenberg *et al.*[15] and Khorana *et al.*,[16] are presented in Figure 1*a*. RNA codons given in Figure 1*b* were selected on the basis of the assumption that each amino acid replacement results from a single base-pair change. The RNA codon for the Ileu of mutant A223 is taken as AUU or AUC, since single mutational events lead to the replacement of Ileu by Thr, Ser, or Asp. The fact that Asp is found in revertant proteins rather than the expected Asn cannot be explained, although it is conceivable that the particular Asn was deamidated during purification of the appropriate peptide. The Cys and Asp codon assignments (A78 and A58) are the only ones possible. Since the A23 Arg is replaced by Thr, Gly, Ser, or Ileu, the A23 Arg RNA codon must be AGA,[17] and AUA is probably the codon for the Ileu derived from the A23 Arg. As yet, *in vitro* studies have not led to the assignment of an amino acid to the AUA triplet. Since the A23 Arg is related to the A46 Glu by mutation from the same Gly, the corresponding Glu and Gly codons must be GAA and GGA. The other assignments follow from these. It is clear from the assignments presented in Figure 1*b* that the Treffers mutator gene caused the same base-pair change in the DNA of all five mutants—a change of AT → CG.[18] All other possible base-pair changes, but one (Table 6), are represented by the amino acid replacements which occur spontaneously at these positions; nevertheless, the AT → CG transversion appears to be the only change favored by the mutator gene. The reverse change, CG → AT, if it were caused by the mutator, should have given Arg → Ileu changes at the A23 position, but these were not observed. Thus, the mutator gene appears to be unidirectional in preference, a conclusion also supported by the finding that mutator-induced auxotrophs are generally not reverted by the mutator gene. The few cases in which the reversion frequencies of mutator-induced auxotrophs were high (Table 5) could be due to mutator-induced suppressor mutations or second-site reversions. This marked mutational specificity of the mutator gene was noted in earlier studies by Bacon and Treffers.[19]

We can postulate on the basis of these results that mutator genes with unidirectional preferences could have been responsible for the very different AT/GC ratios that are now evident in the DNA's of different organisms. The genetic code would appear to permit nucleotide changes in the third position of many codons without concomitant amino acid changes. Consequently, AT/GC ratios could vary considerably without any change in the amino acid content of the proteins of an organism. Furthermore, since amino acids with similar properties appear to be equally acceptable at many positions in many different proteins, changes in the amino acid content of specific proteins could also be tolerated. Thus, the result of the presence of a mutator gene analogous to the one studied here would be an

TABLE 6

BASE-PAIR CHANGES AND AMINO ACID REPLACEMENTS THAT WERE NOT
FAVORED BY THE MUTATOR GENE

Base-pair change in DNA	Amino acid changes expected and not observed[8-12]
$A \rightarrow G$ $T \rightarrow C$	Arg → Gly (A23); Asp → Gly (A58) Glu → Gly (A46); Ileu → Thr (A223)
$A \rightarrow T$ $T \rightarrow A$	Ileu → Asp (A223); Gly → Val (A46)
$C \rightarrow G$ $G \rightarrow C$	Arg → Thr (A23)
$C \rightarrow A$ $G \rightarrow T$	Arg → Ileu (A23)
$C \rightarrow T$ $G \rightarrow A$	None expected

altered GC content in the DNA of the organism, accompanied by changes in the amino acid content of its proteins.   Studies by Sueoka[20] have, in fact, shown that the amino acid content of the total protein of microorganisms does vary, in a definite manner, as the AT/GC ratio varies.   On the basis of these considerations and the unidirectional mutagenic preference of the Treffers mutator gene, it might be expected that the AT/GC ratio of the DNA of the mutator stock would be abnormal and that the amino acid content of the tryptophan synthetase A protein of the mutator stock might be altered.   These possibilities are presently under investigation.

The authors are indebted to D. Vinicor, J. Ito, and G. Pegelow for assistance with this investigation.

* This investigation was supported by grants from the National Science Foundation and the U.S. Public Health Service.

† Postdoctoral trainee of the U.S. Public Health Service.

[1] Demerec, M., *Genetics*, **22**, 469 (1937).

[2] McClintock, B., in *Cold Spring Harbor Symposia on Quantitative Biology*, vol. 16 (1951), p. 13.

[3] Treffers, H. P., V. Spinelli, and N. O. Belser, these PROCEEDINGS, **40**, 1064 (1954).

[4] Miyake, T., *Genetics*, **45**, 11 (1960).

[5] Kirchner, C. E. J., *J. Mol. Biol.*, **2**, 331 (1960).

[6] Goldstein, A., and J. S. Smoot, *J. Bacteriol.*, **70**, 588 (1955).

[7] Speyer, J. F., *Biochem. Biophys. Res. Commun.*, **21**, 6 (1965).

[8] Yanofsky, C., in *Cold Spring Harbor Symposia on Quantitative Biology*, vol. 28 (1963), p. 581.

[9] Yanofsky, C., *Biochem. Biophys. Res. Commun.*, **18**, 899 (1965).

[10] Guest, J. R., and C. Yanofsky, *J. Mol. Biol.*, **12**, 793 (1965).

[11] Carlton, B. C., and C. Yanofsky, *J. Biol. Chem.*, **238**, 2390 (1963).

[12] Yanofsky, C., J. Ito, and V. Horn, in preparation.

[13] Allen, M. K., and C. Yanofsky, *Genetics*, **48**, 1065 (1963).

[14] Henning, U., D. R. Helinski, F. C. Chao, and C. Yanofsky, *J. Biol. Chem.*, **237**, 1523 (1962).

[15] Nirenberg, M., P. Leder, M. Bernfield, R. Brimacombe, J. Trupin, F. Rottman, and C. O'Neal, these PROCEEDINGS, **53**, 1161 (1965).

[16] Söll, D., E. Ohtsuka, D. S. Jones, R. Lohrmann, H. Hayatsu, S. Nishimura, and H. G. Khorana, these PROCEEDINGS, **54**, 1378 (1965).

[17] In a previous paper,[9] two Arg codons were considered, AGA and CGA, and CGA was preferred for the Arg of A23.   On the basis of the ordered triplets that are now known, only AGA is possible.

[18] In the case of the A23 Arg codon, AGA, it cannot be stated whether it is replaced by AGC (Ser) or AGU (Ser), and thus this change might not involve an A→C transversion.

[19] Bacon, D. F., and H. P. Treffers, *J. Bacteriol.*, **81**, 786 (1961).

[20] Sueoka, N., in *Cold Spring Harbor Symposia on Quantitative Biology*, vol. 26 (1961), p. 35.

37

Reprinted from *J. Biol. Chem.*, **247**(22), 7116–7122 (1972)

Studies on the Biochemical Basis of Spontaneous Mutation

I. A COMPARISON OF THE DEOXYRIBONUCLEIC ACID POLYMERASES OF MUTATOR, ANTIMUTATOR, AND WILD TYPE STRAINS OF BACTERIOPHAGE T4*

(Received for publication, June 2, 1972)

Nicholas Muzyczka,‡ Ronald L. Poland, and Maurice J. Bessman

From the McCollum-Pratt Institute, The Johns Hopkins University, Baltimore, Maryland 21218

SUMMARY

The deoxyribonucleic acid polymerases induced by two mutator, two antimutator, and one neutral temperature-sensitive strain of T4D have been partially purified and compared to each other and to the wild type polymerase. Measurements of polymerase, polymerase-associated exonuclease, and deoxynucleoside triphosphate turnover indicate that the mutators and antimutators may be readily distinguished from wild type by the ratios of these activities. The enzymes prepared from mutators have a much lower exonuclease to polymerase ratio than wild type, and the antimutators have a higher ratio of exonuclease to polymerase and a higher rate of nucleotide turnover. Similar measurements with the neutral temperature-sensitive mutant indicate a close resemblance to wild type.

On the basis of these experiments we propose that the spontaneous mutation rate is related to the relative rates of insertion and removal of nucleotides during synthesis of deoxyribonucleic acid. Mutations in the structural gene of the polymerase which perturb this exonuclease to polymerase ratio are expressed as mutator or antimutator genes.

In bacteriophage T4, gene 43 (1) has been shown to be the structural gene for the only DNA polymerase identified with this organism (2–4). In a couple of important papers, Speyer and his colleagues (5, 6) reported that several temperature-sensitive mutants of T4 have greatly increased mutation frequencies. Since all of these mutators mapped in gene 43, it was hypothesized that a missense mutation in the structural gene for the phage-specific DNA polymerase produced an altered enzyme with a reduced base selectivity. Thus, the noncomplimentary base could be inserted more frequently into the DNA of these mutants during replication, leading to increased mutation rates. Supportive evidence for this idea was provided by Hall and Lehman

(7) who demonstrated a greater error frequency in DNA synthesized *in vitro* by a mutant polymerase purified from cells infected with L56, a mutant of phage T4 carrying a mutator gene. On the other hand, Drake *et al.* (8–10) have recently described a series of temperature-sensitive mutants in gene 43 which have markedly reduced spontaneous mutation rates and have designated these as antimutators. In order to explain how different mutations in the same gene which codes for the DNA polymerase can lead to opposite phenotypes, that is, mutators and antimutators, we propose that spontaneous mutation rates reflect the relative rates of insertion and removal of nucleotides at 3′ ends of the growing chain during DNA synthesis. This hypothesis emphasizes the postulated editing function (11) of the 3′-exonuclease resident in the DNA polymerase molecule (12) and predicts that mutations which perturb the normal exonuclease to polymerase ratio may produce either mutators or antimutators depending on whether this ratio is, respectively, decreased or increased.

In this paper we have examined partially purified DNA polymerases induced by five temperature-sensitive mutants in gene 43. The biochemical evidence obtained from this group of mutants is consistent with the hypothesis that the exonuclease to polymerase ratio of the gene 43 product determines the spontaneous mutation rate.

EXPERIMENTAL PROCEDURE

Materials

Nucleotides—Nonradioactive nucleotides were purchased from P-L Biochemicals and radioactive nucleotides from Schwarz-Mann. Diethylaminoethyl cellulose (DEAE), DE23, and cellulose phosphate, P11, were products of Whatman.

Nucleic Acids—Salmon sperm DNA was obtained from Calbiochem. It was dissolved in 0.02 M Tris, pH 7.5, 0.02 M NaCl, 10^{-3} M EDTA at a concentration of approximately 7 mM. Denatured DNA refers to such a solution made 0.05 N in NaOH. *Escherichia coli* DNA was prepared essentially as described by Anraku *et al.* (13) using lysozyme to lyse the cells, Pronase and RNase digestion and phenol extraction. [^{32}P]DNA was prepared the same way from cells grown on [^{32}P]orthophosphate. Partially digested salmon sperm DNA was prepared according to Oleson and Koerner (14).

Polynucleotides—We are grateful to Dr. F. J. Bollum for providing us with the following polynucleotides used in these

* This work was supported by Grant GM 18649 from the National Institutes of Health. This is contribution No. 695 from the McCollum-Pratt Institute.

‡ Supported by National Institutes of Health Predoctoral Training Grant GM-57 to the Department of Biology.

studies: $d(A)_{500} \cdot d(T)_{10}$, $d(A)_{500} \cdot d(T)_{46}$ — [³H]$d(T)_{1.1}$[1]; $d(A)_{500} \cdot d(T)_{46}$ — [³H]$d(C)_{0.80}$; $d(A)_{500} \cdot d(T)_{46}$ — [³H]$d(G)_{0.75}$; $d(A)_{500} \cdot d(T)_{46}$ — [³H]$d(A)_{1.53}$. The ratio of dA to dT residues was approximately 5 in all cases except where noted in the text, and the specific radioactivity of the terminal nucleotide attached to the 3′ end of $d(T)_{46}$ ranged between 56 and 70 cpm per pmole depending on the individual nucleotide. Concentrations are expressed in terms of nucleotide equivalents.

Microorganisms—We are indebted to Dr. John W. Drake for all of the temperature-sensitive mutants in gene 43 used in these studies. They are described in References 9 and 10. Phage T4D and *E. coli* Cr63 were those used previously (15).

Methods

Assay of DNA Polymerase—Incubation mixtures contained in 0.3 ml, 66 mM Tris, pH 8.8; 1 mM NaF; 16 mM (NH₄)₂SO₄; 10 mM β-mercaptoethanol; 6.7 mM MgCl₂; 0.16 mM each of dATP, dCTP, and dGTP; 0.16 mM [³H]dTTP, 1 to 2 × 10⁶ cpm per μmole; 50 μg of serum albumin; 1 mM denatured salmon sperm DNA or partially digested DNA as indicated; and 0.01 to 0.5 unit of enzyme. After 30 min at 30°, the reaction was stopped with 0.2 ml of 0.2 M sodium pyrophosphate and 1 ml of 15% trichloroacetic acid. The precipitate was centrifuged, dissolved in 0.5 ml of 0.1 N NaOH, reprecipitated with 1 ml of 15% trichloroacetic acid, collected on glass fiber filter disks, washed, dried, and counted by liquid scintillation in a toluene fluor. A unit of enzyme converts 10 nmoles of dTTP to an acid-insoluble product in 30 min at 30°. NaF was only used in assays of the crude extracts.

Nuclease Assay—This procedure measures the conversion of ³²P-labeled *E. coli* DNA to an acid-soluble form. The reaction mixtures contained in 0.3 ml: Tris, pH 8.8, 66 mM; (NH₄)₂SO₄, 16 mM; β-mercaptoethanol, 10 mM; MgCl₂, 6.7 mM; denatured [³²P]DNA, 33 μM, 1 to 2 × 10⁶ cpm per μmole; bovine serum albumin, 50 μg; and enzyme, 0.05 to 0.3 unit. After 30 min at 30°, 0.2 ml of carrier DNA (2.5 mg per ml) and 0.5 ml of 0.5 N HClO₄ were added, the precipitate centrifuged, and 0.2 ml of the acid-soluble fraction plated. Three drops of 1 N KOH were added to the planchet which was then dried and counted in a thin window gas-flow counter. A unit of enzyme converts 10 nmoles of denatured [³²P]DNA to an acid-soluble form in 30 min at 30°.

Disk Assay for Polymerase or Nuclease—When the synthetic polymers were used as templates or substrates, the filter disk technique of Bollum (16) was used to measure polymerase or nuclease activity. Aliquots of the reaction mixture (20 to 30 μl) were spotted on glass fiber disks (Whatman GF/C) and transferred to 10 ml of cold 5% trichloroacetic acid and 1% sodium pyrophosphate. The liquid was decanted and replaced by 10 ml of cold H₂O and then the disks were washed on a filter device as in the normal polymerase assay, dried, and counted in a toluene fluor. The constituents of the incubation mixtures are reported in the legends to the figures and tables.

Turnover Assay—Turnover is the DNA-dependent conversion of a deoxynucleoside triphosphate to the corresponding deoxynucleoside monophosphate. The reaction mixture for this assay was the same in all respects as that in the polymerase assay.

---

[1] In this notation, $d(A)_{500} \cdot d(T)_{46}$ — [³H]$d(T)_{1.1}$ is a polydeoxyadenylate chain with an average length of 500 nucleotides hydrogen bonded to a polythymidylate chain labeled at the 3′ terminus where 46 is the average number of unlabeled thymidylate residues in the polymer and 1.1 is the average number of ³H-labeled residues per chain.

After 30 min at 30°, the tubes were chilled and 0.5 ml of H₂O, 0.01 ml of serum albumin (25 mg per ml), and 0.05 ml of 5 N HClO₄ were added to precipitate the acid-insoluble material. The precipitate was centrifuged and the supernatant fraction was carefully drawn off with a Pasteur pipette and transferred to another tube. The precipitate was dissolved in NaOH and worked up as in the polymerase assay to measure incorporated counts. To the supernatant fraction was added 0.083 ml of 5 N KOH to precipitate the perchloric acid as KClO₄, and the tubes were centrifuged again. The clear supernatant fraction was transferred to a short column (Nestor-Faust) of Dowex 1-Cl, (0.6 × 9 cm) and washed with 13 ml of 0.02 N HCl. The columns were eluted with three successive applications of 8 ml each of a solution containing 0.025 N HCl and 0.05 M LiCl. The successive effluents were collected separately directly into scintillation vials and counted in 9 ml of a scintillation mixture containing 2 liters of toluene, 1 liter of Triton X-100, 16.5 g of 2,5-diphenyloxazole (PPO) and 0.375 g of 1,4-bis[2-4-methyl-(5-phenyloxazolyl)]benzene. The sum of the counts in the three successive 8-ml effluents with 0.025 N HCl and 0.05 N LiCl represent >90% of the total dTMP applied to the Dowex 1-Cl column. No thymidine di- or triphosphates were eluted under these conditions. Thymidine was eluted in the 13-ml wash with 0.02 N HCl.

Partial Purification of Enzymes—*E. coli* cR63 was grown at 37° in a carboy equipped with a sintered glass sparger in 10 liters of M-9 medium (17) containing 50 g of Difco casamino acids. When the cell density reached approximately 2.5 × 10⁸ per ml, the culture was transferred to a 30° room and growth was continued to a cell density of 5 × 10⁸ (Klett reading = 115). The phage were added at a multiplicity of five, and aeration was continued for 20 min after which the culture was poured into previously chilled flasks and then harvested in a Sharples centrifuge. Between 20 and 30 g of gummy paste were obtained. Fractions I and II were obtained essentially according to Nossal (4). The fragile cells were disrupted by grinding with alumina (2 g per g of cells) and extracted with 2 volumes of extraction buffer (25% glycerol, 50 mM Tris, pH 7.4, 0.5 mM β-mercaptoethanol, 10 mM MgCl₂, 1 mM EDTA). The alumina and cellular debris were removed by centrifugation for 30 min at 30,000 × g. The turbid, viscous supernatant fraction was centrifuged for 90 min at 105,000 × g and the clear, viscous supernatant solution was designated Fraction I. Fraction I was subjected to chromatography on DEAE-cellulose and this eluate (Fraction II) was in turn chromatographed on cellulose phosphate. The bulk of the DNA polymerase activity eluting from this column was dialyzed against extraction buffer containing 50% glycerol and no MgCl₂. This preparation (Fraction III) was stable when stored at −20° for at least 3 months and was used for most of the studies reported here. Further details and steps in the purification will be reported later. A summary of the specific activities of the enzymes used in the study is presented in Table I. Assayed under our optimum conditions, Fraction III for T4D has a specific activity which is about 50% of the specific activity reported for the pure T4-polymerase (12) when corrections are made for temperature and the difference in the definition of a polymerase unit.

RESULTS

Comparison of Chromatographic Profiles of Polymerases—When Fractions I, prepared from cells infected with wild type or mutant phage, were chromatographed on columns of DEAE-cellulose, the nuclease to polymerase ratios were markedly differ-

TABLE I

Enzyme preparations

Strain	Phenotype	Fraction I	Fraction III	Relative purification
		polymerase units/mg[a]		
T4D	Wild type	6.7	2100	300
L141	Antimutator	3.1	2300	750
L42	Antimutator	2	1100	550
L56	Mutator	2	880	450
L98	Mutator	1	700	700
G40	Neutral	2.7	2185	800

[a] Using partially digested salmon sperm DNA as template-primer.

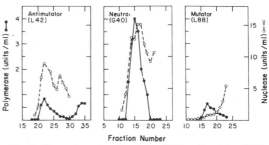

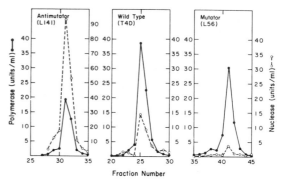

FIG. 2. Chromatography of mutator (L98), antimutator (L42), and neutral (G40) extracts on DEAE-cellulose. Conditions were the same as those described in Fig. 1.

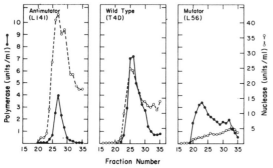

FIG. 1. Chromatography of mutator (L56), antimutator (L141), and wild type (T4D) extracts on DEAE-cellulose. Approximately 20 ml of Fraction I were chromatographed as a column of DEAE-cellulose (2.5 × 20 cm). A linear gradient between 0 and 0.67 M KCl in 0.02 M Tris, pH 7.4, and 25% glycerol was established (total volume, 1000 ml), and 10-ml fractions were collected. Fractions were assayed under standard conditions for polymerase and nuclease. The primer-template for the polymerase assays was denatured salmon sperm DNA.

FIG. 3. Chromatography of DEAE fractions on cellulose phosphate. Pooled fractions from the DEAE column (approximately 50 ml) were applied to a column of cellulose phosphate (1 × 20 cm) and eluted with 100 ml of a linear gradient between 50 and 500 mM potassium phosphate buffer, pH 6.5, in 0.1 mM β-mercaptoethanol and 25% glycerol. Two-milliliter fractions were collected and assayed as in Fig. 1.

ent. This is shown in Figs. 1 and 2 where it can be seen that the nuclease to polymerase ratios in the mutator strains were much lower than wild type, whereas the antimutators showed much higher nuclease to polymerase ratios than wild type. In Fig. 2, the mutator and antimutator may be compared to another temperature-sensitive mutation in gene 43 (G40) which is neutral in respect to its effect on spontaneous mutation rates. It is apparent that this neutral mutation does not perturb the nuclease to polymerase ratio which resembles wild type (Fig. 1). These differences are also seen in more highly purified fractions. In Fig. 3, the profiles of chromatograms on phosphocellulose columns are compared. These enzymes represent a 300- to 700-fold purification over Fraction I and the relationships are still evident. The mutator, L56, has a lower and the antimutator, L141, a higher nuclease to polymerase ratio than the wild type.

Hydrolysis of Synthetic Polynucleotides—The enzymes were tested for their relative ability to remove terminal nucleotides from specific polynucleotides. In this series of experiments, polymerase activities were first measured on the poly d(A)$_{500}$·d(T)$_{10}$ polymer using the disk assay technique. Reaction mixtures were adjusted to contain the same number of polymerase units and the rate of hydrolysis of the terminal labeled nucleotide was measured. In Fig. 4, a comparison of the mutator, antimutator, and wild type polymerases in respect to hydrolysis of the matched nucleotide revealed a marked difference. The

antimutator had a much higher exonuclease activity, and the mutator a much lower nuclease activity than wild type. The same type of experiment is shown in Fig. 5 except the substrates in this case were polynucleotides containing a mismatched terminal nucleotide. Two additional mutant polymerases were included. Again, both mutators hydrolyzed the mismatched nucleotides at a slower rate than wild type, and both antimutators were faster.

The influence of deoxynucleoside triphosphates on the rate of hydrolysis of the matched terminal nucleotide is shown in Fig. 6. In all three cases, addition of a noncomplimentary deoxynucleoside triphosphate, in this case dCTP, had little or no effect on the rate of hydrolysis. However, addition of the complimentary triphosphate, dTTP, markedly inhibited the rate of hydrolysis. It is interesting that dTTP almost completely inhibited the removal of the terminal nucleotide by the mutator and wild type enzymes, whereas appreciable hydrolysis did occur with the antimutator.

Turnover of Deoxynucleoside Triphosphates—The exonuclease activity resident in the polymerase molecule may be visualized by another technique. When a deoxynucleoside triphosphate is added to a DNA primer terminus and then removed, the net reaction is the conversion of a triphosphate to a monophosphate. Thus the turnover of triphosphates is another parameter of interest in comparing the wild type and mutant polymerases.

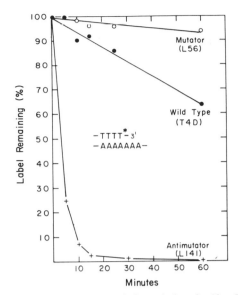

FIG. 4. Hydrolysis of a matched terminal nucleotide. Incubation mixtures (0.2 ml) contained 50 mM Tris, pH 8.8; 15 mM $(NH_4)_2SO_4$; 10 mM β-mercaptoethanol; 10 mM $MgCl_2$; 50 μg of serum albumin; 75 μM d(A)$_{500}$; 15 μM d(T)$_{46}$-[^{3}H]d(T)$_{1.1}$, and 0.3 unit of enzyme (Fraction III). The units[2] refer to the disk assay primed by d(A)$_{500}$·d(T)$_{10}$. The course of hydrolysis was followed by treating 30-μl aliquots according to the disk assay for nuclease.

In Fig. 7, the rate of dTMP formation was plotted as a function of polymerase activity. It is evident that the antimutators turned over dTTP at a much higher rate than either the wild type or neutral mutation (G40) and the mutators turned over dTTP at a lower rate than the latter two enzymes. In this experiment, partially digested salmon sperm DNA was used as primer-template and dTTP was the only nucleoside triphosphate present during the incubation. Therefore, extended synthesis of polymer was not possible. We may ask whether there is turnover of deoxynucleoside triphosphates during synthesis of polymer. Table II indicates the rate of turnover of dTTP during synthesis on the synthetic template-primer d(A)$_{500}$·d(T)$_{10}$. Both the incorporation of dTMP into an acid-insoluble product and the formation of dTMP from dTTP were measured. The results indicated that there was a very marked difference in turnover during poly(dT) synthesis by the mutant and wild type polymerases. Whereas 1 in 25 nucleotides was removed during synthesis by the wild type enzyme, this value was reduced to 1 in 200 for L56 and 1 in 100 for L98, both mutators. On the other hand, the antimutators had quite the opposite effect. Here the rate of turnover of dTTP to dTMP was greatly increased over the wild type and in fact in the case of L141 almost all of the nucleotides inserted were removed during synthesis. It should be emphasized that both incorporation and turnover are linear with respect to time over the duration of the experiment reported in Table II and in no case was more than 30% of

[2] "Total units of enzyme" refers to the sum of the nucleotides incorporated and the nucleotides turned over. For mutators and wild type, this is essentially equal to the nucleotides incorporated, but with the antimutators, a substantial fraction of the total dTTP utilized is not incorporated. This correction was significant only when synthetic templates were used (see Table II).

the available triphosphate utilized. Also, the hydrolysis of dTTP (formation of dTMP) was absolutely dependent on the presence of synthetic polymer or DNA for all of the enzymes, indicating the absence of any nonspecific triphosphatase or dTTPase in these preparations.

DISCUSSION

The DNA polymerase induced by five temperature-sensitive mutants of phage T4 have been compared to wild type in respect to their individual capacities for DNA synthesis and hydrolysis. As seen in Figs. 1 and 2, the crude extracts give very dissimilar chromatographic profiles on ion exchange columns, which we have observed to be highly reproducible and characteristic of the particular mutant. This biochemical "phenotype" does not seem to be absolutely dependent on the host (E. coli CR63) since similar experiments using an endonuclease I minus strain of E. coli gave very similar patterns. We have observed, however, that the procedure for disrupting the cells in the preparation of the crude extract may be important, since substitution of sonic oscillation for alumina grinding has led to variable and aberrant chromatograms. The five mutants studied so far may be relegated to three classes in respect to the wild type (T4D). Both mutators (L56 and L98) have nuclease to polymerase ratios lower than wild type; both antimutators (L141 and L42) have nuclease to polymerase ratios higher than wild type; the neutral temperature-sensitive mutant (G40), which has neither mutator nor antimutator characteristics, has a nuclease to polymerase ratio similar to wild type. Upon subsequent chromatography of the DEAE fractions on phosphocellulose (Fig. 3), the relationships between mutator, antimutator, and wild type persisted although there was a large drop in the absolute nuclease to polymerase ratios. This decrease in the ratios can be explained by preliminary experiments in our laboratory which indicate that there is an endonuclease which co-chromatographs with the polymerase on DEAE but is separated by subsequent chromatography on phosphocellulose. We have located this endonuclease as a discrete peak in early fractions of the phosphocellulose column and are presently examining its properties. Its association with the phage-induced polymerase in the early stages of purification may be fortuitous, or it may augur a specific role in the functional polymerase complex. We realize that slight differences in the isolation and handling of the mutant polymerases might affect their nuclease to polymerase ratios. With this in mind, we have taken care to isolate the polymerases from the different strains in exactly the same way.

Experiments on the excision of terminal nucleotides from specific polynucleotides are also consistent with the notion that the exonuclease to polymerase ratio is correlated with the spontaneous mutation rate. The data in Figs. 4 and 5 indicate that the antimutators are able to excise terminal nucleotides (matched or mismatched) at a higher rate than wild type, and the converse is true for the mutators. If the exonuclease is to serve a corrective function during DNA synthesis, it might be expected to show a bias in favor of the removal of a mispaired rather than a correctly paired terminal nucleotide. Our results (Figs. 4 and 5) as well as those of Brutlag and Kornberg (11) reveal a much higher rate of removal of mismatched nucleotides. Approximately $\frac{1}{40}$ as much enzyme is required to remove mismatched nucleotides at rates comparable to the matched nucleotides. The relative order of mutator, antimutator, and wild type enzymes remained the same on both the paired and mispaired substrates.

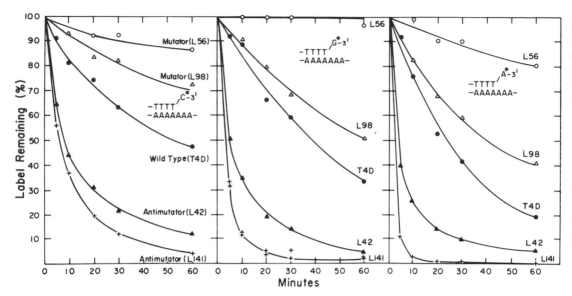

FIG. 5. Hydrolysis of mispaired terminal nucleotides. Reaction conditions were the same as in Fig. 4 with the following modifications. Each incubation contained 0.018 unit of enzyme (Fraction III) and polynucleotide substrates are designated in the panels. The *asterisk* denotes the labeled nucleotide at the 3' terminus of the poly d(T). For a more detailed description of the polynucleotide substrates, see "Materials."

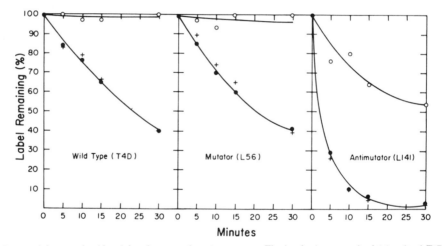

FIG. 6. Influence of deoxynucleoside triphosphates on the rate of hydrolysis of terminal nucleotides. The incubation mixture was similar to that in Fig. 4 and contained 0.25 mM dCTP or dTTP where indicated. As in Fig. 4, $d(A)_{500} \cdot d(T)_{46} - [^3H]d(T)$ was the substrate, and nuclease was measured with the standard disk assay. The incubations contained 0.9 unit of T4D (wild type), 4.3 units of L56 (mutator) or 0.3 unit of L141 (antimutator) polymerase Fraction III. ●——●, no addition; +——+, plus dCTP; ○——○, plus dTTP.

Brutlag and Kornberg (11) showed that a deoxynucleoside triphosphate capable of forming a correct base pair prevented the hydrolysis of a base-paired terminal nucleotide. Our similar experiments with the mutator or wild type enzymes (Fig. 6) showed that the terminal, base-paired nucleotide was fully protected by dTTP against hydrolysis, but dCTP (which cannot form the correct base pair) did not protect. However, it is noteworthy that the antimutator, which has a very high exo-

nuclease to polymerase ratio, hydrolyzed a substantial quantity of the terminal nucleotide even in the presence of dTTP. This is consistent with the high turnover of dTTP by L141 on poly(d(A)·d(T)) (Table II).

Further differences between the three classes of enzymes were revealed by studies of the turnover of deoxynucleoside triphosphates in the absence or presence of DNA synthesis. The antimutators converted deoxynucleoside triphosphates to the mon_

Table II

Insertion and removal of dTMP during polymerization

Incubation mixtures (0.75 ml) contained 66 mM Tris, pH 8.8, 16 mM $(NH_4)_2SO_4$, 10 mM β-mercaptoethanol, 6.7 mM $MgCl_2$, 0.16 mM [^{14}C]dTTP, 2.1×10^6 cpm per μmole, 125 μg of serum albumin, 0.23 mM d(A)$_{500}$, 0.03 mM d(T)$_{10}$, and approximately 0.2 μg of enzyme (Fraction III). At 0, 15, 30, and 45 min, 0.15-ml aliquots were removed and carried through the polymerase and turnover assay. Both incorporation and turnover were linear for 45 min. The values reported are calculated from the 30-min time points and are normalized to 100 moles of total dTTP utilized. For the two mutators, L56 and L98, it was necessary to use four times as much enzyme in order to measure the slow rate of dTMP formation.

Strain	Phenotype	dTMP formed (a)	dTMP incorporated (b)	Turnover (a/a + b)
		moles	*moles*	
L56	Mutator	0.5	99.5	1/200
L98	Mutator	1	99	1/100
T4D	Wild type	4	96	1/25
L42	Antimutator	61	39	1/1.6
L141	Antimutator	93	7	1/1.1

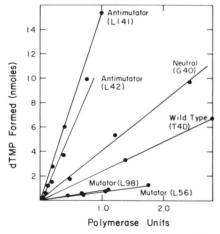

FIG. 7. Turnover of dTTP on partially digested DNA. Standard reaction mixtures contained (in 0.3 ml) 0.68 mM partially digested salmon sperm DNA. The only deoxynucleoside triphosphate present was 0.16 mM [^{14}C]dTTP, 1.7×10^6 cpm per μmole. After 60 min at 30°, the reaction mixtures were treated as described under "Methods." Fraction III enzymes were used.

ophosphates at a much higher rate than wild type, and the mutators were much lower (Fig. 7, Table II). Although only experiments with dTTP have been reported here, we have similar evidence showing that dCTP is also turned over on partially digested DNA. One might ask how a phage like L141 can survive when it turns over almost all (93%) of the dTTP utilized (Table II) and only incorporates 7% into DNA. This exaggerated turnover may be related to the unusual structural characteristics of the synthetic polynucleotide used for these studies and might be quite different for the actual replicating DNA complex *in vivo*. It is interesting however, that phage L141 does

have a low burst size suggesting a possible aberration in DNA synthesis (10).

It will be of interest to examine these mutant enzymes in detail to determine whether any of their catalytic properties may be correlated with specific genetic observations. For example, there is a marked reduction in the effect of base analogues on mutation rates in strains carrying an antimutator gene as shown by Drake *et al.* (10). This might be predicted from our observations on the increased nuclease to polymerase ratios of the antimutator enzymes. A misincorporated base might be expected to be more vulnerable to an enzyme which had an enriched nuclease to polymerase ratio. On the other hand, Freese and Freese have reported (18) that mutagenesis by base analogues in the mutator L56 is not substantially different from wild type. Our finding of a reduced nuclease to polymerase ratio in L56 might have predicted otherwise if the removal of a mispaired nucleotide during DNA synthesis is the rate-limiting step in mutagenesis. Furthermore, the specific effect of antimutators on AT → GC transition and their lack of effect on GC → AT transitions and on transversions (10) is also of interest. Studies with specific polymers and these mutant enzymes might clarify some of the biochemical bases for these genetic observations. For example, Fig. 5 reveals that L56 has a very low activity on a mispaired G compared to a mispaired C. L98, on the other hand, removes G faster than C. These results, although preliminary, suggest that further investigations of specificities under carefully controlled conditions may be profitable.

On the basis of both genetic (10) and biochemical (11) evidence, it has been suggested that polymerase-associated exonuclease contributes to the fidelity of DNA synthesis. Our observations with enzymes induced by phages showing mutator, antimutator, and neutral phenotypes provide a rational explanation for spontaneous mutation rates which emphasizes the importance of both the exonuclease and polymerase functions of the enzyme. Thus, the apparent paradox of how different missense mutations in the same gene can lead to opposite effects on spontaneous mutation rates may be explained by differential effects on regions of the protein molecule which alter the ratio of exonuclease to polymerase activity.

Acknowledgments—We should like to express our gratitude to Dr. John W. Drake for the temperature-sensitive mutants used in these studies. Dr. F. J. Bollum supplied all of the synthetic polynucleotides. John Siegle, Dr. Ko-yu Lo, Joan Lunney, and Doris Singer provided valuable help in several aspects of the work. Mrs. Dorothy Regula is responsible for the preparation of the manuscript, and Anthony V. Zaccaria and Anthony Stetka materially aided us.

REFERENCES

1. STAHL, F. W., EDGAR, R. S., AND STEINBERG, J. (1964) *Genetics* **50**, 539–552
2. DeWAARD, A., PAUL, A. V., AND LEHMAN, I. R. (1965) *Proc. Nat. Acad. Sci. U. S. A.* **54**, 1241–1248
3. WARNER, H. R., AND BARNES, J. E. (1966) *Virology* **28**, 100–107
4. NOSSAL, N. G. (1969) *J. Biol. Chem.* **244**, 218–220
5. SPEYER, J. F. (1965) *Biochem. Biophys. Res. Commun.* **21**, 6–8
6. SPEYER, J. F., KARAM, J. D., AND LENNY, A. B. (1966) *Cold Spring Harbor Symp. Quant. Biol.* **31**, 693–697
7. HALL, Z. W., AND LEHMAN, I. R. (1968) *J. Mol. Biol.* **36**, 321–333
8. DRAKE, J. W., AND ALLEN, E. F. (1968) *Cold Spring Harbor Symp. Quant. Biol.* **33**, 339–344

9. Drake, J. W., Allen, E. F. Forsberg, S. A., Preparata, R. M., and Greening E. O. (1969) *Nature* **221**, 1128–1131

10. Allen, E. F., Albrecht, I., and Drake, J. W. (1970) *Genetics* **65**, 187–200

11. Brutlag, D., and Kornberg, A. (1972) *J. Biol. Chem.* **247**, 241–248

12. Goulian, M., Lucas, Z. J., and Kornberg, A. (1968) *J. Biol. Chem.* **243**, 627–638

13. Anraku, N., Anraku, Y., and Lehman, I. R. (1969) *J. Mol. Biol.* **46**, 481–492

14. Oleson, A. E., and Koerner, J. F. (1964) *J. Biol. Chem.* **239**, 2935–2943

15. Duckworth, D. H., and Bessman, M. J. (1967) *J. Biol. Chem.* **242**, 2877–2885

16. Bollum, F. J. (1966) in *Procedures in Nucleic Acid Research* (Cantoni, G. L., and Davies, D. R., eds) pp. 296–300, Harper and Row, New York

17. Adams, M. H. (1959) *Bacteriophages*, p. 446, Interscience, New York

18. Freese, E. B., and Freese, E. (1967) *Proc. Nat. Acad. Sci. U. S. A.* **57**, 650–657

38

Comparative Rates of Spontaneous Mutation

RATES of spontaneous mutation suitable for making comparisons between different species should consist of averages of rates at many sites, and should reflect all changes in DNA sequence in the tested regions. Average rates can be obtained by measurements of forward mutation, preferably within several different cistrons which should be totally dispensable for growth in permissive conditions, but sensitive even to missense mutations under restrictive or selective conditions. There are very few systems which fulfil most or all of these conditions (Table 1).

Table 1. COMPARATIVE FORWARD MUTATION RATES

Organism	Base pairs per genome	Mutation rate per base pair replication	Total mutation rate
Bacteriophage λ	4.8×10^4	2.4×10^{-8}	1.2×10^{-3}
Bacteriophage T4	1.8×10^5	1.7×10^{-8}	3.0×10^{-3}
Salmonella typhimurium	4.5×10^6	2.0×10^{-10}	0.9×10^{-3}
Escherichia coli	4.5×10^6	2.0×10^{-10}	0.9×10^{-3}
Neurospora crassa	4.5×10^7	0.7×10^{-11}	2.9×10^{-4}
Drosophila melanogaster	2.0×10^8	7.0×10^{-11}	0.8
Man	2.0×10^9	?	?

For all entries up to and including *N. crassa*, mutations arise predominantly during chromosomal replication[1], and the mutation rate was calculated from the formula[2]

$$m = 0.4343 \, (f - f_0)/\log (N/N_0)$$

where m is the mutation rate per chromosomal or nuclear replication, f the mutant frequency and N is the population size. Where necessary, an average amino-acid was assumed to weigh 110 daltons, as calculated from average amino-acid frequencies among *E. coli* proteins[3]; and a gene was assumed to contain 1,000 base pairs, corresponding to a polypeptide weighing 36,700 daltons. The data for bacteriophage λ are collected in ref. 4; the mutation rate was calculated for the cI cistron, the protein of which weighs about 30,000 daltons. The data for bacteriophage T4 are collected in ref. 2; the mutation rate was calculated for the rII cistrons (0·6 of the r mutants), which are estimated to contain about 1,770 base pairs[5,6]. The bacterial data are collected in refs. 2 and 7; the *E. coli* mutation rate was calculated for 5 *ara* cistrons, and the *S. typhimurium* mutation rate was calculated for 10 *his* cistrons. The mutation rate for the *ad*3 locus of *N. crassa* was calculated from the mutant frequency[8] in a growth experiment encompassing about 60 generations (de Serres, personal communication; and ref. 9).

Mutations in *Drosophila* arise both during and independently of cell division[10]. An estimate of the total mutation rate per sexual generation per haploid set of chromosomes[11] is 0·4, and an approximate mutation rate per base pair[12] was calculated by assuming that all mutations arise during cell replication, and that an average of forty cell generations separate the fertilized egg from the progeny gametes. The human genome[13] contains about 40 per cent of redundant species of nucleotide sequences[14], the contribution of which to the mutational target size is unclear. Redundancy is not observed in the procaryotes, and is not yet reported in *Drosophila*.

Spontaneous mutation, expressed per base pair per replication, clearly proceeds at strikingly different rates in different organisms. This is also illustrated by the observation[4] that the temperate bacteriophage λ assumes a mutation rate 20–100 times lower when multiplying as prophage (a set of bacterial genes) than when multiplying lytically (under autonomous control). Among the procaryotes, however, species differences in mutation rates largely disappear when rates are expressed per genome per replication : the mutation rate per base pair is inversely related to the size of the genome. Among the eucaryotes, very few data exist which can be transformed into average mutation rates, and it is not yet clear how these rates will vary with genome size.

If the constant mutation rate per genome suggested for procaryotes by Table 1 is general, then very small genomes should show very high mutation rates per cistron or per base pair. Comparisons should be sought in organisms which encode their own DNA polymerases and which are insulated from any but their own repair systems. Two likely candidates are *Bacillus subtilis* bacteriophage φ29 (ref. 15) and animal cell mitochondria[16], the chromosomes of both of which are composed of double stranded DNA and are less than one tenth of the size of the chromosome of bacteriophage T4.

Observed mutation rates depend on several factors: the primary rate of damage or error insertion, the efficacy of repair, and the probability of detection of an altered phenotype. Decreased local mutation rates associated with increased genome size could be achieved by reducing the error rate, and also by increasing the sophistication of repair systems. (It is less clear whether mutations will tend to produce less drastically altered phenotypes in more complex organisms.) The spontaneous mutation rates listed in Table 1 are chiefly the result of point mutations, which may be divided into base pair substitutions and frameshift mutations. About 80 per cent of T4rII mutants contain frameshifts (although the leaky nature of many rII base pair substitution mutants prevents their recovery)[2]. The frequencies of frameshift mutants are about 20–40 per cent among *S. typhimurium his* mutants, about 15 per cent among *N. crassa ad*3 mutants (where missense mutants are efficiently detected), and an indeterminant but significant percentage among *E. coli lac* mutants (see refs. in preceding article). Species specific differences in average mutation rates therefore reflect changes in both of the elementary types of point mutations. Because these clearly arise by a number of very different mechanisms[2], alterations in average mutation rates must reflect changes in many enzymatic systems.

The frequent appearance of heritable changes in spontaneous mutation rates within a species emphasizes the extent to which mutation rates are genetically determined. Most of these heritable changes consist of mutator genes. In procaryotes, however, and particularly in those with small genomes, there exists the possibility of antimutator genes. These have now been described in bacteriophage T4 (preceding article).

The reasons for the approximately constant mutation rate per genome in the procaryotes are not yet clear. Mutation rates may be adjusted by natural selection to achieve a balance among the mostly deleterious effects of mutation, the need for variation, and the cost of suppressing mutation (for example, by slowing DNA synthesis, by introducing energy-consuming repair processes or by adopting informational redundancy). It can be imagined that procaryotic genomes are all so small that the intrinsic difficulties of achieving a sufficiently low total mutation rate are minor, and that more general factors (such as the ability to adapt rapidly versus the mutational load) are sufficient to fix the rate. At the other end of the scale, however, genome sizes may have expanded to the point where the cost of achieving the necessarily very small mutation rates per base pair becomes very large. It is therefore possible that the human total mutation rate is even larger than the very large *Drosophila* rate. Were this the case, the modern species, with its generally low numbers of offspring, might be particularly sensitive to increases in its mutation rate, such as may result from chemical modifications of the environment.

JOHN W. DRAKE

Department of Microbiology,
University of Illinois,
Urbana, Illinois 61801.

Received January 20, 1969.

[1] Drake, J. W., *Proc. US Nat. Acad. Sci.*, **55**, 738 (1966).
[2] Drake, J. W., *An Introduction to the Molecular Basis of Mutation* (Holden-Day, San Francisco, in the press).
[3] Sueoka, N., *Proc. US Nat. Acad. Sci.*, **47**, 1141 (1961).
[4] Dove, W. F., *Ann. Rev. Genetics*, **2**, 305 (1968).
[5] Stahl, F. W., Edgar, R. S., and Steinberg, J., *Genetics*, **50**, 539 (1964).
[6] Edgar, R. S., Feynman, R. P., Klein, S., Lielausis, I., and Steinberg, C. M., *Genetics*, **47**, 179 (1962).
[7] Cairns, J., *J. Mol. Biol.*, **6**, 208 (1963).
[8] Brockman, H. E., and de Serres, F. J., *Genetics*, **48**, 597 (1963).
[9] Horowitz, N. H., and MacLeod, H., *Microbial Genetics Bull.*, **17**, 6 (1960).
[10] Muller, H. J., *Prog. Nuclear Energy, Ser. VI*, **2**, 146 (1959).
[11] Mukai, T., *Genetics*, **50**, 1 (1964).
[12] Rudkin, G. T., *Proc. XI Intern. Cong. Genet.*, **2**, 359 (1963).
[13] Davidson, J. N., Leslie, I., and White, J. C., *Lancet*, **1**, 1287 (1951).
[14] Britten, R. J., and Kohne, D. E., *Science*, **161**, 529 (1968).
[15] Anderson, D. L., Hickman, D. D., and Reilly, B. E., *J. Bacteriol.*, **91**, 2081 (1966).
[16] Meyer, R. R., and Simpson, M. V., *Proc. US Nat. Acad. Sci.*, **61**, 130 (1968).

AUTHOR CITATION INDEX

355

SUBJECT INDEX

About the Editors

JOHN W. DRAKE is Professor of Microbiology and Basic Medical Sciences at the University of Illinois at Urbana–Champaign, where he has been since 1958. He obtained his B.S. at Yale University in 1954 and his Ph.D. at the California Institute of Technology in 1958. He held a Fulbright Fellowship at the Weizmann Institute of Science in Rehovet in 1957–1958, a Guggenheim Fellowship at the Medical Research Council Laboratory for Molecular Biology in Cambridge in 1964–1965, and a USPH Special Fellowship at the University of Edinburgh in 1971–1972. From 1969 to 1975 he chaired the Genetics Program at the University of Illinois. He was a founding member of the Environmental Society, EMS Vice-President from 1974 to 1976, and President beginning in 1976. He is the author of numerous papers on molecular mechanisms of mutation and of the monograph *The Molecular Basis of Mutation* (Holden-Day, 1970).

ROBERT E. KOCH obtained his B.S. degree at Muhlenberg College in 1964 and his Ph.D. at the University of Illinois at Urbana–Champaign in 1969. Since that time he has continued his professional association with John W. Drake as a Research Associate and has published a number of papers on mechanisms of mutation. He has also taught courses in general genetics.